Geometry of Vector Sheaves

Mathematics and Its Applications

Volume 439

Geometry of Vector Sheaves

An Axiomatic Approach to Differential Geometry

Volume II: Geometry. Examples and Applications

by

Anastasios Mallios

Department of Mathematics,
University of Athens,
Athens, Greece

SPRINGER-SCIENCE+BUSINESS MEDIA, B.V.

A C.I.P. Catalogue record for this book is available from the Library of Congress.

ISBN 978-94-010-6102-5 ISBN 978-94-011-5006-4 (eBook)
DOI 10.1007/978-94-011-5006-4

Printed on acid-free paper

Contents

General Preface (to both Volumes I, II)

Our aim by this study is to exhibit and also exploit a quite general technique, yet simple, by its very nature, which enables one to formulate, at least, a substantial amount (if not all!) of the fundamental notions of the classical differential geometry of $\mathcal{C}^{\infty}$-manifolds and obtain too several standard results thereof. More interestingly, this can be achieved (we might say here, surprisingly enough!), by no use at all (!) of any concept of tangent vectors or of differential forms, in the classical sense of these words; indeed, no calculus is employed altogether! This is, generally speaking, our program, through the present two-volume work, whose secondary title to the adopted main one might be, of course, *An Introduction to* (an) *Abstract* (treatise of) *Differential Geometry.* Such a perspective, even if subconciously evident, more or less, to the experts, has never been, however, quite explicit, to the extent and generality, at least, which is undertaken by this discussion, so that the popularization of this point of view might be of some interest, much more because, as we shall see, there do exist considerable possibilities for applications.

Treatments aiming at the exposition of several aspects of differential geometry, outside the classical framework of (finite-dimensional) $\mathcal{C}^{\infty}$-manifolds, range, of course, from the standard theory of infinite-dimensional $\mathcal{C}^{\infty}$-manifolds, with a vast already literature thereon, to the more sophisticated aspects of "differential geometry" on topological spaces, *more general than* $\mathcal{C}^{\infty}$*-manifolds.* However, this has always been done, through a suitable formulation of the basic notions of the standard case, as for instance, vector fields, differential forms and the like.

By contrast, our treatment is quite general and, so to speak, *axiomatic.* So the basic tools here are *sheaves of modules,* with respect to an appropriate *sheaf of* **C**-*algebras* (alias, **C**-*algebra sheaf*), our *"domain of coefficients"*, by analogy with the classical case. Yet, the sheaves involved are mostly over an arbitrary *topological space* X, that eventually is decreed to be *paracompact* (Hausdorff). The latter requirement for X, base space of the sheaves considered, is only entered into the discussion, to profit from it *cohomologically,* just as occasion serves; so this is, for instance, the case with the counterpart here of *Weil's integrality theorem,* pertaining, as we know, to the cohomology class, determined by the curvature form of a connection of a given line bundle (sheaf).

Thus, *sheaf cohomology* is the second basic constituent of our armament, in order to carry out a rather ambitious(!) program, as is namely, the transfer, to the present abstract setting, of several fundamental results, of a global and/or local nature, of differential geometry of manifolds. Indeed, it is, at least, interesting (and also encouraging!, for an undertaking, like the one at hand) to realize how a considerable amount of the classical theory can be carried over to the present formalism, especially in what concerns an important differential geometric aspect, as is the theory of connections and its consequences, as e.g. curvature, characteristic classes (à la Chern-Weil) and so on.

Our study is, as already mentioned, quite general and essentially of an *algebraic* (topological) *nature,* based, as actually does, on *sheaf theory.* So it contains, as a very special case, to the extent, of course, that is accomplished herewith, the classical aspect of finite (and, partly, of infinite) dimensional differential geometry of $\mathcal{C}^{\infty}$-manifolds, as well as, several other generalizations thereon, as alluded to at the beginning of this Preface. The latter cases are also discussed, as particular non-standard examples, at the pertinent places in the sequel (see thus Volume II, Part Three).

In this context, another item worth pointing out here is the crucial rôle played, in a number of places, by the corresponding to our case *de Rham complex* of the standard theory. This refers, in particular, to the exactness of that complex (here deemed to be in force), in the classical case being, of course, a consequence of the famous *Poincarè Lemma* (viz. that every closed form is exact). Yet, in the aforementioned (non-standard) examples exactness of the previous complex is still valid, as a consequence again of the same lemma, which thus is proved to hold true in those extended cases, as well. On the other hand, in our abstract setting the situation can be remedied by demanding, for instance, the section algebras of the **C**-algebra sheaf $\mathcal{A}$, our *"domain of coefficients"*, to carry further a topological algebraic structure, being, thus, in effect, appropriate *topological algebras.* So our *"sheaf of coefficients"* becomes now, precisely speaking, a *topological* **C**-*algebra sheaf* on X.

The insert at this place of *topological algebra theory* is both crucial and decisive. Of course, this can be construed as revealing, within the present general framework, the oblique, nontheless, significant rôle, played in the classical case by the important (certainly, *non-normed!*) topological algebra $\mathcal{C}^\infty(X)$ of (say, **C**-valued) $\mathcal{C}^\infty$-functions on a given $\mathcal{C}^\infty$-manifold X. (We refer the reader to A. Mallios [2] for a general and detailed account on topological algebra theory, as well as, for the particular algebra at issue). Thus, by employing a suitable *topological algebra sheaf* $\mathcal{A}$, one is able to get an (exact) *resolution of* (the *constant sheaf*) **C**, of the form

$$(*) \qquad 0 \longrightarrow \mathbf{C} \xrightarrow{\varepsilon} \mathcal{A} \xrightarrow{\partial} \Omega^1 \xrightarrow{d^1} \Omega^2 \xrightarrow{d^2} \cdots,$$

where the Ω^i's stand for appropriately defined $\mathcal{A}$-*modules* (modules of *"differential forms"*) and the d^i's, with $d^0 \equiv \partial$, for suitable also (**C**-linear) *morphisms* of the $\mathcal{A}$-modules concerned. Furthermore, a similar aspect (in fact, more feasible than (∗)), which is still worth remarking here, is the exploitation of a *(short exact) exponential sheaf sequence* of the form

$$(**) \qquad 0 \longrightarrow \mathbf{Z} \longrightarrow \mathcal{A} \xrightarrow{e} \mathcal{A}^\bullet \longrightarrow 1,$$

where $\mathcal{A}^\bullet$ denotes the *sheaf of units* of $\mathcal{A}$, while e represents the standard *exponential* (sheaf) *morphism* of the classical theory. Here again an exact sequence, like (∗∗), is further realizable in some particular cases of *topological algebra sheaves,* as before. Indeed, it was the relevant work of S.A. Selesnick [1; 2], in the context of *Banach(-Šilov) algebra theory,* aiming also at applications to harmonic analysis, which, by its intense geometrical flavour, kindled actually our first thoughts about this subject matter. (See thus the first announcement thereon in A. Mallios [3: (i) – (iv)]). Moreover, the same exponential sequence (∗∗) substitutes in many instances for the more stringent exact sequence (∗), more so, when the $\mathcal{A}$-modules involved are, in particular, the analogues, in our sheaf-theoretic approach, of the usual line bundles.

We mentioned already that the prominent features throughout this discussion are certain sheaves of $\mathcal{A}$-modules, in fact, *vector sheaves;* that is, locally free sheaves of $\mathcal{A}$-modules of finite rank, over an arbitrary, in general, topological space X, while $\mathcal{A}$ is our "sheaf of coefficients", as above. This type of $\mathcal{A}$-modules substitute here for the familiar sheaves (of germs) of sections (continuous or otherwise) of *vector bundles* (of finite rank) in the classical theory. If, in particular, the vector sheaf, under consideration, has rank 1, we then speak of a *line sheaf* on X. (See also the Preface to Volume I, concerning this particular terminology).

Within the same context, we further remark that the exploitation of *"structures more general than manifolds"*, pertaining to concrete problems, as for example, calculation of the cohomology of *quotient spaces,* or even of classifying spaces (à la Milnor) and of semi-simplicial manifolds (see e.g. M.A. Mostow [1] and J.W. Smith [1]), indicates a further potential concrete application of the point of view of this treatise. Thus, all the previous structures are indeed a very particular instance of those considered herein, so that the present technique exhibits a possible further extension, within a similar setting (in fact, much more general), not only of the topology, as is actually the case with the above structures, but also of the *geometry* (e.g. connections and curvature) *of* such *"generalized spaces".*

Finally (but certainly, not least!), another fact that supports the framework and terminology adopted by this study, is the ever growing usage of a *sheaf-theoretic language,* compiled with that one of the standard differential geometry, in what actually concerns the *"geometrization of physical thought",* as expressed, for instance, by the nowdays *"gauge theories".* (See, for example, Yu.I. Manin [1] and/or S.A. Selesnick [3], or even R.S. Ward - R.O. Wells, Jr. [1]). All the preceding, along with an extremely positive reaction towards this language of several audiences, both at the University of Athens and elsewhere, has been a strong further inducement to the undertaking of the present more systematic account of the subject.

To make the exposition sufficiently self-contained, we prefix a pertinent introduction to *sheaf theory,* a theme in itself already classical, including the necessary for the present purpose material from *sheaf cohomology.* Yet, we adequately expose the indispensable elements for this study of what we may call *linear and multilinear algebra of vector sheaves,* or even of *sheaf modules,* in general, a subject for which a systematic account is still lacking. *Topological algebra sheaves* are also discussed, to the extent that we need it here, namely, in Part Two of this work. On the other hand, the individual Prefaces to each one of the two Volumes of this treatise contain more details and comments on the material of the corresponding Volume.

Preface (to Volume II)

As already mentioned hitherto (see the General Preface to both Volumes), the present Volume II of our study contains, in fact, the main material referred to differential geometric notions and results, as these are formulated and proved, within the abstract setting, which is advocated by this work. So this essentially concerns the Part Two of the present Volume, while in Part Three of the same we further discuss Applications and Examples of the technique developed so far, outside, of course, of the standard setting of (finite-dimensional) $\mathcal{C}^\infty$-manifolds and/or of holomorphic ones.

The individial chapters of the entire treatise have been consecutively numbered, through this two-volume work, the present Volume II containing thus Chapters VI – IX (Part Two) and X – XI (Part Three). Thus, in brief, these have, in particular, as follows: In Chapter VI we consider, in full generality, the fundamental notion, for the entire Volume, of an *$\mathcal{A}$-connection,* the latter been referred to a given *differential triad*

$$(i) \qquad (\mathcal{A}, \partial, \Omega)$$

on an arbitrary *topological space* X. Here $\mathcal{A}$ stands for a **C**-*algebra sheaf* on X, Ω for an *$\mathcal{A}$-module* (viz. for a sheaf of $\mathcal{A}$-modules on X, not necessarily locally free, or else a *vector sheaf*) and ∂ for a **C**-*linear morphism of $\mathcal{A}$ into* Ω, satisfying the corresponding to our case *Leibniz rule* (concerning here the product of two local (continuous) sections of $\mathcal{A}$). So ∂, is in a sense, the given *flat $\mathcal{A}$-connection,* which is thus the counterpart here of the basic differential operator, deriving in the classical case from the coordinate-differentials of our (finite-dimensional) flat model (euclidean space). In turn, Ω substitutes thus for the (classical) space of *pfaffians* (or else smooth 1-forms); of course, one has to look at all these objects in a sheaf-theoretic disguise. So we next examine the fundamental concept of an *$\mathcal{A}$-connection* for an *$\mathcal{A}$-module $\mathcal{E}$* on X, in general; this is, namely, a map

$$(ii) \qquad D : \mathcal{E} \longrightarrow \mathcal{E} \otimes_{\mathcal{A}} \Omega \equiv \Omega(\mathcal{E}),$$

which is, in particular, a **C**-*linear morphism* of the $\mathcal{A}$-modules involved, satisfying also the respective *Leibniz rule*

$$(iii) \qquad D(\alpha \cdot s) = \alpha \cdot Ds + s \otimes \partial\alpha,$$

for any given (continuous) *local sections* $\alpha \in \mathcal{A}(U)$ and $s \in \mathcal{E}(U)$. It is certain that such a map does not always exist in that generality, even if $\mathcal{E}$ is a vector sheaf on X. Indeed, it is the *vanishing of the* so-called *Atiyah class of* $\mathcal{E}$, which yields a *necessary and sufficient condition* in order a given *vector sheaf* on X *to have an* $\mathcal{A}$*-connection;* concerning the (cohomology) class in question, one has, by definition,

$$(iv) \qquad \mathfrak{a}(\mathcal{E}) \in \check{H}^1(X, M_n(\Omega)),$$

where $n \in \mathbf{N}$ stands for the *rank* of the vector sheaf $\mathcal{E}$ on X, under consideration. Yet, the same class is always defined, when one is given a differential triad, as in (i) above. The situation is reminiscent of and generalizes, of course, the classical case, pertaining to the existence of holomorphic connections (à la Atiyah), which also was the motivation to the above point of view. So one gets a map (the *Atiyah map*)

$$(v) \qquad \mathfrak{a} : \mathbf{\Phi}^n_{\mathcal{A}}(X) \longrightarrow \check{H}^1(X, M_n(\Omega)),$$

whose domain of definition is the set of *isomorphism classes of vector sheaves on* X, of rank $n \in \mathbf{N}$, and whose *kernel*

$$(vi) \qquad \ker(\mathfrak{a}) \equiv \mathbf{\Phi}^n_{\mathcal{A}}(X)^{\nabla}$$

is thus those elements of the previous set, which admit an $\mathcal{A}$*-connection.* In particular, *for* $n = 1$, viz. for *line sheaves* on X, one gets a *cohomological description* of the corresponding set in (vi), in terms of *(Čech) hypercohomology,* the same set being now an abelian group, in such a manner that one has;

$$(vii) \qquad \mathbf{\Phi}^1_{\mathcal{A}}(X)^{\nabla} = \check{\mathbf{H}}^1(X, \mathcal{A}^{\cdot} \xrightarrow{\tilde{\partial}} \Omega)$$

(see Theorems 18.1, 18.2 of Chapt. VI). Indeed, the proof of (vii) requires results from the *local theory of* $\mathcal{A}$*-connections,* which thus constitutes the subject matter of the next Chapter VII. The *local aspects* of the classical theory of connections can be repeated here too, realizing thus, once more, the *essential absence of the calculus,* in the standard sense, *apart,* of course, *from its formal properties* (!) (viz. the formal local properties of standard derivations). So one can formulate, within the present context, the pertinent notions of a *Riemannian* $\mathcal{A}$*-metric,* as well as, that one of a *hermitian* $\mathcal{A}$*-metric;* indeed, one can prove several results from the standard theory as, for instance, the *Fundamental Lemma of Riemannian Geometry,* pertaining here to the *existence of* an *$\mathcal{A}$-connection compatible with a* given *Riemannian $\mathcal{A}$-metric,* for every *Riemannian vector sheaf,* when the corresponding *differential triad,* as in (i) above, is *suitably adapted* to the present (riemannian) framework.

On the other hand, by contrast with the previous situation, concerning the existence of an $\mathcal{A}$-connection, one realizes that, *given an $\mathcal{A}$-connection, we can always associate with it* what we might consider here as the analogue of the classical notion of *curvature.* Thus, the study of this concept, together with the deduction of several standard results, within the present axiomatic framework, connected with the latter notion, constitute the subject matter of Chapter VIII. In this concern, we further note that, in order to formulate the same notion, one has to supplement the initially given differential triad, as in (i) above, with what we call a *curvature datum*

$$(viii) \qquad (\mathcal{A}, \partial, \Omega^1, d, \Omega^2).$$

Thus, in effect, by another *"differential operator"*

$$(ix) \qquad d \equiv d^1 : \Omega \equiv \Omega^1 \longrightarrow \Omega^2 \equiv \wedge^2\Omega^1,$$

satisfying the appropriate *compatibility conditions,* with respect to ∂. We also speak then of a *curvature space* X. So the curvature of a given $\mathcal{A}$-connection D is now defined by the formula

$$(x) \qquad R(D) \equiv D^1 \circ D,$$

where D^1 is the *1st prolongation of* D, by analogy with the classical case, the previous operator $d^1 \equiv d$ being thus the *1st prolongation of* ∂. So one of the compatibility conditions between d and ∂, referred to above, is the assumption that

$$(xi) \qquad R(\partial) \equiv d^1 \circ \partial \equiv d \circ \partial = 0,$$

that is, the *flatness of the* given *$\mathcal{A}$-connection* ∂ of $\mathcal{A}$ (in the strict sense of the word, or else that the *curvature of* ∂ is *zero*). We also examine *flat $\mathcal{A}$-connections,* in general, in the sense that the corresponding curvature, as in (x) above, vanishes. The notion of *flatness* is further related with the analogues here of other classical concepts, as, for example, with those of *complete parallelism, integrability* and *triviality* of the corresponding $\mathcal{A}$-connections; yet, with the *flatness of* a given *vector sheaf* E, relative to the so-called *co-*

ordinate 1-cocycle of E, consisting in that case of *locally constant* **C**-*valued maps* on X. (We have examined that notion in Volume I; Section 5, as well as, in the present Volume II, Chapters VI; Subsection 11.(a) and VII; Section 6). Several classical *"identities"*, connected with the notion of curvature, as, for instance, *Cartan's structural equation, Bianchi's identities,* including *Ricci's Lemma,* are also extended to the present abstract framework. Furthermore, within the same flavour of ideas, one also obtains, in full generality, *Weil's integrality theorem,* referred, as known, to a *geometrical characterization of* certain *integral cohomology classes* of a given (paracompact Hausdorff) space X. For the case at hand, certain appropriate cohomological conditions on X and $\mathcal{A}$ should also be imposed, being, however, fulfilled in all the important cases (take, for instance, X, as before, and $\mathcal{A}$, as in (i), *fine*). One further application of the same material, presented hitherto, is the examination, in the next Chapter IX, of *characteristic classes,* following *Chern-Weil theory.* The corresponding here *Fundamental Theorem* (Chern-Weil), as well as, an appropriate extension of the classical *Chern-Weil homomorphism* are also in force.

Finally, Part Three of this treatise (Chapters X, XI) is concerned with further examples of the preceding material: Thus, in Chapter X we consider various particular cases, apart, of course, of the classical one, which, as already said, fits well within the present framework, this being, in effect, a natural extension of the former. So the same examples constitute special instances of our abstract context, being also by themselves *concrete generalizations* of the standard case of $\mathcal{C}^{\infty}$-*manifolds.* Thus, the so-called *Satake manifolds,* or even some particular instances of *infinite-dimensional* ($\mathcal{C}^{\infty}$-) *manifolds,* fall into the previous framework; namely, within that one of an abstract *differential triad accompanied with* the pertinent sequence of *differential operators,* so that one has, in effect, an *exact de Rham complex,* or else a *resolution of the constant sheaf* **C** (see e.g. (∗) in the preceding General Preface to both Volumes I, II).

In the final Chapter XI we discuss certain particular examples, which are actually deduced from *topological algebra theory;* of course, it is standard that the same algebra ${}^{\mathbf{C}}\mathcal{C}^{\infty}(X) \equiv \mathcal{C}^{\infty}(X)$ of a $\mathcal{C}^{\infty}$-*manifold* X is a very convenient *topological algebra,* which also *cannot be a Banach algebra*! Thus, for a suitable class of topological algebras, the so-called *geometric (topological) algebras,* one can have, through the corresponding *Gel'fand sheaf* (on the *spectrum* of such an algebra), a situation similar to that one of the classical theory. This, in particular, in what essentially concerns the *structure sheaf* $\mathcal{C}_X^\infty$ of a $\mathcal{C}^\infty$*-manifold* X, as before, and the corresponding *de Rham complex.* We consider, instead, the *de Rham-Kähler complex* (and/or its *sheafification*) that corresponds to a given *topological algebra.* However, this *does not yield a resolution of* **C**, in general. On the other hand, topological algebras, as before, do have a nice behaviour, relative to *sheaf exponential sequences* (see, for instance, the same Chapter, Section 7), a very useful device, when dealing, in particular, with *line sheaves.* Yet, a way out of the above inconvenience is to consider some other differential triad, namely, that one, which we call the *Watts-Selesnick differential triad;* this can be associated with a given topological algebra of the pertinent type, general enough, however, to include several important examples, ranging, for instance, from the standard case of the $\mathcal{C}^\infty$-manifolds to the *Fourier algebra* of a (not necessarily abelian) *compact (Lie) group.* Again the final lesson here is that, one can have, through the previous technique, a potential, eventually useful substitute for the *standard* **C***-algebra of* $\mathcal{C}^\infty$*-functions* by a *more perceptive* (!) *function algebra,* as the case may be.

ANASTASIOS MALLIOS

Athens
June, 1997

Acknowledgements

It is a special pleasure to record here my indebtedness to all those, who, in one way or another, contributed to giving me the indispensable impetus and countenance to bring this project to an end. First, to *S. A. Selesnick* for his continuing interest in my work, in particular, in what concerns its relation with topological algebra theory; thus, his own earlier work in harmonic analysis on compact groups, written, within the context of Banach algebra theory, with a strong geometrical flavour, was, in effect, my initial inspiration for the abstract framework that has finally been adopted by the present study. Second, to *E. E. Rosinger,* who, during my visits to the University of Pretoria (South Africa), gave me great encouragement to finish this project, by his extremely enthusiastic reaction to early versions of this material. The final boost, and while the manuscript was almost ready, came from *S. Negrepontis,* who, as Head of our Department in Athens, at that time, offered me kindly the first financial support for converting the manuscript into LaTeX form. Further financial help to finishing the project has been supplied by the University of Athens, thanks, especially, to the kind willingness of *Y. Dracopoulos,* Vice-Rector of the University of Athens. My special thanks are due here to *Dr. M. Gikas* for his excellent rendering of a quite difficult text in LaTeX form, as well as, for an extremely friendly and helpful cooperation. My singular thanks are also due to *my colleagues* in the *Math. Department* of the *University of Athens*, who allowed me an extended leave of absence, very helpful, indeed, to accomplish this undertaking.

Yet, a quite decisive influence to attempt the present venture, in spite of appearing circumspections, as to its applicability, due mainly to the entirely abstract (sheaf-theoretic) language employed, was the very positive final facing of the subject from *various colleagues in Greece and abroad.* For all those I want to keep my very best and thankful memories. For a similar interest I wished, especially, to mention and thank from this place too, *Prof. Ky Fan,* who, as member of the Editorial Board of JMAA, was very kind, as well as, enthusiastic in handling my earlier papers on the subject, connected, in particular, with topological algebra theory, while his response, at that time, was among the first strong impulses for me to continue working on the topic. For their kind help, for the LaTeX writing, during the final corrections of the camera-ready copies of this two-volume work, I want also to thank my (cousins and) colleagues, at the Math. Dept. in Athens, *M. Papatriantafillou* and *E. Vassiliou*. (For any remaining errors, typing or otherwise, it is me, of course, who is to be blamed, not including, certainly, LATEX itself!).

Last, but not least, I wish to express my sincere appreciation to *Prof. M. Hazewinkel* for accepting this book to his Series, as well as, to *Dr. P. Roos* for his extremely kind interest, enthusiasm and editorial endeavour in bringing this work to the light.

Contents of Volume I

PART TWO

GEOMETRY

CHAPTER VI

Geometry of Vector Sheaves. $\mathcal{A}$-connections

From now on we proceed to the main theme of this treatise

"...motivation for the study of a new concept may, in great part, lie in the clarification and simplification of more familiar notions."

John L. Kelley in J. L. Kelley et al., *"Linear Topological Spaces"* (Springer - Verlag, 1963). p.v.

"There are many advantages to developing a theory in the most general context possible."

R. Hartshorne in *"Algebraic Geometry"* (Springer - Verlag, 1977). p.59.

As already remarked below the heading, the present chapter marks, in effect, the beginning of our *abstract* (viz. *axiomatic*) *approach to the standard Differential Geometry of smooth* (i.e. $\mathcal{C}^{\infty}$-)*manifolds.* A crucial aspect hereby is that *we do not employ* for this *any differential analysis;* that is, *no calculus* is involved at all(!), our tools being just *sheaf theory,* in particular, *vector sheaves and their algebra* (linear and multilinear) and/or *sheaf cohomology,* to the extent that these issues have been developed in the preceding chapters. Nontheless, as we shall realize through the succeeding discussion, several fundamental results of the classical theory, global, as well as, local ones, can still be derived, within the present abstract framework. We start thus right away by displaying the necessary preliminaries, in order next to define (cf. Section 2 below) the important (differential geometric) notion of a *connection,* or even, in our terminology, of an *$\mathcal{A}$-connection,* which essentially lies in the core of all the subsequent material; this, with the exception of Chapter X, which, however, again is just an auxiliary one of the next Chapter XI, where anew *$\mathcal{A}$-connections of a particular type* are further examined.

1. C-derivations. Differential triads

A basic notion for the subsequent discussion is that of a *"derivation";* as we shall presently see, this essentially concerns, in our case, an appropriate *sheaf morphism* (see Chapt. I; Definition 1.2). Yet, we shall refer, as usual, to a general **C**-*algebraized space* $(X, \mathcal{A})$, unless otherwise specified (see Chapt. II; Scholium 1.2, or even Chapt. IV; (0.1)). So we begin with the following.

Definition 1.1. Suppose we have a **C**-algebraized space $(X, \mathcal{A})$, and let $\mathcal{E}$ be an $\mathcal{A}$-module on X. Thus, by a **C**-*derivation of $\mathcal{A}$ in $\mathcal{E}$,* one means a sheaf morphism

$$\partial : \mathcal{A} \longrightarrow \mathcal{E}, \tag{1.1}$$

such that the two following conditions are satisfied:

1) ∂ is a **C**-*linear morphism* (relative to the corresponding **C**-vector sheaves $\mathcal{A}$, $\mathcal{E}$ on X; cf. Chapt. II; (1.38)).

2) ∂ satisfies the following *Leibniz condition*

$$\partial(s \cdot t) = s \cdot \partial(t) + t \cdot \partial(s), \tag{1.2}$$

for any (local) sections s, t in $\mathcal{A}(U)$, with U any open set in X.

Of course, concerning the above rel. (1.2), one essentially employs here the corresponding to (1.1) *section-morphism* (cf. Chapt. I; (12.49), (12.50))

$$\overline{\partial} : \Gamma(\mathcal{A}) \longrightarrow \Gamma(\mathcal{E}); \tag{1.3}$$

however, for simplicity's sake, the latter is *still denoted by* ∂ (by an obvious *abuse of notation,* a practice that we shall often apply in the sequel, as well, without specific mention).

Now, a triplet

$$(1.4)\qquad (\mathcal{A}, \partial, \mathcal{E}),$$

as before, is called a *differential triad* on X.

Thus, by analogy with the classical case, we first have the following easy, however, useful.

Lemma 1.1. *For any given differential triad* $(\mathcal{A}, \partial, \mathcal{E})$ (cf. (1.4)), *the restriction of* ∂ *to the constant (algebra) sheaf* $\mathbf{C} \equiv \mathbf{C}_X \overset{\subseteq}{\rightarrow} \mathcal{A}$ (see also Chapt. II; (1.34)) *is zero. That is, one has*

$$(1.5)\qquad \partial|_{\mathbf{C}} = 0$$

(viz. the *zero* **C**-*linear morphism*).

Proof. The assertion follows from applying ∂, first, to the *identity section* $1 \equiv 1_{\mathcal{A}} \in \mathcal{A}(X)$ (see Chapt. II; (1.22)); indeed, one has (cf. also (1.3), (1.2))

$$(1.6)\qquad \partial(1_{\mathcal{A}}) \equiv \partial(1) = \partial(1 \cdot 1) = 1 \cdot \partial(1) + 1 \cdot \partial(1) = \partial(1) + \partial(1),$$

such that one obtains

$$(1.7)\qquad \partial(1_{\mathcal{A}}) \equiv \partial(1) = 0.$$

Therefore, by the **C**-*linearity of* ∂ (Definition 1.1), one has

$$(1.8)\qquad \partial(\lambda) := \partial(\lambda \cdot 1_{\mathcal{A}}) \equiv \partial(\lambda \cdot 1) = \lambda \cdot \partial(1) = \lambda \cdot 0 = 0,$$

for every *"constant section"* $\lambda \in \mathbf{C} \overset{\subseteq}{\rightarrow} \mathbf{C}_X(X) \overset{\subseteq}{\rightarrow} \mathcal{A}(X)$ (see also Chapt. I; (10.2), (11.15)), and this completes the proof. ∎

Yet, as another consequence of (1.2), one also obtains the following useful analogue of the classical case.

Lemma 1.2. *("Quotient rule"). Suppose we have a differential triad* $(\mathcal{A}, \partial, \mathcal{E})$, *as before. Then, for any local sections* $s \in \mathcal{A}(U)$ *and* $t \in \mathcal{A}(U)^{\bullet} = \mathcal{A}^{\bullet}(U)$ (cf. Chapt. IV; (1.15)), *with* U *open in* X, *one has;*

$$(1.9)\qquad \partial(s \cdot t^{-1}) = t^{-2} \cdot (t \cdot \partial s - s \cdot \partial t).$$

Proof. Indeed, based on (1.2), one has;

$$\partial(s \cdot t^{-1}) = s \cdot \partial t^{-1} + t^{-1} \cdot \partial s. \tag{1.10}$$

But, $t \cdot t^{-1} = 1 \equiv 1_{\mathcal{A}}|_U \in \mathcal{A}^{\bullet}(U)$, so that one obtains (cf. (1.7))

$$\begin{aligned} \partial(t \cdot t^{-1}) &= t \cdot \partial(t^{-1}) + t^{-1} \cdot \partial(t) = \partial(1) \equiv \partial(1_{\mathcal{A}}|_U) \equiv \partial_U(1_{\mathcal{A}}|_U) \\ &\equiv \partial_U \left(\rho_U^X(1_{\mathcal{A}})\right) = \lambda_U^X(\partial_X(1_{\mathcal{A}})) \equiv \lambda_U^X(\partial(1_{\mathcal{A}})) \equiv \lambda_U^X(\partial(1)) = 0 \end{aligned} \tag{1.11}$$

(see also e.g. Chapt. I; (6.5), (12.49) and Chapt. II; (1.46), (1.67), or even (2.21)). Therefore, *one* first *obtains;*

$$\partial(t^{-1}) = -t^{-2} \cdot \partial(t), \tag{1.12}$$

for any given $t \in \mathcal{A}^{\bullet}(U) = \mathcal{A}(U)^{\bullet}$ and *open* $U \subseteq X$, as above. Hence, based on (1.10), (1.12), one has

$$\left(s \cdot t^{-1}\right) = s \cdot \left(-t^{-2} \cdot \partial t\right) + t^{-1} \cdot \partial s, \tag{1.13}$$

which obviously yields the desired relation (1.9). ∎

Now, we next consider for later use a *square-matrix extension of the* **C**-*derivation* ∂, as above (cf. (1.4)):

That is, by looking at the $\mathcal{A}$-*algebra sheaf* $M_n(\mathcal{A})$ (cf. Chapt. IV; (3.24)) and the $M_n(\mathcal{A})$-*module* $M_n(\mathcal{E})$ (see also (1.21) below), one defines a **C**-*derivation,* as follows (cf. also Chapt. V; (8.7));

$$\overline{\partial} \equiv \partial : M_n(\mathcal{A}) \longrightarrow M_n(\mathcal{E}) := M_n(\mathcal{A}) \otimes_{\mathcal{A}} \mathcal{E} \tag{1.14}$$

such that one sets;

$$\overline{\partial}(\alpha) \equiv \partial(\alpha) := (\partial(\alpha_{ij})) \in M_n(\mathcal{E}(U)) \tag{1.15}$$

for any $\alpha \equiv (\alpha_{ij}) \in M_n(\mathcal{A}(U)) = M_n(\mathcal{A})(U)$ and U *open* in X. To explain the notation applied in the last relation, we first remark that one can look at $M_n(\mathcal{E})$ as an $M_n(\mathcal{A})$-*module* by defining it as the *sheafification of the complete* $\Gamma(\mathcal{A})$-*presheaf* $M_n(\Gamma(\mathcal{E}))$; that is, one defines

$$M_n(\mathcal{E}) := \mathbf{S}(M_n(\Gamma(\mathcal{E}))), \tag{1.16}$$

such that one sets, for every *open* $U \subseteq X$ (see also Chapt. II; (3.11))

$$M_n(\Gamma(\mathcal{E}))(U) := M_n(\mathcal{E}(U)) = \mathcal{E}(U)^{n^2} = \mathcal{E}^{n^2}(U) = M_n(\mathcal{E})(U), \tag{1.17}$$

the last three relations above standing for $\mathcal{A}(U)$-*isomorphisms of* the $\mathcal{A}(U)$-*modules* involved. Yet, the last item in (1.17) indicates also an *equivalent definition* of the range of (1.14); namely, one can define $M_n(\mathcal{E})$, *equivalently,* as the *sheafification of the* $\Gamma(\mathcal{A})$*-presheaf* (cf. Chapt. II; (5.5))

$$U \longmapsto M_n(\mathcal{E}(U)) := M_n(\mathcal{A}(U)) \otimes_{\mathcal{A}(U)} \mathcal{E}(U) \cong \mathcal{E}(U)^{n^2}, \tag{1.18}$$

where U *varies over the open subsets of* X, along with the appropriate *restriction morphisms.* (For simplicity, we employed an obvious *abuse of notation,* concerning the range of (1.18), in connection with (1.17)). Now, it is easy to see that

(1.19) $M_n(\Gamma(\mathcal{E}))$, as given by (1.17), together with the corresponding restriction morphisms, *yields a complete* $\Gamma(\mathcal{A})$*-presheaf.*

Furthermore, one defines

(1.20) $M_n(\mathcal{E})$ as an $M_n(\mathcal{A})$*-module,* according to the relation;

$$\alpha \cdot s = (\alpha_{ij}) \cdot (s_{ij}) := \left(\sum_{k=1}^{n} \alpha_{ik} \cdot s_{kj}\right) \in M_n(\mathcal{E}(U)), \tag{1.20.1}$$

for any $\alpha \equiv (\alpha_{ij}) \in M_n(\mathcal{A}(U)) = M_n(\mathcal{A})(U)$ *and* $s \equiv (s_{ij}) \in M_n(\mathcal{E}(U))$, with U *open* in X.

Now, based on the very definitions, it is readily seen that;

(1.21) the map (1.14) yields a **C**-*linear morphism* of the $\mathcal{A}$-modules (in fact, by the hypothesis for $\mathcal{A}$, **C**-*vector space sheaves*) on X concerned, while *it* also *fulfils Leibniz condition (1.2).* Therefore, one obtains a **C**-*derivation of the* **C**-*algebra sheaf* $M_n(\mathcal{A})$ *into the* $M_n(\mathcal{A})$*-module* $M_n(\mathcal{E})$ (see also Chapt. II; (1.38), Chapt. IV; (3.24), along with (1.20) above).

The preceding justify now completely our claim about (1.14). ∎ Yet, the triplet, thus defined, *for any given* $n \in \mathbf{N}$,

$$(M_n(\mathcal{A}), \bar{\partial} \equiv \partial, M_n(\mathcal{E})), \tag{1.22}$$

is called the *n-th square matrix extension of the* given *differential triad* $(\mathcal{A}, \partial, \mathcal{E})$ on X. Finally, to emphasize the *functorial nature* of $M_n(\cdot)$, with $n \in \mathbf{N}$, in (1.22), it is still appropriate to set

$$\bar{\partial} \equiv M_n(\partial), \tag{1.23}$$

as the *n-th square matrix extension of* ∂.

1.(a). Logarithmic derivation.- We examine next another fundamental notion for the sequel, associated with the previously defined one of a **C**-derivation (see Definition 1.1); it will constantly appear throughout the ensuing discussion, together with its matrix extension, analogous to (1.14) above. Thus, we first set the following.

Definition 1.2. Suppose we have a differential triad $(\mathcal{A}, \partial, \mathcal{E})$ (cf. (1.4)), and let $\mathcal{A}^{\cdot}$ be the (group) sheaf of units of $\mathcal{A}$ (cf. Chapt. IV; Lemma 1.1). Then, we call *logarithmic derivation,* associated with the given **C**-derivation ∂ (see (1.1)), the following sheaf morphism

$$\tilde{\partial} : \mathcal{A}^{\cdot} \longrightarrow \mathcal{E}, \tag{1.24}$$

given (in terms of sections, cf. Chapt. I; (13.19)) by the relation;

$$\tilde{\partial}(\alpha) := \alpha^{-1} \cdot \partial(\alpha), \tag{1.25}$$

for any (local) section $\alpha \in \mathcal{A}^{\cdot}(U) = \mathcal{A}(U)^{\cdot}$ (cf. Chapt. IV; (1.15)) and U open in X.

Now, as a first application of (1.25), we remark that

$$\varepsilon(\mathbf{C}^{\cdot}) \subseteq \ker \tilde{\partial} \tag{1.26}$$

(see also (1.5) and Chapt. II; (1.34)). On the other hand, as as immediate consequence of (1.25) and (1.2), one realizes that;

(1.27) $\tilde{\partial}$ *is,* in effect, *a morphism of the group sheaves involved;* that is, one has

$$\tilde{\partial}(s \cdot t) = \tilde{\partial}(s) + \tilde{\partial}(t), \tag{1.27.1}$$

for any $s, t \in \mathcal{A}^{\cdot}(U) = \mathcal{A}(U)^{\cdot}$ and U *open* in X. In particular, one obtains (cf. also (1.26));

$$\tilde{\partial}\left(\alpha^{-1}\right) = -\tilde{\partial}(\alpha), \tag{1.27.2}$$

for any $\alpha \in \mathcal{A}^{\cdot}(U)$, with U *open* in X.

Now, by analogy with (1.14), one also defines, *for any* $n \in \mathbf{N}$, the *n-th (square) matrix extension of* $\tilde{\partial}$ (by abusing notation, we continue to denote it by $\tilde{\partial}$, as well), as the following *sheaf morphism*

$$\tilde{\partial} : \mathcal{GL}(n, \mathcal{A}) \longrightarrow M_n(\mathcal{E}), \tag{1.28}$$

given by the relation;

$$\tilde{\partial}(\alpha) := \alpha^{-1} \cdot \partial(\alpha), \tag{1.29}$$

for any $\alpha \in \mathcal{GL}(n, \mathcal{A})(U) = \mathrm{GL}(n, \mathcal{A}(U))$ and U *open* in X (cf. (1.20.1) and Chapt. IV; (1.26), (1.28)).

Thus, based on (1.29) and (1.5), (1.15), one still obtains that

$$\tilde{\partial}|_{\mathcal{GL}(n,\mathbf{C})} = 0 \tag{1.30}$$

(see also Chapt. IV; (1.31), or even Chapt. V; (1.5) for the notation applied). Yet, in view of (1.29), (1.21) and (1.30), one further infers, extending thus (1.27.2), that;

$$\tilde{\partial}\left(\alpha^{-1}\right) = -\tilde{\partial}(\alpha), \tag{1.31}$$

for every $\alpha \equiv (\alpha_{ij}) \in \mathcal{GL}(n, \mathcal{A})(U) = \mathrm{GL}(n, \mathcal{A}(U))$ and U *open* in X.

In fact, something more is actually true; namely, based on the *Leibniz condition for (1.14)* (cf. (1.2), (1.21)), written now in the form (*equivalent to (1.2), for* $n = 1$, due to the commutativity of $\mathcal{A}$)

$$\partial(s \cdot t) = s \cdot \partial(t) + \partial(s) \cdot t, \tag{1.32}$$

for any $s, t \in M_n(\mathcal{A}(U)) = M_n(\mathcal{A})(U)$ and *open* $U \subseteq X$, *one obtains*

$$\tilde{\partial}(s \cdot t) = \mathrm{Ad}(t^{-1}) \cdot \tilde{\partial}(s) + \tilde{\partial}(t), \tag{1.33}$$

for any $s \equiv (s_{ij}), t \equiv (t_{ij})$ *in* $\mathcal{GL}(n, \mathcal{A})(U) = \mathrm{GL}(n, \mathcal{A}(U))$ and U *open* in X: Indeed, one has (cf. (1.29), (1.32));

$$\begin{aligned} \tilde{\partial}(st) &= (st)^{-1} \cdot \partial(st) = t^{-1}s^{-1}(s \cdot \partial(t) + \partial(s) \cdot t) \\ &= t^{-1} \cdot \partial(t) + t^{-1}\left(s^{-1} \cdot \partial(s)\right)t = \tilde{\partial}(t) + t^{-1} \cdot \tilde{\partial}(s) \cdot t, \end{aligned} \tag{1.34}$$

which thus *proves (1.33),* when setting

$$t^{-1} \cdot \tilde{\partial}(s) \cdot t \equiv \mathrm{Ad}\left(t^{-1}\right) \cdot \tilde{\partial}(s). \tag{1.35}$$

Here the multiplication in the first member of (1.35) stands for the usual (section) *matrix multiplication,* as defined by (1.20.1) (see also (1.14), (1.15)). In this respect, we still note that, by analogy with (1.20.1), one infers that;

(1.36) $M_n(\mathcal{E})$ *is,* in fact, *an* $M_n(\mathcal{A})$*-bimodule* (for every $\mathcal{A}$*-module* $\mathcal{E}$ on X).

The notation in the second member of (1.35) is reminiscent, of course, of the standard *adjoint representation*

$$\mathrm{Ad} : \mathcal{GL}(n, \mathcal{A}) \longrightarrow (\mathcal{E}nd\, \mathcal{E}^n)^{\cdot}, \tag{1.37}$$

provided e.g. (1.36) hold true; see also, for intance, (1.17), as well as, Chapt. II; (6.29) and Chapt. V; Scholium 8.2, for the notation applied in the range of (1.37). Yet, (1.37) has certainly a meaning as a *matrix representation* in the case of a *free* $\mathcal{A}$*-module* $\mathcal{E}$ (of *finite rank* $n \in \mathbf{N}$) or, at least, of a *locally free one* (viz. a *vector sheaf*) on X. On the other hand, in the general case, one still defines an *adjoint representation* (cf. also (1.14))

$$\mathrm{Ad} : \mathcal{GL}(n, \mathcal{A}) \longrightarrow \mathcal{E}nd\, (M_n(\mathcal{E}) \equiv M_n(\mathcal{A}) \otimes_{\mathcal{A}} \mathcal{E})^{\cdot}, \tag{1.38}$$

by the relation;

$$\mathrm{Ad}(\alpha)(s \otimes t) := \alpha s \alpha^{-1} \otimes t, \tag{1.39}$$

for any $\alpha \in \mathcal{GL}(n, \mathcal{A})(U) = \mathrm{GL}(n, \mathcal{A}(U))$, $s \in M_n(\mathcal{A})(U) = M_n(\mathcal{A}(U))$ *and* $t \in \mathcal{E}(U)$, with U *open* in X. So *one sets*

$$\mathrm{Ad} := i \otimes \mathrm{id}_{\mathcal{E}}, \tag{1.40}$$

such that

$$i(\alpha)(s) := \alpha s \alpha^{-1}, \tag{1.42}$$

for any α *and* s, *as above.*

Note.- When, in place of (1.29), we set

$$\tilde{\partial}(\alpha) := \partial(\alpha) \cdot \alpha^{-1}, \tag{1.43}$$

for any α, *as in (1.29)* (cf. also (1.36) and (1.20.1)), one obtains, instead of (1.33), the relation;

$$\tilde{\partial}(s \cdot t) = \tilde{\partial}(s) + \mathrm{Ad}(s)\tilde{\partial}(t), \tag{1.43}$$

for any s, t, *as in (1.33).*

Use of the above will be made later on.

2. A standard example: The smooth triad $(\mathcal{C}^\infty_X, d, \Omega^1_X)$

We consider next a concrete example of the previous abstract setting, which, together with its ramifications, constitute, in fact, our fundamental motivation for all the ensuing discussion. Thus, the *concrete differential triad* $(\mathcal{A}, \partial, \mathcal{E})$, under consideration, as already indicated in the heading, has now as follows:

Consider any (finite dimensional Hausdorff) $\mathcal{C}^\infty$*-manifold* X, along with the pair

$$(X, \mathcal{C}^\infty_X) \tag{2.1}$$

as a **C**-*algebraized space;* namely, one thus sets

$$\mathcal{A} \equiv \mathcal{C}^\infty_X := {}^{\mathbf{C}}\mathcal{C}^\infty_X, \tag{2.2}$$

viz. the *sheaf of germs of* **C**-*valued* $\mathcal{C}^\infty$*-functions on* X (cf. Chapt. I; (4.39)) is now taken as the *structure sheaf* of the concrete pair $(X, \mathcal{A})$ in question (cf. Chapt. II; Scholium 1.2), being certainly a **C**-*algebra sheaf* on X (see Chapt. II; Definition 1.4, (1.67), as well as, Chapt. I; (11.11), ii) and (11.12)). In this respect, see also Chapt. III; (8.27), (8.29). Furthermore, one sets

$$\mathcal{E} \equiv \Omega^1_X, \tag{2.3}$$

that is, one further considers the $\mathcal{C}^\infty_X$*-module* (in fact, *vector sheaf*) on X, as this is given by the *sheaf of germs of* **C**-*valued smooth* (viz. $\mathcal{C}^\infty$-) *1-forms on* X; so one has, by definition,

$$\Omega^1_X := \mathbf{S}(\Gamma(\mathcal{T}^*_{\mathbf{C}}(X))). \tag{2.4}$$

Here $\mathcal{T}^*_{\mathbf{C}}(X)$ stands for the *complexified cotangent bundle of* X, such that $\Gamma(\mathcal{T}^*_{\mathbf{C}}(X))$ denotes, of course, the *(complete) presheaf of sections* of the (**C**-vector) bundle in question, being actually isomorphic to that one of the *sheaf (of germs) of sections of* $\mathcal{T}^*_{\mathbf{C}}(X)$. (In this concern, see also Chapt. III; (8.30), for $p = 1$, as well as, Chapt. XI in the sequel). Yet, we note, in passing, that (2.4) is a *locally free* $\mathcal{C}^\infty_X$*-module* on X, *of rank* $n = \dim X < +\infty$, viz. a *vector sheaf* on X, a consequence, in effect, of the very definitions (see, for instance, Chapt. III; (8.34), still for $p = 1$; yet, cf. Section 4 in the sequel). Finally, as a **C**-*derivation* $\partial : \mathcal{A} \longrightarrow \mathcal{E}$, with $\mathcal{A}$ and $\mathcal{E}$ given by (2.2) and (2.3), respectively, one now sets

$$d : \mathcal{C}_X^\infty \longrightarrow \Omega_X^1, \tag{2.5}$$

where the above *sheaf morphism* is defined, *in terms of* the corresponding *sections* of the sheaves concerned (cf. Chapt. I; (13.19) and (2.17)); namely, to each **C**-valued (local) $\mathcal{C}^\infty$-function of X one assigns its *differential,* being thus, by the same definitions, a (local $\mathcal{C}^\infty$-) 1-form on X. On the other hand, it is still a direct consequence of the very definitions that *(2.5) is* indeed *a* **C**-*derivation,* as indicated, in the sense of Definition 1.1 in the preceding.

Thus, one obtains the following triplet

$$(\mathcal{C}_X^\infty, d, \Omega_X^1), \tag{2.6}$$

called the *standard* $\mathcal{C}^\infty$*-triplet,* or even the *standard differential triad,* associated, as above, with the given $\mathcal{C}^\infty$-manifold X.

Another *concrete* (however, non standard!) example of a *differential triad* will be given in the sequel, taken from *topological algebra theory* (see Chapter XI) yet, general enough to specialize to several standard examples. In particular, the familiar *topological* **C**-*algebra* $\mathcal{C}_{\mathbf{C}}^\infty(X) = \Gamma(X, \mathcal{C}_X^\infty)$ is here included. Furthermore, other examples, similar to (2.6), will also be given in the case of *generalized manifolds* (cf. Chapter X).

3. $\mathcal{A}$-connections. Basic definitions

We come now to consider one of the most important notions throughout the whole treatise. However, since a certain standard notation will be applied in the succeeding discussion, with obvious, for that matter, modifications in certain particular cases, we first set the following.

Convention.– Henceforth, we shall denote by

$$(\mathcal{A}, \partial, \Omega) \tag{3.1}$$

a fixed *differential triad* (cf. (1.4)), with respect to a fixed also **C**-algebraized space $(X, \mathcal{A})$, that was usually the case in the preceding, as well. Thus, by definition (see (1.1)), Ω stands here for a particular *$\mathcal{A}$-module* on X, range of the given **C**-*derivation*

$$\partial : \mathcal{A} \longrightarrow \Omega. \tag{3.2}$$

The notation hereby is, of course, reminiscent of the standard example (2.6), as above. Yet, further suitable alterations of this particular "Ω" will be used in the sequel, as the case may be. (See thus, for instance, Section 4 below, as well as, in particular, the next Chapter VII).

In this context, we shall also employ throughout the sequel the following convenient abreviation, which, in a sense, has already been used in the preceding, as well (see thus e.g. (1.14)). So, for any given *$\mathcal{A}$-module* $\mathcal{E}$ on X, one sets;

$$\Omega(\mathcal{E}) := \mathcal{E} \otimes_{\mathcal{A}} \Omega = \Omega \otimes_{\mathcal{A}} \mathcal{E}, \tag{3.3}$$

the second member of (3.3) being, of course, by definition, an *$\mathcal{A}$-module* on X too (cf. Chapt. II; Definition 5.1); yet, the equality thereby stands, in fact, for an *$\mathcal{A}$-isomorphism of* the *$\mathcal{A}$-modules* concerned (ibid.; (5.14)).

Thus, we come now to the definition of an important notion for all the succeeding discussion, as we have already mentioned in the foregoing. That is, one has the following.

Definition 3.1. Let $(\mathcal{A}, \partial, \Omega)$ be a given differential triad (cf. (3.1)) and $\mathcal{E}$ an arbitrary $\mathcal{A}$-module on X. Then, by an *$\mathcal{A}$-connection* of $\mathcal{E}$, one means a sheaf morphism (see also (3.3))

$$D : \mathcal{E} \longrightarrow \Omega(\mathcal{E}) \equiv \mathcal{E} \otimes_{\mathcal{A}} \Omega, \tag{3.4}$$

such that the two following conditions are fulfilled:

i) D is a **C**-*linear morphism* of the **C**-vector space sheaves concerned; viz. one has

$$D(\lambda s + \mu t) = \lambda \cdot D(s) + \mu \cdot D(t), \tag{3.5}$$

for any (local) sections $s, t \in \mathcal{E}(U)$, with U open in X, and $\lambda, \mu \in \mathbf{C} \equiv \mathbf{C}(U) \subsetneq \mathbf{C}_X(U)$.

ii) D satisfies the following *Leibniz condition;*

$$D(\alpha \cdot s) = \alpha \cdot D(s) + s \otimes \partial(\alpha), \tag{3.6}$$

for any (local) sections $\alpha \in \mathcal{A}(U)$, and $s \in \mathcal{E}(U)$, with U open in X,

We first comment a bit more on the previously applied notation: So the *properties (3.5) and (3.6)* of the given sheaf morphism ($\mathcal{A}$-connection) D, as in (3.4), *are* actually *referred to presheaves, generating* the sheaves and, in particular,, the $\mathcal{A}$-modules concerned. Thus, one has

$$\mathcal{E} = \mathbf{S}(\Gamma(\mathcal{E})), \quad \Omega(\mathcal{E}) = \mathbf{S}(\Gamma(\mathcal{E}) \otimes_{\Gamma(\mathcal{A})} \Gamma(\Omega)) \quad \textit{and} \quad \mathbf{C}_X = \mathbf{S}(\mathbf{C}) \tag{3.7}$$

(see Chapt. I; (9.17), (11.21) and Chapt. II; (5.7)). Therefore, the said properties are further extended to the sheaves in question, based on Chapt. I; (12.42), (8.11) and Chapt. II; (2.1) - (2.3) and (1.44). In this context, we also recall that, since $\mathcal{A}$ is, by assumption, a $\mathbf{C}$-*algebra sheaf* on X (cf. e.g. Chapt. IV; (0.1)), *each $\mathcal{A}$-module on X is,* in particular, *a $\mathbf{C}$-vector space sheaf* on X (see Chapt. II; (1.38)).

Yet, concerning the *tensor product section* appeared in the second member of (3.6), one has, of course (cf. (3.7)), that;

$$s \otimes \partial(\alpha) \in \mathcal{E}(U) \otimes_{\mathcal{A}(U)} \Omega(U). \tag{3.8}$$

That is, we actually employ here, as usual, for simplicity, an obvious *abuse of notation,* by writing just $s \otimes \partial(\alpha)$, whereas the *(decomposable) tensor* at issue is always meant with respect to a certain $\mathbf{C}$-*algebra of coefficients* (here $\mathcal{A}(U)$), different, in general, from $\mathbf{C}$ itself. However, to simplify things, as already said, we steadily follow this practice (see also e.g. Chapt. IV; (6.5) and the ensuing comments thereby). The above clarify now completely the notation applied in the above Definition 3.1.

On the other hand, regarding the employed term $\mathcal{A}$-*connection,* as above, we still remark that, since the latter notion is determined modulo the given differential triad, as in (3.1), the term $(\mathcal{A}, \partial, \Omega)$-*connection* of the $\mathcal{A}$-module $\mathcal{E}$ involved would be more appropriate, nevertheless somewhat awkward, as well. Thus, for simplicity, we shall retain the term, already used above, although the name just *connection* will also be employed, occasionally, in the sequel, the determination "$\mathcal{A}$" being, otherwise, easily understood from the context. Yet, the term ∂-*connection* might be applied for D in (3.4), as well, by referring thus, of course, to the $\mathbf{C}$-derivation, in effect, connection (cf. the next Section 4) ∂ of $\mathcal{A}$ involved; moreover, this would be also in agreement with another usage of the term. (See, for instance, M. Cornalba

- P. Griffiths [1]).

Now, before we proceed further to the study of $\mathcal{A}$-connections, along with the applications, we have in mind, within the present abstract setting, of this fundamental notion, we first consider right below certain particular instances of $\mathcal{A}$-connections. Further examples will be encountered throughout the succeeding discussion.

4. Examples of $\mathcal{A}$-connections

We examine below a few particular examples of $\mathcal{A}$-connections, which we have at our disposal from what has been already considered so far.

4.(a). The standard (flat) $\mathcal{A}$-connection.– As a prelude to more sophisticated instances of $\mathcal{A}$-connections, we first look at *the* given **C**-*derivation* $\partial : \mathcal{A} \longrightarrow \Omega$, *as an $\mathcal{A}$-connection* of $\mathcal{A}$ in Ω (see (3.1)):

Thus, we first note the following (canonical) $\mathcal{A}$-*isomorphism* (cf. also (3.3))

$$\Omega(\mathcal{A}) \equiv \mathcal{A} \otimes_{\mathcal{A}} \Omega = \Omega \tag{4.1}$$

(cf. Chapt. II; (5.15)). Therefore, *the* given **C**-*derivation* (see (3.2))

$$\partial : \mathcal{A} \longrightarrow \Omega = \Omega(\mathcal{A}) \tag{4.2}$$

provides, in effect, *an $\mathcal{A}$-connection of* (the $\mathcal{A}$-module) $\mathcal{A}$ *in* Ω, as this is easily verified by the same definitions. See thus Definition 1.1, while we also note hereby the following obvious relation in (1.2);

$$t \cdot \partial(s) = \partial(s) \cdot t \equiv t \otimes \partial(s), \tag{4.3}$$

due to the (canonical) $\mathcal{A}$-*isomorphism* $\mathcal{A} \otimes_{\mathcal{A}} \mathcal{A} = \mathcal{A}$ (again Chapt. II; (5.15)), which thus justifies completely our claim for (4.2).

On the other hand, the geometrical term *flat,* which has been ascribed to ∂ will also be fully vindicated later on (see Chapter VIII below), referred, in fact, as we shall see (ibid.; (2.6)), to the *(zero) curvature* of the $\mathcal{A}$-connection ∂ in question.

4.(b). $\mathcal{C}^{\infty}_X$-connections.– Consider a finite-dimensional (Hausdorff) $\mathcal{C}^{\infty}$-*manifold* X, and let $\mathbf{E}$ be a $\mathcal{C}^{\infty}$-*complex vector bundle,* or *smooth* $\mathbf{C}$-*vector bundle,* of *rank* n over X. Moreover, let $\mathcal{E}$ be the *sheaf of germs of* $\mathcal{C}^{\infty}$-*sections of* $\mathbf{E}$; that is, one has

$$\mathcal{E} := \mathbf{S}(\Gamma(\mathbf{E})), \tag{4.4}$$

such that

$$\Gamma^{\infty}(\mathbf{E}) \equiv \Gamma(\mathbf{E}) \tag{4.5}$$

stands for the *complete presheaf of sections of* $\mathbf{E}$, as given by the assignement

$$U \longmapsto \mathbf{E}(U) := \mathcal{C}^{\infty}(U, \mathbf{E}|_U)_s, \text{ with } U \text{ open in } X, \tag{4.6}$$

along with the obvious *restriction maps* (morphisms); the index "s" in the range of (4.6) is for *section* (see also (4.8) in the sequel). Thus, concerning the same set, we first remark that;

(4.7) by looking at the *indexed set of proper open coverings of* X (see e.g. Chapt. III; (4.9)), one can always obtain a *cofinal subset* of it *such that*

$$\mathbf{E}|_U = U \times \mathbf{C}^n, \tag{4.7.1}$$

within a *diffeomorphism,* for every open set $U \subseteq X$ of the (open) covering of X, belonging to the particular cofinal set of coverings of X, as before. Something more, we may assume that the previous open sets, as in (4.7.1), belong to a (maximal) $\mathcal{C}^{\infty}$-*atlas of* X. We call such an open covering, a *coordinate open covering of* X.

In this context, we still note, in passing, that in view of (4.4), one has (see also Chapt. I; Proposition 11.1);

$$\Gamma(\mathbf{E})(U) = \mathbf{E}(U) \cong \mathcal{E}(U), \tag{4.7.2}$$

(within an *isomorphism* of the spaces involved), *for every open set* $U \subseteq X$.

Thus, for an open covering of X, as in (4.7), one obtains;

$$
\begin{aligned}
(4.8)\qquad \mathcal{C}^\infty(U,\mathbf{E}|_U)_s &= \mathcal{C}^\infty(U,U\times\mathbf{C}^n)_s = \mathcal{C}^\infty(U,\mathbf{C}^n)\\
&= \mathcal{C}^\infty(U,\mathbf{C})^n \equiv \mathcal{C}^\infty_{\mathbf{C}}(U)^n \equiv \mathcal{C}^\infty(U)^n,
\end{aligned}
$$

the last two equalities denoting, in fact, *isomorphisms of* the $\mathbf{C}$*-vector spaces* concerned, while the last item in (4.8) stands for the standard *free* $\mathcal{C}^\infty(U)$-*module* of *rank* n. Of course, *(4.8) is valid, for any open* $V \subseteq U$. Therefore, in view of (4.8), one concludes that

$$
(4.9)\qquad \Gamma(\mathbf{E})|_U = \Gamma(\mathbf{E}|_U) = \Gamma((\mathcal{C}^\infty_X)^n|_U) = \Gamma((\mathcal{C}^\infty_X)^n)|_U,
$$

for every open set $U \subseteq X$, *as in (4.7.1)* (see also Chapt. I; (11.41) and Chapt. II; (3.7)). Hence, *one obtains* (cf. (4.4), as well as, Chapt. I; (9.23), (9.17));

$$
\begin{aligned}
(4.10)\qquad \mathcal{E}|_U &= \mathbf{S}(\Gamma(\mathbf{E}))|_U = \mathbf{S}(\Gamma(\mathbf{E})|_U) = \mathbf{S}(\Gamma((\mathcal{C}^\infty_X)^n)|_U)\\
&= \mathbf{S}(\Gamma((\mathcal{C}^\infty_X)^n))|_U = (\mathcal{C}^\infty_X)^n|_U)
\end{aligned}
$$

up to a $\mathcal{C}^\infty_X|_U$*-isomorphism of* the $\mathcal{C}^\infty_X|_U$*-modules* involved, *for any open* $U \subseteq X$, *as above.* Thus, in other words, one infers that

(4.11) $\mathcal{E} \equiv \mathbf{S}(\Gamma(\mathcal{E}))$ *is a locally free* $\mathcal{C}^\infty_X$*-module on* X, *of rank* $n = \operatorname{rk}\mathbf{E} \in \mathbf{N}$*;* that is, a *vector sheaf* on X.

In this regard, see also Chapt. XI for an ample generalization of (4.11).

Now, for an *integer* $p \geq 1$, let

$$
(4.12)\qquad \Omega^p(X,\mathbf{E}) \equiv \Omega^p(\mathbf{E})
$$

be the *sheaf of* $\mathbf{E}$*-valued smooth* (viz. $\mathcal{C}^\infty$-) p*-forms* on X; thus, modulo a $\mathcal{C}^\infty_X$*-isomorphism,* one has

$$
(4.13)\qquad \Omega^p(\mathbf{E}) = \mathbf{S}(\Gamma(\Omega^p)\otimes_{\Gamma(\mathcal{C}^\infty_X)}\Gamma(\mathbf{E})),
$$

such that one sets

$$
(4.14)\qquad \Omega^p \equiv \Omega^p_X := \bigwedge^p \Omega^1_X,
$$

this standing for the p*-th exterior power of the* $\mathcal{C}^\infty_X$*-module* Ω^1_X (see (2.4) above, as well as, Chapt. IV; (7.1), (7.3)).

So, *for* $p = 1$, a $\mathcal{C}^\infty_X$*-connection,* for the case at hand, is given by the following *sheaf morphism* (cf. (4.4), (4.13))

$$
(4.15)\qquad D:\mathcal{E}\longrightarrow\Omega^1_X(\mathcal{E})\equiv\Omega^1(\mathcal{E})=\mathcal{E}\otimes_{\mathcal{C}^\infty_X}\Omega^1=\Omega^1\otimes_{\mathcal{C}^\infty_X}\mathcal{E},
$$

which further satisfies the conditions of Definition 3.1. Thus, (4.15) is, in fact, a *$\mathcal{C}^\infty$-connection,* or just a *connection, on the* given *smooth* **C***-vector bundle* **E**, in the standard sense of Differential Geometry. In particular, suppose that **E**, as before, is the *complexified tangent bundle* of X,

$$\mathcal{T}_{\mathbf{C}}(X) := \mathcal{T}(X) \otimes_{\mathbf{R}} \mathbf{C}, \tag{4.16}$$

that is, the result of the *"complexification functor"*, applied on the *tangent bundle* of X, $\mathcal{T}(X)$ (the latter being, of course, a *smooth* **R***-vector bundle* on X. See e.g. M. Karoubi [1: Example 4.8,e) p.21]). So in that case, a $\mathcal{C}^\infty_X$-connection on (4.16), by analogy with (4.15), is a *$\mathcal{C}^\infty$-connection,* or simply a *connection, on the* given *$\mathcal{C}^\infty$-manifold X.* (In this respect, see also, for instance, S. Kobayashi [1: p.1], or R.O. Wells [1: p. 70, Definition 1.5], or even J.W. Milnor - J.D. Stasheff [1: p. 289]).

The above example will often be used in the sequel too, as a link of several abstract results, within the present framework, with corresponding standard ones in Differential Geometry and/or Complex Manifold Theory. Further examples of $\mathcal{A}$-connections, along the lines of the classical theory, are considered right below.

5. Induced $\mathcal{A}$-connections

We examine below certain particular cases of $\mathcal{A}$-connections, which will be of special use throughout the succeeding discussion, corresponding, as already said, to *familiar instances of* the classical theory of *$\mathcal{C}^\infty$-connections,* according to the above Example 4.(b). Thus, we begin with the following standard, otherwise, notion.

5.(a). Whitney sum of $\mathcal{A}$-connections.– We consider, as usual, a basic *differential triad* $(\mathcal{A}, \partial, \Omega)$ over a given *topological space* X, common base space of all the sheaves, involved in our discussion, and let

$$(\mathcal{E}_i)_{1 \leq i \leq n} \tag{5.1}$$

be a finite sequence of *$\mathcal{A}$-modules* on X, along with a corresponding (finite) sequence of *$\mathcal{A}$-connections* on $\mathcal{E}_i$, $i = 1, \ldots, n$, respectively;

(5.2) $$(D_i)_{1 \leqq i \leqq n}.$$

Then, one concludes that;

(5.3) the *sheaf morphism*

(5.3.1) $$\begin{aligned} D := \bigoplus_{i=1}^n D_i : \bigoplus_{i=1}^n \mathcal{E}_i \equiv \mathcal{E} \longrightarrow \mathcal{E} \otimes_{\mathcal{A}} \Omega \equiv \Omega(\mathcal{E}) \\ \equiv \Omega\left(\bigoplus_i \mathcal{E}_i\right) = \bigoplus_{i=1}^n (\mathcal{E}_i \otimes_{\mathcal{A}} \Omega) \equiv \bigoplus_i \Omega(\mathcal{E}_i), \end{aligned}$$

which is given by

(5.3.2) $$D(s) \equiv D(s_1 \oplus \cdots \oplus s_n) := \bigoplus_{i=1}^n D_i(s_i) \in \bigoplus_i \Omega(\mathcal{E}_i) \equiv \Omega(\mathcal{E}),$$

for any given (local) *sections* $s_i \in \mathcal{E}(U)$, $1 \leq i \leq n$, with U *open* in X (so that $s = \bigoplus_{i=1}^n s_i \in \left(\bigoplus_i \mathcal{E}_i\right)(U) \equiv \mathcal{E}(U)$, as above), is an *$\mathcal{A}$-connection* on $\mathcal{E} \equiv \bigoplus_{i=1}^n \mathcal{E}_i$.

Concerning the notation applied above, see Chapt. II; (3.7), (3.14) and (5.17). Now, the assertion in (5.3) is an immediate consequence of (5.3.2) and Definition 3.1. ∎ Yet, the sheaf morphism D, as in (5.3.1), is also called the *Whitney sum of the $\mathcal{A}$-connections D_i*, $1 \leq i \leq n$.

On the other hand, an important *particular example of (5.3),* which will also be applied in the sequel, is the *$\mathcal{A}$-connection*

(5.4) $$\partial^n := \underbrace{\partial \oplus \cdots \oplus \partial}_{n-\text{times}} : \mathcal{A}^n \longrightarrow \mathcal{A}^n \otimes_{\mathcal{A}} \Omega \equiv \Omega(\mathcal{A}^n) = \Omega^n \ (\equiv \Omega^{(n)})$$

(see also Chapt. II; (5.18)), *n-th Whitney sum of the* given *standard (flat) $\mathcal{A}$-connection* ∂ (cf. Example 4.(a); (4.2)), defined thus on the free $\mathcal{A}$-module $\mathcal{A}^n$, according to (5.3.2); that is, one sets

(5.5) $$\partial^n(s \equiv (s_i)) := \bigoplus_{i=1}^n \partial(s_i) \equiv (\partial(s_i))_{1 \leqq i \leqq n} \in \Omega(U)^n = \Omega^n(U),$$

for any $s \equiv (s_i) \in \mathcal{A}^n(U) = \mathcal{A}(U)^n$, and U *open* in X. Yet, due to the particular form of ∂^n (cf. (5.4)), the same is still named the *n-th power* (or even the *n-th dimensional extension) of* ∂.

So, for every *integer* $n \geqq 2$, one considers the *n-th dimensional extension of the* given *differential triad* $(\mathcal{A}, \partial, \Omega)$, that is, the triple

(5.6) $$(\mathcal{A}^n, \partial^n, \Omega^n).$$

Of course, (5.6) has a special bearing on the *square-matrix extension of* ∂, already considered in the preceding by (1.22) (cf. (1.23) too); in this regard, see also (1.15), (1.18), together with Chapt. IV; (3.23) and Chapt. II; (5.18).

Now, more generally, by extending the situation in (5.4), one can look at the *n-th dimensional extension of an $\mathcal{A}$-connection D* on a given *$\mathcal{A}$-module* $\mathcal{E}$ on X. So one has the triple

(5.7) $$(\mathcal{E}^n, D^n, \Omega(\mathcal{E}^n)),$$

extending (5.6), such that one sets;

(5.8) $$\begin{aligned} D^n(s \equiv (s_i)) &:= (D(s_i)) \equiv \bigoplus D(s_i) \in \bigoplus_{i=1}^n (\mathcal{E}_i \otimes_{\mathcal{A}} \Omega) \\ &= (\bigoplus_i \mathcal{E}_i) \otimes_{\mathcal{A}} \Omega = \mathcal{E}^n \otimes_{\mathcal{A}} \Omega \equiv \Omega(\mathcal{E}^n) = \Omega(\mathcal{E})^n, \end{aligned}$$

for any $s = (s_i) \in \mathcal{E}^n(U) = \mathcal{E}(U)^n$, with U *open* in X (see also Chapt. II; (3.7), (5.17) and (3.14)).

Scholium 5.1.– Based on (5.3.2), it is straightforward to consider *any family of summands (not necessarily finite)* in a *Whitney sum of $\mathcal{A}$-connections.*

The interest here lies, among other things, in the fact that, through e.g. such an *infinite Whitney sum of vector sheaves,* one has an example of an *$\mathcal{A}$-module (not necessarily a vector sheaf) endowed with an $\mathcal{A}$-connection.* Further examples of the same sort, as before, are also obtained by considering, for instance, *projective limits of vector sheaves with $\mathcal{A}$-connections* (E. Vassiliou), or *inductive limits* of such. We leave, however, the technical details hereby to the reader.

5.(b). Tensor product of $\mathcal{A}$-connections.– As usual, we tacitly assume, here too, the existence of a *differential triad* $(\mathcal{A}, \partial, \Omega)$ (see (3.1)) on a *topological space* X, while we further suppose that we are given two *$\mathcal{A}$-modules* $\mathcal{E}$ and $\mathcal{F}$ on X, endowed with *$\mathcal{A}$-connections* $D_{\mathcal{E}} \equiv D$ and $D_{\mathcal{F}} \equiv D'$. So our task is to define the pair

(5.9) $$(\mathcal{E} \otimes_{\mathcal{A}} \mathcal{F}, D_{\mathcal{E}\otimes_{\mathcal{A}}\mathcal{F}});$$

that is, to consider on the *tensor product $\mathcal{A}$-module* $\mathcal{E} \otimes_{\mathcal{A}} \mathcal{F}$ (cf. Chapt. II; (5.6)), an *$\mathcal{A}$-connection* $D_{\mathcal{E}\otimes_{\mathcal{A}}\mathcal{F}} \equiv \widetilde{D}$, which is actually given by the relation

(5.10) $$\widetilde{D} \equiv D_{\mathcal{E}\otimes_{\mathcal{A}}\mathcal{F}} := (D_{\mathcal{E}} \otimes 1_{\mathcal{F}}) + (1_{\mathcal{E}} \otimes D_{\mathcal{F}}) \equiv D \otimes 1 + 1 \otimes D',$$

where $1_{\mathcal{E}}, 1_{\mathcal{F}}$ denote the *identity $\mathcal{A}$-automorphisms* of the $\mathcal{A}$-modules concerned. (In this regard, we still note that the notation "$\otimes$" in the second member of (5.10) stands, in effect, for an *"adorned tensor product"*, viz. for "$\otimes_{\mathcal{A}}$"; see also the comments about (3.8) in the preceding).

Now, our claim on (5.10) is, in fact, a straightforward implication of the same definitions (see thus Definition 3.1). So we check, for instance, the corresponding *Leibniz condition* for (5.10) (see (3.6)), by first applying *decomposable tensors (sections);* that is, one has

$$
\begin{aligned}
&\widetilde{D}(\alpha \cdot (s \otimes t)) = \widetilde{D}(s \otimes \alpha t) = [(D \otimes 1) + (1 \otimes D')](s \otimes \alpha t)\\
&= (D \otimes 1)(s \otimes \alpha t) + (1 \otimes D')(s \otimes \alpha t) = Ds \otimes \alpha t + s \otimes D'(\alpha t)\\
&= \alpha \cdot (Ds \otimes t) + s \otimes (\alpha D't + t \otimes \partial\alpha) = \alpha \cdot (Ds \otimes t + s \otimes D't) + s \otimes t \otimes \partial\alpha\\
&= \alpha \cdot \widetilde{D}(s \otimes t) + (s \otimes t) \otimes \partial\alpha,
\end{aligned} \tag{5.11}
$$

for any (local) *sections* $\alpha \in \mathcal{A}(U)$, $s \in \mathcal{E}(U)$ *and* $t \in \mathcal{F}(U)$, with U *open* in X. The general case follows now *extending (5.11) by linearity,* which thus proves our assertion. ∎

5.(c). $\mathcal{A}$-connections vis-à-vis to the functor $\mathcal{H}om_{\mathcal{A}}$.- Suppose we are given the *same data, as in 5.(b)* above. Thus, our claim now is the following;

(5.12) under the previous hypothesis and the supplementary assumption that $\mathcal{E}$ is a *vector sheaf* on X, *one can define an $\mathcal{A}$-connection on the $\mathcal{A}$-module* $\mathcal{H}om_{\mathcal{A}}(\mathcal{E}, \mathcal{F})$ on X, as follows:

That is, the sought $\mathcal{A}$-connection is supplied by the relation

$$D_{\mathcal{H}om_{\mathcal{A}}(\mathcal{E},\mathcal{F})}(\phi) := D_{\mathcal{F}} \circ \phi - (\phi \otimes 1_{\Omega}) \circ D_{\mathcal{E}}, \tag{5.13}$$

where ϕ stands for any (local) section of $\mathcal{H}om_{\mathcal{A}}(\mathcal{E}, \mathcal{F})$; viz. we assume that

$$\phi \in \mathcal{H}om_{\mathcal{A}}(\mathcal{E}, \mathcal{F})(U) = \mathrm{Hom}_{\mathcal{A}|_U}(\mathcal{E}|_U, \mathcal{F}|_U) \tag{5.14}$$

for some *open* $U \subseteq X$ (see Chapt. II; (6.8)), while 1_{Ω} in (5.13) denotes the *identity $\mathcal{A}$-automorphism* of the given *$\mathcal{A}$-module* Ω on X. Yet, applying, for convenience, the following abbreviated notation,

$$\widetilde{D} \equiv D_{\mathcal{H}om_{\mathcal{A}}(\mathcal{E},\mathcal{F})}, \quad D \equiv D_{\mathcal{E}} \text{ and } D' \equiv D_{\mathcal{F}}, \tag{5.15}$$

the defining relation (5.13) is written;

(5.16) $$\widetilde{D}(\phi) := D' \circ \phi - (\phi \otimes 1_{\Omega}) \circ D$$

with ϕ, *as in (5.14)*. In this regard, the following comments on (5.13) (or (5.16)) are still in order:

Namely, suppose that $s \in \mathcal{E}(U)$, with U open in X; then setting, for simplicity,

(5.17) $$\phi(s) \equiv \overline{\phi}(s) := \phi \circ s$$

(see Chapt. I; (12.49), (12.50)), one obtains

(5.18) $$D_{\mathcal{F}}(\phi(s)) \in \mathcal{F}(U) \otimes_{\mathcal{A}(U)} \Omega(U)$$

(see (3.4) and the comments following Definition 3.1. Yet, concerning (5.14), see also Chapt. I; (11.40)). Analogously, since $s \in \mathcal{E}(U)$, one has

(5.19) $$D_{\mathcal{E}}(s) \in \mathcal{E}(U) \otimes_{\mathcal{A}(U)} \Omega(U),$$

so that one further obtains;

(5.20) $$(\phi \otimes 1_{\Omega})(D_{\mathcal{E}}(s)) \in \mathcal{F}(U) \otimes_{\mathcal{A}(U)} \Omega(U).$$

Thus, operating now on $\phi \equiv \overline{\phi}$, via $\widetilde{D}$ (see (5.14), (5.14)), and then on $s \in \mathcal{E}(U)$, that is, setting (see (5.16)),

(5.21) $$\widetilde{D}(\phi)(s) = D'(\phi(s)) - (\phi \otimes 1_{\Omega})(D(s)),$$

one finally obtains a (local) *section over the open* $U \subseteq X$ of the following $\mathcal{A}$-*module* (cf. also (5.14), (5.19), (5.20)), together with Chapt. II; (5.25), (6.24))

(5.22) $$\mathcal{H}om_{\mathcal{A}}(\mathcal{E}, \mathcal{F} \otimes_{\mathcal{A}} \Omega) = \mathcal{H}om_{\mathcal{A}}(\mathcal{E}, \mathcal{F}) \otimes_{\mathcal{A}} \Omega \equiv \Omega(\mathcal{H}om_{\mathcal{A}}(\mathcal{E}, \mathcal{F})),$$

as desired, for the case at issue, according to (3.4). In this regard, one should also notice that the equality (in fact, $\mathcal{A}$-*isomorphism*) in (5.22) *is* actually *based on the supplementary hypothesis for* $\mathcal{E}$(!) (cf. (5.12) and Chapt. IV; (6.8)). The preceding clarify now completely our notation in (5.13), in conjunction with (5.22) and (5.18), (5.20).

Finally, concerning our claim for (5.13), we check, for instance, that *the said sheaf morphism satisfies the Leibniz condition;* thus, *for any* ϕ, s, as above (cf. e.g. (5.21)), *and* $\alpha \in \mathcal{A}(U)$, one has;

$$(5.23)\qquad \begin{aligned}\widetilde{D}(\alpha\cdot\phi)(s) &= D'(\alpha\cdot\phi(s)) - (\alpha\phi\otimes 1_\Omega)(D(s))\\ &= \alpha\cdot D'(\phi(s)) + \phi(s)\otimes\partial\alpha - \alpha\cdot(\phi\otimes 1_\Omega)(D(s))\\ &= \alpha\cdot\widetilde{D}(\phi)(s) + \phi(s)\otimes\partial\alpha,\end{aligned}$$

that was to be proved. (The **C**-*linearity* of (5.13) is still straightforward from (5.21)). This justifies thus completely our assertion in (5.12). ∎

We consider next a particular instance of (5.13), namely, the case $\mathcal{E} = \mathcal{F}$; notice also that, according to our hypothesis in (5.12), $\mathcal{E}$ is a *vector sheaf* on X. Thus, for any given $\mathcal{A}$-*connection* D of $\mathcal{E}$, one obtains from (5.13) (or (5.16)) the following $\mathcal{A}$-*connection* on the *vector sheaf* $\mathcal{H}om_{\mathcal{A}}(\mathcal{E},\mathcal{E}) \equiv \mathcal{E}nd\,\mathcal{E}$ on X (cf. Chapt. II; (6.28) and Lemma 6.1)

$$(5.24)\qquad D_{\mathcal{E}nd\,\mathcal{E}}(\phi) = D\circ\phi - (\phi\otimes 1_\Omega)\circ D,$$

such that (ibid.; (6.30))

$$(5.25)\qquad \phi\in(\mathcal{E}nd\,\mathcal{E})(U) = \mathcal{H}om_{\mathcal{A}|_U}(\mathcal{E}|_U,\mathcal{E}|_U) \equiv \mathrm{End}\,(\mathcal{E}|_U),$$

with U *open* in X. The same relation (5.24) is still written, by an obvious *abuse of notation* (however, useful!), reminiscent of the *"Lie product"*, in the following form;

$$(5.26)\qquad D_{\mathcal{E}nd\,\mathcal{E}}(\phi) = D\circ\phi - \phi\circ D \equiv [D,\phi].$$

On the other hand, based on (5.13), and *taking* $\mathcal{E} = \mathcal{F}$, one obtains, for any two $\mathcal{A}$-connections D, D'

$$(5.27)\qquad D_{\mathcal{E}nd\,\mathcal{E}}(\phi) = D'\circ\phi - (\phi\otimes 1_\Omega)\circ D,$$

with ϕ, as in (5.25). *In particular, taking in (5.27)*

$$(5.28)\qquad \phi\equiv 1_{\mathcal{E}}\in(\mathcal{E}nd\,\mathcal{E})(X) = \mathrm{Hom}_{\mathcal{A}}(\mathcal{E},\mathcal{E}) \equiv \mathrm{End}\,\mathcal{E},$$

that is, the *identity* $\mathcal{A}$-*automorphism* of $\mathcal{E}$ (cf. (3.7) and Chapt. II; (6.30)), one obtains the following useful relation;

$$(5.29)\qquad D_{\mathcal{E}nd\,\mathcal{E}}(1) = D' - D\in\Omega(\mathcal{E}nd\,\mathcal{E})(X) = \mathrm{Hom}_{\mathcal{A}}(\mathcal{E},\mathcal{E}\otimes_{\mathcal{A}}\Omega).$$

Thus, as a consequence of (5.29), one realizes that *the difference of two* $\mathcal{A}$-*connections is* not merely a **C**-linear morphism but, in particular, an $\mathcal{A}$-*morphism,* as well; this will be explained in Section 7 below (see thus Theorem 7.1).

Now, as a particular application of the preceding, we further examine in the next Subsection *the case* $\mathcal{F} = \mathcal{A}$.

5.(d) Dual $\mathcal{A}$-connections.– Suppose we are given an *$\mathcal{A}$-module* $\mathcal{E}$ on X, endowed with an *$\mathcal{A}$-connection* $D_{\mathcal{E}}$, with respect to a given *differential triad* $(\mathcal{A}, \partial, \Omega)$ (cf. (4.1)). Furthermore, let

$$\mathcal{E}^* \equiv \mathcal{H}om_{\mathcal{A}}(\mathcal{E}, \mathcal{A}) \tag{5.30}$$

be the *dual $\mathcal{A}$-module of* $\mathcal{E}$ on X (see Chapt. IV; (5.1)). In particular, *due to (5.12),* assume that $\mathcal{E}$ is a *vector sheaf* on X, hence, applying (5.13), for $\mathcal{F} = \mathcal{A}$ and $D_{\mathcal{F}} = D_{\mathcal{A}} = \partial$ (see (4.2)), one obtains an *$\mathcal{A}$-connection of* $\mathcal{E}^*$ by the relation;

$$D_{\mathcal{E}^*}(u) := \partial \circ u - (u \otimes 1_{\Omega}) \circ D_{\mathcal{E}}, \tag{5.31}$$

such that (cf. (5.30) and Chapt. II; (6.8))

$$u \in \mathcal{E}^*(U) = \mathrm{Hom}_{\mathcal{A}|_U}(\mathcal{E}|_U, \mathcal{A}|_U), \tag{5.32}$$

with U *open* in X. Yet, cf. e.g. (5.21), one has

$$D_{\mathcal{E}^*}(u)(s) = \partial(u(s)) - (u \otimes 1_{\Omega})(Ds), \tag{5.33}$$

for any u, as in (5.32), and $s \in \mathcal{E}(U)$. The last relation is also written, in a more standard way, as;

$$\partial(u(s)) = u(Ds) + (D_{\mathcal{E}^*}(u))(s), \tag{5.34}$$

which is also taken as a *defining relation for* $D_{\mathcal{E}^*}$. We have still employed hereby an obvious *abuse of notation,* by simply writing

$$(u \otimes 1)(Ds) \equiv u(Ds). \tag{5.35}$$

Yet, setting

$$u(s) \equiv \langle s, u\rangle \quad and \quad (u \otimes 1)(Ds) \equiv u(Ds) \equiv \langle Ds, u\rangle,$$

one still writes (5.34) in the form;

$$\partial(\langle s, u\rangle) = \langle Ds, u\rangle + \langle s, D^*u\rangle, \tag{5.35'}$$

taking it as a *defining relation* of $D^* \equiv D_{\mathcal{E}^*}$ (cf. (5.36) below). In this regard, notice also that, by definition, one has;

$$D_{\mathcal{E}^*} : \mathcal{E}^* \longrightarrow \mathcal{E}^* \otimes_{\mathcal{A}} \Omega = \mathcal{H}om_{\mathcal{A}}(\mathcal{E}, \Omega) \tag{5.36}$$

(cf. Chapt. IV; Theorem 6.1), which thus justifies the notation applied in (5.33) or (5.34) (cf. also (5.35)).

The previously defined $\mathcal{A}$-connection $D_{\mathcal{E}^*}$ (cf. (5.31)) is called the *dual $\mathcal{A}$-connection* of the given $\mathcal{A}$-connection $D_{\mathcal{E}}$ of the *vector sheaf* $\mathcal{E}$ on X, under consideration; it is thus, by definition (cf. (5.36)), an *$\mathcal{A}$-connection of the dual vector sheaf $\mathcal{E}^*$ of $\mathcal{E}$* on X (cf. (5.12) and Chapt. IV; (5.19)).

On the other hand, as an application of (5.31), one has the following variant, concerning the definition of *$\mathcal{A}$-connections of $\mathcal{E}nd\,\mathcal{E}$*, where $\mathcal{E}$ is still a *vector sheaf* on X (see (5.12)): So we first remark that;

$$\mathcal{E}nd\,\mathcal{E} \equiv \mathcal{H}om_{\mathcal{A}}(\mathcal{E}, \mathcal{E}) = \mathcal{E}^* \otimes_{\mathcal{A}} \mathcal{E} = \mathcal{E} \otimes_{\mathcal{A}} \mathcal{E}^*, \tag{5.37}$$

within *$\mathcal{A}$-isomorphisms of* the *$\mathcal{A}$-modules* (in fact, *vector sheaves*) involved (see Chapt. II; (6.28), (5.14), Lemma 6.1 and Chapt. IV; Theorem 6.1, in conjunction with (5.12) above). Therefore, if $D_{\mathcal{E}}$ is a given *$\mathcal{A}$-connection of* $\mathcal{E}$, then, based on (5.10), one obtains (cf. also (5.36));

$$D_{\mathcal{E}\otimes_{\mathcal{A}}\mathcal{E}^*} := (D_{\mathcal{E}} \otimes 1_{\mathcal{E}^*}) + (1_{\mathcal{E}} \otimes D_{\mathcal{E}^*}). \tag{5.38}$$

Now, the interesting fact here is that, *due to (5.37),*

(5.39) *the two $\mathcal{A}$-connections (5.38) and (5.27) are, in effect, identical, modulo the $\mathcal{A}$-isomorphism,* as given by (5.37), *say,*

$$\theta : \mathcal{E} \otimes_{\mathcal{A}} \mathcal{E}^* \longrightarrow \mathcal{H}om_{\mathcal{A}}(\mathcal{E}, \mathcal{E}) \equiv \mathcal{E}nd\,\mathcal{E}. \tag{5.39.1}$$

Precisely, one has the following relation;

$$D_{\mathcal{E}nd\,\mathcal{E}} = (\theta \otimes 1_{\Omega}) \circ D_{\mathcal{E}\otimes_{\mathcal{A}}\mathcal{E}^*} \circ \theta^{-1}. \tag{5.40}$$

The above relation (5.40) is explained through the following fundamental notion, whose particular instance is thus (5.40). The same general notion will be employed several times in the sequel; thus, anticipating hereby its use, in the general form, we set the following.

Definition 5.1. Suppose we have a differential triad $(\mathcal{A}, \partial, \Omega)$ and let $\mathcal{E}$,$\mathcal{F}$ be two given $\mathcal{A}$-modules on X, endowed with $\mathcal{A}$-connections $D_{\mathcal{E}}$ and $D_{\mathcal{F}}$, respectively. Moreover, let ϕ be a given $\mathcal{A}$-morphism of $\mathcal{E}$ in $\mathcal{F}$, viz. $\phi \in \mathrm{Hom}_{\mathcal{A}}(\mathcal{E}, \mathcal{F}) = \mathcal{H}om_{\mathcal{A}}(\mathcal{E}, \mathcal{F})(X)$ (cf. Chapt. II; (6.9)). Then, we say that

$D_{\mathcal{E}}$ and $D_{\mathcal{F}}$ are *ϕ-related,* whenever the following diagram is commutative

(5.41)
$$\begin{array}{ccc} \mathcal{E} & \xrightarrow{\;D_{\mathcal{E}}\;} & \Omega(\mathcal{E}) \\ {\scriptstyle \phi}\downarrow & & \downarrow{\scriptstyle \phi\otimes 1_{\Omega}} \\ \mathcal{F} & \xrightarrow{\;D_{\mathcal{F}}\;} & \Omega(\mathcal{F}) \end{array}$$

that is, one has;

(5.42)
$$D_{\mathcal{F}} \circ \phi = (\phi \otimes 1_{\Omega}) \circ D_{\mathcal{E}}.$$

In particular, if ϕ in (5.42) is an *$\mathcal{A}$-isomorphism* of the $\mathcal{A}$-modules involved (cf. (5.41)), then one speaks of *gauge equivalent $\mathcal{A}$-connections;* thus, (5.42) can now be written in the form

(5.43)
$$D_{\mathcal{F}} = (\phi \otimes 1_{\Omega}) \circ D_{\mathcal{E}} \circ \phi^{-1}.$$

Thus, based now on (5.43) and the $\mathcal{A}$-isomorphism (5.39.1), our previous assertion in (5.39) and/or (5.40) says that;

(5.44) given the *vector sheaf* $\mathcal{E}$ on X and an *$\mathcal{A}$-connection* $D_{\mathcal{E}}$ of it, *the* resulting *$\mathcal{A}$-connections* $D_{\mathcal{E}nd\mathcal{E}}$ (cf. (5.27)) *and* $D_{\mathcal{E}\otimes_{\mathcal{A}}\mathcal{E}^*}$ (see (5.38)) *are gauge equivalent,* with respect to the *gauge* (viz. *$\mathcal{A}$-isomorphism*), given by (5.39.1).

The proof of (5.40) is straightforward, by the very definitions; thus, cf. (5.38), (5.33) and (5.24) (or even (5.21) for the case in hand), in conjunction with our argument, throughout the preceding, connected e.g. with (3.7).

6. Induced $\mathcal{A}$-connections (contn'd). Pull-back of $\mathcal{A}$-connections

We continue in this Section our previous study of induced $\mathcal{A}$-connections, by considering the situation that results, when *one changes the base space* of the sheaves involved.

Thus, suppose that we have a *continuous map*

$$f : X \longrightarrow Y \tag{6.1}$$

between the *topological spaces* X and Y, while we also assume that we are given a *differential triad on* Y,

$$(\mathcal{A}_Y \equiv \mathcal{A}, \partial, \Omega). \tag{6.2}$$

So we first remark that:

(6.3) *(6.2) entails a differential triad on X too,* such that one has;

$$f^*(\mathcal{A}_Y, \partial, \Omega) \equiv (f^*(\mathcal{A}), f^*(\partial), f^*(\Omega)) \tag{6.3.1}$$

(see the succeeding discussion for the notation applied). We call (6.3.1) the *pull-back of (6.2).*

Thus, *the pull-back of a differential triad, via a continuous map, is still a differential triad on the source of the given map.*

Now, concerning the *notation in (6.3.1),* we applied hereby our previous terminology in Chapt. I; Subsection 14.(b): So we know (cf. Chapt. II; (2.65) - (2.67)) that $f^*(\mathcal{A}) \equiv f^*(\mathcal{A}_Y)$ (see (6.2)) *is a* **C**-*algebra sheaf on* X (of the *same type, as* $\mathcal{A}_Y$, concerning the **C**-algebras involved, cf., for instance, Chapt. IV; (0.1)), such that the pair

$$(X, \mathcal{A}_X \equiv f^*(\mathcal{A}_Y) \equiv f^*(\mathcal{A})) \tag{6.4}$$

is now our **C**-*algebraized space,* relative to X. Moreover, on the same space X, $f^*(\Omega)$ *is an* $f^*(\mathcal{A})$-*module* (cf. Chapt. II; (2.65)). On the other hand, by considering the given **C**-derivation $\partial : \mathcal{A}_Y \equiv \mathcal{A} \longrightarrow \Omega$, as in (6.2), one also obtains a *sheaf morphism,* as well,

$$f^*(\partial) : f^*(\mathcal{A}) \longrightarrow f^*(\Omega) \tag{6.5}$$

(cf. Chapt. I; (14.41)); indeed, one easily proves that

(6.6) $f^*(\partial)$ *is a* **C**-*derivation,* as indicated (cf. Definition 1.1).

Thus, one proves e.g. that (see also Chapt. I; (14.20))

$$f_V^* : \mathcal{A}(V) \longrightarrow f^*(\mathcal{A})\left(f^{-1}(V)\right) \tag{6.7}$$

is a **C**-*algebra morphism, for every open* $V \subseteq Y$ (apply Chapt. I; (14.23)), while by a similar argument one verifies the *Leibniz condition for (6.5)* (cf. (1.2)); This already justifies our claim in (6.3). ∎

Now suppose that

$$(\mathcal{E}, D) \tag{6.8}$$

is a given pair, consisting of an $\mathcal{A}$-*module* $\mathcal{E}$ on Y (cf. (6.2)) and an $\mathcal{A}$-*connection* D *of* $\mathcal{E}$. Thus, by considering D as a *morphism of the appropriate* (**C**-*vector space) presheaves* (cf. Definition 3.1 and the relevant comments on (3.7)), one gets *a map* (we still retain, for convenience, the same symbol)

$$D : \Gamma(\mathcal{E}) \longrightarrow \Gamma(\mathcal{E}) \otimes_{\Gamma(\mathcal{A})} \Gamma(\Omega) \tag{6.9}$$

satisfying (3.5), (3.6). Therefore, one further obtains a corresponding *presheaf morphism;*

$$f^*(D) : \Gamma(f^*(\mathcal{E})) \longrightarrow \Gamma(f^*(\mathcal{E})) \otimes_{\Gamma(f^*(\mathcal{A}))} \Gamma(f^*(\Omega)), \tag{6.10}$$

such that one has

$$f^*(D)(f_V^*(t)) := f_V^*(D(t)), \tag{6.11}$$

for every (local) section $t \in \mathcal{E}(V)$, with V *open* in Y. (In this context, see Chapt. I; (14.27), (14.20) and (14.40), (14.42). Yet, concerning (6.11), we also note that $D(t)$ thereby is, by definition, a (local) *section,* viz. element, *of the presheaf, range of (6.9),* hence the notation in (6.11), its second member being thus in the range of (6.10); in fact, *we abused notation* here, using, for simplicity, the same symbol for the *two* obvious *different uses of* "f_V^*", according to Chapt. I; (14.20)). So we are now in a position to set the following.

Definition 6.1. Suppose we have the situation described by (6.1), (6.2) and (6.8). Then, the sheaf morphism corresponding to (6.10) (see Chapt. I; (8.4), (8.5) and (8.10)) is called the *pull-back of the* given $\mathcal{A}$-*connection* D, still denoted by $f^*(D)$. That is, one has;

$$\begin{aligned} f^*(D) : f^*(\mathcal{E}) \longrightarrow f^*(\mathcal{E}) \otimes_{f^*(\mathcal{A})} f^*(\Omega) &\equiv f^*(\Omega)(f^*(\mathcal{E})) \\ &\equiv f^*(\Omega(\mathcal{E})). \end{aligned} \tag{6.12}$$

Now, it is, in fact, a routine matter to verify, by actually applying (6.10), that;

(6.13) $f^*(D)$, as given by (6.12), *is an $f^*(\mathcal{A})$-connection of $f^*(\mathcal{E})$.*

On the other hand, concerning the range of (6.12), one obtains, in effect, the following result, an extension indeed of Lemma 5.1 of Chapter II (see thus Subsection 6.(a) in the sequel).

Lemma 6.1. *Suppose we have a continuous map $h : X \longrightarrow Y$, with $(Y, \mathcal{A})$ a $\mathbf{C}$-algebraized space, while $\mathcal{E}$,$\mathcal{F}$ are two given $\mathcal{A}$-modules on Y. Then, one obtains the following relation;*

$$h^*(\mathcal{E} \otimes_{\mathcal{A}} \mathcal{F}) = h^*(\mathcal{E}) \otimes_{h^*(\mathcal{A})} h^*(\mathcal{F}), \tag{6.14}$$

valid, within an $h^*(\mathcal{A})$-isomorphism of the $h^*(\mathcal{A})$-modules on X concerned.

Proof. Applying Theorem 9.1 of Chapter I on the members of (6.14) and then based on the rels. (14.20), (14.23) and (14.27) of the same Chapter, one gets a *presheaf morphism*

$$\Gamma(h^*(\mathcal{E} \otimes_{\mathcal{A}} \mathcal{F})) \longrightarrow \Gamma(h^*(\mathcal{E}) \otimes_{\Gamma(h^*(\mathcal{A}))} \Gamma(h^*(\mathcal{F})); \tag{6.15}$$

now, this in turn yields (same Chapter I; (8.4), (8.5)) an associated to (6.15) *sheaf morphism* of the corresponding sheaves (in fact, $h^*(\mathcal{A})$-*modules* on X), as in (6.14). However, the latter is, in effect, a *sheaf isomorphism,* being such *fiber-wise* (cf. also Chapt. I; (1.21)); indeed, one has the following *(canonical) isomorphisms*

$$h^*(\mathcal{E} \otimes_{\mathcal{A}} \mathcal{F})_x = h^*(\mathcal{E})_x \otimes_{h^*(\mathcal{A})_x} h^*(\mathcal{F})_x = (h^*(\mathcal{E}) \otimes_{h^*(\mathcal{A})} h^*(\mathcal{F}))_x, \tag{6.16}$$

for every point $x \in X$ (cf. Chapt. I; (14.36) and Chapt. II; (5.9)). Yet, the above isomorphisms are, of course, *module structure preserving* (see also Chapt. II; (1.44)), which thus terminates the proof. ∎

Furthermore, referring still to (6.12), by analogy with the classical case (see, for instance, J.W. Milnor - J.D. Stasheff [1: p. 292, Lemma 3]), one has the following *commutative diagram* (cf. also (6.10))

$$
(6.17)\qquad
\begin{array}{ccc}
\Gamma(\mathcal{E}) & \xrightarrow{\quad D \quad} & \Gamma(\mathcal{E}) \otimes_{\Gamma(\mathcal{A})} \Gamma(\Omega) \equiv \Gamma(\Omega(\mathcal{E})) \\
\Gamma(f^*) \Big\downarrow & & \Big\downarrow \Gamma(f^*) \\
\Gamma(f^*(\mathcal{E})) & \xrightarrow{\quad f^*(D) \quad} & \Gamma(f^*(\mathcal{E}) \otimes_{\Gamma(f^*(\mathcal{A}))} \Gamma(f^*(\Omega)) = \Gamma(f^*(\Omega(\mathcal{E}))).
\end{array}
$$

Indeed, our claim can be proved by verifying it, for instance, *fiber-wise*, which thus follows from (6.16). (Yet, for simplicity, we employed in (6.17) an obvious *abuse of notation*, concerning the two *different* actually *morphisms* $\Gamma(f^*)$; see Chapt. I; (14.20)). ∎ Now, applying a similar argument, as for (6.15), one concludes the validity of an analogous *commutative diagram* for the sheaf morphisms, associated with (6.17). So extending, in an obvious way, the terminology of Definition 5.1 above, we still express (6.17) and/or its *sheaf analogue*, as before, by saying that;

(6.18) *D and $f^*(D)$ are f^*-related.*

6.(a). Restriction of $\mathcal{A}$-connections.- The preceding have a special bearing, when one looks at an *open* subspace U of a given *topological space* X, where the latter space is endowed with a *differential triad* $(\mathcal{A}, \partial, \Omega)$ (see (3.1)); yet, we assume that we are given a pair

$$(6.19)\qquad (\mathcal{E}, D),$$

consisting now of an *$\mathcal{A}$-module $\mathcal{E}$ on X* and an *$\mathcal{A}$-connection D* of $\mathcal{E}$. Thus, as an application of our previous argument, our task now is to *transfer* the situation just described *on U, via the* canonical *inclusion map*

$$(6.20)\qquad i_U \equiv i : U \subsetneq X.$$

So applying first (6.3.1), one concludes that;

(6.21) the *restriction on U of the* given *differential triad* $(\mathcal{A}, \partial, \Omega)$ on X, thus, *by definition,* the triple

(6.21.1) $i^*(\mathcal{A}, \partial, \Omega) = (i^*(\mathcal{A}), i^*(\partial), i^*(\Omega)) \equiv (\mathcal{A}|_U, \partial|_U, \Omega|_U)$,

yelds a *differential triad on $U \subseteq X$.*

Therefore, by further employing (6.13), one infers that:

(6.22) $i^*(D) \equiv D|_U$, called the *restriction of D on U,* (viz. the *pull-back of D via (6.20)*), is an *$\mathcal{A}|_U$-connection* of the *$\mathcal{A}|_U$-module* $i^*(\mathcal{E}) \equiv \mathcal{E}|_U$, restriction of $\mathcal{E}$ to U.

Of course, one sets directly;

$$(6.23) \qquad (D|_U)(s|_U) := (Ds)|_U,$$

such that

$$(6.24) \qquad (D|_U)(s|_U)(x) = (Ds)(x),$$

for any $s \in \mathcal{E}(U)$ and $x \in U$ (cf. Chapt. I; (14.23), (14.27), as well as, Chapt. II; (5.25)).

The preceding are going to be used right below (see thus Section 8). Yet, we can still examine the *local expression of an $\mathcal{A}$-connection, pull-back.* (Cf. also Chapt. VIII and Chapt. I; (14.23), (14.34)).

7. Affine space of $\mathcal{A}$-connections

Suppose that we are given a *differential triad* $(\mathcal{A}, \partial, \Omega)$ on a *topological space* X (cf. e.g. (3.1)), and let $\mathcal{E}$ be an *$\mathcal{A}$-module* on X. Then, we denote by

$$(7.1) \qquad Conn_{\mathcal{A}}(\mathcal{E})$$

the *set of $\mathcal{A}$-connections of $\mathcal{E}$*; of course, the set (7.1) might be just the *empty one!,* even if $\mathcal{E}$ was a vector sheaf on X. (We shall discuss the relevant situation later on, giving also conditions guaranteeing the non-emptyness of the set (7.1). Thus, we make the *basic assumption,* throughout this Section, that;

$$(7.2) \qquad Conn_{\mathcal{A}}(\mathcal{E}) \neq \emptyset.$$

Now, suppose that we are given an *$\mathcal{A}$-connection D of $\mathcal{E}$,* as before, viz. let

(7.3) $$D \in \mathcal{C}onn_{\mathcal{A}}(\mathcal{E}).$$

Then, according to Definition 3.1 (cf. (3.5)), one has

(7.4) $$D \in \mathrm{Hom}_{\mathbf{C}}(\mathcal{E}, \Omega(\mathcal{E})) = \mathcal{H}om_{\mathbf{C}}(\mathcal{E}, \Omega(\mathcal{E})(X),$$

where one looks at $\mathbf{C}$ as the *constant* $\mathbf{C}$-*algebra sheaf* on X (cf. Chapt. II; (1.27), (1.32), (1.34), along with (6.9), for $\mathcal{A} = \mathbf{C}_X \equiv \mathbf{C}$. Yet, see ibid.; (1.38)). Consequently, one obtains so far that;

(7.5) $$\mathcal{C}onn_{\mathcal{A}}(\mathcal{E}) \subseteq \mathrm{Hom}_{\mathbf{C}}(\mathcal{E}, \mathcal{E} \otimes_{\mathcal{A}} \Omega) \equiv \mathrm{Hom}_{\mathbf{C}}(\mathcal{E}, \Omega(\mathcal{E})),$$

while *the last set in (7.5) is* further *characterized by (3.6) (Leibniz condition).* In this context, we next make the easy, however, very useful observation, that

(7.6) *given two* $\mathcal{A}$-*connections* D *and* D' *of* $\mathcal{E}$, as above, *one has;*

(7.6.1) $$D' - D \in \mathrm{Hom}_{\mathcal{A}}(\mathcal{E}, \Omega(\mathcal{E})) \subseteq \mathrm{Hom}_{\mathbf{C}}(\mathcal{E}, \Omega(\mathcal{E})).$$

Indeed, we have already seen (7.6.1) in the preceding (cf. (5.29)), in the particular case that $\mathcal{E}$ is a vector sheaf on X. But, the same assertion is still valid, *for any* $\mathcal{A}$-*module* $\mathcal{E}$, as above, as an immediate consequence of (3.6) (viz. the *Leibniz condition*), applied to each one of the two summands in $D - D'$. ∎

Now, our next objective is to show that such differences constitute, in fact, the elements of the *"model"* (viz. the first set in (7.6.1)), which the previous set (7.1) can be built on, in a sense that we are going to specify right away (see thus (7.15) below). So our next result has now as follows.

Lemma 7.1. *Let* $(\mathcal{A}, \partial, \Omega)$ *be a differential triad on a topological space* X *and* $\mathcal{E}$ *an* $\mathcal{A}$-*module on* X. *Then, for every* $\mathcal{A}$-*connection* D *of* $\mathcal{E}$ (see also (7.2)), *one obtains a bijection*

(7.7) $$\tau_D : \mathrm{Hom}_{\mathcal{A}}(\mathcal{E}, \Omega(\mathcal{E})) \longrightarrow \mathcal{C}onn_{\mathcal{A}}(\mathcal{E})$$

given by the relation;

(7.8) $$\tau_D(u) := D + u.$$

What amounts to the same, given an $\mathcal{A}$-connection D of $\mathcal{E}$, as above, *for every* $u \in \mathrm{Hom}_{\mathcal{A}}(\mathcal{E}, \Omega(\mathcal{E}))$, *there exists a uniquely defined* $\mathcal{A}$-*connection* D' *of* $\mathcal{E}$, *such that*

$$u = D' - D; \tag{7.9}$$

viz. one has

$$D' := D + u. \tag{7.10}$$

Proof. We first remark that *the map (7.7)*, as given by (7.8), *is well-defined*, viz. *one has;*

$$D + u \in Conn_{\mathcal{A}}(\mathcal{E}), \tag{7.11}$$

for any $D \in Conn_{\mathcal{A}}(\mathcal{E})$ *and* $u \in \mathrm{Hom}_{\mathcal{A}}(\mathcal{E}, \Omega(\mathcal{E}))$. It is clear that it suffices only to check the *Leibniz condition for* $D + u \in \mathrm{Hom}_{\mathbf{C}}(\mathcal{E}, \Omega(\mathcal{E}))$ (cf. (3.6)). Thus, *one has;*

$$\begin{aligned} (D+u)(\alpha \cdot s) &= D(\alpha s) + u(\alpha s) = \alpha \cdot D(s) + s \otimes \partial\alpha + \alpha \cdot u(s) \\ &= \alpha \cdot (D+u)(s) + s \otimes \partial(\alpha), \end{aligned} \tag{7.12}$$

for any $\alpha \in \mathcal{A}(U)$ *and* $s \in \mathcal{E}(U)$, with U *open* in X, as was to be proved. Moreover, as already noted, it is clear that;

$$D + u \in \mathrm{Hom}_{\mathbf{C}}(\mathcal{E}, \Omega(\mathcal{E})), \tag{7.13}$$

since this holds true for each one of the summands of $D + u$.

On the other hand, it is still obvious by (7.8), that *(7.7) is one-to-one.* Yet, for every $D' \in Conn_{\mathcal{A}}(\mathcal{E})$, one defines (see (7.6.1))

$$u := D' - D \in \mathrm{Hom}_{\mathcal{A}}(\mathcal{E}, \Omega(\mathcal{E})), \tag{7.14}$$

so that, in view of (7.8), the map (7.7) *is onto, as well,* which, in effect, terminates the proof. ∎

Now, as a result of the previous lemma, one obtains the following *basic relation;*

$$Conn_{\mathcal{A}}(\mathcal{E}) = D + \mathrm{Hom}_{\mathcal{A}}(\mathcal{E}, \Omega(\mathcal{E})), \tag{7.15}$$

within a *bijection,* defined through (7.8), *for any given* $D \in Conn_{\mathcal{A}}(\mathcal{E})$ (cf. (7.3)). Accordingly, it is instructive to look now at (7.15), in terms of the (geometric) language of *"affine spaces"*. So, for convenience, we first comment a bit on the relevant terminology through the following.

Scholium 7.1.- The succeeding terminological comments will facilitate the statement of our next result (see thus Theorem 7.1 below).

So assume that we are given a (commutative unital) *ring* R and a (left) *R-module* M.

Thus, by an *affine space, modelled on the R-module M,* one means a non-empty set E and a map

$$\phi : E \times E \longrightarrow M : (x, y) \longmapsto \phi(x, y) \equiv y - x \equiv \overrightarrow{xy} \tag{7.16}$$

such that the two following conditions are fulfilled:

i) *For every $x \in E$*, the partial map of ϕ,

$$\phi_x : E \longmapsto M : y \longmapsto \phi_x(y) := \phi(x, y), \tag{7.17}$$

is one-to-one.

ii) *For any $x, y, z \in E$, one has* (cf. (7.18));

$$\overrightarrow{xy} + \overrightarrow{yz} + \overrightarrow{zx} = 0. \tag{7.18}$$

(We call (7.18) the *triangle axiom,* or even the *Chasles law*).

We come now to the statement of the following basic conclusion of the previous discussion. That is, one has.

Theorem 7.1. *Let $(\mathcal{A}, \partial, \Omega)$ be a given differential triad on a topological space X and $\mathcal{E}$ an $\mathcal{A}$-module on X. Then, the set of $\mathcal{A}$-connections of $\mathcal{E}$, $Conn_{\mathcal{A}}(\mathcal{E})$* (cf. also (7.2)), *is an affine space, modelled on the $\mathcal{A}(X)$-module* $\mathrm{Hom}_{\mathcal{A}}(\mathcal{E}, \Omega(\mathcal{E}))$.

In particular, if $\mathcal{E}$ is a vector sheaf on X, then (7.15) becomes;

$$Conn_{\mathcal{A}}(\mathcal{E}) = D + \Omega(\mathcal{E}nd\, \mathcal{E})(X), \tag{7.19}$$

within a bijection (see (7.7)), *for every $D \in Conn_{\mathcal{A}}(\mathcal{E})$. That is, $Conn_{\mathcal{A}}(\mathcal{E})$ is now an affine space, modelled on the $\mathcal{A}(X)$-module $\Omega(\mathcal{E}nd\, \mathcal{E})(X)$.*

Proof. Specializing the terminology of the previous Scholium 7.1 to the case, under consideration, E (viz. the *affine space,* we look for) is *the set (7.1), assuming, of course, (7.2),* while M (the *"model"*) the *$\mathcal{A}(X)$-module* $\mathrm{Hom}_{\mathcal{A}}(\mathcal{E}, \Omega(\mathcal{E}))$; the latter reduces to $\Omega(\mathcal{E}nd\, \mathcal{E})(X)$, whenever $\mathcal{E}$ is a *vector sheaf* on X (see (7.22) below). Furthermore, the map corresponding to (7.16) is that one, defined by the *difference of two* given $\mathcal{A}$-*connections* of $\mathcal{E}$, that is,

$$\mathit{Conn}_{\mathcal{A}}(\mathcal{E}) \times \mathit{Conn}_{\mathcal{A}}(\mathcal{E}) \longrightarrow \mathrm{Hom}_{\mathcal{A}}(\mathcal{E}, \Omega(\mathcal{E})), \tag{7.20}$$

such that (cf. (7.6.1))

$$(D, D') \longmapsto D' - D. \tag{7.21}$$

Thus, (7.18) is obvious, by the very definition of (7.21), while (7.17) is just (7.15); so this proves already our assertion, concerning the *affine space* $\mathit{Conn}_{\mathcal{A}}(\mathcal{E})\ (\neq \emptyset)$.

Now, if $\mathcal{E}$ is a *vector sheaf* on X, one has (cf. Chapt. IV; Corollary 6.1)

$$\begin{aligned} \mathcal{H}om_{\mathcal{A}}(\mathcal{E}, \Omega(\mathcal{E})) &\equiv \mathcal{H}om_{\mathcal{A}}(\mathcal{E}, \mathcal{E} \otimes_{\mathcal{A}} \Omega) = \mathcal{H}om_{\mathcal{A}}(\mathcal{E}, \mathcal{E}) \otimes_{\mathcal{A}} \Omega \\ &\equiv (\mathcal{E}nd\, \mathcal{E}) \otimes_{\mathcal{A}} \Omega \equiv \Omega(\mathcal{E}nd\, \mathcal{E}), \end{aligned} \tag{7.22}$$

so that one obtains (see Chapt. II; (6.9));

$$\mathrm{Hom}_{\mathcal{A}}(\mathcal{E}, \Omega(\mathcal{E})) = \mathcal{H}om_{\mathcal{A}}(\mathcal{E}, \Omega(\mathcal{E})(X) = \Omega(\mathcal{E}nd\, \mathcal{E})(X), \tag{7.23}$$

which thus yields our assertion, concerning (7.19), and this completes the proof. ∎

Corollary 7.1. *Let $\mathcal{L}$ be a line sheaf on X, together with an $\mathcal{A}$-connection D on $\mathcal{L}$. Then, one has*

$$\mathit{Conn}_{\mathcal{A}}(\mathcal{L}) = D + \Omega(X), \tag{7.24}$$

within a bijection, given by (7.8). Thus, $\mathit{Conn}_{\mathcal{A}}(\mathcal{L})$ is an affine space, modelled on the $\mathcal{A}(X)$-module $\Omega(X)$.

Proof. The assertion is, of course, a specialization of the second part of Theorem 7.1 to the case of a *line sheaf* $\mathcal{L}$ on X; thus, (7.24) follows from (7.19), in view of the $\mathcal{A}$-*isomorphism* (cf. Chapt. II; Lemma 6.2)

$$\mathcal{E}nd\, \mathcal{L} = \mathcal{A}, \tag{7.25}$$

so that one then obtains $\Omega(\mathcal{A}) \equiv \mathcal{A} \otimes_{\mathcal{A}} \Omega = \Omega$ (still up to an $\mathcal{A}$-*isomorphism*, ibid.; (5.15)), and this terminates the proof. ∎

Yet, as a further application of the classical *theory of affine spaces*, one obtains the following.

Lemma 7.2. *Let $(\mathcal{A}, \partial, \Omega)$ be a given differential triad on a topological space X and $\mathcal{E}$ an $\mathcal{A}$-module on X. Then, for any two different and non-trivial $\mathcal{A}$-connections D and D' of $\mathcal{E}$, the corresponding $\mathcal{A}(X)$-line, they determine, that is, the set*

$$L_{D,D'}^{\mathcal{A}(X)} = \left\{\widetilde{D} \equiv \alpha \cdot D + (1-\alpha) \cdot D' : \alpha \in \mathcal{A}(X)\right\} \tag{7.26}$$

is contained in $Conn_{\mathcal{A}}(\mathcal{E})$.

Proof. First, one defines;

$$(\alpha \cdot D)(s) := \alpha \cdot D(s) \in \Omega(\mathcal{E})(U), \tag{7.27}$$

for any $\mathcal{A}$-connection D of $\mathcal{E}$, $\alpha \in \mathcal{A}(X)$ and $s \in \mathcal{E}(U)$, with U *open* in X. Hence, by hypothesis for D and D', the set (7.26) *is a subset of the* **C**-*vector space* $\mathrm{Hom}_{\mathbf{C}}(\mathcal{E}, \Omega(\mathcal{E}))$ (cf. (7.6)). Now, the assertion about (7.26) is straightforward by the very definitions. ∎

In particular, one concludes that;

(7.28) *for any two points $D \neq D'$ in $Conn_{\mathcal{A}}(\mathcal{E})$* (cf. also (7.2)), the corresponding **C**-*line,* which *they determine* in the **C**-*vector space* $\mathrm{Hom}_{\mathbf{C}}(\mathcal{E}, \Omega(\mathcal{E}))$, *lies actually in $Conn_{\mathcal{A}}(\mathcal{E})$.* That is, *one has;*

$$\lambda D + (1-\lambda) D' \in Conn_{\mathcal{A}}(\mathcal{E}), \tag{7.28.1}$$

for any $\lambda \in \mathbf{C}$ and $D, D' \in Conn_{\mathcal{A}}(\mathcal{E})$, with $D \neq D'$.

For the classical case of Differential Geometry, see, for instance, W.A. Poor [1: p.85, and p.86, Proposition 2.71], or even J.-L. Brylinski [1: p.71, Proposition 2.2.2].

So, according to the classical terminology, the $\mathcal{A}$-*morphism*

$$u = D' - D \in \mathrm{Hom}_{\mathcal{A}}(\mathcal{E}, \Omega(\mathcal{E})) \tag{7.29}$$

(see (7.14)), or in the case of a *vector sheaf* $\mathcal{E}$ on X, that one given by

$$u = D' - D \in \Omega(\mathcal{E}nd\, \mathcal{E})(X) \tag{7.30}$$

(see (7.22)), is, in effect, the so-called *connection difference (*End $\mathcal{E}$*-valued) 1-form* on X, a nomenclature justified by the *standard example* in Section 2 above (cf. also, for instance, (4.15)). In particular, this is really the case, viz. one has

(7.31) $$u = D' - D \in \Omega(X),$$

namely, a *"1-form"* on X (the same example, as before, especially, (2.3), (2.4)), when we are given a *line sheaf* $\mathcal{L}$ on X (cf. (7.24)).

On the other hand, the preceding rels. (7.15) and/or (7.19) permit one to employ, for convenience, even in the classical theory, an obvious *abuse of language,* by *identifying,* namely, *the two sets* appeared in either one of the relations in question (yet, in particular, of (7.19)). This happens, especially with (7.24), when one speaks usually of an *identification of $\mathcal{A}$-connections of* the given line sheaf $\mathcal{L}$ *with 1-forms on X.* (In this concern, see also, however, the next Chapter VII).

Scholium 7.2.- Before we leave this Section, we also note for later use the following more general aspect of the above Lemma 7.2:

Namely, suppose that we are given a *family*

(7.32) $(\alpha_i)_{i\in I} \subseteq \mathcal{A}(X)$, with *finite support;*

viz. one has $\alpha_i = 0 \in \mathcal{A}(X)$, at the exception only of finite many indices $i \in I$. Yet, we assume that

(7.33) $$\sum_{i\in I} \alpha_i = 1 \in \mathcal{A}(X).$$

(The latter sum has, of course, a meaning, in view of (7.32)). Finally, consider any family of *$\mathcal{A}$-connections* of the given *$\mathcal{A}$-module* $\mathcal{E}$ on X

(7.34) $$(D_i)_{i\in I} \subseteq Conn_{\mathcal{A}}(\mathcal{E})$$

(indexed by the same I, as in (7.32), but not necessarily of finite support). Thus, under the above circumstances, *one* now *concludes that;*

(7.35) $$\sum_{i\in I} \alpha_i D_i \in Conn_{\mathcal{A}}(\mathcal{E}).$$

The previous sum is taken, in principle, in the *$\mathcal{A}(X)$-module* $\mathrm{Hom}_{\mathbf{C}}(\mathcal{E}, \Omega(\mathcal{E}))$ (see also (7.27)). Then, (7.33) contributes to prove that *(7.35) is,* in effect, *an $\mathcal{A}$-connection of $\mathcal{E}$*, as well. ▮

8. Localization of $\mathcal{A}$-connections

We come by the present Section to one of the main points of our discussion, as is the *existence of $\mathcal{A}$-connections for a given $\mathcal{A}$-module $\mathcal{E}$* on X. Thus, the suitable tool hereby, as we shall presently see, is *sheaf cohomology* (Chapter III), while we shall also restrict ourselves to *vector sheaves* on X. So, following the classical pattern (of *Differential Geometry*), our first task is to show that:

(8.1) every *vector sheaf* on X, can always be endowed with a *"connection 1-cocycle"*.

Now, the importance of the previous information lies in the fact that;

(8.2) *the vanishing of the* corresponding *cohomology class of the previous 1-cocycle guarantees the existence of an $\mathcal{A}$-connection for the vector sheaf* in question.

Indeed, we further note, in anticipation, that the aforementioned *vanishing* (of the *"connection cohomology class"*) yields, in fact, a *criterion* (the counterpart in our case of the standard *Atiyah's criterion*) *for the existence of an $\mathcal{A}$-connection.* However, this second information, as in (8.2), along with its consequences, will be our object of study in the ensuing few Sections. Thus, we start with fixing the relevant terminology, which we shall use throughout the sequel.

So assume, as usual, that we are given a *differential triad* $(\mathcal{A}, \partial, \Omega)$ on a *topological space* X, base space of a (fixed) $\mathbf{C}$-*algebraized space* $(X, \mathcal{A})$.

On the other hand, we have already seen that:

i) we can always consider the *n-th dimensional extension of the* given *differential triad* $(\mathcal{A}, \partial, \Omega)$, as above, viz. the triple

$$(\mathcal{A}^n, \partial^n, \Omega^n), \tag{8.3}$$

for every $n \in \mathbf{N}$ (cf. (5.6)), getting thus a *free $\mathcal{A}$-module,* of *rank* n, equipped with an *$\mathcal{A}$-connection;* that is, one has

$$\partial^n : \mathcal{A}^n \longrightarrow \mathcal{A}^n \otimes_{\mathcal{A}} \Omega \equiv \Omega(\mathcal{A}^n) = \Omega(\mathcal{A})^n = \Omega^n \tag{8.4}$$

(see also (5.5) and (4.1)). Furthermore,

ii) one can always take the *localization,* or even *restriction,* of the previous pair $(\mathcal{A}^n, \partial^n)$, *to any open* set $U \subseteq X$, according to our argument in Section 6 in the preceding. That is, one has (cf. (6.21), (6.22));

$$(\mathcal{A}^n, \partial^n)|_U = (\mathcal{A}^n|_U, \partial^n|_U) = ((\mathcal{A}|_U)^n, (\partial|_U)^n) \tag{8.5}$$

(see also e.g. Chapt. II; (4.5)), getting thus an $\mathcal{A}|_U$*-module* (in fact, a *free one,* of *rank* n), endowed with an $\mathcal{A}|_U$*-connection*

$$\begin{aligned} \partial^n|_U &= (\partial|_U)^n : \mathcal{A}^n|_U = (\mathcal{A}|_U)^n \longrightarrow \mathcal{A}^n|_U \otimes_{\mathcal{A}|_U} \Omega|_U \\ &= (\mathcal{A}^n \otimes_{\mathcal{A}} \Omega)|_U = ((\mathcal{A} \otimes_{\mathcal{A}} \Omega)|_U)^n = \Omega^n|_U = (\Omega|_U)^n \end{aligned} \tag{8.6}$$

(*commutativity of Whitney sum and localization,* referred to a given *differential triad.* See also Chapt. II; Lemma 5.1 and (5.18)).

Now, it is clear that one can *transfer the previous data in ii), via any* given *local isomorphism of* the *sheaf modules* involved. This is, in effect, the *key idea* behind our next result, lying at the basis of all that follows, as well.

On the other hand, although the $\mathcal{A}$*-connection of* $\mathcal{A}^n$, which we shall use in the sequel is (8.4), *for the sake of generality* (but, no extra cost is paid for that, anyway!) we do consider right below an arbitrary $\mathcal{A}$*-connection* d *of* $\mathcal{A}^n$. However, due to (7.19), one actually obtains;

$$d = \partial^n + u, \tag{8.7}$$

for some *uniquely defined*

$$\begin{aligned} u &\in \Omega(\mathcal{E}nd\, \mathcal{A}^n)(X) = \Omega(\mathcal{A}^{n^2})(X) = \Omega(\mathcal{A})^{n^2}(X) \\ &= \Omega^{n^2}(X) \equiv M_n(\Omega)(X) = M_n(\Omega(X)) \equiv \Omega(X)^{n^2} \end{aligned} \tag{8.8}$$

(see Chapt. IV; (3.23) and Chapt. II; (5.18), (3.14)). In this regard, we still note that the *existence of* ∂ *entails that one of* ∂^n, *for every* $n \in \mathbf{N}$ (cf. (5.4)), *hence, of* d *too,* for some unique u, as in (8.8).

Thus, we are now in the position to start with our first basic information, pertaining to the *effect of the data in ii),* as above, *to a given vector sheaf on* X (this, of course, *to a local extent!*). Indeed, as was already hinted at in the preceding, *the situation can be copied* (locally!), *through any* given *local frame of the vector sheaf* concerned. Precisely, one thus obtains the next.

Lemma 8.1. *Suppose we are given a differential triad* $(\mathcal{A}, \partial, \Omega)$ *on a topological space* X *and a vector sheaf* $\mathcal{E}$ *on* X *of rank* $n \in \mathbf{N}$. *Moreover,* *assume that we are given an $\mathcal{A}$-connection of* (the free $\mathcal{A}$-module) $\mathcal{A}^n$,

$$d : \mathcal{A}^n \longrightarrow \mathcal{A}^n \otimes_{\mathcal{A}} \Omega \equiv \Omega(\mathcal{A}^n) = \Omega^n \tag{8.9}$$

(we can take, for instance, the *$\mathcal{A}$-connection* ∂^n, cf. (8.4), (8.7)), *while let*

$$\mathcal{U} = (U_\alpha)_{\alpha \in I} \tag{8.10}$$

be a local frame of $\mathcal{E}$. Then, for every $\alpha \in I$, one defines the following commutative diagram;

$$\begin{array}{ccc} \mathcal{E}|_{U_\alpha} & \xrightarrow{\quad D_\alpha \quad} & \mathcal{E} \otimes_{\mathcal{A}} \Omega|_{U_\alpha} \equiv \Omega(\mathcal{E})|_{U_\alpha} = \mathcal{E}|_{U_\alpha} \otimes_{\mathcal{A}|_{U_\alpha}} \Omega|_{U_\alpha} \\ \eta_\alpha \Big\downarrow & & \Big\uparrow \eta_\alpha^{-1} \otimes 1_{\Omega|_{U_\alpha}} \\ \mathcal{A}^n|_{U_\alpha} & \xrightarrow{\quad d_\alpha \equiv d|_{U_\alpha} \quad} & \mathcal{A}^n \otimes_{\mathcal{A}} \Omega|_{U_\alpha} = \Omega^n|_{U_\alpha} \end{array} \tag{8.11}$$

Here

$$\eta_\alpha : \mathcal{E}|_{U_\alpha} \longrightarrow \mathcal{A}^n|_{U_\alpha} = (\mathcal{A}|_{U_\alpha})^n, \quad \alpha \in I, \tag{8.12}$$

denotes the *(canonical) $\mathcal{A}|_{U_\alpha}$-isomorphism,* provided by our hypothesis for $\mathcal{E}$ and $\mathcal{U}$ (cf. Chapt. II; (4.7), (4.8)), *so that one then sets;*

$$D_\alpha := \left(\eta_\alpha^{-1} \otimes 1_{\Omega|_{U_\alpha}}\right) \circ d_\alpha \circ \eta_\alpha, \quad \alpha \in I. \tag{8.13}$$

Then, for any $\alpha \in I$, D_α yields an $\mathcal{A}|_{U_\alpha}$-connection of the (free) $\mathcal{A}|_{U_\alpha}$-module $\mathcal{E}|_{U_\alpha}$.

Scholium 8.1.– As already remarked in the foregoing, the idea of the previous lemma is that *any $\mathcal{A}$-connection of $\mathcal{A}^n$*, for which *one has, of course* (cf. (7.1)),

$$Conn_{\mathcal{A}}(\mathcal{A}^n) \neq \emptyset, \quad n \in \mathbf{N}, \tag{8.14}$$

whenever we are given ∂ (cf. (5.4)), *can be transferred locally,* modulo a local $\mathcal{A}$-isomorphism, *to any other $\mathcal{A}$-module on X.* Therefore, naturally, to a given *vector sheaf* on X, through any *local frame* of it. So, roughly speaking, the lesson thus far is that;

(8.15) *given a differential triad* $(\mathcal{A}, \partial, \Omega)$ on a topological space X, *$\mathcal{A}$-connections always exist locally!*

(for *vector sheaves* on X), a fact, however, that, as we shall see, through the ensuing discussion, *does not hold, in general, true globally.*

Yet, another issue of (8.11), which should be clarified at this point, is that, *whenever one has an $\mathcal{A}$-isomorphism $\phi : \mathcal{E} \longrightarrow \mathcal{F}$ between* two given *$\mathcal{A}$-modules $\mathcal{E}$ and $\mathcal{F}$* on X, and an *$\mathcal{A}$-connection D of $\mathcal{E}$*, then *the following diagram can be completed* (through the dashed map, becoming thus commutative), so as to give an *$\mathcal{A}$-connection, say D', of $\mathcal{F}$*; i.e., one has

(8.16)
$$\begin{array}{ccc} \mathcal{E} & \xrightarrow{\ \phi\ } & \mathcal{F} \\ {\scriptstyle D}\downarrow & & \downarrow{\scriptstyle D'} \\ \Omega(\mathcal{E}) & \xrightarrow[\phi \otimes 1_{\Omega}]{} & \Omega(\mathcal{F}). \end{array}$$

The assertion, as already noted in the foregoing, is an immediate consequence of the very definitions and the hypothesis for $\phi \in \mathrm{Hom}_{\mathcal{A}}(\mathcal{E}, \mathcal{F})$. So one has, *by definition,*

(8.17)
$$D' := (\phi \otimes 1_{\Omega}) \circ D \circ \phi^{-1};$$

that is, D' is thus forced to be *gauge equivalent to D, with respect to* ϕ (see (5.43)). Said otherwise, one thus infers that;

(8.18) *elements of* $\mathrm{Isom}_{\mathcal{A}}(\mathcal{E}, \mathcal{F})$ (set of *$\mathcal{A}$-isomorphisms* between the *$\mathcal{A}$-modules* $\mathcal{E}$ and $\mathcal{F}$) *preserve $\mathcal{A}$-connections,* such that *the resulting $\mathcal{A}$-connections of $\mathcal{F}$*, through the element of $\mathrm{Isom}_{\mathcal{A}}(\mathcal{E}, \mathcal{F})$, under consideration, *are gauge equivalent to a given $\mathcal{A}$-connection of $\mathcal{E}$.*

Indeed, we are going to use (8.18), *suitably localized,* straight away. So we come next to the

Proof of Lemma 8.1. The only thing we have to check is that D_α, $\alpha \in I$, as given by (8.13), is an *$\mathcal{A}|_{U_\alpha}$-connection of $\mathcal{E}|_{U_\alpha}$*: Now, based on the same relation, it is clear that (see e.g. Chapt. II; (6.8))

$$D_\alpha \in \mathrm{Hom}_{\mathbf{C}}(\mathcal{E}|_{U_\alpha}, \Omega(\mathcal{E})|_{U_\alpha}) = \mathcal{H}om_{\mathbf{C}}(\mathcal{E}, \Omega(\mathcal{E}))(U_\alpha),\ \alpha \in I, \tag{8.19}$$

hence, it remains to prove the corresponding *Leibniz condition for* D_α (see Definition 3.1). However, taking into account that η_α (cf. (8.12)) is an $\mathcal{A}|_{U_\alpha}$-*isomorphism,* the desired property follows already from (8.18) *localized to* $U_\alpha \in \mathcal{U}$ (see (8.10), along with (8.17)) and the fact that

$$d_\alpha := d|_{U_\alpha},\quad \alpha \in I, \tag{8.20}$$

is an $\mathcal{A}|_{U_\alpha}$*-connection of* $\mathcal{A}^n|_{U_\alpha}$, according to the hypothesis for d (cf. (8.9)) and to (6.22), which thus terminates the proof. ∎

Thus, following (8.17), we still conclude, in view of (8.13), that:

(8.21) the $\mathcal{A}|_{U_\alpha}$*-connections* d_α *and* D_α are *gauge equivalent,* through the $\mathcal{A}|_{U_\alpha}$*-isomorphism*

$$\eta_\alpha \in \mathrm{Isom}_{\mathcal{A}|_{U_\alpha}}(\mathcal{E}|_{U_\alpha}, \mathcal{A}^n|_{U_\alpha}) = \mathcal{I}som_{\mathcal{A}}(\mathcal{E}, \mathcal{A}^n)(U_\alpha) \tag{8.21.1}$$

(cf. (8.23) in the sequel), *for every* $\alpha \in I$.

In this context, it is still useful, for later use too, to comment a bit on the following *(physics oriented!)* terminology: Namely, every element of the set

$$\mathrm{Isom}_{\mathcal{A}}(\mathcal{E}, \mathcal{F}) = \mathcal{I}som_{\mathcal{A}}(\mathcal{E}, \mathcal{F})(X) \tag{8.22}$$

(see Chapt. V; (2.33) for the notation applied) is still a *gauge transformation* between the given $\mathcal{A}$*-modules* $\mathcal{E}$ and $\mathcal{F}$. (Of course, the terminology fits, in particular, well, when either one of $\mathcal{E}$,$\mathcal{F}$ is a *free* $\mathcal{A}$*-module,* of *finite rank*). Now, by a *local gauge transformation* between $\mathcal{E}$ and $\mathcal{F}$, one means any *local section of the sheaf* $\mathcal{I}som_{\mathcal{A}}(\mathcal{E}, \mathcal{F})$ (ibid.; (2.31)); viz. any element of the set

$$\mathcal{I}som_{\mathcal{A}}(\mathcal{E}, \mathcal{F})(U) = \mathrm{Isom}_{\mathcal{A}|_U}(\mathcal{E}|_U, \mathcal{F}|_U) \subseteq \mathrm{Hom}_{\mathcal{A}|_U}(\mathcal{E}|_U, \mathcal{F}|_U), \tag{8.23}$$

with U any *open* set in X (see e.g. (8.21.1) above).

Thus, the obvious, however, *fundamental fact* at this point is that (another formulation actually of (8.18));

(8.24) *(local) gauge transformations preserve* ($\mathcal{A}$-)*connections,* the *gauge transform of* a given $\mathcal{A}$*-connection* being supplied by (an analogous relation to) (8.17) (cf., for instance, (8.13)).

On the other hand, supplementing hereby our previous terminology in Section 5, we further remark that (5.42), in conjunction with (5.13), *can* also *be written in the form;*

$$D_{\mathcal{H}om_{\mathcal{A}}(\mathcal{E},\mathcal{F})}(\phi) = 0, \tag{8.25}$$

with ϕ, *as in (5.14).* We still say that ϕ is a *horizontal section* of $\mathcal{H}om_{\mathcal{A}}(\mathcal{E}, \mathcal{F})$ (a terminology that will be justified later on).

Accordingly, based on (8.13), one concludes now that;

> *the local gauge transformation* (or even *"coordinate" local isomorphism*) η_α (see (8.12)) of the given *vector sheaf* $\mathcal{E}$ is always a *horizontal section* of the *vector sheaf* (actually *free* $\mathcal{A}|_{U_\alpha}$*-module;* see also Chapt. II; (6.23))

(8.26) (8.26.1) $\mathcal{F}_\alpha \equiv \mathcal{H}om_{\mathcal{A}|_{U_\alpha}}(\mathcal{E}|_{U_\alpha}, \mathcal{A}^n|_{U_\alpha}) = \mathcal{H}om_{\mathcal{A}}(\mathcal{E}, \mathcal{A}^n)|_{U_\alpha}, \ \alpha \in I.$

> *That is, one has*
>
> (8.26.2) $$D_{\mathcal{F}_\alpha}(\eta_\alpha) = 0, \quad \alpha \in I.$$

8.(a). Local Levi-Civita $\mathcal{A}$-connections.- The connections in the title of this Subsection refer to the particular (however, *substantial!*) case, one takes in (8.13), when

$$d = \partial^n : \mathcal{A}^n \longrightarrow \Omega^n = \Omega(\mathcal{A}^n) \tag{8.27}$$

(see also (8.9), (8.4), (5.5), (8.20), as well as, (8.7); viz. whenever one defines D_α in (8.13), by employing the *standard flat* $\mathcal{A}$*-connection* ∂ *of* $\mathcal{A}$ (cf. Subsection 4.(a) in the preceding). Note, in passing, that the $\mathcal{A}$-connection ∂^n *is* also *flat,* in a sense that will be specified later on (see thus Chapter VIII in the sequel).

Thus, setting now (see e.g. (8.6), along with (8.10)),

$$\partial^n_\alpha \equiv \partial^n|_{U_\alpha} = (\partial|_{U_\alpha})^n, \quad \alpha \in I, \tag{8.28}$$

one obtains the following (local) $\mathcal{A}|_{U_\alpha}$*-connection of* $\mathcal{E}|_{U_\alpha}$ ($\alpha \in I$), which *there does exist!* (in view of our hypothesis for ∂; see also (8.15)). Namely, one now defines;

$$D_\alpha := \left(\eta_\alpha^{-1} \otimes 1_{\Omega|_{U_\alpha}}\right) \circ \partial^n_\alpha \circ \eta_\alpha, \quad \alpha \in I. \tag{8.29}$$

Thus, stated otherwise, according to our previous terminology,

(8.30) *for every* $\alpha \in I$ (cf. (8.10)), one considers the *(local) gauge transform, through* η_α, *of the standard flat (local)* $\mathcal{A}$*-connection of* $\mathcal{A}^n$, with $n = \operatorname{rk} \mathcal{E}$.

This also justifies already, formally, our previous claim in (8.15).

Now, extending the classical terminology, we call (8.29) the *local Levi-Civita* $\mathcal{A}|_{U_\alpha}$*-connection* of the given *vector sheaf* $\mathcal{E}$, the latter being restricted to the *local gauge* U_α of it, under consideration. Yet, for simplicity's sake, we also refer to (8.29), by abusing language, just as a *local Levi-Civita* $\mathcal{A}$*-connection of* $\mathcal{E}$ (always existing, given ∂, $\mathcal{E}$ and U_α, as before). Finally, we call (8.11), in the particular case at issue (cf. (8.27)), the *Levi-Civita diagram,* associated with a given *vector sheaf* $\mathcal{E}$ on X and a *local gauge* of it.

Therefore, according to (8.21), and taking (8.27) into account, one has that:

(8.31) *every local Levi-Civita* $\mathcal{A}$*-connection,* associated with a given vector sheaf $\mathcal{E}$ on X and a local gauge of it, *is gauge equivalent to the standard flat (local)* $\mathcal{A}$*-connection of* $\mathcal{A}^n$ (cf. (8.28)), with $n = \operatorname{rk} \mathcal{E} \in \mathbf{N}$.

Now, as another consequence of the previous discussion, we already deduce that:

(8.32) given a *vector sheaf* $\mathcal{E}$ on X and a *local frame* of it, $\mathcal{U} = (U_\alpha)_{\alpha \in I}$, one obtains, through (8.29) (see also e.g. (8.19)), the following *Levi-Civita 0-cochain*

(8.32.1) $$(D_\alpha) \in C^0(\mathcal{U}, \mathcal{H}om_{\mathbf{C}}(\mathcal{E}, \Omega(\mathcal{E})))$$

being *gauge equivalent* to the *standard flat* $\mathcal{A}$*-connection 0-cochain* (in fact, *0-cocycle,* cf. (8.33) below) *of* $\mathcal{A}^n$;

(8.32.2) $$\begin{aligned}(\partial_\alpha^n) &\in C^0(\mathcal{U}, \mathcal{H}om_{\mathbf{C}}(\mathcal{A}^n, \Omega(\mathcal{A}^n))) \\ &= C^0(\mathcal{U}, \mathcal{H}om_{\mathbf{C}}(\mathcal{A}^n, \Omega^n)).\end{aligned}$$

That is, *one has*

(8.32.3) $$(D_\alpha) \underset{\alpha}{\sim} (\partial_\alpha^n),$$

in the sense that *(8.29) holds true, for every* $\alpha \in I$.

Yet, as already said, *(8.32.2) is,* actually *a 0-cocycle,* a consequence, in fact, of (8.28). Thus, one obtains

(8.33)
$$\begin{aligned}\partial^n = (\partial^n_\alpha) \in Z^0(\mathcal{U}, \mathcal{H}om_{\mathbf{C}}(\mathcal{A}^n, \Omega^n)) &= \mathcal{H}om_{\mathbf{C}}(\mathcal{A}^n, \Omega^n)(X) \\ &= \mathrm{Hom}_{\mathbf{C}}(\mathcal{A}^n, \Omega^n) = \mathrm{Hom}_{\mathbf{C}}(\mathcal{A}^n, \Omega(\mathcal{A}^n))\end{aligned}$$

(cf. Chapt. III; Lemma 4.1, or even (3.11) therein).

On the other hand, taking, for instance, (7.22) and (7.30) into account, one concludes that (see also Chapt. III; (4.58));

(8.34) *the 1-cocycle, which corresponds to (8.32.1),* is given by

(8.34.1)
$$\begin{aligned}\delta(D_\alpha) = (D_\beta - D_\alpha) &\in Z^1(\mathcal{U}, \mathcal{H}om_{\mathcal{A}}(\mathcal{E}, \Omega(\mathcal{E}))) \\ &= Z^1(\mathcal{U}, \Omega(\mathcal{E}nd\,\mathcal{E})).\end{aligned}$$

We call it the *Levi-Civita ($\mathcal{A}$-connection) 1-cocycle* of the *vector sheaf $\mathcal{E}$*, which is thus associated with the given *local frame $\mathcal{U}$* of $\mathcal{E}$. We also call it the *standard $\mathcal{A}$-connection 1-cocycle of $\mathcal{E}$*, (canonically) *associated with $\mathcal{U}$.*

In this context, we still remark, in anticipation, that the *$\mathcal{A}$-module* on X, $\Omega(\mathcal{E}nd\,\mathcal{E})$, appeared in (8.34.1) is going to play a significant rôle throughout the sequel. (It will deliberately be called *"sheaf of $\mathcal{A}$-connection,* or just *connection, coefficients"* right below).

Note 8.1.– As a result of the preceding discussion one infers that:

(8.35) Given a *differential triad* $(\mathcal{A}, \partial, \Omega)$ on a topological space X, a *vector sheaf* $\mathcal{E}$ on X and a *local frame* $\mathcal{U}$ of $\mathcal{E}$, one always gets the *standard $\mathcal{A}$-connection 1-cocycle of $\mathcal{E}$.*

Thus, its existence simply follows from the hypothesis that *∂ is given* and that *$\mathcal{E}$ is locally free.* Yet, as we shall see (cf. Section 10 below), there is an *intimate (natural) relation between the above 1-cocycle and* the vector sheaf *$\mathcal{E}$ itself.*

However, the extent to which the same 1-cocycle provides an $\mathcal{A}$-connection of $\mathcal{E}$ (*in general, does not!*) depends, so to speak, on our *"arithmetic"*(!); viz. on the *cohomological behaviour,* as we shall see in the sequel, of the *sheaf of coefficients $\mathcal{A}$* and, in turn, of that one of *connection coefficients* $\Omega(\mathcal{E}nd\,\mathcal{E})$ (see the next Section 9, for the terminology applied). So the appeared obstruction to the classical case of Differential Geometry diversifies exactly the present abstract setting.

Yet, one *final remark,* of theoretical character only, is that the preceding machinery, through which one arrives at (8.35), *is,* of course, *always, in force, locally;* that is, once we are given just an $\mathcal{A}$-*module* $\mathcal{E}$ on X, for which there exists an *open set* $U \subseteq X$, such that, *for some* $n \in \mathbf{N}$, one has

(8.36) $$\mathcal{E}|_U = \mathcal{A}^n|_U,$$

within an $\mathcal{A}|_U$-*isomorphism of* the $\mathcal{A}|_U$-*modules* involved. (*Finite* n here is still *not crucial*).

9. Sheaf of connection coefficients

We consider now the particular $\mathcal{A}$-*module* on X, appeared in (8.34.1), as the *"sheaf of coefficients"* of the relevant cochains (in fact, cocycles), which are naturally involved, when employing cohomology arguments. Indeed, such methods are going to be used systematically in the succeeding discussion. So our purpose here is to show that:

(9.1) by using *cohomology* $\mathcal{A}(X)$-*modules* (cf. Chapter III in the preceding), *the "sheaf of coefficients", that is finally involved, is actually independent of the vector sheaf* $\mathcal{E}$, under consideration, *depending only on the given differential triad* $(\mathcal{A}, \partial, \Omega)$; *precisely,* on the range of the standard flat $\mathcal{A}$-connection ∂ of $\mathcal{A}$, viz. *on the* $\mathcal{A}$-*module* Ω.

To simplify notation, we also accept the following terminology (see also the comments at the end of the previous Section 8):

(9.2) Given a differential triad $(\mathcal{A}, \partial, \Omega)$, we say that *an* $\mathcal{A}$-*module* $\mathcal{E}$ on X *has a local gauge,* whenever (8.36) *holds true, for some open* $U \subseteq X$ (U being thus the *local gauge of* $\mathcal{E}$, in question, of *rank* $n \in \mathbf{N}$).

Of course, it is clear that any *vector sheaf* $\mathcal{E}$ on X *has always*, by its very definition (see e.g. Chapt. II; (4.1.1)), a *local gauge* U, of *rank* $n = \mathrm{rk}_{\mathcal{A}}(\mathcal{E}) \in \mathbf{N}$ (ibid.; (4.2)), *for every point* $x \in X$, *with* $x \in U$. So our first relevant result is now the following.

Lemma 9.1. *Suppose we are given an* $\mathcal{A}$*-module* $\mathcal{E}$ *on* X, *having a local gauge* U, *of rank* $n \in \mathbf{N}$ (cf. (9.2) and (8.36)). *Then, one obtains* (see also Chapt. V; (8.7));

$$(9.3)\quad \mathcal{H}om_{\mathcal{A}}(\mathcal{E}, \mathcal{E} \otimes_{\mathcal{A}} \Omega)(U) \equiv \mathcal{H}om_{\mathcal{A}}(\mathcal{E}, \Omega(\mathcal{E}))(U) = M_n(\Omega)(U) = M_n(\Omega(U)),$$

within an $\mathcal{A}(U)$*-isomorphism of the* $\mathcal{A}(U)$*-modules involved.*

In particular, if $\mathcal{E}$ *is a vector sheaf on* X, *of rank* $n \in \mathbf{N}$, *and* U *a local gauge of it, then one has;*

$$(9.4)\qquad \Omega(\mathcal{E}nd\,\mathcal{E})(U) = M_n(\Omega)(U),$$

up to an $\mathcal{A}(U)$*-isomorphism.*

Proof. Indeed, one obtains;

$$\begin{aligned}
&\mathcal{H}om_{\mathcal{A}}(\mathcal{E}, \Omega(\mathcal{E}))(U) \equiv \mathcal{H}om_{\mathcal{A}}(\mathcal{E}, \mathcal{E} \otimes_{\mathcal{A}} \Omega)(U) = \mathrm{Hom}_{\mathcal{A}|_U}(\mathcal{E}|_U, (\mathcal{E} \otimes_{\mathcal{A}} \Omega)|_U)\\
&= \mathrm{Hom}_{\mathcal{A}|_U}(\mathcal{E}|_U, \mathcal{E}|_U \otimes_{\mathcal{A}|_U} \Omega|_U) = \mathrm{Hom}_{\mathcal{A}|_U}(\mathcal{A}^n|_U, \mathcal{A}^n|_U \otimes_{\mathcal{A}|_U} \Omega|_U)\\
(9.5)\quad &= \mathcal{H}om_{\mathcal{A}}(\mathcal{A}^n, \mathcal{A}^n \otimes_{\mathcal{A}} \Omega)(U) = (\mathcal{H}om_{\mathcal{A}}(\mathcal{A}^n, \mathcal{A}^n) \otimes_{\mathcal{A}} \Omega)(U)\\
&\equiv (\mathcal{E}nd\,\mathcal{A}^n \otimes_{\mathcal{A}} \Omega)(U) \equiv \Omega(M_n(\mathcal{A}))(U) = M_n(\Omega)(U)\\
&\equiv \Omega^{n^2}(U) = \Omega(U)^{n^2} \equiv M_n(\Omega(U)),
\end{aligned}$$

which thus proves our claim, concerning (9.3). In this context, see also (8.36) above, together with Chapt. II; (6.8), (5.25), (3.14) and Chapt. IV; (6.8). On the other hand, (9.4) follows now from the preceding and (7.22), and this terminates the proof. ∎

Now, as a consequence of the above, which also motivates the terminology we apply in the sequel, one obtains the next.

Corollary 9.1. *Suppose we are given a differential triad* $(\mathcal{A}, \partial, \Omega)$ *on a topological space* X, $\mathcal{E}$ *a vector sheaf on* X, *of rank* $n \in \mathbf{N}$, *and* $\mathcal{U} = (U_\alpha)_{\alpha \in I}$ *a local frame of* $\mathcal{E}$. *Then, the Levi-Civita* ($\mathcal{A}$*-connection*) *1-cocycle of* $\mathcal{E}$, *associated with* $\mathcal{U}$ (see (8.34)),

(9.6) $$(D_\beta - D_\alpha) \in Z^1(\mathcal{U}, \Omega(\mathcal{E}nd\,\mathcal{E})),$$

is such that (cf. (9.4))

(9.7) $$D_\beta - D_\alpha \in \Omega(\mathcal{E}nd\,\mathcal{E})(U_{\alpha\beta}) = M_n(\Omega)(U_{\alpha\beta}) = M_n(\Omega(U_{\alpha\beta})),$$

for any $\alpha, \beta \in I$, *with* $U_{\alpha\beta} \equiv U_\alpha \cap U_\beta \neq \emptyset$. *Therefore, one has;*

(9.8) $$(D_\beta - D_\alpha) \in Z^1(\mathcal{U}, M_n(\Omega)).$$

Proof. The only thing we have to check, by applying (9.4) to derive (9.7), is that *the intersection of two local gauges of* $\mathcal{E}$ *is still such a one,* and this is certainly clear. The rest is obvious, by the very definitions (see e.g. Chapt. III; (4.11), (4.12), or even (11.3) therein). ∎

In addition to the above relation (9.8), where $M_n(\Omega)$ appears (in fact, just local sections of it (cf. (9.7)), as we shall see in the sequel (cf. the next Chapter VII), *any given* $\mathcal{A}$*-connection of a vector sheaf* $\mathcal{E}$ on X *can be expressed locally through (local) sections of the* $\mathcal{A}$*-module* $M_n(\Omega)$; viz. via elements of the *complete presheaf*

(9.9) $$\Gamma(M_n(\Omega)) \equiv M_n(\Gamma(\Omega)).$$

So, as is also the case in the classical theory, this is a good reason to call the above $\mathcal{A}$*-module* on X,

(9.10) $$M_n(\Omega) \equiv M_n(\mathcal{A}) \otimes_{\mathcal{A}} \Omega \equiv \Omega(M_n(\mathcal{A})),$$

the *sheaf of* $\mathcal{A}$*-connection coefficients* of the vector sheaf $\mathcal{E}$, associated thus with the given differential triad $(\mathcal{A}, \partial, \Omega)$. Yet, for simplicity, we call (9.10) the *sheaf of connection coefficients of* $\mathcal{E}$.

The preceding establish thus completely, by now, our initial objective, as stated in (9.1). In this context, we still remark that the conclusion about (9.10) was actually emerged by employing local gauges, hence, *local frames,* of the vector sheaf involved. As a matter of fact, the point here is that:

(9.11) *by using,* for instance, *Čech cohomology* (see Chapt. III; Section 4), *one can always argue in terms of local frames of* $\mathcal{E}$, when considering (Čech) *cohomology groups* (in effect, $\mathcal{A}(X)$*-modules;* ibid.)

(In this regard, see also Chapt. V; Section 3, in particular, (3.10)). Now, concerning our claim in (9.11), this is due, indeed, to the fact that (Chapt. III; (4.9), in conjunction with Chapt. II; (4.9)):

(9.12) *the local frames of a given vector sheaf $\mathcal{E}$ on X constitute a cofinal subset of the set of all (proper) open coverings of X.*

So taking, for instance, the *first (Čech) cohomology group* (in fact, $\mathcal{A}(X)$-*module*) of X with coefficients in the $\mathcal{A}$-*module* (9.10), as above, one obtains;

$$\check{H}^1(X, M_n(\Omega)) = \varinjlim_{\mathcal{U}} \check{H}^1(\mathcal{U}, M_n(\Omega)), \tag{9.13}$$

where $\mathcal{U}$ *ranges over the local frames of* $\mathcal{E}$. (Here (9.13) is valid, within an $\mathcal{A}(X)$-*isomorphism of* the $\mathcal{A}(X)$-*modules* involved; cf. Chapt. III; (4.53), as well as, e.g. N. Bourbaki [1: p.95, Proposition 8]).

Accordingly, as a consequence of (9.8) and (9.13), in conjunction with our previous argument in (8.34), we conclude now the following;

(9.14) given a *vector sheaf* $\mathcal{E}$ on X and a *local frame* $\mathcal{U}$ of it (cf. (8.10)), by using the *Levi-Civita 1-cocycle of* $\mathcal{E}$, associated with $\mathcal{U}$ (cf. (8.34.1)), one defines a *1-dimensional (Čech) cohomology class*

$$[(D_\beta - D_\alpha)] \in \check{H}^1(X, M_n(\Omega)). \tag{9.14.1}$$

Thus, we set

$$\mathfrak{a}(\mathcal{E}) := [(D_\beta - D_\alpha)]. \tag{9.14.2}$$

The latter class is, of course, *uniquely determined by* $\mathcal{U}$ (see (9.8) and (9.13). We call $\mathfrak{a}(\mathcal{E})$ the *1-dimensional $\mathcal{A}$-connection (Čech) cohomology class of* $\mathcal{E}$; yet, by extending the classical terminology (cf., for instance, J.L. Koszul [2: p.119]), we simply call it too, the *Atiyah class of* $\mathcal{E}$.

Furthermore, it is still clear that *(9.14.1) depends only on the isomorphism class of a given vector sheaf* $\mathcal{E}$ on X (cf. also Chapt. II; (4.15)). Therefore, the same definition (9.14.2) yields further, *for every* $n \in \mathbf{N}$, a map

$$\mathbf{\Phi}^n_{\mathcal{A}}(X) \longrightarrow \check{H}^1(X, M_n(\Omega)), \tag{9.15}$$

called the *Atiyah map* (or even *Atiyah operator*) of X; its definition is thus associated simply with the given *differential triad* $(\mathcal{A}, \partial, \Omega)$ on X.

Thus, one obtains an "exact sequence" of the form;

$$(9.16) \qquad 0 \longrightarrow \ker \mathfrak{a} \longrightarrow \mathbf{\Phi}^n_{\mathcal{A}}(X) \xrightarrow{\mathfrak{a}} \check{H}^1(X, M_n(\Omega)).$$

As we shall see in the sequel (cf., for instance, Corollary 12.1), *the kernel of the Atiyah map,* as above,

$$(9.17) \qquad \ker \mathfrak{a} \equiv \mathbf{\Phi}^n_{\mathcal{A}}(X)^{\nabla},$$

is exactly the set of those vector sheaves (of rank $n \in \mathbf{N}$) *on* X, *which admit an* $\mathcal{A}$*-connection* (*"Yang-Mills fields"* on X; of course, we do not distinguish here between *isomorphic vector sheaves* (see e.g. (18.6) below). In the particular *case* $n = 1$ (*line sheaves* on X), we further give below an appropriate *cohomological interpretation of* the previous set (9.17), in terms of (Čech) *hypercohomology* (see Theorem 18.2). Namely, as we are going to see right below, the above cohomology class, as given by (9.14.1), provides thus a means to examine the existence of an $\mathcal{A}$-connection for a given *vector sheaf* $\mathcal{E}$ on X. Precisely, by complete analogy with the classical case (cf. M.F. Atiyah [1]), one obtains through it a *criterion,* what, namely, is called in the sequel the *Atiyah criterion,* for the existence of an $\mathcal{A}$-connection of $\mathcal{E}$. This will actually occupy us in the succeeding few Sections. However, it is instructive first to consider (we shall use it, for that matter) an existing relationship between (8.34.1) and the *coordinate 1-cocycle of* $\mathcal{E}$, which, as we know (cf. Chapt. V; (1.8)), can be associated with any given *local frame* $\mathcal{U}$ of $\mathcal{E}$.

10. Levi-Civita 1-cocycles, as related to coordinate 1-cocycles of vector sheaves

As already said, we consider below a useful relation, that naturally emerges between the 1-cocycles in the heading of this Section. Yet, the latter cocycles are, as already known, canonically associated with any given local frame of a vector sheaf (cf. the following discussion).

So to fix the terminology applied, assume that we are given a *vector sheaf* $\mathcal{E}$ on a *topological space* X, the latter being, as usual, the base space of a

C-*algebraized space* $(X, \mathcal{A})$. Yet, we consider a *differential triad* $(\mathcal{A}, \partial, \Omega)$. Finally, suppose that

(10.1) $$\mathcal{U} = (U_\alpha)_{\alpha \in I}$$

is a given *local frame of* $\mathcal{E}$.

Thus, the two following *1-cocycles of* $\mathcal{U}$ *with coefficients in* $M_n(\Omega)$ *and* $\mathcal{GL}(n, \mathcal{A})$, respectively, are canonically associated with the given vector sheaf $\mathcal{E}$, where *we set* $n = \mathrm{rk}\,(\mathcal{E}) \in \mathbf{N}$:

1) The *Levi-Civita 1-cocycle of* $\mathcal{E}$, canonically associated with $\mathcal{U}$, i.e., one has (see (8.34), (9.4) and (9.8))

(10.2) $$(D_\beta - D_\alpha) \in Z^1(\mathcal{U}, \Omega(\mathcal{E}nd\,\mathcal{E})) = Z^1(\mathcal{U}, M_n(\Omega)).$$

2) The *coordinate 1-cocycle of* $\mathcal{E}$, as defined by $\mathcal{U}$; namely, one has (see Chapt. V; (1.7) and Definition 1.1),

(10.3) $$(\eta_{\alpha\beta}) \in Z^1(\mathcal{U}, \mathcal{GL}(n, \mathcal{A})).$$

Here we have set;

(10.4) $$\eta_{\alpha\beta} := \eta_\alpha \circ \eta_\beta^{-1},$$

for any $\alpha, \beta \in I$, *with* $U_{\alpha\beta} \equiv U_\alpha \cap U_\beta \neq \emptyset$, while the $\mathcal{A}|_{U_\alpha}$-*isomorphism* η_α, $\alpha \in I$, is given by (8.12) (see also Chapt. V; (1.3.2) and (1.5.1)).

Now, based on (8.29), *one obtains,* concerning (10.2);

(10.5) $$D_\beta - D_\alpha = \left(\eta_\beta^{-1} \otimes 1\right) \partial_\beta^n \eta_\beta - \left(\eta_\alpha^{-1} \otimes 1\right) \partial_\alpha^n \eta_\alpha,$$

for any $\alpha, \beta \in I$, *as in (10.4)*. For convenience, *we* still *set* here

(10.6) $$1_{\Omega|U_\alpha} \equiv 1, \quad \alpha \in I,$$

while an *analogous notation will* also *be applied,* in the case one considers *any* $\alpha, \beta \in I$, *as in (10.4)*. Thus, in view of (10.4) and (10.5), one has;

(10.7) $$(\eta_\alpha \otimes 1)(D_\beta - D_\alpha)\eta_\alpha^{-1} = (\eta_{\alpha\beta} \otimes 1)\partial_\beta^n \eta_{\alpha\beta}^{-1} - \partial_\alpha^n.$$

Now, based e.g. on our argument in (9.5), we understand that the *first member of (10.7)* is an *element of the* following *set;*

(10.8) $$M_n(\Omega)(U_{\alpha\beta}) = \mathrm{Hom}_{\mathcal{A}|_{U_{\alpha\beta}}}(\mathcal{A}^n|_{U_{\alpha\beta}}, \Omega^n|_{U_{\alpha\beta}})$$

(it concerns, in fact, an $\mathcal{A}(U_{\alpha\beta})$-*module*). Therefore (cf. Chapt. I; Proposition

2.1, yet (2.17)), the same element is known, *whenever we know its value at the "constant" section*

$$\mathbf{1} \in \mathcal{A}^n(X) = \mathcal{A}(X)^n, \tag{10.9}$$

suitably *localized,* as the case may be. Of course, we have set in (10.9)

$$\mathbf{1} \equiv \underbrace{1 \oplus \cdots \oplus 1}_{n-\text{times}}, \tag{10.10}$$

where $\mathbf{1} \equiv 1_{\mathcal{A}} \in \mathcal{A}(X)$ denotes the (global) *identity section of* (the $\mathbf{C}$*-algebra sheaf*) $\mathcal{A}$; see also Chapt. II; Definition 1.4 and (1.34), (1.34$'$). Thus, in other words, one may *replace the element in question, by its value at* (the *section*) 1, as before. (In this context, see also e.g. Chapt. I; (12.50)). On the other hand, for convenience, we still note that, if u belongs, for instance, to (9.5), then, *for every* $\alpha \in \mathcal{A}^n(U)$, *one has* $u(\alpha) \equiv u(\alpha \cdot 1) = \alpha \cdot u(1)$, which thus vindicates completely our last assertion. ▮ Thus, in view now of (10.7), one obtains:

$$\begin{aligned}(\eta_\alpha \otimes 1)(D_\beta - D_\alpha)\eta_\alpha^{-1} &\equiv ((\eta_\alpha \otimes 1)(D_\beta - D_\alpha)\eta_\alpha^{-1})(\mathbf{1}) \\ &= \left((\eta_{\alpha\beta} \otimes 1) \circ \partial_\beta^n\right)\left(\eta_{\alpha\beta}^{-1}(\mathbf{1})\right) \equiv \left((\eta_{\alpha\beta} \otimes 1)\partial_\beta^n \eta_{\alpha\beta}^{-1}\right)(\mathbf{1}),\end{aligned} \tag{10.11}$$

given that, one has, of course,

$$\partial_\alpha^n(\mathbf{1}) = 0; \tag{10.12}$$

the last relation follows from (5.6) and (5.4) (cf. also (8.28)), in conjunction with (the obvious *extension to* $\mathcal{A}^n$ *of*) Lemma 1.1. Thus, here one considers, *fiber-wise,* $\mathcal{A}^n$ as an $\mathcal{A}$*-algebra sheaf* on X, with "$\mathbf{1}$" given by (10.10); see also Chapt. II; Definition 6.2). ▮

In a similar manner, one may consider, *equivalently,* $\eta_{\alpha\beta}(\mathbf{1})$, with $\mathbf{1}$ given by (10.10), *in place of*

$$\begin{aligned}\eta_{\alpha\beta} \in \mathcal{GL}(n, \mathcal{A})(U_{\alpha\beta}) &\subseteq \mathcal{H}om_{\mathcal{A}}(\mathcal{A}^n, \mathcal{A}^n)(U_{\alpha\beta}) \\ &= \mathrm{Hom}_{\mathcal{A}|_{U_{\alpha\beta}}}(\mathcal{A}^n|_{U_{\alpha\beta}}, \mathcal{A}^n|_{U_{\alpha\beta}}).\end{aligned} \tag{10.13}$$

So *we* further *set;*

$$g_{\alpha\beta} := \eta_{\alpha\beta}^{-1}(\mathbf{1}) \in \mathcal{A}^n(U_{\alpha\beta}) = \mathcal{A}(U_{\alpha\beta})^n, \tag{10.14}$$

for any $\alpha, \beta \in I$, *with* $U_{\alpha\beta} \equiv U_\alpha \cap U_\beta \neq \emptyset$. Yet, applying the same argument, as before (cf. (10.11)), and based on (10.14), one also obtains;

(10.15) $$\eta_{\alpha\beta} \otimes 1 \equiv \eta_{\alpha\beta}(\mathbf{1}) \cdot 1_{\Omega|_{U_{\alpha\beta}}} \equiv \eta_{\alpha\beta}(\mathbf{1}) = g_{\alpha\beta}^{-1}.$$

Therefore, one concludes from (10.11) that;

(10.16) $$(\eta_\alpha \otimes 1)(D_\beta - D_\alpha)\eta_\alpha^{-1} = g_{\alpha\beta}^{-1} \cdot \partial_\beta^n(g_{\alpha\beta}),$$

that is, in view of (1.29) (cf. also (1.28), for $\mathcal{E} = \Omega$, and (3.1)), along with (8.28), one finally obtains;

(10.17) $$(\eta_\alpha \otimes 1)(D_\beta - D_\alpha)\eta_\alpha^{-1} = \tilde{\partial}(g_{\alpha\beta}).$$

On the other hand, concerning the *1-cochain* of $\mathcal{U}$

(10.18) $$(g_{\alpha\beta}) \in C^1(\mathcal{U}, \mathcal{H}om_{\mathcal{A}}(\mathcal{A}^n, \mathcal{A}^n)),$$

as defined by (10.14), (10.13), one easily sees, according to the very definitions, that the same yields, in fact, a *1-cocycle*

(10.19) $$(g_{\alpha\beta}) \in Z^1(\mathcal{U}, \mathcal{GL}(n, \mathcal{A}));$$

moreover, the latter actually constitutes the *coordinate 1-cocycle of the vector sheaf* $\mathcal{E}$ at issue, which is *associated with the given local frame* $\mathcal{U}$ *of* $\mathcal{E}$ (see (10.1), and Chapt. V; (1.2), (1.3.2), together with the above rels. (8.12), (10.4) and (10.13), (10.14)).

So we can now summarize the preceding discussion into the form of the following.

Theorem 10.1. *Consider a differential triad* $(\mathcal{A}, \partial, \Omega)$ *on a topological space* X, $\mathcal{E}$ *a vector sheaf on* X, *of rank* $n \in \mathbf{N}$, *and* $\mathcal{U} = (U_\alpha)_{\alpha \in I}$ *a local frame of* $\mathcal{E}$. *Then, the Levi-Civita (*$\mathcal{A}$*-connection) 1-cocycle of* $\mathcal{E}$, *relative to* $\mathcal{U}$ (see (8.34), (9.8)),

(10.20) $$(D_\beta - D_\alpha) \in Z^1(\mathcal{U}, M_n(\Omega)),$$

is gauge equivalent (relative to the $\mathcal{A}|_{U_\alpha}$*-isomorphism* η_α, $\alpha \in I$, cf. (8.12) or even (8.21.1)) *to the logarithmic derivative of the coordinate 1-cocycle of* $\mathcal{E}$, *associated with* $\mathcal{U}$, *say,*

(10.21) $$(g_{\alpha\beta}) \in Z^1(\mathcal{U}, \mathcal{GL}(n, \mathcal{A}))$$

(cf. (10.19)). *That is, one has* (see (10.17));

(10.22) $$(\eta_\alpha \otimes 1)(D_\beta - D_\alpha)\eta_\alpha^{-1} = \tilde{\partial}(g_{\alpha\beta}).$$

for any $\alpha, \beta \in I$, *with* $U_{\alpha\beta} \equiv U_\alpha \cap U_\beta \neq \emptyset$. *Yet, we write (10.22) in the following abbreviated form* (see also (8.32.3));

(10.23) $$\delta(D_\alpha) \underset{\alpha}{\sim} \tilde{\partial}(g_{\alpha\beta}), \quad \alpha \in I. \blacksquare$$

Now, in view of (9.5) and (9.7) (see also e.g. (10.8)), one has;

(10.24) $$\begin{aligned}(\eta_\alpha \otimes 1)(D_\beta - D_\alpha)\eta_\alpha^{-1} &\in Z^1(\mathcal{U}, \mathcal{H}om_{\mathcal{A}}(\mathcal{A}^n, \Omega(\mathcal{A}^n))) \\ &= Z^1(\mathcal{U}, M_n(\Omega)).\end{aligned}$$

Furthermore, concerning the *coordinate 1-cocycle of the vector sheaf* $\mathcal{E}$, under consideration (cf. (10.19)), *one* still *obtains* that;

(10.25) $$\tilde{\partial}((g_{\alpha\beta})) \equiv \left(\tilde{\partial}(g_{\alpha\beta})\right) \in Z^1(\mathcal{U}, M_n(\Omega)).$$

Indeed, the last assertion follows from the relation;

(10.26) $$\tilde{\partial}(g_{\alpha\gamma}) = \mathrm{Ad}\left(g_{\beta\gamma}^{-1}\right)\tilde{\partial}(g_{\alpha\beta}) + \tilde{\partial}(g_{\beta\gamma}),$$

where $\alpha, \beta, \gamma \in I$, *such that* $U_\alpha \cap U_\beta \cap U_\gamma \neq \emptyset$, which, in turn, in an immediate consequence of (10.21) and (1.33). ∎

Accordingly, as a result of (10.22), (10.24) and (10.25), *one obtains the following,* basic for the sequel, *relation* in (Čech) *cohomology;*

(10.27) $$[(D_\beta - D_\alpha)] = \left[\left(\tilde{\partial}(g_{\alpha\beta})\right)\right] \in \check{H}^1(X, M_n(\Omega)).$$

Therefore, by virtue of (9.14), one concludes, by now, that;

(10.28) *the Atiyah class of a vector sheaf* $\mathcal{E}$ on X *is given by*

(10.28.1) $$\mathfrak{a}(\mathcal{E}) := [(D_\beta - D_\alpha)] = \left[\left(\tilde{\partial}(g_{\alpha\beta})\right)\right] \in \check{H}^1(X, M_n(\Omega)).$$

Hence, by considering the Atiyah map (9.15), together with the (set-theoretic) *bijection*

(10.29) $$\Phi^n_{\mathcal{A}}(X) = \check{H}^1(X, \mathcal{GL}(n, \mathcal{A}))$$

(see (9.14.2) and Chapt. V; Theorem 2.1), one obtains the following *commutative diagram;*

(10.30)
$$\begin{array}{ccc} \check{H}^1(X, \mathcal{GL}(n, \mathcal{A})) & \xrightarrow{\tilde{\partial}} & \check{H}^1(X, M_n(\Omega)) \\ \Vert & \nearrow_{\mathfrak{a}} & \\ \Phi^n_{\mathcal{A}}(X) & & \end{array}$$

The maps appeared in the previous diagram are, of course, the relevant *extensions to cohomology* (due to the *functorial behaviour*(!) of the cohomology functors, see e.g. Chapt. III; Theorem 3.1 and Lemma 8.1, along with (8.3)) *of the maps (1.28)* (for $\mathcal{E} = \Omega$, cf. (3.1) and (9.15)). Yet, the *commutativity of (10.30)* holds certainly true, by virtue of (10.28.1), so that one actually obtains the following relation;

(10.31) $$\mathfrak{a} = \tilde{\partial},$$

a fact that will be of use right in the next Section 11.

Note 10.1.– Concerning the above calculations about the *basic relation (10.22),* we still note, from a historical perspective, that the motive thereby was initially a seminal (for our whole subject matter(!), in effect) paper of S.A. Selesnick [2], which we shall also use later on. The same was actually referred to certain relevant questions, within the framework of *Banach algebra theory*! Yet, we shall have the opportunity to look at it, more closely, in terms also of *topological algebra theory,* subsequently (see thus Chapter XI below).

11. Existence of $\mathcal{A}$-connections

As already promised, we start with this Section to seek for conditions guaranteeing the *existence of $\mathcal{A}$-connections of a* given *vector sheaf* on X, an inquiry that deviates from our standard experience, pertaining e.g. to smooth (viz. $\mathcal{C}^\infty$-) **C**-vector bundles. The relevant reason, classically exhibited already by M.F. Atiyah [1], attains also, within the present abstract framework, an analogous formulation; so this will be our objective in this and the two following Sections.

Thus, to begin with, suppose that we are given a vector sheaf $\mathcal{E}$ on X, and let

$$\mathfrak{a}(\mathcal{E}) \equiv [(D_\beta - D_\alpha)] \in \check{H}^1(X, M_n(\Omega)) \tag{11.1}$$

be the *Atiyah class of* $\mathcal{E}$, determined (uniquely) through a given local frame of $\mathcal{E}$ (cf. (9.14.2)). So our program now is to show that:

(11.2) *a vector sheaf* $\mathcal{E}$ *has an* $\mathcal{A}$*-connection if, and only if, one has;*

$$\mathfrak{a}(\mathcal{E}) = 0. \tag{11.2.1}$$

Thus, we start right away, with considering first the *"if"* part (in fact, the easy one) of our assertion, while the rest (viz. the *"only if"* part) will be treated in the sequel (see Sections 12, 13 below). So we now have the following.

Theorem 11.1. *Let* $\mathcal{E}$ *be a given vector sheaf on* X*, such that the Atiyah class of it, associated with a local frame of* $\mathcal{E}$*, vanishes. That is, we assume that*

$$\mathfrak{a}(\mathcal{E}) = 0. \tag{11.3}$$

Then, $\mathcal{E}$ *admits an* $\mathcal{A}$*-connection.*

Proof. Suppose that

$$\mathfrak{a}(\mathcal{E}) \equiv [(D_\beta - D_\alpha)] = 0 \in \check{H}^1(X, M_n(\Omega)) \tag{11.4}$$

is the Atiyah class of the given vector sheaf $\mathcal{E}$ on X, with $n = \operatorname{rk} \mathcal{E}$, corresponding to a *local frame of* $\mathcal{E}$,

$$\mathcal{U} = (U_\alpha)_{\alpha \in I} \tag{11.5}$$

(see (9.14.2)). Thus, in view of (11.4) (see also, for instance, Chapt. II; (11.26.1), (11.16), along with Remark 11.1 and Lemma 8.1), one has

$$(D_\beta - D_\alpha) = \delta(\theta_\alpha) \in \delta^0(\check{C}^0(\mathcal{U}, M_n(\Omega))), \tag{11.6}$$

for some *0-cochain*

$$\theta \equiv (\theta_\alpha) \in \check{C}^0(\mathcal{U}, M_n(\Omega)). \tag{11.7}$$

Therefore, one obtains (cf. Chapt. III; (4.16))

$$\delta(D_\alpha) = (D_\beta - D_\alpha) = \delta(\theta_\alpha), \tag{11.8}$$

so that one has

$$\delta(D_\alpha) - \delta(\theta_\alpha) = \delta(D_\alpha - \theta_\alpha) = 0. \tag{11.9}$$

Furthermore, one has

$$(D_\alpha - \theta_\alpha) \in \check{C}^0(\mathcal{U}, \mathcal{H}om_{\mathbf{C}}(\mathcal{E}, \Omega(\mathcal{E}))), \tag{11.10}$$

which follows from (8.32.1) and (11.7), in conjunction with (9.3). Yet, we still note, in passing, that from the same rel. (9.3), *one obtains;*

$$\check{C}^0(\mathcal{U}, M_n(\Omega)) = \check{C}^0(\mathcal{U}, \mathcal{H}om_{\mathcal{A}}(\mathcal{E}, \Omega(\mathcal{E}))). \tag{11.11}$$

Now, by virtue of (11.9), *the 0-cochain (11.10) is,* in effect, *a 0-cocycle,* so that one has;

$$\begin{aligned}(D_\alpha) - (\theta_\alpha) = (D_\alpha - \theta_\alpha) &\in \check{Z}^0(\mathcal{U}, \mathcal{H}om_{\mathbf{C}}(\mathcal{E}, \Omega(\mathcal{E}))) \\ &= \mathcal{H}om_{\mathbf{C}}(\mathcal{E}, \Omega(\mathcal{E}))(X) = \mathrm{Hom}_{\mathbf{C}}(\mathcal{E}, \Omega(\mathcal{E}))\end{aligned} \tag{11.12}$$

(see Chapt. III; Lemma 8.1). Therefore, let

$$\widetilde{D} := (D_\alpha - \theta_\alpha) \tag{11.13}$$

be the corresponding $\mathbf{C}$*-linear morphism, as given by (11.12).* On the other hand, based on (11.13), (11.7) and (9.4), as well as, on (7.19), *one* still *obtains;*

$$\begin{aligned}\widetilde{D}_\alpha := \widetilde{D}|_{U_\alpha} = D_\alpha - \theta_\alpha &\in D_\alpha + \Omega(\mathcal{E}nd\,\mathcal{E})(U_\alpha) \\ &= \mathcal{C}onn_{\mathcal{A}|_{U_\alpha}}(\mathcal{E}|_{U_\alpha}) \equiv \mathcal{C}onn_{\mathcal{A}}(\mathcal{E})|_{U_\alpha},\end{aligned} \tag{11.14}$$

for every $\alpha \in I$. Thus, $\widetilde{D}_\alpha$ *is an* $\mathcal{A}|_{U_\alpha}$*-connection of* $\mathcal{E}|_{U_\alpha}$*, for every* $U_\alpha \in \mathcal{U}$ (cf. (11.5)); accordingly, in view of (11.13) and (11.14),

$$\widetilde{D} = \left(\widetilde{D}_\alpha\right) = (D_\alpha - \theta_\alpha) \tag{11.15}$$

is an $\mathcal{A}$*-connection of* $\mathcal{E}$ too, since the remaining yet (see thus (11.12)) *Leibniz condition for* $\widetilde{D}$ is now fulfilled, this holding true *locally,* according to (11.14), which thus terminates the proof. ∎

Now, by analogy with the terminology, already employed for (8.29) (see the comments following (8.30)), we set the next.

Definition 11.1. Let $\mathcal{E}$ be a vector sheaf on X. Then, the $\mathcal{A}$-connection of $\mathcal{E}$, which is provided by (11.15), under the cond. (11.3), is called a *Levi-Civita $\mathcal{A}$-connection of $\mathcal{E}$*. Yet, for simplicity, the term *Levi-Civita connection of $\mathcal{E}$* will occasionally be employed for the same notion, as before.

In this context, we still note, in anticipation (see also (11.2)), that;

(11.16) given a *vector sheaf $\mathcal{E}$* on X, *the condition* (cf. (7.1))

(11.16.1) $$Conn_{\mathcal{A}}(\mathcal{E}) \neq \emptyset$$

is necessary and sufficient in order $\mathcal{E}$ to have a Levi-Civita $\mathcal{A}$-connection. In other words, *a vector sheaf $\mathcal{E}$* on X, *having an $\mathcal{A}$-connection, has also a Levi-Civita connection* (*and conversely,* of course).

Full justification of the previous claim will be given in the sequel (see thus Theorem 12.1, as well as, Corollary 12.1 below along with (12.31)).

11.(a). Integrable $\mathcal{A}$-connections.– The notion in the title of this Subsection will be fully clarified later on (see Chapter VIII in the sequel). However, we do look at it hereby, in anticipation, giving its formal definition, while, as we shall see, right away, the same notion is naturally connected with the previous one of a Levi-Civita connection. So we start with the following.

Definition 11.2. Suppose we have a vector sheaf $\mathcal{E}$ on X. Then, by an *integrable $\mathcal{A}$-connection* of $\mathcal{E}$, one means an $\mathcal{A}$-connection $\widetilde{D}$ of $\mathcal{E}$, for which there exists a local frame $\mathcal{U} = (U_\alpha)_{\alpha \in I}$ of $\mathcal{E}$, such that one has

(11.17) $$\widetilde{D}|_{U_\alpha} \equiv \widetilde{D}_\alpha = D_\alpha, \quad \alpha \in I,$$

where D_α, $\alpha \in I$, is given by (8.13).

As already said, $\mathcal{A}$-connections of the previous type will be amply considered in the sequel, namely, when the notion of *curvature* of a given $\mathcal{A}$-connection will be available. Nonetheless, we do examine in the meantime the above kind of $\mathcal{A}$-connections in conjunction with our argument in Section 8, a fact which also explains the reason for the succeeding discussion.

Thus, suppose that the *Levi-Civita 0-cochain*

(11.18) $$(D_\alpha) \in \check{C}^0(\mathcal{U}, \mathcal{H}om_{\mathbf{C}}(\mathcal{E}, \Omega(\mathcal{E}))),$$

which can be defined for any given *local frame* $\mathcal{U} = (U_\alpha)_{\alpha \in I}$ of a *vector sheaf* $\mathcal{E}$ on X, is, in particular, a *0-cocycle* of $\mathcal{U}$; that is, *assume that:*

(11.19) $$(D_\alpha) \in \check{Z}^0(\mathcal{U}, \mathcal{H}om_{\mathbf{C}}(\mathcal{E}, \Omega(\mathcal{E}))) = \mathrm{Hom}_{\mathbf{C}}(\mathcal{E}, \Omega(\mathcal{E}))$$

(see also, for instance, (11.12)). Hence, in view of our hypothesis for the individual sections D_α, $\alpha \in I$, in (11.19) (cf. Lemma 8.1, in particular, (8.13)), one concludes that;

(11.20) $$D := (D_\alpha),$$

as in (11.19), *is an $\mathcal{A}$-connection of $\mathcal{E}$*, such that, by definition, one has

(11.21) $$D|_{U_\alpha} = D_\alpha, \quad \alpha \in I.$$

Consequently,

(11.22) *(11.20) is an integrable $\mathcal{A}$-connection* of $\mathcal{E}$.

Yet, in this context, we further remark that, according to the very definitions (see (11.15) and (11.17)),

(11.23) *every integrable $\mathcal{A}$-connection* of a given *vector sheaf $\mathcal{E}$ is,* in particular, *a Levi-Civita $\mathcal{A}$-connection* of $\mathcal{E}$, *in such a manner that*

(11.23.1) $$(\theta_\alpha) = 0 \in C^0(\mathcal{U}, M_n(\Omega)).$$

Concerning (11.23.1), see also (11.7), (11.14) and (11.17). ∎ The same relation (11.23.1), for an integrable $\mathcal{A}$-connection, will be of use later on (cf. Chapter VIII in the sequel). Finally, based on the preceding, one concludes that:

(11.24) *the Levi-Civita 0-cochain (11.18) is a 0-cocycle,* yielding *thus a (Levi-Civita) $\mathcal{A}$-connection* (cf. (11.20)), *if, and only if, the latter is an integrable $\mathcal{A}$-connection.*

11.(b). Levi-Civita $\mathcal{A}$-connections of $\mathcal{A}^n$, $n \in \mathbf{N}$.– Our purpose by the following discussion is to show that;

(11.25) *the existence of an $\mathcal{A}$-connection of a* given *vector sheaf* $\mathcal{E}$ on X *implies that one too of a Levi-Civita $\mathcal{A}$-connection of* $\mathcal{A}^n$, *where* $n = \operatorname{rk}\mathcal{E}$.

The *converse* of the above claim is, of course, *not always true, unless (11.3) is fulfilled,* as this is easily seen by the standard *$\mathcal{A}$-connection ∂^n of $\mathcal{A}^n$*; see (8.4), (11.2) and (11.16), along with the succeeding comments to the two last quotations. Yet, cf. also Lemma 8.1.

Thus, suppose we are given a *vector sheaf* $\mathcal{E}$ on X, of *rank* $n \in \mathbf{N}$, admitting an *$\mathcal{A}$-connection* $\widetilde{D}$; moreover, let $\mathcal{U}$ be a *local frame of* $\mathcal{E}$, as in (11.5). Then, by setting

$$\widetilde{D}_\alpha := \widetilde{D}|_{U_\alpha}, \quad \alpha \in I, \tag{11.26}$$

one defines, *for every* $\alpha \in I$, by analogy with (8.11), a *commutative diagram,* as follows;

$$\begin{array}{ccc} \mathcal{E}|_{U_\alpha} & \xrightarrow{\ \widetilde{D}_\alpha\ } & \mathcal{E} \otimes_{\mathcal{A}} \Omega|_{U_\alpha} \\ \Big\downarrow{\scriptstyle \eta_\alpha^{-1}} & & \Big\downarrow{\scriptstyle \eta_\alpha \otimes 1_{\Omega|_{U_\alpha}}} \\ \mathcal{A}^n|_{U_\alpha} & \xrightarrow{\ d_\alpha\ } & \mathcal{A}^n \otimes_{\mathcal{A}} \Omega|_{U_\alpha} \end{array} \tag{11.27}$$

That is, we set

$$d_\alpha := \left(\eta_\alpha \otimes 1_{\Omega|_{U_\alpha}}\right) \circ \widetilde{D} \circ \eta_\alpha^{-1}, \quad \alpha \in I. \tag{11.28}$$

As a result of the preceding, one thus infers, so far, that;

(11.29) *for any $\mathcal{A}$-connection $\widetilde{D}$ of the vector sheaf $\mathcal{E}$ and any given local frame $\mathcal{U} = (U_\alpha)_{\alpha \in I} \in \mathcal{E}$, one obtains a 1-cocycle*

$$\delta(d_\alpha) = (d_\beta - d_\alpha) \in Z^1(\mathcal{U}, M_n(\Omega)) \tag{11.29.1}$$

(see also e.g. (10.8)). We still refer to (11.29.1) as the *Levi-Civita 1-cocycle of $\mathcal{A}^n$, associated with $\mathcal{U}$*, while the *0-cochain*

$$\begin{aligned}(d_\alpha) &\in C^0(\mathcal{U}, \mathcal{H}om_{\mathbf{C}}(\mathcal{A}^n, \Omega(\mathcal{A}^n))) \\ &= C^0(\mathcal{U}, \mathcal{H}om_{\mathbf{C}}(\mathcal{A}^n, \Omega^n)),\end{aligned} \tag{11.29.2}$$

as defined by (11.28), is still named the *Levi-Civita 0-cochain of $\mathcal{A}^n$, corresponding to* the given $\mathcal{U}$, as before.

Now, the question *whether the* above *0-cochain (11.29.2) yields an $\mathcal{A}$-connection of $\mathcal{A}^n$*, is treated by the following.

Theorem 11.2. *Let $\mathcal{E}$ be a vector sheaf on X, of rank $n \in \mathbf{N}$, $\mathcal{U} = (U_\alpha)_{\alpha \in I}$ a local frame of $\mathcal{E}$ and $\widetilde{D}$ an $\mathcal{A}$-connection of $\mathcal{E}$. Then, the Levi-Civita 0-cochain of $\mathcal{A}^n$, as derived from $\widetilde{D}$ and $\mathcal{U}$* (see (11.29.2)), *yields an $\mathcal{A}$-connection of $\mathcal{A}^n$, if, and only if, the respective Levi-Civita 1-cocycle of $\mathcal{A}^n$* (cf. (11.29.1)) *is a coboundary; i.e., equivalently, if, and only if, one has*

$$[(D_\beta - D_\alpha)] = 0 \in \check{H}^1(X, M_n(\Omega)). \tag{11.30}$$

Furthermore, in that case too, one obtains (see (9.14))

$$\mathfrak{a}(\mathcal{E}) = [(D_\beta - D_\alpha)] = 0 \in \check{H}^1(X, M_n(\Omega)); \tag{11.31}$$

hence (Theorem 11.1, along with Definition 11.1), *the given vector sheaf $\mathcal{E}$ admits a Levi-Civita $\mathcal{A}$-connection, as well.*

Proof. Assuming (11.30), one obtains (see also e.g. (11.6))

$$\delta(d_\alpha) = \delta(\psi_\alpha), \tag{11.32}$$

for some *0-cochain*

$$\begin{aligned}\psi \equiv (\psi_\alpha) \in C^0(\mathcal{U}, \mathcal{H}om_{\mathcal{A}}(\mathcal{A}^n, \mathcal{A}^n \otimes_{\mathcal{A}} \Omega)) &= C^0(\mathcal{U}, \mathcal{H}om_{\mathcal{A}}(\mathcal{A}^n, \Omega^n)) \\ &= C^0(\mathcal{U}, M_n(\Omega))\end{aligned} \tag{11.33}$$

(cf. also, for instance, (10.8) or even the argument in (9.5)). Hence, one obtains from (11.32),

(11.34) $$\delta(d_\alpha) - \delta(\psi_\alpha) = \delta(d_\alpha - \psi_\alpha) = 0,$$

(see Chapt. III; (4.15)), that is, we thus obtain a *0-cocycle,*

(11.35) $$\begin{aligned}(d_\alpha - \psi_\alpha) \in Z^0(\mathcal{U}, \mathcal{H}om_{\mathbf{C}}(\mathcal{A}^n, \mathcal{A}^n \otimes_{\mathcal{A}} \Omega)) &= \mathcal{H}om_{\mathbf{C}}(\mathcal{A}^n, \mathcal{A}^n \otimes \Omega)(X) \\ &= \mathrm{Hom}_{\mathbf{C}}(\mathcal{A}^n, \mathcal{A}^n \otimes_{\mathcal{A}} \Omega) = \mathrm{Hom}_{\mathbf{C}}(\mathcal{A}^n, \Omega^n).\end{aligned}$$

Therefore, setting

(11.36) $$\tilde{d} := (d_\alpha - \psi_\alpha),$$

or, *equivalently,*

(11.37) $$\tilde{d}|_{U_\alpha} \equiv \tilde{d}_\alpha := d_\alpha - \psi_\alpha, \quad \alpha \in I,$$

one concludes, by (11.28) and (8.24), that

(11.38) $\tilde{d}$, as given by (11.36), *yields an $\mathcal{A}$-connection of $\mathcal{A}^n$.*

Conversely, suppose that the 0-cochain (d_α), as in (11.29.2), provides an $\mathcal{A}$-connection of $\mathcal{A}^n$; so, in particular, one has that

(11.39) $$(d_\alpha) \in Z^0(\mathcal{U}, \mathcal{H}om_{\mathbf{C}}(\mathcal{A}^n, \Omega^n)) = \mathrm{Hom}_{\mathbf{C}}(\mathcal{A}^n, \Omega^n),$$

so that one obtains that $\delta(d_\alpha) = (d_\beta - d_\alpha) = 0$, which obviously implies (11.30). So *this completes the proof of the first part of the assertion.*

On the other hand, *to prove (11.31),* we first remark that, by virtue of (11.30) and since ∂^n *is an $\mathcal{A}$-connection of $\mathcal{A}^n$* (cf. (8.4)), one concludes for $\tilde{d}$ (cf. (11.36), (11.38)), in view of Theorem 7.1, that:

(11.40) $$\tilde{d} = \partial^n + \phi,$$

for some *uniquely defined*

(11.41) $$\begin{aligned}\phi \equiv (\phi_\alpha) \in \Omega(\mathcal{E}nd\, \mathcal{A}^n)(X) \equiv M_n(\Omega)(X) &= M_n(\Omega(X)) \\ &= Z^0(\mathcal{U}, M_n(\Omega)) \subseteq C^0(\mathcal{U}, M_n(\Omega))\end{aligned}$$

(see also Chapt. IV; (3.24) and Chapt. V; (8.7), for the notation applied). Therefore, by (11.37) and (11.40), we obtain;

(11.42) $$d_\alpha = \tilde{d}_\alpha + \psi_\alpha = \partial^n_\alpha + (\phi_\alpha + \psi_\alpha), \quad \alpha \in I,$$

such that, by (11.33) and (11.41), we have

(11.43) $$\chi \equiv (\chi_\alpha) := \phi + \psi \equiv (\phi_\alpha + \psi_\alpha) \in C^0(\mathcal{U}, M_n(\Omega)).$$

Furthermore, based now on our hypothesis for $\widetilde{D}$ (cf. also (11.26)), on (11.28) and on (10.4), (10.6) and (10.7), along with (11.42) and (11.43), we obtain;

(11.44) $$\begin{aligned} 0 &= (\eta_\alpha \otimes 1)(\widetilde{D}_\beta - \widetilde{D}_\alpha)\eta_\alpha^{-1} = (\eta_\alpha \otimes 1)\left(\eta_\beta^{-1} \otimes 1\right) d_\beta \eta_\beta \eta_\alpha^{-1} - d_\alpha \\ &= (\eta_{\alpha\beta} \otimes 1)\partial_\beta^n \eta_{\alpha\beta}^{-1} - \partial_\alpha^n + (\eta_{\alpha\beta} \otimes 1)\chi_\beta \eta_{\alpha\beta}^{-1} - \chi_\alpha \\ &= (\eta_\alpha \otimes 1)(D_\beta - D_\alpha)\eta_\alpha^{-1} + (\eta_{\alpha\beta} \otimes 1)\chi_\beta \eta_{\alpha\beta}^{-1} - \chi_\alpha, \end{aligned}$$

for any $\alpha, \beta \in I$, with $U_\alpha \cap U_\beta \equiv U_{\alpha\beta} \neq \emptyset$. Thus, one has (still cf. (10.4))

(11.45) $$D_\beta - D_\alpha = \left(\eta_\beta^{-1} \otimes 1\right) \chi_\beta \eta_\beta - \left(\eta_\alpha^{-1} \otimes 1\right) \chi_\alpha \eta_\alpha,$$

for any $\alpha, \beta \in I$, as above. Therefore, setting now

(11.46) $$\psi'_\alpha := \left(\eta_\alpha^{-1} \otimes 1\right) \circ \chi_\alpha \circ \eta_\alpha, \quad \alpha \in I,$$

such that (see also (9.4))

(11.47) $$(\psi'_\alpha) \in C^0(\mathcal{U}, \Omega(\mathcal{E}nd\,\mathcal{E})) = C^0(\mathcal{U}, M_n(\Omega)),$$

one thus obtains from (11.45);

(11.48) $$\delta(D_\alpha) = \delta(\psi'_\alpha).$$

In other words, one infers that (cf. e.g. Chapt. III; (4.22))

(11.49) $$\begin{aligned} \delta(D_\alpha) = (D_\beta - D_\alpha) \in \operatorname{im}\delta^0 &= B^1(\mathcal{U}, \Omega(\mathcal{E}nd\,\mathcal{E})) \\ &= B^1(\mathcal{U}, M_n(\Omega)), \end{aligned}$$

so that (ibid.; (4.23)) one finally concludes that;

(11.50) $$[(D_\beta - D_\alpha)] = 0 \in \check{H}^1(X, M_n(\Omega)),$$

that was actually *to be proved* (cf. (11.31)), and this, in effect, terminates the proof of the theorem. ∎

Note.– The question that was formulated before the statement of Theorem 11.2 is actually treated by the same theorem in a *round-about way.* So it would be interesting, at least, to have a *direct attack* to the same question, *in terms,* namely, *of the $\mathcal{A}$-connections $\widetilde{D}$ and $\tilde{d}$* (cf. (11.36)) *themselves.*

We come next to handle, as already said, by the following Section 12, the *"only if"* part of (11.2).

12. $\mathcal{A}$-connections as splittings of jet-line sheaves

As a result of the succeeding discussion and in conjunction with the above Theorem 11.1, we are going actually to see that:

(12.1) *the Atiyah class* of a given *vector sheaf* $\mathcal{E}$, which is associated with a *local frame* $\mathcal{U} = (U_\alpha)_{\alpha \in I}$ of $\mathcal{E}$,

(12.1.1) $$\mathfrak{a}(\mathcal{E}) \equiv [(D_\beta - D_\alpha)] \in \check{H}^1(X, M_n(\Omega))$$

(see (9.14)), *constitutes, in effect, the obstruction to the existence of an $\mathcal{A}$-connection of $\mathcal{E}$.*

Indeed, see Theorem 12.1 and Corollary 12.1 below. ∎ So we start with fixing again the terminology we are going to apply.

Thus, suppose, as usual, that we are given a *differential triad* $(\mathcal{A}, \partial, \Omega)$, with respect to a fixed **C**-*algebraized space* $(X, \mathcal{A})$, and let $\mathcal{E}$ be an $\mathcal{A}$-*module* on X. Moreover, consider the following **C**-*vector space sheaf* on X

(12.2) $$\mathcal{J}^1_{\mathcal{A}}(\mathcal{E}) := \mathcal{E} \oplus (\mathcal{E} \otimes_{\mathcal{A}} \Omega) \equiv \mathcal{E} \oplus \Omega(\mathcal{E})$$

(cf. also Chapt. II; (1.38)). Hereby we have, of course, employed the fact that $\mathcal{J}^1_{\mathcal{A}}(\mathcal{E})$ *is,* according to our hypothesis for $\mathcal{E}$ and Ω, *an $\mathcal{A}$-module* on X (see Chapt. II; (3.6)). However, for reasons that will become apparent right below, we do define *another $\mathcal{A}$-module structure* on the above **C**-vector space sheaf $\mathcal{J}^1_{\mathcal{A}}(\mathcal{E})$, as follows;

(12.3) $$\alpha \cdot (s \oplus t) := \alpha \cdot s \oplus (\alpha \cdot t + s \otimes \partial(\alpha)),$$

for any (local) sections $\alpha \in \mathcal{A}(U)$, $s \in \mathcal{E}(U)$ and $t \in \Omega(\mathcal{E})(U)$, with U *open* in X.

Now, the $\mathcal{A}$-*module*

$$\mathcal{J}^1_{\mathcal{A}}(\mathcal{E}) \equiv \mathcal{J}^1(\mathcal{E}), \tag{12.4}$$

as defined by (12.3), is called the *jet-line sheaf* (or even *jet sheaf of order 1*) of the *$\mathcal{A}$-module $\mathcal{E}$*, relative to the given differential triad $(\mathcal{A}, \partial, \Omega)$. The terminology employed extends, of course, the classical one of *vector bundle* (differential) *geometry;* see, for instance, M. Karoubi - G. Leruste [1: p. 338, §B7].

Thus, our purpose now, by the ensuing discussion, is to relate the above *$\mathcal{A}$-module* $\mathcal{J}^1(\mathcal{E})$ with certain *$\mathcal{A}$-extensions of $\mathcal{E}$* (cf. Chapt. III; Section 10 and Chapt. V; Sections 7, 8), in such a manner that *splittings of the latter assure the existence of $\mathcal{A}$-connections of $\mathcal{E}$, and conversely:*

Indeed, by considering $\mathcal{J}^1(\mathcal{E})$ *as an $\mathcal{A}$-module* on X, *according to (12.3),* one obtains the following *short exact $\mathcal{A}$-sequence*

$$0 \longrightarrow \Omega(\mathcal{E}) \xrightarrow{\ i\ } \mathcal{J}^1(\mathcal{E}) \xrightarrow{\ \pi\ } \mathcal{E} \longrightarrow 0, \tag{12.5}$$

where π denotes the obvious *projection $\mathcal{A}$-morphism* (with the *$\mathcal{A}$-module structure of $\mathcal{I}^1(\mathcal{E})$, as before!*); yet, i stands again for the natural *injection $\mathcal{A}$-morphism* of $\Omega(\mathcal{E})$ in the *Whitney sum of* the **C**-*vector space sheaves* concerned, as in (12.2) (see also (12.3), in conjunction with Lemma 1.1). Therefore (cf. Chapt. III; Definition 10.1),

(12.6) given a *differential triad* $(\mathcal{A}, \partial, \Omega)$ and an *$\mathcal{A}$-module* $\mathcal{E}$ on X, *the* corresponding *jet-line sheaf of $\mathcal{E}$*, $\mathcal{J}^1(\mathcal{E})$ (cf. (12.4)), *provides an $\mathcal{A}$-extension of $\mathcal{E}$ by $\Omega(\mathcal{E}) \equiv \mathcal{E} \otimes_{\mathcal{A}} \Omega$.*

(*Warning!* not the trivial one; cf. (12.3) and Chapt. III; (10.2)). In other words, *(12.5) defines an element of the* respective (abelian) *Baer group*

$$\mathrm{ext}_{\mathcal{A}}(\mathcal{E}, \Omega(\mathcal{E})) = \mathrm{Ext}^1_{\mathcal{A}}(\mathcal{E}, \Omega(\mathcal{E})) \tag{12.7}$$

(see Chapt. III; Theorem 10.1 and (10.36)).

On the other hand, for use right below, we further note that, in the case that $\mathcal{E}$ is, in particular, a *vector sheaf* on X, *with* $\mathrm{rk}\,\mathcal{E} = n \in \mathbf{N}$, one obtains the following (abelian) *group isomorphisms;*

$$H^1(X, \mathcal{H}om_{\mathcal{A}}(\mathcal{E}, \Omega(\mathcal{E}))) = H^1(X, \Omega(\mathcal{E}nd\,\mathcal{E})) = H^1(X, M_n(\Omega)). \tag{12.8}$$

Indeed, based on Chapt. III; Lemma 8.1 and (4.53), as well as, on our previous remark in (9.12), along with (9.13), (9.4) and (7.22) in the preceding, one infers that;

$$\begin{aligned} & H^1(X, \mathcal{H}om_{\mathcal{A}}(\mathcal{E}, \Omega(\mathcal{E}))) = H^1(X, \Omega(\mathcal{E}nd\,\mathcal{E})) = \check{H}^1(X, \Omega(\mathcal{E}nd\,\mathcal{E})) \\ (12.9) \quad & = \varinjlim_{\mathcal{U}} \check{H}^1(\mathcal{U}, \Omega(\mathcal{E}nd\,\mathcal{E})) = \varinjlim_{\mathcal{U}} \check{H}^1(\mathcal{U}, M_n(\Omega)) = \check{H}^1(X, M_n(\Omega)) \\ & = H^1(X, M_n(\Omega)), \end{aligned}$$

as was actually to be proved, according to (12.8). ∎ Thus, based now on Chapt. V; Theorem 7.2, one concludes that;

$$\begin{aligned} (12.10) \quad \mathrm{ext}_{\mathcal{A}}(\mathcal{E}, \Omega(\mathcal{E})) &= \mathrm{Ext}^1_{\mathcal{A}}(\mathcal{E}, \Omega(\mathcal{E})) = H^1(X, \mathcal{H}om_{\mathcal{A}}(\mathcal{E}, \Omega(\mathcal{E}))) \\ &= H^1(X, \Omega(\mathcal{E}nd\,\mathcal{E})) = H^1(X, M_n(\Omega)), \end{aligned}$$

up to (abelian) *group isomorphisms,* as indicated, for every *vector sheaf* $\mathcal{E}$ on X, *with* $\operatorname{rk}\mathcal{E} = n \in \mathbf{N}$, and a given *differential triad* $(\mathcal{A}, \partial, \Omega)$.

The previous rels. (12.10) will be applied presently below in the proof of (12.1). But, first we have to state the following fundamental lemma, on which the same proof is based too. So we have.

Lemma 12.1. *Let* $(\mathcal{A}, \partial, \Omega)$ *be a given differential triad on a topological space* X, $\mathcal{E}$ *an* $\mathcal{A}$*-module on* X *and* $\mathcal{J}^1(\mathcal{E})$ *the corresponding jet-line sheaf of* $\mathcal{E}$ (cf. (12.4)). *Moreover, let*

$$(12.11) \qquad \pi : \mathcal{J}^1(\mathcal{E}) \longrightarrow \mathcal{E}$$

be the (canonical) projection $\mathcal{A}$*-morphism* (see (12.5)). *Then,* π *admits a section, viz. an* $\mathcal{A}$*-morphism*

$$(12.12) \qquad D' : \mathcal{E} \longrightarrow \mathcal{J}^1(\mathcal{E}),$$

satisfying the relation

$$(12.13) \qquad \pi \circ D' = 1_{\mathcal{E}},$$

or, equivalently (cf. Chapt. III; Proposition 10.1 and Definition 10.3), *one has a splitting of (12.5), if, and only if, there exists a uniquely defined* $\mathcal{A}$*-connection* D *of* $\mathcal{E}$, *such that one has* (cf. (12.2));

$$(12.14) \qquad D' = 1_{\mathcal{E}} \oplus D.$$

Proof. Indeed, suppose that

$$(12.15) \qquad D : \mathcal{E} \longrightarrow \mathcal{E} \otimes_{\mathcal{A}} \Omega \equiv \Omega(\mathcal{E})$$

is a given $\mathcal{A}$*-connection* of the $\mathcal{A}$-module $\mathcal{E}$. Then, the relation

(12.16) $$D' := 1_{\mathcal{E}} \oplus D \equiv (1_{\mathcal{E}}, D) : \mathcal{E} \longrightarrow \mathcal{E} \oplus \Omega(\mathcal{E}) \equiv \mathcal{J}^1(\mathcal{E})$$

provides a *splitting of (12.5)*; that is, *equivalently* (cf. the statement of the lemma), *a section of* π (cf. (12.11)). Furthermore,

(12.17) *D' is an $\mathcal{A}$-morphism* of the $\mathcal{A}$-modules involved, as in (12.16) (cf. also (12.3)), *satisfying (12.13).*

Thus, *for any* (local) *section* $s \in \mathcal{E}(U)$, with U *open* in X, one has, by (12.16),

(12.18) $$D'(s) = s \otimes D(s) \equiv (s, D(s)),$$

so that one obtains

(12.19) $$(\pi \circ D')(s) = \pi(D'(s)) \equiv \pi_{\mathcal{E}}(s, Ds) = s,$$

as desired. (We set here $\pi \equiv \pi_{\mathcal{E}}$, as in (12.5)). Yet, for any $\alpha \in \mathcal{A}(U)$ and $s \in \mathcal{E}(U)$, as before, one has, *by virtue of (12.3)*;

(12.20) $$\begin{aligned} D'(\alpha \cdot s) &= \alpha s \oplus D(\alpha s) = \alpha s \oplus (\alpha \cdot Ds + s \otimes \partial \alpha) \\ &= \alpha \cdot (s \oplus Ds) = \alpha \cdot D'(s), \end{aligned}$$

given that $D(s) \in \Omega(\mathcal{E})(U)$, by (12.15). (We still note here the *significance of the definition (12.3)*). So *D' is an $\mathcal{A}$-morphism,* which thus completely justifies our claim about it in (12.17).

Conversely, we shall prove now that;

(12.21) *any splitting $D' : \mathcal{E} \longrightarrow \mathcal{J}^1(\mathcal{E})$ of (12.5)* (in other words, any *section of* π, such that $\pi \circ D' = 1_{\mathcal{E}}$) *is of the form (12.16), for some uniquely defined $\mathcal{A}$-connection D of $\mathcal{E}$.*

Namely, by first considering the following diagram

(12.22) $$\begin{array}{ccc} \mathcal{E} & \xrightarrow{\quad D' \quad} & \mathcal{E} \oplus \Omega(\mathcal{E}) \equiv \mathcal{J}^1(\mathcal{E}) \\ & {\scriptstyle D} \searrow & \downarrow {\scriptstyle p_2} \\ & & \Omega(\mathcal{E}), \end{array}$$

one makes it *commutative,* by defining,

$$D := p_2 \circ D', \tag{12.23}$$

where p_2 stands for the (canonical) *projection of* $\mathcal{J}^1(\mathcal{E}) \equiv \mathcal{E} \oplus \Omega(\mathcal{E})$ *onto the second factor;* as already remarked in the preceding, the latter sheaf is the *Whitney sum of the* **C**-*vector space sheaves involved,* so that p_2 *is* certainly *a* **C**-*linear morphism.* Hence, *the same holds true for* D, by virtue of (12.23) and of our hypothesis for D' (cf. (12.21), as well as, Chapt. III; (10.10), along with Definition 10.3 and Chapt. II; (2.3)).

Thus, we further prove that D, as given by (12.23), *satisfies the Leibniz condition (3.6):* First, by hypothesis for D' (cf. (12.21)), along with (12.13) and (12.23), one has

$$D' = \pi \circ D' \oplus p_2 \circ D' = 1_{\mathcal{E}} \oplus D. \tag{12.24}$$

Therefore, *for any* given $\alpha \in \mathcal{A}(U)$ *and* $s \in \mathcal{E}(U)$, with U *open* in X, one obtains (cf. also (12.3));

$$\begin{aligned} D(\alpha \cdot s) &= (p_2 \circ D')(\alpha s) = p_2(D'(\alpha s)) = p_2(\alpha \cdot D'(s)) \\ &= p_2(\alpha \cdot (s \oplus Ds)) = p_2(\alpha s \oplus (\alpha \cdot Ds + s \otimes \partial\alpha)) = \alpha \cdot Ds + s \otimes \partial\alpha, \end{aligned} \tag{12.25}$$

that was actually to be proved. Finally, the same relation (12.24) *proves,* of course, *the uniqueness of* D, and this terminates the proof of the lemma. ∎

Now, as an application of the previous lemma, we come next, as already promised, to the *vindication of (12.1).* However, we do state first the following more convenient (although less precise) formulation of the same lemma. That is, one has.

Theorem 12.1. *Let* $(\mathcal{A}, \partial, \Omega)$ *be a given differential triad on a topological space* X, $\mathcal{E}$ *an* $\mathcal{A}$*-module on* X *and* $\mathcal{J}^1(\mathcal{E})$ *the respective jet-line sheaf of* $\mathcal{E}$ (cf. (12.2)). *Moreover, consider the following short exact* $\mathcal{A}$*-sequence*

$$0 \longrightarrow \Omega(\mathcal{E}) \xrightarrow{i} \mathcal{J}^1(\mathcal{E}) \xrightarrow{\pi} \mathcal{E} \longrightarrow 0; \tag{12.26}$$

viz. an $\mathcal{A}$*-extension of* $\mathcal{E}$ *by* $\Omega(\mathcal{E})$, *as defined by* $\mathcal{J}^1(\mathcal{E})$ (cf. (12.5)).

Then, the given $\mathcal{A}$*-module* $\mathcal{E}$ *admits an* $\mathcal{A}$*-connection if, and only if, (12.26) splits.*

Proof. The assertion is, in effect, an immediate consequence of the above Lemma 12.1. (We do not actually refer here to the *uniqueness of the connection of* $\mathcal{E}$ supplied by the splitting of (12.26)). ∎

Thus, as a result of the preceding, we now have the following reformulation of our claim in (12.1).

Corollary 12.1. *Suppose we are given a differential triad* $(\mathcal{A}, \partial, \Omega)$ *on a topological space* X, *and let* $\mathcal{E}$ *be a vector sheaf on* X, *of rank* $n \in \mathbf{N}$. *Then, the two following assertions are equivalent:*

i) The given vector sheaf $\mathcal{E}$ *admits an* $\mathcal{A}$*-connection.*

ii) The Atiyah class of $\mathcal{E}$, *relative to a given local frame*

$$\mathcal{U} = (U_\alpha)_{\alpha \in I} \tag{12.27}$$

of $\mathcal{E}$ *vanishes; viz. one has* (cf. (9.14))

$$\mathfrak{a}(\mathcal{E}) = [(D_\beta - D_\alpha)] = 0 \in \check{H}^1(X, M_n(\Omega)). \tag{12.28}$$

Proof. We already know that *ii) implies i)* (cf Theorem 11.1). Thus, *conversely,* assume that $\mathcal{E}$ *admits an* $\mathcal{A}$*-connection,* so that, according to Theorem 12.1, *(12.26) splits.* Now, this actually means that the equivalence class, defined by $\mathcal{J}^1(\mathcal{E})$, in the Baer group $\mathrm{ext}_{\mathcal{A}}(\mathcal{E}, \Omega(\mathcal{E}))$ is the trivial one; that is, the zero element of the latter group or, equivalently (cf. also (12.7)), of the (abelian) group (in fact, $\mathcal{A}(X)$-module) $\mathrm{Ext}^1_{\mathcal{A}}(\mathcal{E}, \Omega(\mathcal{E}))$ (cf. Chapt. III; (10.30), (10.36)). Hence, based on (12.10), the same class corresponds to the zero element of (the $\mathcal{A}(X)$-module) $H^1(X, M_n(\Omega))$. However, due to the same rel. (12.10), *the* $\mathcal{A}$*-extension at issue* (viz. of $\mathcal{E}$ by $\Omega(\mathcal{E})$, as given by $\mathcal{J}^1(\mathcal{E})$) *defines uniquely an element* of $H^1(X, M_n(\Omega))$, while *the same element is uniquely determined, as well, by the Atiyah class of* $\mathcal{E}$, as the latter is defined by any given local frame of $\mathcal{E}$ (cf. (12.27)), according to the relation (see also (9.14), (10.28), along with (10.29), (10.31));

$$\mathfrak{a}(\mathcal{E}) = [(D_\beta - D_\alpha)] = [\tilde{\partial}(g_{\alpha\beta})] \in \check{H}^1(X, M_n(\Omega)). \tag{12.29}$$

Therefore, by what has been said before, one concludes that $\mathfrak{a}(\mathcal{E}) = 0$, that was actually to be proved, so that *i) implies ii), as well,* and this terminates the proof. ∎

Scholium 12.1.– As a clarification of our argument in the last part of the preceding proof, we still set out the following comments:

Given a *vector sheaf* $\mathcal{E}$ on X, as before, the *Levi-Civita (*$\mathcal{A}$*-connection) 1-cocycle of* $\mathcal{E}$, associated with any given *local frame of* $\mathcal{E}$ (cf. (12.27) and (8.34.1)), *yields* further *an* $\mathcal{A}$*-extension of* $\mathcal{E}$ *by* $\Omega(\mathcal{E})$, as determined e.g.

by the *jet-line sheaf of* $\mathcal{E}$, $\mathcal{J}^1(\mathcal{E})$ (see (12.5)), according to the following (abelian) *group isomorphisms;*

$$\begin{aligned} &\operatorname{ext}_{\mathcal{A}}(\mathcal{E},\Omega(\mathcal{E})) = \operatorname{Ext}^1_{\mathcal{A}}(\mathcal{E},\Omega(\mathcal{E})) = \operatorname{Ext}^1_{\mathcal{A}}(\mathcal{E}\otimes_{\mathcal{A}}\mathcal{A},\Omega(\mathcal{E})) \\ &= \operatorname{Ext}^1_{\mathcal{A}}(\mathcal{A},\mathcal{E}^*\otimes_{\mathcal{A}}\Omega(\mathcal{E})) = \operatorname{Ext}^1_{\mathcal{A}}(\mathcal{A},\mathcal{H}om_{\mathcal{A}}(\mathcal{E},\Omega(\mathcal{E}))) \\ &= \operatorname{Ext}^1_{\mathcal{A}}(\mathcal{A},\Omega(\mathcal{E}nd\,\mathcal{E})) = H^1(X,\Omega(\mathcal{E}nd\,\mathcal{E})) = H^1(X,M_n(\Omega)) \end{aligned} \tag{12.30}$$

(see (12.7), (12.8), (12.10) and Chapt. V; (7.7)).

Thus (cf. Theorem 12.1), *the aforementioned 1-cocycle yields,* in fact, *an $\mathcal{A}$-connection of $\mathcal{E}$ if, and only if, the $\mathcal{A}$-extension at issue* (cf. (12.5)) *splits. Equivalently* (cf. Chapt. III; (10.30), (10.36)), *if, and only if, the* same *1-cocycle is a coboundary,* or yet (cf. (10.28.1)) *if, and only if, the Atiyah class of $\mathcal{E}$* (defined by the given local frame of $\mathcal{E}$, as above) *vanishes.*

In this context, see also Note 8.1 in the foregoing, along with Theorem 14.1 and (14.10) below. Yet, the above picture is supplemented by further *specializing,* for instance, *on* Ω, by requiring this to be a *vector sheaf* on X, as well (cf. the next Section 13), or even *on the* base *space* X, assuming it now *(Hausdorff) paracompact* (see Section 14 below).

Finally, as another consequence of the preceding, we still infer that;

(12.31) a given *vector sheaf* $\mathcal{E}$ on X *admits an $\mathcal{A}$-connection if, and only if, it also admits a Levi-Civita $\mathcal{A}$-connection.*

Indeed, if $\mathcal{E}$ has an $\mathcal{A}$-connection, then (Corollary 12.1) $\mathfrak{a}(\mathcal{E}) = 0$, so that (cf. Theorem 11.1 and Definition 11.1) $\mathcal{E}$ admits a Levi-Civita $\mathcal{A}$-connection, as well, the rest of (12.31) being clear. ∎ This explains also completely (11.16) in the preceding.

13. $\mathcal{A}$-connections as splitting extensions (contn'd)

We examine in this Section another short exact $\mathcal{A}$-sequence, save (12.5) above, that is, another *$\mathcal{A}$-extension,* (yet, naturally) connected with a given *vector sheaf* $\mathcal{E}$ on X (we always refer, of course, to a given fixed, as usual, differential triad $(\mathcal{A}, \partial, \Omega)$) and whose splitting is still equivalent with the existence of an *$\mathcal{A}$-connection* of $\mathcal{E}$. However, now, under an extra hypothesis for the $\mathcal{A}$-module Ω; yet, a further restriction on the topological space X considered, will involve, as we shall see, but in the next Section 14, the so-called *"coordinate 1-cocycle"* of the extension concerned.

So assume that we are given a *differential triad* $(\mathcal{A}, \partial, \Omega)$ on an arbitrary *topological space* X (cf. (3.1)), such that

(13.1) Ω is a *vector sheaf* on X, as well.

Therefore, by considering the *dual $\mathcal{A}$-module* of Ω,

$$\Omega^* := \mathcal{H}om_{\mathcal{A}}(\Omega, \mathcal{A}), \tag{13.2}$$

one gets another *vector sheaf* on X *with* $\operatorname{rk}\Omega^* = \operatorname{rk}\Omega$ (cf. Chapt. IV; Definition 5.1, and (5.19)). Furthermore, due to (13.1), Ω *is reflexive;* that is, one has

$$\Omega^{**} = \Omega, \tag{13.3}$$

up to an *$\mathcal{A}$-isomorphism of* the *$\mathcal{A}$-modules* (in fact, *vector sheaves*) involved (cf. Chapt. IV; Theorem 5.1). In this context, we also note that one has, of course, a similar situation to (13.1) in the classical case of differential geometry (see thus Section 2 in the preceding, as well as, Section 16 below).

On the other hand, we still recall that, given a *vector sheaf* $\mathcal{E}$ on X, *with* $\operatorname{rk}\mathcal{E} = n \in \mathbf{N}$, the set

$$\mathcal{H}om_{\mathcal{A}}(\mathcal{E}, \mathcal{E}) = \mathcal{E}nd\,\mathcal{E} \tag{13.4}$$

is a *vector sheaf* on X too, *with* $\operatorname{rk}(\mathcal{E}nd\,\mathcal{E}) = n^2$ (cf. Chapt. II; Lemma 6.1).

Thus, under the above circumstances, consider further the *short exact $\mathcal{A}$-sequence,*

$$0 \longrightarrow \mathcal{E}nd\,\mathcal{E} \longrightarrow \mathcal{S} \longrightarrow \Omega^* \longrightarrow 0; \tag{13.5}$$

that is, an *$\mathcal{A}$-extension of Ω^* by $\mathcal{E}nd\,\mathcal{E}$* (cf. Chapt. III; Definition 10.1), or even when looking at the corresponding *isomorphism class of (13.5),* an element of the *Baer group*

$$\operatorname{ext}_{\mathcal{A}}(\Omega^*, \mathcal{E}nd\,\mathcal{E}) = \operatorname{Ext}^1_{\mathcal{A}}(\Omega^*, \mathcal{E}nd\,\mathcal{E}), \tag{13.6}$$

while the latter group is also suitably expressed through the functor $\mathrm{Ext}^1_{\mathcal{A}}$, as above (cf. Chapt. III; Theorem 10.1). Therefore, based on Chapt. V; Proposition 8.1, we conclude that;

(13.7) under the assumption (13.1), *the* previous *$\mathcal{A}$-extensions (12.5) and (13.5) have the same* (in fact *isomorphic*) *Baer groups.* That is, one has the following (abelian) *group isomorphisms;*

$$(13.7.1)\quad \mathrm{ext}_{\mathcal{A}}(\mathcal{E},\Omega(\mathcal{E})) = \mathrm{ext}_{\mathcal{A}}(\Omega^*, \mathcal{E}nd\,\mathcal{E}) = H^1(X, M_n(\Omega)).$$

However, for convenience, we do give the relevant calculations, concerning (13.7.1); see thus (12.8), (12.20) *and* (13.3) above, along with Chapt. IV; (6.1). So we have;

$$\begin{aligned}(13.8)\quad &\mathrm{ext}_{\mathcal{A}}(\mathcal{E},\Omega(\mathcal{E})) = \mathrm{Ext}^1_{\mathcal{A}}(\mathcal{E},\Omega(\mathcal{E})) = H^1(X, \mathcal{H}om_{\mathcal{A}}(\mathcal{E},\Omega(\mathcal{E})))\\ &= H^1(X,\Omega(\mathcal{E}nd\,\mathcal{E})) = H^1(X, M_n(\Omega)) = H^1(X, \Omega^{**}(\mathcal{E}nd\,\mathcal{E}))\\ &= H^1(X, \mathcal{H}om_{\mathcal{A}}(\Omega^*, \mathcal{E}nd\,\mathcal{E})) = \mathrm{Ext}^1_{\mathcal{A}}(\Omega^*, \mathcal{E}nd\,\mathcal{E}) = \mathrm{ext}_{\mathcal{A}}(\Omega^*, \mathcal{E}nd\,\mathcal{E}),\end{aligned}$$

which thus justifies (13.7). ∎ Accordingly (see also Chapt. V; (8.60)), one further concludes that;

(13.9) under the hypothesis (13.1), *the splitting of either one of the two $\mathcal{A}$-extensions (12.5) and (13.5) is tantamount to the existence of an $\mathcal{A}$-connection of $\mathcal{E}$.*

Indeed, the two extensions, under consideration, have the same splittings, corresponding to the zero elements of the isomorphic Baer groups in (13.7.1), respectively (see Chapt. III; (10.30) and (10.36)). Thus, our claim in (13.9) follows already from Theorem 12.1. ∎

For convenience, we summarize the information, we gathered hitherto, into the following.

Theorem 13.1. *Suppose we have a differential triad $(\mathcal{A}, \partial, \Omega)$ on a topological space X, with Ω being, in particular, a vector sheaf on X. Moreover, let also $\mathcal{E}$ be another given vector sheaf on X. Then, all the next four assertions are equivalent:*

1) The given vector sheaf $\mathcal{E}$ admits an $\mathcal{A}$-connection.

2) $\mathfrak{a}(\mathcal{E}) = [(D_\beta - D_\alpha)] = 0$; viz. *the Atiyah class of $\mathcal{E}$*, relative to a given local frame $\mathcal{U} = (U_\alpha)_{\alpha\in I}$ of $\mathcal{E}$ (cf. (9.14)), *vanishes.*

3) The following $\mathcal{A}$-extension splits;

$$0 \longrightarrow \Omega(\mathcal{E}) \longrightarrow \mathcal{J}^1(\mathcal{E}) \longrightarrow \mathcal{E} \longrightarrow 0 \tag{13.10}$$

(see also (12.5)), *and*

4) the next $\mathcal{A}$-extension splits

$$0 \longrightarrow \mathcal{E}nd\,\mathcal{E} \longrightarrow \mathcal{S} \longrightarrow \Omega^* \longrightarrow 0. \tag{13.11}$$

Proof. First, according to (13.9), the assertions *1), 3) and 4) are* all *equivalent* to one another. On the other hand, *1) and 2) are* already *equivalent,* by Corollary 12.1 (even without any restriction on Ω), and this completes the proof. ∎

We come now, by the succeeding Section 14, to relate the above $\mathcal{A}$-*extensions,* in effect, their respective *coordinate 1-cocycles,* with another familiar such cocycle, connected with a given *vector sheaf* $\mathcal{E}$ on X and a *local frame* of it, viz. the *Levi-Civita $\mathcal{A}$-connection 1-cocycle of $\mathcal{E}$.* However, this under the further assumption of *paracompactness* of X.

14. Coordinate 1-cocycles of $\mathcal{A}$-extensions, as related to $\mathcal{A}$-connections

To begin with, suppose that we are given a *differential triad* $(\mathcal{A}, \partial, \Omega)$ on a topological space X, such that the two following supplementary conditions are fulfilled:

(14.1)
i) X is a *paracompact (Hausdorff)* space.
ii) Ω is a *vector sheaf* on X.

Furthermore, suppose that $\mathcal{E}$ is a *vector sheaf* on X, *with* $\operatorname{rk}\mathcal{E} = n \in \mathbf{N}$, and consider again the $\mathcal{A}$-*extension,* given by (13.5) above,

$$0 \longrightarrow \mathcal{E}nd\,\mathcal{E} \longrightarrow \mathcal{S} \longrightarrow \Omega^* \longrightarrow 0. \tag{14.2}$$

Thus, by looking now at the *characteristic class of (14.2)* (see Chapt. III; (10.39)), one obtains ((14.1); ii) is needed here, *only!*);

$$\begin{aligned}(14.3)\qquad \delta(1_{\Omega^*}) &\in \mathrm{Ext}^1_{\mathcal{A}}(\Omega^*, \mathcal{E}nd\,\mathcal{E}) = \mathrm{Ext}^1_{\mathcal{A}}(\mathcal{A}, \Omega^{**}\otimes_{\mathcal{A}} \mathcal{E}nd\,\mathcal{E})\\ &= \mathrm{Ext}^1_{\mathcal{A}}(\mathcal{A}, \Omega\otimes_{\mathcal{A}} \mathcal{E}nd\,\mathcal{E}) = \mathrm{Ext}^1_{\mathcal{A}}(\mathcal{A}, \Omega(\mathcal{E}nd\,\mathcal{E}))\\ &= H^1(X, \Omega(\mathcal{E}nd\,\mathcal{E})) = H^1(X, M_n(\Omega)).\end{aligned}$$

Concerning the above calculations, cf. Chapt. V; (7.1), or even (7.7), for $n = 1$, along with (13.8) above. Yet, as already known (Chapt. III; (10.39)), $\delta(1_{\Omega^*})$ *measures the obstruction to the splitting of (14.2).* Therefore (cf. Theorem 13.1), one further concludes that;

(14.4) by assuming (14.1); ii) *only* (hence, for any space X), one infers that $\delta(1_{\Omega^*})$ *measures the obstruction to the existence of an $\mathcal{A}$-connection of $\mathcal{E}$.*

In this context, see also (14.12) below.

By the same token, one can actually get a better *criterion for the existence of an $\mathcal{A}$-connection* of the *vector sheaf $\mathcal{E}$* involved, patterned after Theorem 12.1, that is, by considering the *$\mathcal{A}$-extension* of $\mathcal{E}$ by $\Omega(\mathcal{E})$, as given by (12.5);

$$(14.5)\qquad 0 \longrightarrow \Omega(\mathcal{E}) \longrightarrow \mathcal{J}^1(\mathcal{E}) \longrightarrow \mathcal{E} \longrightarrow 0.$$

Thus, one now concludes that;

(14.6) *without* any *restriction on Ω or X,* one infers that *the characteristic class $\delta(1_{\mathcal{E}})$ of (14.5) measures the obstruction to the existence of an $\mathcal{A}$-connection of the vector sheaf $\mathcal{E}$.*

It is clear, of course, in view of Chapt. III; (10.39), that (14.6) is, in effect, another formulation of the previous Theorem 12.1.

On the other hand, *assuming* now (14.1) and based on Chapt. V; (8.61), we further conclude that:

(14.7) *the $\mathcal{A}$-extensions (14.2) and (14.5) are* (cohomologically) *classified by the same characteristic class, or even the same coordinate 1-cocycle;* viz. one has

$$(14.7.1)\qquad \delta(1_{\mathcal{E}}) = \delta(1_{\Omega^*}) = [(h_\beta - h_\alpha)] \in H^1(X, M_n(\Omega))$$

(see also Chapt. V; (8.27), (8.29)).

Now, the last conclusion with (12.29), in concert, provide already the proof of the following.

Theorem 14.1. *Let $(\mathcal{A}, \partial, \Omega)$ be a given differential triad on a paracompact (Hausdorff) space X, with Ω being also a vector sheaf on X. Moreover, let $\mathcal{E}$ be a vector sheaf on X, of rank $n \in \mathbf{N}$, together with the following 1-cocycles of a local frame $\mathcal{U} = (U_\alpha)_{\alpha \in I}$ of $\mathcal{E}$:*

i) A *coordinate 1-cocycle of $\mathcal{E}$*, associated with $\mathcal{U}$, say,

$$(g_{\alpha\beta}) \in Z^1(\mathcal{U}, \mathcal{GL}(n, \mathcal{A})) \tag{14.8}$$

(see e.g. (10.19)), along with *its image via the logarithmic derivation* $\tilde{\partial}$ (cf. (10.25)); i.e.,

$$\tilde{\partial}(g_{\alpha\beta}) \in Z^1(\mathcal{U}, M_n(\Omega)). \tag{14.9}$$

ii) The *Levi-Civita ($\mathcal{A}$-connection) 1-cocycle of $\mathcal{E}$* (cf. (9.8));

$$(D_\beta - D_\alpha) \in Z^1(\mathcal{U}, M_n(\Omega)). \tag{14.10}$$

iii) The *coordinate 1-cocycle of an $\mathcal{A}$-extension of Ω^** (dual vector sheaf of Ω) *by $\mathcal{E}nd\,\mathcal{E}$*;

$$\delta(h_\alpha) = (h_\beta - h_\alpha) \in Z^1(\mathcal{U}, M_n(\Omega)) \tag{14.11}$$

(see Chapt. V; (8.27), for $\mathcal{E} = \Omega^*$ and $\mathcal{F} = \mathcal{E}nd\,\mathcal{E}$, along with Chapt. IV; (6.1) and (5.11)). *The latter coincides* (through a (canonical) *group isomorphism;* see e.g. (13.8)) *with that one of (14.5)* (cf. also (14.7.1)).

Then, the 1-dimensional cohomology classes of X, which correspond to all the preceding $M_n(\Omega)$-valued 1-cocycles of $\mathcal{U}$ (viz. to (14.9), (14.10) and (14.11)) *are all equal; that is, one has* (see also (14.7.1), concerning the first of the succeeding equalities)

$$\begin{aligned} \delta(1_{\mathcal{E}}) = \delta(1_{\Omega^*}) &= [(h_\beta - h_\alpha)] = [(D_\beta - D_\alpha)] \\ &= [\tilde{\partial}(g_{\alpha\beta})] \in H^1(X, M_n(\Omega)). \end{aligned} \tag{14.12}$$

(In this regard, cf. also Chapt. V; (8.54), *with $\mathcal{E}$ and $\mathcal{F}$, as in iii) above,* along with (8.59), (8.61) therein).

Consequently, once more (see also (14.6)), the *triviality of any one of the previous 1-cocycles, as* appeared *in (14.12),* viz. if any one of them is cohomologous to zero, then this *is a necessary and sufficient condition for the existence of an $\mathcal{A}$-connection of $\mathcal{E}$.* ∎

The above beautiful and all embracing result constitutes, in effect, the *echo,* within the present abstract framework, of a classical one due to M.F.

Atiyah [1] and referred to (holomorphic) *vector bundle geometry.*

Now, by further applying the terminology of Chapt. V; (8.48), and based also on (8.59) (cf. (8.51.1) too) of the same Chapter, as well as, for instance, on (9.3), (9.4) in the preceding, we can say that:

(14.13) the *sheaf of $\mathcal{A}$-connection coefficients*

(14.13.1) $$M_n(\Omega) := M_n(\mathcal{A}) \otimes_{\mathcal{A}} \Omega \equiv \Omega(M_n(\mathcal{A}))$$

(cf. (9.10)), in fact, another *vector sheaf* on X too, in view of (14.1) (cf. also Chapt. II; (5.27)), *is* still (cohomologically) *the structure sheaf of any $\mathcal{A}$-extension of the form (14.2), or even of (14.5).* In particular, one concludes that;

(14.13.2) *every element of* $H^1(X, M_n(\Omega))$, viz. *any 1-cocycle* with coefficients from the sheaf in question, *determines uniquely* (cf. (13.7.1)) *an $\mathcal{A}$-extension of any one of the previous two equivalent forms.*

15. Examples (contn'd)

We consider below certain particular important instances from the classical theory, on which our previous considerations have a special bearing, while the same classical examples often motivated the present abstract context. Now, to put things in perspective, we first describe the relevant abstract setting, in accordance with the hitherto general framework.

Thus, suppose that we have a *differential triad* $(\mathcal{A}, \partial, \Omega)$ on a *topological space* X, as usual, for which, however, we further assume the following cohomological condition;

(15.1) $$H^1(X, M_n(\Omega)) = 0, \quad n \in \mathbf{N}.$$

An important *special case,* for which (15.1) is fulfilled is, for example, that one which occurs when,

(15.2) concerning a given $\mathbf{C}$-algebraized space $(X, \mathcal{A})$, the structure sheaf $\mathcal{A}$ is *soft* (cf. Chapt. III; (8.25)), while the base space X is *paracompact* (Hausdorff).

Of course, the important thing here, in connection with (15.4) below, is the fact that;

(15.3) *if $\mathcal{A}$ is soft over a paracompact* (Hausdorff) *space X,* then *every $\mathcal{A}$-module $\mathcal{E}$* on X *is fine* (cf. Chapt. III; Definition 8.1), while, *every fine sheaf* on such a space *is soft* too.

Concerning (15.3), see, for instance, G.E. Bredon [1: p.50, Theorem 9.12]. ▮ Furthermore, one obtains that:

(15.4) *every soft sheaf over a paracompact* (Hausdorff) *space is acyclic;* that is, *it has trivial cohomology.*

(ibid.; p.49, Theorem 9.8). ▮ Thus, by virtue of (15.3) and (15.4), it is now clear that;

(15.5) *if (15.2) holds true,* then, *every $\mathcal{A}$-module* on X, therefore $M_n(\Omega)$ itself, *with $n \in \mathbf{N}$* (cf. (14.3.1) and Chapt. V; (8.7)) *is fine,* hence, soft and *thus acyclic.* So, *(15.1) is in force.*

Now, the fundamental fact that interests us here is, of course, the following.

(15.6) *Assume that (15.2) holds true, so* that, in particular, *(15.1) is valid.* Then, *every vector sheaf* on X *admits an $\mathcal{A}$-connection.*

This assertion is certainly clear, in view of (15.1) and Theorem 11.1 (cf. (11.4)). ▮ On the other hand, as we shall presently see, by the ensuing Subsections, (15.6) represents the situation, one actually meets in the classical theory:

15.(a). $\mathcal{C}^\infty$-manifolds.– In Section 2 in the preceding we have already considered the *standard differential triad*

$$(\mathcal{C}_X^\infty, d, \Omega_X^1), \tag{15.7}$$

which is naturally associated with any given (finite dimensional) $\mathcal{C}^\infty$ (or *smooth*)*-manifold* X (see (2.6)). In particular, assuming that X is *paracompact* (Hausdorff) (cf. Chapt. III; (8.54), along with the subsequent comments there), one concludes that the respective *structure sheaf*

$$\mathcal{A} \equiv \mathcal{C}_X^\infty \tag{15.8}$$

is fine (see (2.2), as well as, Chapt. III; (8.27) and (8.55.1)). Therefore, due to the hypothesis for X and (15.3), the sheaf $\mathcal{C}^\infty_X$ is soft, as well, so that (yet, (15.3) and (15.4)) one concludes that;

(15.9) *every $\mathcal{C}^\infty_X$-module on X is* a fine sheaf on X and, therefore, *acyclic.*

In particular, *the last conclusion is valid for the sheaf of connection coefficients $M_n(\Omega^1_X)$, $n \in \mathbf{N}$,* where now Ω^1_X is given by (2.4) (cf. also (14.13.1) and Chapt. III; (8.30), for $p = 1$). Thus, (15.1) *is in force,* concerning now the triple (15.7).

On the other hand, as already mentioned (cf. (4.11) above), for any given (finite dimensional) *smooth* (viz. $\mathcal{C}^\infty$-) **C**-*vector bundle* **E** on X, the corresponding *sheaf $\mathcal{E}$ of germs of smooth sections of* **E** is a *vector sheaf* on X, with respect to (15.8), and whose *rank* equals *the dimension of* **E**. Furthermore, a $\mathcal{C}^\infty_X$-*connection of* $\mathcal{E}$, as above, in the sense of Definition 3.1, is (still, by definition, cf. (4.15)) a standard (linear) *connection* (viz. *law of derivation* of smooth sections (again!)) *of* **E**, in the usual sense of differential geometry. Therefore, as a consequence of (15.6) and the preceding, one thus arrives at the classical result, according to which:

(15.10) *every smooth* **C**-*vector bundle* **E** *over a paracompact $\mathcal{C}^\infty$-manifold X admits a connection.*
In particular, *this holds true for $\mathcal{T}(X)$, the* (complexified) *tangent bundle of X* (cf. (4.16)); in that case we also say that *the manifold X itself has a* (smooth) *connection* (i.e., a *law of derivation* of the *smooth vector fields* on X, the latter being the *smooth sections of $\mathcal{T}(X)$*).

In this context, see e.g. W.A. Poor [1: p.61, Theorem 2.35], or even J.W. Milnor - J.D. Stasheff [1: p.291, Lemma 2]. Yet, see also the next Section 16 for another proof of the same result. ∎

Now, it is still standard (Atiyah) that *(15.10) does not hold, in general, true* in the case of *holomorphic connections* (cf. Definition 15.1 below). Thus, we do discuss it briefly right away in the following Subsection.

15.(b). Complex manifolds. Holomorphic connections.– Suppose that we have a *complex manifold X*, such that let

$$(X, \mathcal{O}) \tag{15.11}$$

be the corresponding **C**-*algebraized space,* where $\mathcal{O}$ stands for the *sheaf of germs of* **C**-*valued holomorphic functions on* X (cf. also Chapt. III; (8.65)).

Now, as already noted (ibid.; (8.66)), $\mathcal{O}$ *is not a fine sheaf* on X, the latter property being violated, by the principle of analytic continuation. Therefore, (15.4) *cannot be applied:* Precisely, supposing further that our (complex) manifold X is Hausdorff and 2nd countable, hence, *paracompact,* we remark that *the* (*unital* **C**-*algebra,* cf. Chapt. III; comments following (8.65)) *sheaf* $\mathcal{O}$, being an $\mathcal{O}$-module over itself, *is fine if, and only if, it is soft* (cf. (15.3). See also G.E. Bredon [1: p.267; (j)]). ∎

However, thanks to the classical *Cartan's Theorem B* (see (15.16) below), the previous situation, about the non validity of (15.4), can be remedied, when restricting ourselves to a *Stein manifold* X:

In this context, we first recall, for convenience (see, for instance, L. Hörmander [1: p. 109, Definition 5.1.3]), that by the last term one means a (Hausdorff, 2nd countable, hence, *paracompact*) *complex manifold* X, satisfying the following properties;

i) $\mathcal{O}(X) = \Gamma(X, \mathcal{O})$ (viz. the algebra of **C**-*valued holomorphic functions* on X) *separates points* in X and *supplies global-local coordinates at every point of* X.

ii) X is *holomorphically convex,* viz. the holomophically convex hull of any compact subset of X is (relatively) compact, as well.

> **Scholium 15.1.**– Note that the condition of *separation of points* in X by $\mathcal{O}(X)$ in the above definition of a Stein manifold *is* actually *redundant;* cf. H. Grauert [1], as well as, S.G. Krantz [1: p. 276; ex. 14,(iv)].
>
> There are, of course, other equivalent, more sophisticated (nevertheless useful!) definitions of this important class of complex manifolds, through e.g. functional analysis methods (cf., for instance, A. Mallios [4: p. 230; Theorem 3.1]) or cohomological ones (see L. Kaup - B. Kaup [1: p. 93, Theorem 63.2]).

Now, given a *complex manifold* X, one considers the $\mathcal{O}$-*module,* in effect, *vector sheaf,* on X (with respect to the structure sheaf $\mathcal{O}$, cf. (15.11))

$$\Omega^1 := \Omega^1_X \tag{15.12}$$

of (germs of) **C**-*valued holomorphic 1-forms* on X, *with*

(15.13) $$\mathrm{rk}_{\mathcal{O}}(\Omega^1) = n = \dim_{\mathbf{C}} X.$$

Thus, we have that;

(15.14) Ω^1 is a *coherent* (analytic) *sheaf* on X.

See, for instance, H. Grauert - R. Remmert [1: p.67, Corollary]. ∎ Hence,

(15.15) *the same holds true for the vector sheaf* (cf. (15.31))

(15.15.1) $$M_n(\Omega^1) := M_n(\mathcal{O}) \otimes_{\mathcal{O}} \Omega^1 = (\Omega^1)^{n^2} \equiv \Omega^{(n^2)},$$

with $n \in \mathbf{N}$.

(See also Chapt. II; (5.27)). Here, as in (15.14) too, one actually applies the fact that, since $\mathcal{O}$ *is a coherent sheaf (Oka's Coherence Theorem), every vector sheaf* on X *is coherent,* as well. (See e.g. H. Grauert - R. Remmert [2: pp. 12, 16]).

Thus, according to *Cartan's Theorem B,* one now concludes that

(15.16) *every coherent* (analytic) *sheaf* $\mathcal{S}$ *on a Stein manifold* X *is acyclic; viz. one has*

(15.16.1) $$H^p(X, \mathcal{S}) = 0,$$

Cf. e.g. L. Hörmander [1: p.182, Theorem 7.4.3]. ∎ Consequently, based on (15.15) and (15.16), one infers that;

(15.17) *for every Stein manifold* X, *one has*

(15.17.1) $$H^1(X, M_n(\Omega^1)) = 0, \quad \textit{with} \;\; n \in \mathbf{N}.$$

Therefore, in view of (15.1) and the last part of Theorem 14.1, one infers that (see also (15.32) in the sequel);

(15.18) *Let* X *be a Stein manifold. Then, every* (principal holomorphic fiber bundle, hence, any) *holomorphic vector bundle on* X *admits a holomorphic connection.*

However, it is instructive to clarify at this place, a bit more, the term *holo-*

morphic connection, which has been used before. So, according to our general notion, as given by Definition 3.1 in the foregoing, the term is, in principle, referred to a (*differential* or, rather, for the case at hand, to a) *holomorphic triad;* in particular, this concerns here the triple

$$(\mathcal{O}, d, \Omega^1) \tag{15.19}$$

(cf. (15.11), (15.12)), where, of course,

$$d : \mathcal{O} \longrightarrow \Omega^1 \tag{15.20}$$

stands for the usual *differential of holomorphic functions* (viz. *sections of* $\mathcal{O}$; see also Chapt. I; (10.14), (11.11); iii), along with Proposition 2.1 and (12.49), (12.50) therein). In this context, see also the next Chapter VII, concerning a *local description of any given* $\mathcal{A}$*-connection,* hence, of a *holomorphic connection* too. However, following M.F. Atiyah [1], we do display right away a *holomorphic connection of an* $\mathcal{O}$*-extension,* relative to the particular **C**-*algebraized space* $(X, \mathcal{O})$, under consideration, according to the general pattern, as already presented in the foregoing:

Thus, suppose that we are given a *principal holomorphic fiber bundle* P over a *complex manifold* X (the latter being, as usual, Hausdorff and 2nd countable, hence *paracompact*), having as *structure group* a (connected) *complex Lie group* G. Furthermore, let $T(X)$ be the *(holomorphic) tangent bundle of* X; hence, denoting by $\mathcal{T}(X)$ the *sheaf of* (germs of) *holomorphic vector fields* on X (viz. the respective *sheaf of germs of (holomorphic) sections of* $T(X)$, cf. e.g. (4.4)), one has

$$\mathcal{T}(X) = (\Omega^1)^* \equiv \mathcal{H}om_{\mathcal{O}}(\Omega^1, \mathcal{O}), \tag{15.21}$$

that is, the *dual* (vector) *sheaf of* Ω^1, as above (cf. also Chapt. IV; (5.1)). Yet, let $L(P)$ be the *holomorphic vector bundle* over X, associated to P, via the *adjoint representation of* G on its Lie algebra; so $L(P)$ is, in fact, an *algebra bundle* over X, its fiber being isomorphic to $\mathcal{G}$, the *Lie algebra of* G. On the other hand, one has a natural *action of* G on $T(P)$, the (holomorphic) *tangent bundle of* P, induced on it by the given action of G on P; hence, one can further consider the *quotient vector bundle*

$$Q := T(P)/G. \tag{15.22}$$

(In this concern, see e.g. S. Kobayashi - K. Nomizu [1: p.50ff], [2: Chapt. IX], along with M.F. Atiyah [2: p.10]). So the first relevant result here is the following *short exact sequence of the holomorphic vector bundles* concerned (*Atiyah exact sequence*);

(15.23) $$0 \longrightarrow L(P) \longrightarrow Q \longrightarrow T(X) \longrightarrow 0$$

(see M.F. Atiyah [1: p.187, Theorem 1]).

Therefore, by considering the respective sheaves of germs of (holomorphic) sections of the vector bundles in (15.23), one gets the following *short exact $\mathcal{O}$-sequence of vector sheaves* on X;

(15.24) $$0 \longrightarrow \mathcal{L}(P) \longrightarrow \mathcal{Q} \longrightarrow \mathcal{T}(X) \longrightarrow 0$$

(see M.F. Atiyah [1: p.183, Proposition 1]. Yet, cf. (4.11) and (4.10) in the foregoing). So, by applying the terminology of $\mathcal{A}$-extensions (Chapt. III; Section 10), we first conclude that;

(15.25) *every principal holomorphic fiber bundle P* over a complex manifold X *entails an $\mathcal{O}$-extension of $\mathcal{T}(X)$ by $\mathcal{L}(P)$* (see (15.24)).

Consequently, *the same $\mathcal{O}$-extension is classified by the* respective *characteristic class;*

(15.26) $$\delta(1_{\mathcal{T}(X)}) \equiv \mathfrak{a}(P) \in H^1(X, \mathcal{H}om_{\mathcal{O}}(\mathcal{T}(X), \mathcal{L}(P))$$

(cf. Chapt. III; (10.39.2) and Chapt. V; (8.4). See also (10.28) and (14.12) in the preceding, concerning the above applied notation). Thus, by still following M.F. Atiyah [1: p.188, Definition], one sets the following.

Definition 15.1. Let P be a holomorphic principal fiber bundle on a complex manifold X. Then, we call a *holomorphic connection* of P, any splitting of (15.24).

Now, as we presently show, the previous terminology can also be embodied in the hitherto applied framework: Indeed, one first obtains (see also (15.21));

(15.27) $$\begin{aligned}\mathcal{H}om_{\mathcal{O}}(\mathcal{T}(X), \mathcal{L}(P)) &= \mathcal{T}(X)^* \otimes_{\mathcal{O}} \mathcal{L}(P) = \mathcal{L}(P) \otimes_{\mathcal{O}} \mathcal{T}(X)^* \\ &= \mathcal{L}(P) \otimes_{\mathcal{O}} (\Omega^1)^{**} = \mathcal{L}(P) \otimes_{\mathcal{O}} \Omega^1(\mathcal{L}(P)).\end{aligned}$$

(Concerning the previous calculations, cf. Chapt. IV; (6.1), (5.11), Chapt. II; (5.14), along with (15.12) above). On the other hand, by employing any *local gauge U* of the *vector sheaf* $\mathcal{L}(P)$, one obtains;

$$
\begin{aligned}
&\Omega^1(\mathcal{L}(P))(U) = (\Omega^1(\mathcal{L}(P))|_U)(U) \equiv (\mathcal{L}(P) \otimes_{\mathcal{O}} \Omega^1|_U)(U) \\
(15.28)\qquad &= (\mathcal{L}(P)|_U \otimes_{\mathcal{O}|_U} \Omega^1|_U)(U) = (\mathcal{O}^m|_U \otimes_{\mathcal{O}|_U} \Omega^1|_U) \\
&= (\mathcal{O}^m \otimes_{\mathcal{O}} \Omega^1)(U) = (\Omega^1)^{(m)}(U) \equiv \Omega^{(m)}(U),
\end{aligned}
$$

such that

$$
(15.29)\qquad m = \operatorname{rk} \mathcal{L}(P) = \operatorname{rk} L(P) = \dim_{\mathbf{C}} \mathcal{G} = \dim_{\mathbf{C}} G
$$

(see thus (4.10), (4.11) above, in conjunction with Chapt. I; (11.40) and Chapt. II; Lemma 5.1, for the calculations in (15.28)). Therefore, one has (see also, for instance, the argument used for (12.8) above: (12.9), in conjunction with Lemma 9.1)

$$
(15.30)\qquad H^1(X, \mathcal{H}om_{\mathcal{O}}(\mathcal{T}(X), \mathcal{L}(P))) = H^1(X, \Omega^1(\mathcal{L}(P))) = H^1(X, \Omega^{(m)}).
$$

Now, given that the *vector sheaf* (Whitney sum)

$$
(15.31)\qquad \Omega^{(m)} \equiv (\Omega^1)^m := \underbrace{\Omega^1 \oplus \cdots \oplus \Omega^1}_{m-\text{times}} \quad (m \in \mathbf{N}),
$$

being, by its definition (cf. (15.29), as well as, Chapt. II; (3.7)), another *$\mathcal{O}$-module, is coherent* (see the comments after (15.15) above), one concludes that (see also (15.16));

(15.32) for every *Stein manifold* X, one has

$$
(15.32.1)\qquad H^1(X, \Omega^{(m)}) = 0, \quad m \in \mathbf{N}.
$$

Therefore (cf. (15.26) and Definition 15.1), *every holomorphic principal fiber bundle P over X admits a holomorphic connection.*

The above contributes to the vindication of (15.18) (see also (15.36) below).

On the other hand, given a *holomorphic vector bundle* $\mathbf{E}$ on a *complex manifold* X, one can further consider the corresponding *principal holomorphic fiber bundle* P on X of the *complex linear frames of* $\mathbf{E}$, having *structure group* $G = \mathrm{GL}(n, \mathbf{C})$, *where* $n = \operatorname{rk} \mathbf{E}$. Hence, by considering the respective *vector sheaf of* (germs of holomorphic) *sections of* $\mathbf{E}$, $\mathcal{E}$ (see, for instance, (4.4)), one obtains, according to the very definitions;

$$
(15.33)\qquad \mathcal{L}(P) = \mathcal{E}nd\, \mathcal{E}.
$$

Therefore, (15.24) takes now the form

(15.34) $$0 \longrightarrow \mathcal{E}nd\,\mathcal{E} \longrightarrow \mathcal{Q} \longrightarrow (\Omega^1)^* \longrightarrow 0,$$

whose *sructure sheaf* (cf. Chapt. V; Theorem 8.1, in particular, (8.50)), is given by

(15.35) $$\mathcal{H}om_{\mathcal{O}}((\Omega^1)^*, \mathcal{E}nd\,\mathcal{E}) = \Omega^1(\mathcal{E}nd\,\mathcal{E})$$

(cf. (15.27), (15.21) and (15.33)). Accordingly, one infers that;

(15.36) *a holomorphic vector bundle* **E** *on a complex manifold* X *has a holomorphic connection* (viz., by definition, the respective *vector sheaf* $\mathcal{E}$, as above, *has* an $\mathcal{O}$-*connection,* with respect to (15.19)) *if, and only if, this happens for the* corresponding *principal holomorphic fiber bundle of (complex) linear frames of* **E**.

In particular, *this is always the case, if* X is a *Stein manifold.*

The above, in conjunction with (15.32), justifies now completely our assertion in (15.18).

Finally, concerning the $\mathcal{O}(X)$-*module* ("cohomology group"), *which classifies* (15.34) (cf. Chapt. V; Theorem 8.1, in particular, (8.51)), one has

(15.37) $$\begin{aligned} H^1(X, \Omega^1(\mathcal{L}(P))) &= H^1(X, \Omega^1(\mathcal{E}nd\,\mathcal{E})) = H^1(X, M_n(\Omega^1)) \\ &= H^1(X, \Omega^{(k)}), \end{aligned}$$

such that

(15.38) $$k = n^2 = \operatorname{rk}_{\mathcal{O}}(\mathcal{E}nd\,\mathcal{E}), \quad \textit{with} \quad n = \operatorname{rk}\mathcal{E} = \operatorname{rk}\mathbf{E}$$

(see (15.15) and Chapt. II; Lemma 6.1).

16. Fine vector sheaves

The sheaves in the title of this Section provide another important instance of *sheaves admitting* $\mathcal{A}$-*connections,* once the base space X is *paracompact* and, of course, a standard *(law of) derivation,* viz. a *differential triad* is available.

Now, *fine sheaves,* in fact, *fine $\mathcal{A}$-modules,* have already been considered in the preceding (see Chapt. III; Definition 8.1); thus, we still refer to Chapt. III; Subsection 8.(a) for eventual unexplained terminology, we shall use below.

So assume that we are given a *differential triad* $(\mathcal{A}, \partial, \Omega)$ on a *topological space* X, and let $\mathcal{E}$ be an *$\mathcal{A}$-module* on X. Moreover, let

$$\mathcal{U} = (U_\alpha)_{\alpha \in I} \tag{16.1}$$

be an *open covering* of X, and

$$(D_\alpha)_{\alpha \in I} \tag{16.2}$$

a family of *local $\mathcal{A}$-connections of $\mathcal{E}$*. That is, each D_α, $\alpha \in I$, is an $\mathcal{A}|_{U_\alpha}$-*connection of* $\mathcal{E}|_{U_\alpha}$, so that one has

$$\begin{aligned}(D_\alpha)_{\alpha \in I} \in \textstyle\prod_\alpha \mathrm{Hom}_{\mathbf{C}}(\mathcal{E}|_{U_\alpha}, \mathcal{E} \otimes_{\mathcal{A}} \Omega|_{U_\alpha}) &= C^0(\mathcal{U}, \mathcal{H}om_{\mathbf{C}}(\mathcal{E}, \mathcal{E} \otimes_{\mathcal{A}} \Omega)) \\ &\equiv C^0(\mathcal{U}, \mathcal{H}om_{\mathbf{C}}(\mathcal{E}, \Omega(\mathcal{E})))\end{aligned} \tag{16.3}$$

(see, for example, Chapt. III; (11.12)); thus, (16.2) is, by definition, an *$\mathcal{O}$-cochain of $\mathcal{U}$* with coefficients in the **C**-*vector space sheaf* $\mathcal{H}om_{\mathbf{C}}(\mathcal{E}, \Omega(\mathcal{E}))$, satisfying (coordinate-wise) the corresponding *Leibniz condition* (cf. (3.6)).

Thus, one has, for instance, the previous situation, by supposing that $\mathcal{E}$ is, in particular, a *vector sheaf* on X, while (16.1) stands for a *local frame of $\mathcal{E}$*. Then, one gets a family, like (16.2), by considering the *Levi-Civita $\mathcal{A}$-connection 0-cochain, associated with $\mathcal{U}$* (cf. (8.32.1)).

In this context, we first have the following general result, which will be of use presently below; yet, we refer to Chapt. III; (8.23), (8.23′) for the terminology applied.

Lemma 16.1. *Let $(\mathcal{A}, \partial, \Omega)$ be a given differential triad on a topological space X and $\mathcal{E}$ an $\mathcal{A}$-module on X. Moreover, suppose that*

$$\mathcal{U} = (U_\alpha)_{\alpha \in I} \tag{16.4}$$

is a locally finite open covering of X, while

$$(\phi_\alpha)_{\alpha \in I} \subseteq \mathrm{End}\,\mathcal{E} \tag{16.5}$$

is a partition of $1 \equiv 1_{\mathcal{E}}$ ($\in \mathrm{Aut}\,\mathcal{E}$, identity automorphism of $\mathcal{E}$) and

$$(D_\alpha)_{\alpha \in I} \in C^0(\mathcal{U}, \mathcal{H}om_{\mathbf{C}}(\mathcal{E}, \Omega(\mathcal{E}))) \tag{16.6}$$

a given family of local $\mathcal{A}$-connections of $\mathcal{E}$ (cf. (16.2), (16.3)). *Then, the relation;*

$$\widetilde{D} := \sum_{\alpha\in I}(\phi_\alpha \otimes 1) \circ D_\alpha \tag{16.7}$$

provides an $\mathcal{A}$-connection of $\mathcal{E}$.

Scholium 16.1.– As we shall see right away (see the proof of Theorem 16.1 below), one has the situation of the previous lemma by taking a *fine vector sheaf $\mathcal{E}$* over a *paracompact (Hausdorff)* space X, along with a *locally finite local frame $\mathcal{U}$ of $\mathcal{E}$.*

Proof of Lemma 16.1. Of course, for convenience, we have employed in (16.7) the obvious abuse of notation;

$$1 \equiv 1_{\Omega|_{U_\alpha}}, \quad \alpha \in I. \tag{16.8}$$

Thus, each one of the $\mathbf{C}$-*linear morphisms*

$$\widetilde{D}_\alpha := (\phi_\alpha \otimes 1) \circ D_\alpha : \mathcal{E}|_{U_\alpha} \longrightarrow \Omega(\mathcal{E})|_{U_\alpha}, \quad \alpha \in I \tag{16.9}$$

(cf. (16.3), (16.6)), may be viewed as a *local $\mathcal{A}$-connection of $\mathcal{E}$*, when the latter (sheaf) is *restricted to $U_\alpha \subseteq X$* (along, of course, with a similar restriction of the given differential triad $(\mathcal{A}, \partial, \Omega)$; cf. (6.21)). However, what one really needs here is just the fact that

$$\widetilde{D}_\alpha \in \mathrm{Hom}_{\mathbf{C}}(\mathcal{E}|_{U_\alpha}, \Omega(\mathcal{E})|_{U_\alpha}), \quad \alpha \in I. \tag{16.10}$$

Yet, based on the hypothesis for the family (16.5) (cf. Chapt. III; (8.22), (8.23′)), one still obtains that

$$\mathrm{Supp}\,(\phi_\alpha \otimes 1) = \mathrm{Supp}\,(\phi_\alpha) \subseteq U_\alpha, \quad \alpha \in I; \tag{16.11}$$

Therefore, in view of (16.9), and *extending $\widetilde{D}_\alpha$, $\alpha \in I$, in $\complement\,\mathrm{Supp}\,(\phi_\alpha \otimes 1)$ by zero,* one may further assume that

$$\widetilde{D}_\alpha \in \mathrm{Hom}_{\mathbf{C}}(\mathcal{E}, \Omega(\mathcal{E})), \quad \alpha \in I, \tag{16.12}$$

which, of course, implies, in turn, that;

$$\widetilde{D} := \sum_{\alpha\in I} \widetilde{D}_\alpha \equiv \sum_{\alpha}(\phi_\alpha \otimes 1) \circ D_\alpha \in \mathrm{Hom}_{\mathbf{C}}(\mathcal{E}, \Omega(\mathcal{E})). \tag{16.13}$$

(Here the *local finiteness* of $\mathcal{U}$ guarantees, as usual, the meaning of the "infinite sum" in (16.13)).

Now, based on (16.11)) and the hypothesis for (16.5) (see Chapt. III; (8.23′), 2)), one concludes that *the family*

(16.14) $$\phi_\alpha \otimes 1_\Omega \equiv \phi_\alpha \otimes 1 \in \mathrm{End}(\Omega(\mathcal{E})), \quad \alpha \in I,$$

yields a *partition of* $1_{\Omega(\mathcal{E})} \equiv 1$, so that, in addition to (16.11), one obtains;

(16.15) $$\sum_{\alpha \in I} (\phi_\alpha \otimes 1) = 1.$$

Thus, the last relation contributes now, substantially (see, for example, Scholium 7.2 in the preceding), in proving that *the* **C**-*linear morphism* $\widetilde{D}$, as defined by (16.13), *fulfils the* corresponding *Leibniz condition,* as well, which thus completes the proof. ∎

As already noted, we come now to clarify, by the next more concrete result, the framework adopted by the previous Lemma 16.1. So we have.

Theorem 16.1. *Let $(\mathcal{A}, \partial, \Omega)$ be a given differential triad on a (Hausdorff) paracompact space X. Then, every fine vector sheaf $\mathcal{E}$ on X admits an $\mathcal{A}$-connection.*

Proof. We first remark that, due to the hypothesis for X, for any given *local frame* of $\mathcal{E}$, one can consider a *locally finite open refinement* of it, which thus will provide a *locally finite local frame of* $\mathcal{E}$ too. However, for any *(locally finite) local frame* $\mathcal{U}$ of $\mathcal{E}$, one can further consider the *Levi-Civita $\mathcal{A}$-connection 0-cochain, associated with $\mathcal{U}$*, as in (8.32.1) in the preceding. Thus, the proof of our claim is now an immediate consequence of the previous Lemma 16.1, due again to the hypothesis for X and now for $\mathcal{E}$, as well (cf. thus Chapt. III; Definition 8.1). ∎

Scholium 16.2.– In connection with our argument in the previous proof, we still remark here, in passing, that:

(16.16) given a *vector sheaf* $\mathcal{E}$ on a *paracompact (Hausdorff)* space X, *the locally finite local frames of* $\mathcal{E}$ *constitute a cofinal subset of the set of all* (proper) *open coverings of* X, the latter set being upwards directed by refinement.

See also Chapt. II; (4.9) and Chapt. III; (4.9).

As already remarked in the preceding (cf. the comments following (15.10) above), the situation described by Theorem 16.1, is what one actually applies in the classical case of $\mathcal{C}^\infty$*-manifolds.* Thus, due to its significance in that important context, we restate Theorem 16.1, for convenience, less formally, as follows:

(16.17) *every fine vector sheaf $\mathcal{E}$ over a paracompact (Hausdorff) space X admits an $\mathcal{A}$-connection, once we are given the standard* (flat) *$\mathcal{A}$-connection $\partial : \mathcal{A} \longrightarrow \Omega$* (viz. a *differential triad* $(\mathcal{A}, \partial, \Omega)$).

So in the case of the usual *differential geometry of smooth* (viz. $\mathcal{C}^\infty$-)*manifolds,* the standard result, as expressed by (15.10), has also the following, more classical justification, as an application now of (16.17):

Namely, as already remarked in (15.8), when taking the base $\mathcal{C}^\infty$-manifold X *paracompact* (Hausdorff), *the structure sheaf* $\mathcal{C}^\infty_X$ *is fine,* so that (cf. (15.9)) *this is the case for every* $\mathcal{C}^\infty_X$*-module* too; in particular, the same holds true for the *vector sheaves* appeared as *sheaves of* (germs of) $\mathcal{C}^\infty$*-sections of smooth* $\mathbf{C}$*-vector bundles* on X (see, for instance, (4.11)). As a consequence of the above discussion, one thus arrives at the following conclusion, as desired. That is,

(16.18) *every vector sheaf, as above, admits a* $\mathcal{C}^\infty_X$*-connection; hence* (by definition), *any* given *smooth vector bundle,* as well.

Yet, concerning the situation in (16.18), one can further employ, of course, the existing one-to-one (and onto) correspondence between (isomorphism classes of) the two objects, under consideration. Indeed, we actually have the following *bijections;*

$$\Phi^n_{\mathcal{A}}(X) = H^1(X, \mathcal{GL}(n, \mathcal{A})) = H^1\left(X, \mathcal{A}^{\mathrm{GL}(n,\mathbf{C})}\right), \qquad (16.19)$$

for every $n \in \mathbf{N}$, *with* $\mathcal{A} \equiv \mathcal{C}^\infty_X$. (See Chapt. V; Theorem 2.1 and (2.65), ii); yet, Chapt. IV; (2.11), for $\mathbf{A} = \mathbf{C}$ and $\mathcal{C}^\infty_X$ insted of $\mathcal{C}_X$. Finally, cf. also Chapter XI in the sequel, for a more general point of view than in (16.19)).

17. Moduli space of $\mathcal{A}$-connections

Our aim in this Section is to show the way one can group together *gauge equivalent connections* of a given *$\mathcal{A}$-module $\mathcal{E}$*, a procedure, which, along with its consequences, is of a particular importance in the classical context *(gauge field theories)*, the respective technique amounting thus to the title of the heading. The relevant basic idea has already been tackled in the preceding, in its general form (cf. Definition 5.1), while here we restrict ourselves to the case of one and the same $\mathcal{A}$-module or, in particular, vector sheaf $\mathcal{E}$.

So to start with, suppose, as usual, that we are given a *differential triad*

$$(\mathcal{A}, \partial, \Omega) \tag{17.1}$$

on an arbitrary *topological space* X, while let $\mathcal{E}$ be an *$\mathcal{A}$-module* on X, for which we further assume that

$$\mathcal{C}onn_{\mathcal{A}}(\mathcal{E}) \neq \emptyset, \tag{17.2}$$

viz. we still assume that $\mathcal{E}$ *admits an $\mathcal{A}$-connection* (cf. (7.2), (7.15)).

On the other hand, let

$$\mathcal{A}ut_{\mathcal{A}}(\mathcal{E}) \equiv \mathcal{A}ut\,(\mathcal{E}) := (\mathcal{E}nd\,\mathcal{E})^{\cdot} \tag{17.3}$$

be the *group sheaf of $\mathcal{A}$-automorphisms* (or just *automorphisms*) *of* $\mathcal{E}$ (i.e., *$\mathcal{A}$-isomorphisms of $\mathcal{E}$ onto itself;* cf. also Chapt. II; (6.29) and Chapt. V; (2.31) together with Scholium 8.1). Of course, (17.3) is a *non-abelian group sheaf,* in general, unless $\mathcal{E}$ is a *line sheaf* on X (see Chapt. II; Lemma 6.2 and Chapt. IV; Lemma 1.1).

Thus, extending the classical terminology, we call (17.3) the *gauge group of* $\mathcal{E}$. (Concerning the standard theory, see e.g. T. Petrie - J. Randall [1: p.23ff]). Yet, it is still usual to employ (by abusing language again!) the previous term, *gauge group of* $\mathcal{E}$, for

$$\Gamma(\mathcal{A}ut\,\mathcal{E}), \tag{17.4}$$

viz. for the *(complete) presheaf of* (local) *sections of* $\mathcal{A}ut\,\mathcal{E}$ (see also Chapt. V; Scholium 8.2). Finally, the same term, as above, is ascribed to the *group of global sections of* $\mathcal{A}ut\,\mathcal{E}$,

$$\mathrm{Aut}\,\mathcal{E} := (\mathcal{A}ut\,\mathcal{E})(X) = (\mathcal{E}nd\,\mathcal{E})^{\cdot}(X) = (\mathcal{E}nd\,\mathcal{E})(X)^{\cdot} = (\mathrm{End}\,\mathcal{E})^{\cdot} \tag{17.5}$$

(see (17.3) and Chapt. V; (8.44)).

So, as an application of Definition 5.1 in the foregoing, our next objective is now to show, under the assumption (17.2), that;

(17.6) *the gauge group Aut* $\mathcal{E}$ (in fact, *the group* Aut $\mathcal{E}$, cf. (17.5)) *acts on the (affine) space of* $\mathcal{A}$*-connections of* $\mathcal{E}$, $Conn_{\mathcal{A}}(\mathcal{E})$,

in a sense that will be made precise. Thus, we first set the following definition, a particular case, in effect (for $\mathcal{F} = \mathcal{E}$), of our previous Definition 5.1. So one has.

Definition 17.1. Assuming the previous data (cf. (17.1)), two given $\mathcal{A}$-connections D and D' of an $\mathcal{A}$-module $\mathcal{E}$ are said to be *gauge equivalent,* if there exists an element $\phi \in \mathrm{Aut}\,\mathcal{E}$ (see (17.5)), making the following diagram commutative;

(17.7)
$$\begin{array}{ccc} \mathcal{E} & \xrightarrow{\quad D \quad} & \Omega(\mathcal{E}) \\ \phi \downarrow & & \downarrow \phi \otimes 1_{\Omega} \equiv \phi \otimes 1 \\ \mathcal{E} & \xrightarrow{\quad D' \quad} & \Omega(\mathcal{E}) \end{array}$$

That is, one has

(17.8) $$D' \circ \phi = (\phi \otimes 1) \circ D,$$

or, equivalently,

(17.9) $$D' = (\phi \otimes 1) \circ D \circ \phi^{-1}.$$

Scholium 17.1.– Concerning the above relation (17.9), it is still convenient to employ here a simpler (however, *abused*) notation, by just *writing, instead of (17.9);*

(17.10) $$D' = \phi \circ D \circ \phi^{-1} \equiv \phi D \phi^{-1} \equiv \mathrm{Ad}(\phi) D.$$

In this context, it is clear now that;

(17.11) *(17.9) defines an equivalence relation on the set of $\mathcal{A}$-connections of the given $\mathcal{A}$-module $\mathcal{E}$, $Conn_{\mathcal{A}}(\mathcal{E})$* ($\neq \emptyset$, cf. (17.2)). We still denote it by

(17.11.1) $$D \underset{\phi}{\sim} D', \quad \textit{where} \quad \phi \in \operatorname{Aut}\mathcal{E}.$$

On the other hand, one further concludes that;

(17.12) *the* above *equivalence relation (17.11.1) in the set $Conn_{\mathcal{A}}(\mathcal{E})$ is,* in effect, *that one defined by the action of the group* $\operatorname{Aut}\mathcal{E}$ *on* $Conn_{\mathcal{A}}(\mathcal{E})$.

Indeed, one has the following correspondence

(17.13) $$\tau : \operatorname{Aut}\mathcal{E} \times Conn_{\mathcal{A}}(\mathcal{E}) \longrightarrow Conn_{\mathcal{A}}(\mathcal{E}),$$

defined by (cf. (17.10))

(17.14) $$(\phi, D) \longmapsto \tau(\phi, D) \equiv \phi \cdot D \equiv \phi(D) := \phi D \phi^{-1} \equiv \operatorname{Ad}(\phi) D.$$

That is, we first note that *(17.13) is well-defined,* according to (17.14): Thus, by hypothesis *for ϕ and D, as in (17.14),* we first have, of course, that (see also (17.9), (17.10));

(17.15) $$\phi D \equiv \phi D \phi^{-1} \equiv (\phi \otimes 1) \circ D \circ \phi^{-1} \in \operatorname{Hom}_{\mathbf{C}}(\mathcal{E}, \Omega(\mathcal{E})).$$

Furthermore, *for any* (local) *sections $\alpha \in \mathcal{A}(U)$ and $s \in \mathcal{E}(U)$,* with *U open* in X, one obtains;

(17.16)
$$\begin{aligned}
&(\phi D)(\alpha \cdot s) \equiv (\phi D \phi^{-1})(\alpha \cdot s) = (\phi D)(\phi^{-1}(\alpha s)) = (\phi D)(\alpha \cdot \phi^{-1}(s)) \\
&\equiv (\phi D)(\alpha \cdot t) = \phi(D(\alpha \cdot t)) = \phi(\alpha \cdot Dt + t \otimes \partial \alpha) \\
&= \phi(\alpha \cdot Dt) + \phi(t \otimes \partial\alpha) \equiv \alpha \cdot \phi D(\phi^{-1}(s)) + (\phi \otimes 1)(t \otimes \partial \alpha) \\
&\quad = \alpha\, (\phi D \phi^{-1})(s) + \phi(t) \otimes \partial\alpha \equiv \alpha \cdot (\phi D)(s) + s \otimes \partial \alpha,
\end{aligned}$$

that is, $D' \equiv \phi \cdot D$ (cf. (17.14)) satisfies also the respective *Leibniz condition,* which thus, along with (17.15), justifies now our claim about (the *range of*) the map (17.13). ▮ Now, it is easy to see that;

(17.17) *the map (17.13), as given by (17.14),* yields an *action of the group* $\operatorname{Aut}\mathcal{E}$ *on the set* $Conn_{\mathcal{A}}(\mathcal{E})$. *That is, one has:*

(17.17.1) $$1_{\mathcal{E}} \cdot D = D, \quad D \in Conn_{\mathcal{A}}(\mathcal{E}),$$

as well as,

(17.17.2) $$\phi(\psi D) = (\phi\psi)(D),$$

for any $\phi, \psi \in \operatorname{Aut}\mathcal{E}$ *and* $D \in Conn_{\mathcal{A}}(\mathcal{E})$.

Thus, *for every* $D \in Conn_{\mathcal{A}}(\mathcal{E})$, the corresponding *orbit of* D is given by the set (see also (17.10), (17.11) and (17.14));

(17.18) $$\mathcal{O}_D := \{\phi \cdot D \in Conn_{\mathcal{A}}(\mathcal{E}) : \phi \in \operatorname{Aut}\mathcal{E}\} = \Big\{D' \in Conn_{\mathcal{A}}(\mathcal{E}) : D' \underset{\phi}{\sim} D, \ \text{for some } \phi \in \operatorname{Aut}\mathcal{E}\Big\}.$$

In other words,

(17.19) the *orbit* of an element $D \in Conn_{\mathcal{A}}(\mathcal{E})$, under the action of (the *gauge group*) $\operatorname{Aut}\mathcal{E}$ on the latter set, consists of *those* $\mathcal{A}$*-connections of* $\mathcal{E}$, which are *gauge equivalent to* D.

The above yields now a complete justification of our previous claim, concerning (17.11.1). ∎

Now, the corresponding *orbit space* of the action considered, i.e., the set;

(17.20) $$M(\mathcal{E}) \equiv Conn_{\mathcal{A}}(\mathcal{E})/\operatorname{Aut}\mathcal{E} := \bigcup_{D \in Conn_{\mathcal{A}}(\mathcal{E})} \mathcal{O}_D = \sum_D \mathcal{O}_D,$$

is called the (global) *moduli space of the* $\mathcal{A}$*-connections* of $\mathcal{E}$.

On the other hand, the specification *"global"* in the last definition indicates the possibility of considering the *local aspect of (17.20)* by restricting ourselves to an *open* $U \subseteq X$: So consider the *restriction to* U of the preceding data; thus, we first have that one of (17.1), viz. the *differential triad* on U (see (6.21))

(17.21) $$(\mathcal{A}, \partial, \Omega)|_U = (\mathcal{A}|_U, \partial|_U, \Omega|_U),$$

yet the $\mathcal{A}|_U$*-module* $\mathcal{E}|_U$, restriction to U of the given $\mathcal{A}$*-module* $\mathcal{E}$, as well as, the restriction to U of the set $Conn_{\mathcal{A}}(\mathcal{E})$ (*non-void,* by hypothesis, cf. (17.2)), so that one has

$$\mathcal{C}onn_{\mathcal{A}}(\mathcal{E})|_U \subseteq \mathcal{C}onn_{\mathcal{A}|_U}(\mathcal{E}|_U) \tag{17.22}$$

(see (6.22) – (6.24)). Finally, we also consider a similar restriction of the *group sheaf* $\mathcal{A}ut\,\mathcal{E}$ and then (guided by (17.5)) take sections over U of the same restriction; thus, we have (cf. Chapt. I; (11.40))

$$\begin{aligned}(\mathcal{A}ut\,\mathcal{E})|_U(U) &= (\mathcal{A}ut\,\mathcal{E})(U) = \mathrm{Isom}_{\mathcal{A}|_U}(\mathcal{E}|_U, \mathcal{E}|_U) \\ &= \mathcal{I}som_{\mathcal{A}|_U}(\mathcal{E}|_U), \mathcal{E}|_U)(U) \equiv \mathcal{A}ut\,(\mathcal{E}|_U)(U) = \mathrm{Aut}\,(\mathcal{E}|_U)\end{aligned} \tag{17.23}$$

(cf. also e.g. Chapt. V; (2.32)). Thus, one finally obtains

$$(\mathcal{A}ut\,\mathcal{E})(U) = \mathrm{Aut}\,(\mathcal{E}|_U), \tag{17.24}$$

for every open set $U \subseteq X$ (up to an *isomorphism of* the *groups* concerned). So one has an *action of the last group on the set(s) in (17.22)*; viz. one gets the following map

$$\mathrm{Aut}\,(\mathcal{E}|_U) \times \mathcal{C}onn_{\mathcal{A}|_U}(\mathcal{E}|_U) \longrightarrow \mathcal{C}onn_{\mathcal{A}|_U}(\mathcal{E}|_U), \tag{17.25}$$

defined *analogously to (17.14)*.

In this context, it is not difficult to see that one actually has here a *presheaf of orbit spaces,* by analogy with (17.20), when U ranges over the open subsets of X, together with the obvious restriction maps. Accordingly, one can further speak of the *sheaf (of germs) of moduli spaces of $\mathcal{A}$-connections* of the given *$\mathcal{A}$-module $\mathcal{E}$*, denoted by

$$\mathcal{M}(\mathcal{E}) = \mathcal{C}onn_{\mathcal{A}}(\mathcal{E})/\mathcal{A}ut\,(\mathcal{E}). \tag{17.26}$$

However, we are not going to use (17.26) in the sequel, formally (at least!). Nevertheless, the following special case, that mostly concerns the applications, as well, has an interest per se:

Thus, assume that $\mathcal{E}$ is, in particular, a *vector sheaf* on X, *with*

$$\mathrm{rk}\,\mathcal{E} = n \in \mathbf{N}. \tag{17.27}$$

Moreover, consider an *open* $U \subseteq X$, which is a *local gauge of* $\mathcal{E}$, viz. one has

$$\mathcal{E}|_U = \mathcal{A}^n|_U, \tag{17.28}$$

up to an *$\mathcal{A}|_U$-isomorphism of* the *$\mathcal{A}|_U$-modules* involved. (Of course, we could consider here, more generally, just an *$\mathcal{A}$-module* $\mathcal{E}$ on X, together with a *local gauge* of it *of rank* $n \in \mathbf{N}$; cf., for instance, the comments for (8.36) in the foregoing). Thus, one now obtains

$$(17.29)\qquad \begin{aligned}(\mathcal{A}ut\,\mathcal{E})(U) &= \mathrm{Aut}\,(\mathcal{E}|_U) = \mathrm{Aut}\,(\mathcal{A}^n|_U) = (\mathcal{A}ut\mathcal{A}^n)(U)\\ &= M_n(\mathcal{A})^{\cdot}(U) \equiv \mathcal{GL}(n,\mathcal{A})(U) = \mathrm{GL}(n,\mathcal{A}(U))\end{aligned}$$

(cf. (17.24), (17.28), along with Chapt. IV; (1.28)). Consequently, one thus concludes that;

(17.30) *any local automorphism of a* given *vector sheaf* $\mathcal{E}$ *over a local gauge* U of $\mathcal{E}$ *is* (given, in effect, by) *a local automorphism of* $\mathcal{A}^n$, *where* $n = \mathrm{rk}\,\mathcal{E}$, *viz. an element of*

$$(17.30.1)\qquad \mathrm{GL}(n,\mathcal{A}(U)) = \mathcal{GL}(n,\mathcal{A})(U) \equiv \mathrm{GL}(n,\mathcal{A}|_U).$$

Thus, in the special case of a *vector sheaf* $\mathcal{E}$, as mostly happens, of course, in the applications, the respective *gauge group* $\mathcal{A}ut\,\mathcal{E}$, or $\mathrm{Aut}\,\mathcal{E}$ (see (17.5)) is actually reduced *locally* (precisely, in terms of a *local frame* $\mathcal{U}$ of $\mathcal{E}$) to that one of $\mathcal{A}^n$ ($n = \mathrm{rk}\,\mathcal{E}$), viz. to the *group sheaf* $\mathcal{GL}(n,\mathcal{A})$, or its *(complete) presheaf of sections* $\Gamma(\mathcal{GL}(n,\mathcal{A}))$ (cf. Chapt. IV; (1.29)). Therefore, the terminology one usually employs hereby, referring rather to the latter as the respective *gauge groups.* This also in conjunction with the particular rôle, the *local frames* of a given *vector sheaf* play, with respect to the open coverings of X. (See e.g. Chapt. II; (4.9) and Chapt. V; (3.1), as well as, Scholium 16.2 in the preceding).

Applications of the above will be considered by the ensuing discussion. See thus the succeeding Chapters VII and VIII.

18. Cohomological classification of line sheaves, admitting $\mathcal{A}$-connections

Given an arbitrary $\mathbf{C}$-*algebraized space* $(X,\mathcal{A})$ (cf., for instance, Chapt. IV; (0.1)), we already know that the *line sheaves* on X are classified, within $\mathcal{A}$-*isomorphisms,* by what we have called the Picard group of X; thus, one first has

$$(18.1)\qquad \mathbf{\Phi}^1_{\mathcal{A}}(X) = H^1(X,\mathcal{A}^{\cdot}),$$

within a (set-theoretic) *bijection* (cf. Chapt. V; (2.59), (2.60)). Now, the previous identification becomes an *abelian group isomorphism,* when the set of isomorphism classes of line sheaves on X, $\mathbf{\Phi}^1_{\mathcal{A}}(X)$, is further endowed with the *abelian group structure,* defined on it by the *tensor product of line sheaves* (see Chapt. V; (4.3)). The resulting abelian group, viz. the *Picard group of* X, denoted by $\mathrm{Pic}_{\mathcal{A}}(X) \equiv \mathrm{Pic}\,(X)$ (ibid.; Definition 4.1), is now isomorphic to the *1st cohomology group* in (18.1); i.e., one has

$$(\mathbf{\Phi}^1_{\mathcal{A}}(X), \otimes_{\mathcal{A}}) \equiv \mathrm{Pic}\,(X) = H^1(X, \mathcal{A}^{\cdot}) = \check{H}^1(X, \mathcal{A}), \tag{18.2}$$

up to *isomorphisms of* the (abelian) *groups* involved (ibid.; Theorem 4.1, along with Chapt. III; Lemma 8.1, regarding the last equality).

On the other hand, suppose now that we are given on X a *differential triad*

$$(\mathcal{A}, \partial, \Omega) \tag{18.3}$$

(cf. (3.1)). So let further

$$\mathbf{\Phi}^1_{\mathcal{A}}(X)^{\nabla} \tag{18.4}$$

be the *set of equivalence classes of pairs*

$$(\mathcal{L}, D), \tag{18.5}$$

consisting of a *line sheaf* $\mathcal{L}$ on X and an *$\mathcal{A}$-connection* D *of* $\mathcal{L}$, where we set

$$(\mathcal{L}, D) \sim (\mathcal{L}', D') \tag{18.6}$$

for two such pairs, whenever there exists an *$\mathcal{A}$-isomorphism*

$$\phi \in \mathrm{Isom}_{\mathcal{A}}(\mathcal{L}, \mathcal{L}'), \tag{18.7}$$

making the $\mathcal{A}$-connections D and D' of $\mathcal{L}$ and $\mathcal{L}'$, respectively, *gauge equivalent;* viz. one has

$$D' = (\phi \otimes 1_{\Omega}) \circ D \circ \phi^{-1} \equiv (\phi \otimes 1) D \phi^{-1} \equiv \phi D \phi^{-1} \tag{18.8}$$

(see (5.43), as well as, e.g. (17.9), (17.10), for the notation applied).

Thus, in view of (18.6), it is easy to see that every element

$$[(\mathcal{L}, D)] \in \mathbf{\Phi}^1_{\mathcal{A}}(X)^{\nabla} \tag{18.9}$$

constitutes, indeed, an *equivalence class* of pairs $(\mathcal{L}, D)$, as in (18.5). In this context, we can further say that:

(18.10) *(18.4) is that subset* (possibly empty) *of* $\mathbf{\Phi}^1_{\mathcal{A}}(X)$, *the elements of which have a representative admitting an* $\mathcal{A}$*-connection.*

Indeed, if $[\mathcal{L}] \in \mathbf{\Phi}^1_{\mathcal{A}}(X)$ and $\mathcal{L}$ admits an $\mathcal{A}$-connection D, then *every* $\mathcal{L}' \in [\mathcal{L}]$, ϕ*-isomorphic* to $\mathcal{L}$, with ϕ given by (18.7), *has an* $\mathcal{A}$*-connection, as well,* say D' defined by (18.8), so that $(\mathcal{L}, D) \sim (\mathcal{L}', D')$, in the sense of (18.6) (see also (8.18)). ∎

Now, for reasons that will become clear later on (see thus Chapter X in the sequel), one employs for (18.5) a physics originated nomenclature; thus, a pair

$$(\mathcal{L}, D), \tag{18.11}$$

as above, consisting of a line sheaf $\mathcal{L}$ on X and an $\mathcal{A}$-connection D of $\mathcal{L}$, is called a *Maxwell field* on X. Yet, the same term is employed, occasionally, to the respective equivalence class,

$$[(\mathcal{L}, D)] \in \mathbf{\Phi}^1_{\mathcal{A}}(X)^{\nabla}, \tag{18.12}$$

as well. Thus, we also speak of (18.4), as the *set of Maxwell fields* on X.

In this regard, as an application of our considerations in previous Sections, our first result says that *the above set inherits from the Picard group of* X *the respective abelian group structure, becoming,* in fact, a *subgroup* of the latter. That is, one has the next.

Theorem 18.1. *Let* $(\mathcal{A}, \partial, \Omega)$ *be a differential triad on a topological space* X. *Then, the set of Maxwell fields on* X,

$$\mathbf{\Phi}^1_{\mathcal{A}}(X)^{\nabla} \overset{\subseteq}{\rightarrow} \operatorname{Pic}(X) \equiv (\mathbf{\Phi}^1_{\mathcal{A}}(X), \otimes_{\mathcal{A}}) \tag{18.13}$$

(cf. also (18.10, (18.2)), *endowed with the relative (abelian) group structure from the Picard group of* X, *becomes a subgroup of the latter, thus an abelian group itself, as well.*

Proof. First, supplementing our argument in (18.10), by justifying thus too the (set-theoretic) *injection* appeared in (18.13), we remark that the map in question is, in effect, the correspondence,

$$[(\mathcal{L}, D)] \longmapsto [\mathcal{L}] : \mathbf{\Phi}^1_{\mathcal{A}}(X)^{\nabla} \longrightarrow \mathbf{\Phi}^1_{\mathcal{A}}(X), \tag{18.14}$$

as already follows from the comments after (18.10); yet, the same map is certainly *one-to-one.*

Now, given the elements $(\mathcal{L}, D)$ and $(\mathcal{L}', D')$ of (18.4) (in fact, two *representatives,* respectively, of the elements in question), we further consider the pair

$$(\mathcal{L} \otimes_{\mathcal{A}} \mathcal{L}', D \otimes D'), \tag{18.15}$$

which is again a *representative of an element of the same set:* Indeed, this follows from Chapt. V; (4.3), by further setting

$$D \otimes D' \equiv D_{\mathcal{L} \otimes_{\mathcal{A}} \mathcal{L}'} := (D \otimes 1_{\mathcal{L}'}) + (1_{\mathcal{L}} \otimes D') \equiv D \otimes 1 + 1 \otimes D', \tag{18.16}$$

according to (5.10) in the preceding.

On the other hand, based further on Chapt. V; (4.3) and on Subsection 5.(d) of the present Chapter, one defines the *inverse of an element* $[(\mathcal{L}, D)] \in \mathbf{\Phi}^1_{\mathcal{A}}(X)^{\nabla}$ (in fact, *of a representative* of the latter) by the relation

$$(\mathcal{L}, D)^{-1} := (\mathcal{L}^*, D^*), \tag{18.17}$$

where, of course, $\mathcal{L}^*$ stands for the *dual line sheaf* of $\mathcal{L}$ and D^* for the respective *dual $\mathcal{A}$-connection of the* given *$\mathcal{A}$-connection D of $\mathcal{L}$*, being thus an *$\mathcal{A}$-connection of the line sheaf $\mathcal{L}^*$*; so one has (cf. (5.31))

$$D^*(u) := \partial \circ u - (u \otimes 1_{\Omega}) \circ D \equiv \partial \circ u - (u \otimes 1) \circ D, \tag{18.18}$$

for every

$$u \in \mathcal{L}^*(U) \equiv \mathcal{H}om_{\mathcal{A}}(\mathcal{L}, \mathcal{A})(U) = \mathrm{Hom}_{\mathcal{A}|_U}(\mathcal{L}|_U, \mathcal{A}|_U) \tag{18.19}$$

and any *open $U \subseteq X$*. Therefore, *for every (local) section $s \in \mathcal{L}(U)$*, one has, in view of (18.18);

$$D^*(u)(s) = \partial(u(s)) - (u \otimes 1)(Ds). \tag{18.20}$$

So one obtains (cf. also (18.17))

$$[(\mathcal{L}, D)]^{-1} := \left[(\mathcal{L}, D)^{-1}\right] \equiv [(\mathcal{L}^*, D^*)] \in \mathbf{\Phi}^1_{\mathcal{A}}(X)^{\nabla}, \tag{18.21}$$

which thus, together with (18.15), completes now the proof. ∎

Thus, based on Theorem 18.1, one may speak of the *abelian group of Maxwell fields* (in fact, equivalence classes of such) on X, *a subgroup,* in effect (viz. modulo the injection in (18.13), see (18.14)), *of the corresponding Picard group of X*. So our next objective is now to identify the former group with an appropriate cohomology group, by analogy with the already familiar situation in (18.1). Indeed, one has the following.

Theorem 18.2. *Suppose that we have a differential triad* $(\mathcal{A}, \partial, \Omega)$ *on topological space* X *and let*

$$\tilde{\partial} : \mathcal{A}^{\cdot} \longrightarrow \Omega \tag{18.22}$$

be the logarithmic derivation, induced by ∂ *on* $\mathcal{A}^{\cdot}$*, the group sheaf of units of* $\mathcal{A}$*. Then, the group of Maxwell fields on* X *coincides, within an (abelian) group isomorphism, with the 1st (Čech) hypercohomology group of* X *with coefficients in the 2-term* $\mathcal{A}$*-complex on* X*, defined by (18.22). That is, one has* (cf. Chapt. III; (7.43))

$$\mathbf{\Phi}^1_{\mathcal{A}}(X)^{\nabla} = \check{\mathbf{H}}^1(X, \mathcal{A}^{\cdot} \xrightarrow{\tilde{\partial}} \Omega), \tag{18.23}$$

up to an isomorphism of the abelian groups concerned.

The proof of the above theorem is actually based on material related with the *"local theory" of* $\mathcal{A}$*-connections,* a subject that we are dealt with right in the next Chapter (see thus ibid.; Section 11). Notwithstanding, the previous statement was aiming, in effect, to give a particular application of Theorem 18.1, the latter result belonging, for that matter, to the global theory of $\mathcal{A}$-connections, as this has already been exposed in the foregoing.

Concerning the classical counterpart of the previous result, see, for instance, J.-L. Brylinski [1: p.76, Theorem 2.2.11, (2)]. In this context, we also note that, by analogy with the classical pattern (ibid.), one can further obtain the corresponding result to the previous theorem, by employing *Deligne cohomology* (see, for instance, ibid.; p.77, Theorem 2.2.12). However, we have not treated the cohomology in question in the preceding, hence we omit here the relevant discussion, concerning the group (18.4) (cf. Theorem 18.1) in that context.

CHAPTER VII

$\mathcal{A}$-connections. Local Theory

" ...to understand what's what ... is a vital aspect of Mathematics."

S. Mac Lane in "Mathematics Form and Function" (Springer - Verlag, 1986). p. 288.

Our aim in this Chapter is to study the *local aspects* of certain fundamental concepts, which we were dealt with in the previous Chapter VI, where we mainly examined the *global point of view* of the notion of an $\mathcal{A}$-connection. As we shall presently see, several standard results of the classical *(local) differential geometry of C^∞-manifolds* are still in force, within the present abstract (axiomatic) framework. This means, at least, that *even in the classical setting no calculus* (notwithstanding, its *formal properties,* yes!) *is actually needed* for the results in question.

Furthermore, various applications of this local point of view of $\mathcal{A}$-connections are also considered, according to the classical patterns, in this and the following Chapter VIII; yet, as another application of this same aspect, we can still mention our treatment of *characteristic classes* of vector sheaves *(Chern-Weil theory),* as this is given in Chapter IX in the sequel.

1. Local form of $\mathcal{A}$-connections

We start with looking at the form, which is taken by an $\mathcal{A}$-connection D of an $\mathcal{A}$-module $\mathcal{E}$, when we restrict ourselves to a given local gauge of $\mathcal{E}$, as this latter notion is specified in the sequel (see thus (1.15) below; but, cf. also, for instance, Chapt. VI; (8.36), (9.2)). Of course, our argument applies when, in particular, one looks at a given vector sheaf and a local frame of it.

Thus, to begin with, suppose, as usual, that we are given a *differential triad*

$$(\mathcal{A}, \partial, \Omega) \tag{1.1}$$

on a *topological space* X (see Chapt. VI; (3.1)). Moreover, let $\mathcal{E}$ be an *$\mathcal{A}$-module* on X, along with an *open* $U \subseteq X$, being a *local gauge of* $\mathcal{E}$, of *rank* $n \in \mathbf{N}$; viz. one has, by definition,

$$\mathcal{E}|_U = \mathcal{A}^n|_U, \tag{1.2}$$

within an *$\mathcal{A}|_U$-isomorphism of* the *$\mathcal{A}|_U$-modules* involved.

Now, assume further that

$$(e_1, \ldots, e_n) \equiv (e_i)_{1 \leqq i \leqq n} \subseteq \mathcal{E}(U) \tag{1.3}$$

is a (finite) sequence of *local sections* of $\mathcal{E}$ over the *open* $U \subseteq X$, as above, providing a *basis* of the *free* $\mathcal{A}(U)$-*module* $\mathcal{E}(U)$; indeed, one obtains

$$\mathcal{E}(U) = (\mathcal{E}|_U)(U) = (\mathcal{A}^n|_U)(U) = \mathcal{A}^n(U) = \mathcal{A}(U)^n \tag{1.4}$$

(see (1.2), and Chapt. I; (11.40). Of course, all of the equalities in (1.4) stand for *$\mathcal{A}(U)$-isomorphisms of* the *$\mathcal{A}(U)$-modules* involved).

> **Scholium 1.1.**– Concerning the *$\mathcal{A}(U)$-basis* (1.3) of $\mathcal{E}(U)$, as above, we remark that this is, in effect, a *consequence* of the *$\mathcal{A}|_U$-isomorphism* (1.2). Indeed, one gets such a basis by composing, for instance, the *inverse* of the previous isomorphism with the (canonical) *Kronecker sections* of $\mathcal{A}^n$, restricted to U; see Chapt. IV; (3.3) and (3.11.1).

On the other hand, for convenience of reference, we adopt, heretofore, the following succinct terminology, concerning the previous situation: Namely, we shall speak of a *local gauge* of an *$\mathcal{A}$-module $\mathcal{E}$* over an *open $U \subseteq X$*, of *rank $n \in \mathbf{N}$*, and write

$$e^U \equiv \{U; (e_1, \ldots, e_n)\}, \tag{1.5}$$

whenever *one has (1.2), together with a given basis (1.3),* as before. Yet, the term *local coordinate system* (or simply *local coordinates*) *of $\mathcal{E}$ over U* are applied, occasionally, for (1.5).

Thus, based on the preceding, one now concludes that (cf. also Chapt. IV; (3.11)):

(1.6) given a *local gauge of the $\mathcal{A}$-module $\mathcal{E}$*, as in (1.5), any *local section $s \in \mathcal{E}(U)$* is written in the form;

$$s = \sum_{i=1}^{n} s_i e_i, \tag{1.6.1}$$

such that (cf. (1.4))

$$(s_i)_{1 \leqq i \leqq n} \in \mathcal{A}(U)^n = \mathcal{A}(U)^n. \tag{1.6.2}$$

We call the (local) sections $s_i \in \mathcal{A}(U)$, $1 \leqq i \leqq n$, the *coordinates of* the given $s \in \mathcal{E}(U)$, with respect to e^U, as in (1.5). Thus, when (1.5) is given, they are, of course, in view of (1.4), *uniquely defined, for any $s \in \mathcal{E}(U)$.*

Now, assume further that we are given an *$\mathcal{A}$-connection D* of the *$\mathcal{A}$-module $\mathcal{E}$*, under consideration (cf. Chapt. VI; Definition 3.1). Thus, by applying D to (1.6.1) above, one obtains;

$$D(s) = D\left(\sum_{i=1}^{n} s_i e_i\right) = \sum_i D(s_i \cdot e_i) = \sum_i (s_i De_i + e_i \otimes \partial s_i). \tag{1.7}$$

So this is the *local form of the* given *$\mathcal{A}$-connection D on U*, precisely speaking on $\mathcal{E}|U$, when U is a local gauge of $\mathcal{E}$, as in (1.5).

Remark 1.1.– In fact, one considers in (1.7) the *restriction of the $\mathcal{A}$-connection D to the open $U \subseteq X$*, in the sense of Chapt. VI; (6.23).

Accordingly, by virtue of (1.7), one concludes the following:

(1.8) Suppose we are given a *local gauge* e^U of an *$\mathcal{A}$-module* $\mathcal{E}$, as in (1.5), and an *$\mathcal{A}$-connection D of $\mathcal{E}$*. Then, *the* given *$\mathcal{A}$-connection D is known on U* (cf. Remark 1.1), *when we know its values at the basis (1.3); viz. once we know*

(1.8.1) $D(e_i) \in (\mathcal{E} \otimes_{\mathcal{A}} \Omega)(U) \equiv \Omega(\mathcal{E})(U) = \mathcal{E}(U) \otimes_{\mathcal{A}(U)} \Omega(U),$

with $1 \leqq i \leqq n$. (See also (1.10) below).

Now, the following argument justifies (1.8.1); thus, by (1.2) and Chapt. II; Lemma 5.1, along with (5.18), one first has the next *$\mathcal{A}|_U$-isomorphisms,*

$$\begin{aligned}\Omega(\mathcal{E})|_U \equiv (\mathcal{E} \otimes_{\mathcal{A}} \Omega)|_U = \mathcal{E}|_U \otimes_{\mathcal{A}|_U} \Omega|_U = \mathcal{A}^n|_U \otimes_{\mathcal{A}|_U} \Omega|_U \\ = (\mathcal{A}^n \otimes_{\mathcal{A}} \Omega)|_U = \Omega^n|_U = (\Omega|_U)^n.\end{aligned} \tag{1.9}$$

Therefore, one obtains (cf. also (1.4), Chapt. I; (11.40) and Chapt. II; (5.25));

$$\begin{aligned}\Omega(\mathcal{E})(U) &\equiv (\mathcal{E} \otimes_{\mathcal{A}} \Omega)(U) = (\mathcal{E} \otimes_{\mathcal{A}} \Omega)|_U(U) \\ &= (\Omega^n|_U)(U) = \Omega^n(U) = \Omega(U)^n = \mathcal{A}(U)^n \otimes_{\mathcal{A}(U)} \Omega(U) \\ &= \mathcal{E}(U) \otimes_{\mathcal{A}(U)} \Omega(U),\end{aligned} \tag{1.10}$$

which thus completely explains (1.8.1). ∎

On the other hand, in view of (1.4), one has that ($\mathcal{E}(U)$ *is a free $\mathcal{A}(U)$-module of rank* $n \in \mathbf{N}$. Accordingly, *the values of the* given $\mathcal{A}$-connection D of $\mathcal{E}$ at the elements of the basis (1.3), being, by virtue of (1.8.1), *elements of the $\mathcal{A}(U)$-module of the last term in (1.10),* are now determined by the following relations (cf. also, for instance, W. Greub [1: p.7, Lemma 1.5.2]);

$$D(e_j) = \sum_{i=1}^{n} e_i \otimes \omega_{ij}, \quad 1 \leqq j \leqq n, \tag{1.11}$$

for some *uniquely defined local sections*

$$\omega_{ij} \in \Omega(U), \quad 1 \leqq i, j \leqq n. \tag{1.12}$$

Therefore, *one has*

$$\omega \equiv (\omega_{ij}) \in M_n(\Omega(U)) = M_n(\Omega)(U) \tag{1.13}$$

(cf. also Chapt. V; (8.7) and Chapt. VI; (9.3)).

We call (1.13) the *local $\mathcal{A}$-connection matrix of D*, which is thus associated with the given *local gauge* e^U of the *$\mathcal{A}$-module* $\mathcal{E}$ (cf. (1.5)).

Thus, as a consequence of the preceding discussion, we now have the following basic result.

Theorem 1.1. *(Local form of an $\mathcal{A}$-connection). Suppose we are given a differential triad $(\mathcal{A}, \partial, \Omega)$ on a topological space X, an $\mathcal{A}$-module $\mathcal{E}$ on X and an $\mathcal{A}$-connection D of $\mathcal{E}$. Moreover, let*

$$e^U \equiv \{U; (e_1, \ldots, e_n)\} \tag{1.14}$$

be a local gauge of $\mathcal{E}$ over an open set $U \subseteq X$, of rank $n \in \mathbf{N}$ (cf.(1.5)). *Then, D* (restricted to U) *is given by the relation*

$$D(s) = \sum_{i=1}^{n} e_i \otimes \left(\partial s_i + \sum_{j=1}^{n} s_j \omega_{ij}\right), \tag{1.15}$$

for any (local) section of $\mathcal{E}$ over U,

$$s = \sum_{i=1}^{n} s_i e_i \in \mathcal{E}(U) \tag{1.16}$$

(cf. (1.6.1)), *while*

$$\omega \equiv (\omega_{ij}) \in M_n(\Omega(U)) = M_n(\Omega)(U) \tag{1.17}$$

is a uniquely determined (given D and e^U) *$n \times n$ matrix with entries in $\Omega(U)$* (alias *"1-forms"* on U).

Conversely, given (1.16), by further assuming (1.13), the relation (1.15) determines uniquely an $\mathcal{A}$-connection of $\mathcal{E}$ over U, viz. precisely speaking, an $\mathcal{A}|_U$-connection of $\mathcal{E}|_U$.

Proof. First, we prove (1.15): Thus, for any local section $s \in \mathcal{E}(U)$, as in (1.16), one obtains, *in view of (1.11);*

$$\begin{aligned} D(s) &= D\left(\textstyle\sum_{j=1}^{n} s_j e_j\right) = \textstyle\sum_j D(s_j e_j) = \textstyle\sum_j (s_j De_j + e_j \otimes \partial s_j) \\ &\textstyle\sum_j (s_j (\sum_{i=1}^{n} e_i \otimes \omega_{ij}) + e_j \otimes \partial s_j) = \textstyle\sum_j (s_j (\sum_i e_i \otimes \omega_{ij})) + \textstyle\sum_j e_j \otimes \partial s_j \\ &= \textstyle\sum_j (\sum_i e_i \otimes s_j \omega_{ij}) + \textstyle\sum_j e_j \otimes \partial s_j = \textstyle\sum_i (e_i \otimes \sum_j s_j \omega_{ij}) + \textstyle\sum_i e_i \otimes \partial s_i \\ &= \textstyle\sum_{i=1}^{n} e_i \otimes (\partial s_i + \sum_{j=1}^{n} s_j \omega_{ij}), \end{aligned} \tag{1.18}$$

as was actually to be proved. Now, this, in conjunction also with our previous conclusion about the *uniqueness of (1.12),* already establishes the first part of the assertion.

Conversely, we prove now that, *given the* $n \times n$ *matrix* (1.17), the relation (1.15) *yields,* in fact, *an* $\mathcal{A}|_U$*-connection of the* $\mathcal{A}|_U$ *module* $\mathcal{E}|_U$ (cf. e.g. Chapt. VI; (6.21), (6.22)): First, the **C**-*linearity* of D (in fact, of $D|_U$), with respect to $s \in \mathcal{E}(U)$, as the latter is given by (1.16), is certainly clear by the same form of (1.15). So, for convenience, we check the corresponding *Leibniz condition* (see Chapt. VI; (3.6)); thus, *for any (local) sections* $s \in \mathcal{E}(U)$, as given by (1.16) (see also (1.6.1)), *and* $\alpha \in \mathcal{A}(U)$, one obtains, by virtue of (1.15) (cf. also Chapt. VI; (1.2)),

$$\begin{aligned} D(\alpha \cdot s) &= \textstyle\sum_{i=1}^n e_i \otimes (\partial(\alpha s_i) + \sum_{j=1}^n (\alpha s_j)\omega_{ij}) \\ &= \textstyle\sum_i e_i \otimes (\alpha \cdot \partial s_i + s_i \cdot \partial\alpha + \alpha \cdot \sum_j s_j \omega_{ij}) = \alpha \cdot (\sum_i \otimes (\partial s_i + \sum_j s_j \omega_{ij})) \\ &\quad + \textstyle\sum_i e_i \otimes s_i \partial \alpha = \alpha \cdot D(s) + (\sum_i s_i e_i) \otimes \partial\alpha = \alpha \cdot Ds + s \otimes \partial\alpha, \end{aligned} \tag{1.19}$$

as desired. Furthermore, it is clear that *assuming (1.11),* the relation (1.15) *defines* $D|_U$ *uniquely* (cf. also (1.7)), which thus completes the proof. ∎

Now, by analogy with (1.11), it is still useful to put (1.15) into the following form;

$$D(s) = \sum_{i=1}^n e_i \otimes (Ds)_i, \tag{1.20}$$

where, of course, we have set

$$(Ds)_i := \partial(s_i) + \sum_{j=1}^n s_j \omega_{ij} \in \Omega(U), \quad 1 \leqq i \leqq n, \tag{1.21}$$

for any $s \in \mathcal{E}(U)$*, given by (1.6.1),* in view always of our hypothesis for e^U. As a consequence of the above, one can further look at (1.20), as given by the relation (cf. also (1.10))

$$D(s) = ((Ds)_i)_{1 \leqq i \leqq n} \in \Omega(U)^n = \Omega(\mathcal{E})(U). \tag{1.22}$$

This also explains the obvious abuse of language, one usually applies to this point, by referring to the elements (1.21) of $\Omega(U)$, as the *(local) coordinates of*

$$D(s) \in \Omega(\mathcal{E})(U) = \mathcal{E}(U) \otimes_{\mathcal{A}(U)} \Omega(U) \tag{1.23}$$

(cf. (1.10)), *with respect to the given local gauge* e^U *of* $\mathcal{E}$ (see (1.5)). They are, of course, *uniquely defined, for any* given $s \in \mathcal{E}(U)$, as before, once D *and* e^U are kept *fixed* (cf. e.g. (1.11), (1.12)).

Furthermore, the same relation (1.22), in conjunction with (1.21), permits also another, *very convenient expression of* D, in fact, of $D|_U$, *in matrix form;* that is, we have

$$(1.24)\qquad D = \partial + \omega \ (\textit{locally on } U),$$

such that one has (cf. also (1.29) below)

$$(1.25)\qquad D(s) = \partial(s) + \omega \cdot s \equiv \partial(s) + \omega(s),$$

for every (local) section

$$(1.26)\qquad s \equiv (s_i)_{1 \leqq i \leqq n} \in \mathcal{A}(U)^n = \mathcal{E}(U)$$

(cf. (1.4)). So concerning (1.25), we have also set

$$(1.27)\qquad \partial(s) \equiv (\partial(s_i))_{1 \leqq i \leqq n} \in \Omega(U)^n;$$

that is, we abused notation here, for convenience, by setting ∂, in place actually of its *n-dimensional extension, restricted to* U, viz.

$$(1.28)\qquad (\partial \equiv)\, \partial^n|_U : \mathcal{A}^n|_U \longrightarrow \Omega^n|_U,$$

given exactly by (1.27) (cf. Chapt. VI; (5.4), (5.5) and (8.6)). Yet, concerning the last item in (1.25), we have also set (cf. (1.21))

$$(1.29)\qquad \left(\sum_{j=1}^{n} s_j \cdot \omega_{ij}\right) = \left(\sum_{j=1}^{n} \omega_{ij} \cdot s_j\right) = \omega \cdot s \equiv \omega(s),$$

for any $s \in \mathcal{E}(U)$, *as given by (1.26).*

On the other hand, by hypothesis for D and (1.2), (1.9), one obtains

$$(1.30)\qquad D|_U \in \mathrm{Hom}_{\mathbf{C}}(\mathcal{E}|_U, \Omega(\mathcal{E})|_U) = \mathrm{Hom}_{\mathbf{C}}(\mathcal{A}^n|_U, \Omega^n|_U),$$

such that *one has,* in effect, *that*

$$(1.31)\qquad D|_U \in \mathcal{C}onn_{\mathcal{A}|_U}(\mathcal{A}^n|_U).$$

Now, this along with (1.28), serve to *justify* (1.24): Namely, the resulting $\mathcal{A}|_U$*-connection difference*

$$(1.32)\qquad D - \partial\, (= \omega)$$

is exactly an element, indeed, an $n \times n$ *section matrix,* as follows;

$$
\begin{aligned}
&\omega \in \mathrm{Hom}_{\mathcal{A}|_U}(\mathcal{A}^n|_U, \Omega^n|_U) = \mathcal{H}om_{\mathcal{A}}(\mathcal{A}^n, \Omega^n)(U) \\
(1.33)\qquad &= \mathcal{H}om_{\mathcal{A}}(\mathcal{A}^n, \mathcal{A}^n \otimes_{\mathcal{A}} \Omega)(U) = (\mathcal{H}om_{\mathcal{A}}(\mathcal{A}^n, \mathcal{A}^n) \otimes_{\mathcal{A}} \Omega)(U) \\
&\equiv (M_n(\mathcal{A}) \otimes_{\mathcal{A}} \Omega)(U) \equiv M_n(\Omega)(U) = M_n(\Omega)(U)),
\end{aligned}
$$

which is actually the case with (1.24), *according to (1.17).* In this context, cf. also Chapt. VI; (7.14), (7.30), as well as, e.g. (8.7), (8.8)). So this, in conjunction with (1.21), (1.22) and (1.28), (1.29), *explains* now *completely (1.24).* ∎

Scholium 1.2.– The above *(local) matrix form of D*, as given by (1.24), leads often to a convenient *abuse of language* (even in the standard theory of differential geometry), by saying that:

(1.34) *the $\mathcal{A}$-connection D is (uniquely) determined by ω*, a fact *holding true,* of course, *only locally*(!) (see thus Theorem 1.1 above), viz. for every *local gauge e^U* of the *$\mathcal{A}$-module $\mathcal{E}$*, under consideration.

Thus, by further looking at the things *locally,* as before, we still realize that:

(1.35) a given *$\mathcal{A}$-connection D* of an *$\mathcal{A}$-module $\mathcal{E}$* on X, *restricted to a local gauge e^U of $\mathcal{E}$* (see (1.5)), becomes, in fact, an *$\mathcal{A}|_U$-connection of $\mathcal{A}^n|_U$* (cf. (1.31)), *determining* thus *a change (perturbation) of the standard (flat) $\mathcal{A}|_U$-connection of $\mathcal{A}^n|_U$*, $\partial^n|_U \equiv \partial$ (the latter being already supplied by the given *differential triad* $(\mathcal{A}, \partial, \Omega)$ on X; cf. Chapt. VI; (8.6)). Yet, *this perturbation is expressed* (*locally,* always!) *by* (the *potential*) $\omega \in M_n(\Omega(U))$, according to (1.24) (see also (1.31)).

So, putting it differently, we can say that;

(1.36) ω in (1.24) measures the *deviation from "differentiating flatly"*, viz. in terms of $\partial^n|_U \equiv \partial$ (again (1.31)).

A word of *terminology,* before we proceed: So we recall that we have already named in the preceding ω, the *local $\mathcal{A}$-connection matrix* of D. Yet, we call ω, as in (1.24) (cf. also (1.13)), the *(local) connection 1-form* of D, or just *(local) connection form.* Moreover, the terms *potential form* (*of the connection* involved), or simply *potential* (as applied at the end of (1.35) above), are in use, as well, especially when referring to applications in *physics*.

So, by terminating this discussion, we can still say that, as a result of (1.35), one can repeat, here too (along with the classical theory) that, roughly speaking, *(locally)*

(1.37) *a connection is a potential.*

Finally, as we shall see later on (cf. Chapter X in the sequel), the same *(local)* relation (1.24) admits a nice *physical interpretation,* as well.

On the other hand, by still referring to the terminology of the last Scholium, we also note, in anticipation, that a means of estimating the deviation at issue (cf. (1.36)), or even the way one realizes the above *potential* (see (1.37)), is the respective *"field strength"* of the potential, under consideration, in other words, the corresponding *curvature* of the connection involved. Indeed, this fundamental notion is going to be treated in the subsequent Chapter VIII.

Now, the preceding have, of course, a natural application when one is concerned with *vector sheaves* on X, a fact that will also be considered in the sequel. In particular, the previous formulae about (1.24) become extremely simple in the case one is dealt with a *Maxwell field* $(\mathcal{L}, D)$ on X (see Chapt. VI; (18.11), (18.12), for the relevant terminology). Thus, by looking at a *local gauge* of (the *line sheaf*) $\mathcal{L}$, say

$$e^U \equiv (U; e) \tag{1.38}$$

(see (1.5), for $n = 1$), one has

$$e \in \mathcal{L}(U) = \mathcal{A}(U) \tag{1.39}$$

(see (1.4), for $n = 1$, the equality in (1.39) standing, of course, for an $\mathcal{A}(U)$-*isomorphism*).

Accordingly, for the case in question, one obtains for the (local) *potential of the $\mathcal{A}$-connection D of $\mathcal{L}$,*

$$\omega \in \Omega(U), \tag{1.40}$$

viz. a *"1-form"* on U; yet, concerning (1.25), *one has*

$$(D|_U)(s) \equiv D(s) = \partial(s) + \omega \cdot s = \partial(s) + s \cdot \omega, \tag{1.41}$$

for any local section

$$s \in \mathcal{L}(U) = \mathcal{A}(U) \tag{1.42}$$

(see (1.39); yet, concerning (1.41), we recall that, according to our hypothesis for $\mathcal{A}$, $\mathcal{A}(U)$ is a *commutative* **C**-algebra. The same argument has been applied in (1.29)).

Now, we are interested next in the way the various local $\mathcal{A}$-connection matrices, as above, vary when (the local gauge) *U is running over a local frame $\mathcal{U}$ of a* given *vector sheaf* on X. Indeed, this point of view is dealt with straight off in the following Section.

2. Change of local gauges

We examine in this Section the manner that the local connection forms of a given $\mathcal{A}$-connection of a vector sheaf are transformed when a change of a local gauge involved is on foot (see thus Theorem 2.1 below).

So to start with, suppose first that we are given an *$\mathcal{A}$-module $\mathcal{E}$* on X, as in the preceding Section 1, along with the *local gauges* of it

$$e^U \equiv \{U; (e_1, \ldots, e_n)\} \text{ and } f^V \equiv \{V; (f_1, \ldots, f_n)\} \tag{2.1}$$

over the *open* sets U, V of X, of the *same dimension* (or *rank*) $n \in \mathbf{N}$ (see (1.5)). Thus, by further assuming that

$$U \cap V \neq \emptyset, \tag{2.2}$$

one obtains, by *restriction of* the (local) gauges in (2.1) *to $U \cap V$, two n-dimensional local gauges of $\mathcal{E}$, with common domain of definition the open set $U \cap V \subseteq X$.*

Therefore, by first applying (1.6.1), one gets

$$f_j = \sum_{i=1}^{n} g_{ij} \cdot e_i, \quad 1 \leqq j \leqq n, \tag{2.3}$$

(of course, every thing here is *restricted to $U \cap V$*; cf. the comments above, following (2.2)). So one actually obtains

$$(2.4) \qquad g \equiv (g_{ij}) \in \mathrm{GL}(n, \mathcal{A}(U \cap V)) = \mathcal{GL}(n, \mathcal{A})(U \cap V)$$

(cf. also (1.6.2) and Chapt. IV; (1.28)). Accordingly,

(2.5) *the $n \times n$ matrix g* in (2.4), as given by (2.3), is the *matrix of change between the local gauges e^U and f^V*. Thereby, we call g a *local gauge transformation;* yet, in addition to the comments, following (17.30) in the previous Chapter VI, the form of g in (2.4) explains also the reason that we still apply, occasionally, for $\mathcal{GL}(n, \mathcal{A})$ the mame *gauge group (sheaf).*

Now, our next objective is to determine the *form, that is taken by a local $\mathcal{A}$-connection matrix ω* (see (1.13)), *under the action of a local gauge transformation g*, as in (2.4) (cf. also (2.5)):

First, for such a transformed ω, we employ the notation;

$$(2.6) \qquad g^*(\omega) \equiv \omega' \equiv (\omega'_{ij}) \in M_n(\Omega(U \cap V)) = M_n(\Omega)(U \cap V).$$

So, *by definition,* ω' is the *local $\mathcal{A}$-connection matrix* of D (the latter being a given $\mathcal{A}$-connection of the $\mathcal{A}$-module $\mathcal{E}$ involved), which is associated with the *local gauge of $\mathcal{E}$ over $U \cap V$*, given by

$$(2.7) \qquad f^{U\cap V} \equiv f^V|_{U\cap V} \equiv \{U \cap V; (f_1, \ldots, f_n)|_{U\cap V}\}$$

(viz. the *restriction to $U \cap V$ of the* given *local gauge f^V* of $\mathcal{E}$, as in (2.1)).

Therefore, according to the general pattern, as given by (1.11), we further consider the corresponding *defining relation for ω'*, with respect to the local gauge (2.7); that is, we have

$$(2.8) \qquad D(f_j) = \sum_{i=1}^{n} f_i \otimes \omega'_{ij}, \quad 1 \leqq j \leqq n,$$

such that one has

$$(2.9) \qquad \omega'_{ij} \in \Omega(U \cap V), \quad 1 \leqq i, j \leqq n.$$

Moreover, the latter sections are *uniquely defined,* given D and the local gauge $f^{U\cap V}$ of $\mathcal{E}$, as in (2.7) (see the comments about (1.12) above). So this also explains the notation adopted by (2.6).

Thus, in view of (2.3) and (2.8), one now obtains

$$(2.10)\qquad \begin{aligned} D(f_j) &= \textstyle\sum_i f_i \otimes \omega'_{ij} = \sum_i (\sum_k g_{ki} \cdot e_k) \otimes \omega'_{ij} \\ &= \textstyle\sum_k e_k \otimes (\sum_i g_{ki} \cdot \omega'_{ij}). \end{aligned}$$

On the other hand, according to the same relation (2.3) and the very properties of the given $\mathcal{A}$-connection D of $\mathcal{E}$ (cf. thus Chapt. VI; (3.5), (3.6)), one further obtains;

$$(2.11)\qquad \begin{aligned} D(f_j) &= D(\textstyle\sum_i g_{ij} e_i) = \sum_i D(g_{ij} e_i) = \sum_i (g_{ij} De_i + e_i \otimes \partial g_{ij}) \\ &= \textstyle\sum_i ((\sum_k e_k \otimes \omega_{ki}) g_{ij} + e_i \otimes \partial g_{ij}) = \sum_i (\sum_k e_k \otimes \omega_{ki}) g_{ij} + \sum_i e_i \otimes \partial g_{ij} \\ &= \textstyle\sum_k e_k \otimes (\sum_i \omega_{ki} g_{ij} + \partial g_{kj}). \end{aligned}$$

Yet, concerning both of the previous relations (2.10) and (2.11), one should also remark that, in addition to (2.7), we do consider here, in effect, an analogous restriction to $U \cap V$ of the given local gauge e^U of $\mathcal{E}$, as in (2.1); that is, one has the following *local gauge of $\mathcal{E}$ over $U \cap V$*

$$(2.12)\qquad e^{U\cap V} \equiv e^U|_{U\cap V} \equiv \{U \cap V; (e_1, \ldots, e_n)|_{U\cap V}\},$$

which thus, along with (2.7), appeared, in fact, in the above two relations.

Remark 2.1.- In this context, by rectifying our previous argument in (2.5), we should say, precisely speaking, that the (section) matrix $g \equiv (g_{ij})$ in (2.4) is actually the *matrix of change of the* given *local gauges e^U and f^V* of $\mathcal{E}$, as in (2.1), when both are *restricted to* $U \cap V$, as in (2.7) and (2.12), respectively. Thus, one essentially has the *change of (local) gauges*

$$(2.13)\qquad e^{U\cap V} \equiv (e_i)|_{U\cap V} \xrightarrow[g\equiv(g_{ij})]{} (f_j)|_{U\cap V} \equiv f^{U\cap V},$$

as expressed by (2.3) above.

So, based on the same relations (2.10) and (2.11), we finally conclude that;

$$(2.14)\qquad \sum_{i=1}^{n} g_{ki}\omega'_{ij} = \sum_{i=1}^{n} \omega_{ki} g_{ij} + \partial g_{kj} \in \Omega(U \cap V).$$

(In this regard, see also, for instance, W. Greub [1: p. 6, Lemma 1.5.1], or even our previous argument on (1.11), concerning the uniqueness of the ω_{ij}'s). *Equivalently,* one can write (2.14) in the following *matrix form* (cf. also (2.4) and (2.6));

$$g \cdot \omega' = \omega \cdot g + \partial(g). \tag{2.15}$$

We still *abused notation* above, given that ∂ stands here, in fact, for the respective *extension to* $M_n(\mathcal{A})$ (cf. Chapt. VI; (1.22), (1.23)). Thus, from (2.15) and (2.4) one now obtains

$$\omega' = g^{-1}\omega g + g^{-1}\partial g, \tag{2.16}$$

or by employing the notation of Chapt. VI; (1.29), (1.35) and (1.41), one finally has the relation;

$$\begin{aligned} \omega' = g^{-1}\omega g + \tilde{\partial}(g) \equiv \mathrm{Ad}(g^{-1})\omega + \tilde{\partial}(g) \in M_n(\Omega(U \cap V)) \\ = M_n(\Omega)(U \cap V). \end{aligned} \tag{2.17}$$

We can now summarize the preceding into the form of the following.

Lemma 2.1. *Suppose we are given a differential triad* $(\mathcal{A}, \partial, \Omega)$ *on a topological space* X, *an* $\mathcal{A}$*-module* $\mathcal{E}$ *on* X, *as well as, two local gauges of* $\mathcal{E}$,

$$e^U \equiv \{U; (e_1, \ldots, e_n)\} \quad and \quad f^V \equiv \{V; (f_1, \ldots, f_n)\} \tag{2.18}$$

over the open sets U *and* V *of* X, *respectively, such that*

$$U \cap V \neq \emptyset, \tag{2.19}$$

both being of the same rank $n \in \mathbf{N}$. *Moreover, let*

$$g \equiv (g_{ij}) \in \mathcal{GL}(n, \mathcal{A})(U \cap V) = \mathrm{GL}(n, \mathcal{A}(U \cap V)) \tag{2.20}$$

be the respective local gauge transformation over $U \cap V$ (cf. (2.19)) *between the above two local gauges of* $\mathcal{E}$ (see (2.4), (2.5) and (2.13)). *Finally, let* D *be a given* $\mathcal{A}$*-connection of* $\mathcal{E}$, *whose local* $\mathcal{A}$*-connection matrices over* U, V *are*

$$\omega \equiv (\omega_{ij}) \in M_n(\Omega(U)) = M_n(\Omega)(U) \tag{2.21}$$

and

$$\omega' \equiv (\omega'_{ij}) \in M_n(\Omega(V)) = M_n(\Omega)(V), \tag{2.22}$$

respectively (cf. (1.13)).

Then, ω *and* ω' *are gauge equivalent through* g; *viz. their restrictions to* $U \cap V$ (we keep the same notation) *are related by*

$$\omega' = \mathrm{Ad}(g^{-1})\omega + \tilde{\partial}(g). \blacksquare \tag{2.23}$$

Thus, as a consequence of the preceding, one further realizes that;

(2.24) the above relation (2.23) *is a necessary condition for being two local matrices* ω *and* ω', as before (see (2.21), (2.22)), *the* (local) *$\mathcal{A}$-connection matrices* of an $\mathcal{A}$-connection D of the $\mathcal{A}$-module $\mathcal{E}$, *resulted from a change of local gauges* of $\mathcal{E}$, as this is expressed by a section matrix g, *as in (2.4)* (cf. also (2.13)).

Yet, as we shall see in the succeeding Section 3, the same relation (2.23) is still a *sufficient condition for the existence of an $\mathcal{A}$-connection D* of a given *vector sheaf* $\mathcal{E}$ (cf. Theorem 3.1 in the sequel).

Scholium 2.1..- By analogy with the *"global case"* (cf. (2.25) below, together with (17.10) in Chapt. VI;), two *local (matrix) forms* ω *and* ω', as above, are said to be *gauge equivalent,* whenever a relation like (2.23) is in force, while g is then called a *local gauge transformation* (see also (2.5) above); yet, we say that ω' is the *gauge transform* of ω. In this concern, it is still easy to see that;

(2.25) *(2.23) defines, in fact, an equivalence relation among the various local* (matrix) *forms with common domain of definition.*

Thus, denoting by

$$\omega \underset{g}{\sim} \omega' \tag{2.26}$$

the *condition that (2.23) hold true,* we check, for instance, that (2.26) *is transitive:* That is, *if* $\omega \underset{g}{\sim} \omega'$ *and* $\omega' \underset{f}{\sim} \omega''$, *then* $\omega \underset{gf}{\sim} \omega''$, such that

$$gf := g \circ f \in \mathrm{GL}(n, \mathcal{A}(U \cap V \cap W)), \tag{2.27}$$

the notation applied having the obvious meaning (cf. (2.20)); indeed, one has, by hypothesis (see also Chapt. VI; (1.33)),

$$\begin{aligned} \omega'' &= \mathrm{Ad}(f^{-1})\omega' + \tilde{\partial}(f) = \mathrm{Ad}(f^{-1})(\mathrm{Ad}(g^{-1})\omega + \tilde{\partial}(g)) + \tilde{\partial}(f) \\ &= \mathrm{Ad}((gh)^{-1})\omega + \mathrm{Ad}(f^{-1})\tilde{\partial}(g) + \tilde{\partial}(f) = \mathrm{Ad}((gf)^{-1})\omega + \tilde{\partial}(gf), \end{aligned} \tag{2.28}$$

as desired. (For the *symmetry* of (2.26), cf. also Chapt. VI; (1.31)). ∎

In this context, another relevant terminology in connection with (2.23), is to call the latter relation the *gauge transformation law,* or even the *transformation law of potentials,* a nomenclature of obvious origin, of course (see thus Scholium 1.2 in the preceding, as well as, Theorem 3.1 in the sequel).

Now, in conclusion, we also remark that *the above constitute,* in fact, a *localization of our previous considerations in Chapt. VI; Section 17,* when, in the particular case considered, $\mathcal{E}$ is a *vector sheaf* on X (see especially Chapt. VI; (17.30)). Thus,

(2.29) the *effect of a (global) automorphism* $\phi \in \operatorname{Aut}\mathcal{E}$ *on a* given $\mathcal{A}$*-connection* D *of* $\mathcal{E}$ (see e.g. Chapt. VI; (17.10)) is locally expressed by an *analogous transformation of* the respective *local* $\mathcal{A}$*-connection matrix.* Thus, g in (2.4) can be conceived as but the *(local) action of* ϕ *on the* particular *local gauge of* $\mathcal{E}$ involved (cf. Chapt. VI; (17.29)).

2.(a). Vector sheaves.– The foregoing will be applied in particular, in the special case one considers a *vector sheaf* $\mathcal{E}$ on X, say, of *rank* $n \in \mathbf{N}$ (cf. thus also Section 3 in the sequel). Now, assume further that

$$\mathcal{U} = \{U_\alpha\}_{\alpha \in I} \tag{2.30}$$

is a *local frame* of $\mathcal{E}$. So it is clear that anything, has been said above holds, in particular, true for any *local gauge* $U_\alpha \in \mathcal{U}$ of $\mathcal{E}$; thus, precisely speaking (see e.g. (1.5)), the latter is given by

$$e^{(\alpha)} \equiv \left\{U_\alpha; (e_1^{(\alpha)}, \ldots, e_n^{(\alpha)})\right\} \tag{2.31}$$

such that

$$(e_i^{(\alpha)})_{1 \leqq i \leqq n = \mathrm{rk}\,\mathcal{E}} \in \mathcal{E}(U_\alpha)^n = \mathcal{A}(U_\alpha)^{n^2} \tag{2.32}$$

stands for the respective *basis (of sections)* of the *free* $\mathcal{A}(U_\alpha)$*-module* (see also (1.3), (1.4));

$$\mathcal{E}(U_\alpha) = (\mathcal{E}|_{U_\alpha})(U_\alpha) = (\mathcal{A}^n|_{U_\alpha})(U_\alpha) = \mathcal{A}^n(U_\alpha) = \mathcal{A}(U_\alpha)^n. \tag{2.33}$$

Now, suppose further that we are given an $\mathcal{A}$*-connection* of $\mathcal{E}$

$$D : \mathcal{E} \longrightarrow \Omega(\mathcal{E}) \equiv \mathcal{E} \otimes_{\mathcal{A}} \Omega; \tag{2.34}$$

so, based on the preceding (cf. (1.17), (1.35)), one defines, *for every* $\alpha \in I$, the respective *local* $\mathcal{A}$*-connection matrix* (or even the *potential*) *of* D *over* $U_\alpha \subseteq X$,

(2.35) $$\omega^{(\alpha)} \equiv (\omega_{ij}^{(\alpha)}) \in M_n(\Omega(U_\alpha)) = M_n(\Omega)(U_\alpha),$$

according to the relation,

(2.36) $$\omega^{(\alpha)} = D_\alpha - \partial_\alpha^n.$$

Here we have set, of course,

(2.37) $$D_\alpha := D|_{U_\alpha},$$

so that one has

(2.38) $$D_\alpha \equiv D|_{U_\alpha} \in Conn_{\mathcal{A}|_{U_\alpha}}(\mathcal{E}|_{U_\alpha}) = Conn_{\mathcal{A}|_{U_\alpha}}(\mathcal{A}^n|_{U_\alpha});$$

in this context, see (1.31) – (1.33), (1.28) and (1.30) above, as well as, Chapt. VI; (8.28). Yet, cf. (2.47) below. In conclusion, one thus infers that:

(2.39) given an *$\mathcal{A}$-connection D* of a *vector sheaf $\mathcal{E}$* on X, then, *for every local frame $\mathcal{U} = (U_\alpha)_{\alpha \in I}$ of $\mathcal{E}$*, one obtains a *0-cochain of matrix-valued "1-forms"*,

(2.39.1) $$(\omega^{(\alpha)}) \in C^0(\mathcal{U}, M_n(\Omega)).$$

We call it the *$\mathcal{A}$-connection 0-cochain matrix form,* associated with the given local frame $\mathcal{U}$ of $\mathcal{E}$.

On the other hand, suppose now that

(2.40) $$g \equiv (g_{\alpha\beta}) \in Z^1(\mathcal{U}, \mathcal{GL}(n, \mathcal{A}))$$

is a *coordinate 1-cocycle* of the *vector sheaf $\mathcal{E}$*, associated with the given *local frame $\mathcal{U}$ of $\mathcal{E}$*, as in (2.30) (cf. Chapt. V; (1.8), along with Definition 1.1). Thus, *for any $\alpha, \beta \in I$, with*

(2.41) $$U_{\alpha\beta} \equiv U_\alpha \cap U_\beta \neq \emptyset,$$

the respective *section-matrix*

(2.42) $$g_{\alpha\beta} \in \mathcal{GL}(n, \mathcal{A})(U_{\alpha\beta}) = \mathrm{GL}(n, \mathcal{A}(U_{\alpha\beta}))$$

entails, in effect, the *(local) gauge transformation between the local gauges U_α and U_β of $\mathcal{E}$* (cf. Chapt. V; (1.5), (1.18), in conjunction with (2.13) above).

Consequently, by considering now the respective *transformation law of potentials,* as the latter are defined by (2.35), one obtains, by virtue of (2.23) (see also (2.40)), the following *basic relation;*

$$\omega^{(\beta)} = \mathrm{Ad}(g_{\alpha\beta}^{-1})\omega^{(\alpha)} + \tilde{\partial}(g_{\alpha\beta}). \tag{2.43}$$

In this regard, it is instructive (yet, in anticipation for later applications) to give here a bit further *clarification of the notation applied in (2.43):*

Thus, looking at the *local gauge* (2.31) of $\mathcal{E}$, one has

$$e_i^{(\alpha)} := \eta_\alpha^{-1} \circ \varepsilon_i^{(\alpha)}, \quad 1 \leqq i \leqq n = \mathrm{rk}\,\mathcal{E}, \quad \alpha \in I, \tag{2.44}$$

with $(\varepsilon_i^{(\alpha)}) \subseteq \mathcal{A}^n(U_\alpha)$ the *Kronecker gauge of* $\mathcal{A}^n$ *over* U_α; cf. Chapt. IV; (3.2), (3.3) and (3.11.1), along with Chapt. VI; (8.12). Yet, see Chapt. V; (1.13), (1.15). Moreover, we also set (cf. Chapt. VI; (10.4))

$$g_{\alpha\beta} := \eta_\alpha \circ \eta_\beta^{-1}, \textit{ for any } \alpha, \beta \in I, \textit{ satisfying (2.41).} \tag{2.45}$$

So, based now on (1.11), one obtains (cf. also (2.35), (2.37))

$$D_\alpha(e_j^{(\alpha)}) \equiv (D|_{U_\alpha})(e_j^{(\alpha)}) \equiv D(e_j^{(\alpha)}) = \sum_{i=1}^{n} e_i^{(\alpha)} \otimes \omega_{ij}^{(\alpha)}, \tag{2.46}$$

with $1 \leqq j \leqq n$ *and* $\alpha \in I$. From this and from (1.15), (1.21) and (1.32), one now obtains;

$$D_\alpha(e_j^{(\alpha)})_i = \omega_{ij}^{(\alpha)} = (D_\alpha - \partial_\alpha^n)(e_j^{(\alpha)}), \quad 1 \leqq i, j \leqq n \ \textit{ and } \ \alpha \in I, \tag{2.47}$$

which also *justifies* now *completely* (2.36) in the foregoing.

On the other hand, one still obtains (see also (2.42), (2.45))

$$e_j^{(\beta)} = \sum_{i=1}^{n} g_{ij}^{(\alpha\beta)} e_i^{(\alpha)}, \quad 1 \leqq j \leqq n, \tag{2.48}$$

where we set

$$g_{\alpha\beta} \equiv (g_{ij}^{(\alpha\beta)}) \in \mathrm{GL}(n, \mathcal{A}(U_{\alpha\beta}) = \mathcal{GL}(n, \mathcal{A})(U_{\alpha\beta}), \tag{2.49}$$

for any $\alpha, \beta \in I$, *as in (2.41).*

The preceding can be summarized now into the following.

Theorem 2.1. *Suppose that we are given a differential triad* $(\mathcal{A}, \partial, \Omega)$ *on a topological space* X. *Moreover, let* $\mathcal{E}$ *be a vector sheaf on* X, *of rank* $n \in \mathbf{N}$, D *an* $\mathcal{A}$*-connection of* $\mathcal{E}$ *and* $(g_{\alpha\beta}) \in Z^1(\mathcal{U}, \mathcal{GL}(n, \mathcal{A}))$ *a coordinate 1-cocycle of* $\mathcal{E}$, *with respect to a given local frame* $\mathcal{U} = (U_\alpha)_{\alpha \in I}$ *of* $\mathcal{E}$.

Then, the local $\mathcal{A}$-connection matrices of D, which are associated with $\mathcal{U}$ (cf. (2.35), along with (2.39))

$$\omega^{(\alpha)} \in M_n(\Omega)(U_\alpha), \quad \alpha \in I, \tag{2.50}$$

are pairwise gauge equivalent, through $g_{\alpha\beta} \in \mathrm{GL}(n, \mathcal{A}(U_{\alpha\beta}))$ (cf. (2.49)); *that is, one has the following relation (transformation law of potentials), valid, for any $\alpha, \beta \in I$, satisfying (2.41)* (cf. (2.43):

$$\omega^{(\beta)} = \mathrm{Ad}(g_{\alpha\beta}^{-1})\omega^{(\alpha)} + \tilde{\partial}(g_{\alpha\beta}). \tag{2.51}$$

In particular, for every point $x \in U_{\alpha\beta} \equiv U_\alpha \cap U_\beta$, one obtains;

$$\omega^{(\beta)}(x) = \mathrm{Ad}(g_{\alpha\beta}^{-1}(x))\omega^{(\alpha)}(x) + \tilde{\partial}(g_{\alpha\beta})(x). \ \blacksquare \tag{2.52}$$

The converse of the previous statement is also true, as we shall presently see in the next Section 3 (see thus Theorem 3.1 below). Indeed, one can realize it straightaway, at least, for *line sheaves.*

So it is still worth looking at the *transformation law of potentials,* as given above by (2.51), in the particular case of a *line sheaf* $\mathcal{L}$ on X, endowed with an $\mathcal{A}$*-connection* D, viz. for a *Maxwell field* $(\mathcal{L}, D)$ on X (see Chapt. VI; (18.11)):

Thus, by taking again a *local frame* $\mathcal{U}$ of $\mathcal{L}$ (cf. (2.30)), one now obtains for the corresponding $\mathcal{A}$*-connection 0-cochain matrix form,* the relation (see (2.39.1), *for* $n = 1$);

$$(\omega_\alpha) \in C^0(\mathcal{U}, \Omega) = \prod_{\alpha \in I} \Omega(U_\alpha) \tag{2.53}$$

(cf. also Chapt. III; (11.12)). So, for any *coordinate 1-cocycle of* $\mathcal{L}$,

$$(g_{\alpha\beta}) \in Z^1\mathcal{U}, \mathcal{A}^{\cdot}), \tag{2.54}$$

so that, for any $\alpha, \beta \in I$, with (2.41) in force, one has

$$g_{\alpha\beta} \in \mathcal{A}^{\cdot}(U_{\alpha\beta}) = \mathcal{A}(U_{\alpha\beta})^{\cdot} \tag{2.55}$$

(see (2.4), (2.49), *for* $n = 1$, *one obtains for (2.51);*

$$\omega_\beta = g_{\alpha\beta}^{-1}\omega_\alpha g_{\alpha\beta} + \tilde{\partial}(g_{\alpha\beta}) \tag{2.56}$$

(cf. also e.g. (2.16)). However, due to (2.55), along with the *commutativity of the* $\mathbf{C}$*-algebra* $\mathcal{A}(U)$, for any *open* $U \subseteq X$, according to our hypothesis for $\mathcal{A}$, one has from (2.56);

$$\omega_\beta = \omega_\alpha + \tilde{\partial}(g_{\alpha\beta}). \tag{2.57}$$

Therefore, by further applying the *Bockstein operator* δ (Chapt. III; (4.16), (4.29)), so that one has

$$\omega_\beta - \omega_\alpha = \delta(\omega_\alpha) \in B^1(\mathcal{U}, \Omega) \subseteq Z^1(\mathcal{U}, \Omega), \tag{2.58}$$

one finally *obtains* from (2.57) *the relation;*

$$\delta(\omega_\alpha) = \tilde{\partial}(g_{\alpha\beta}). \tag{2.59}$$

In other words, one thus infers that:

(2.60) *the 1-cocycle*

$$\tilde{\partial}(g_{\alpha\beta}) \in Z^1(\mathcal{U}, \Omega) \tag{2.60.1}$$

(cf. Chapt. VI; (10.25), *for* $n = 1$) *is,* by virtue of (2.59), *a (1-)coboundary,* so that one has;

$$\left[\tilde{\partial}(g_{\alpha\beta})\right] = 0 \in H^1(X, \Omega). \tag{2.60.2}$$

On the other hand, we already know (Chapt. VI; (10.28.1) and Corollary 12.1) that (2.60.2) *is a necessary and sufficient condition for a line sheaf to admit an* $\mathcal{A}$*-connection (Atiyah's criterion).* So one concludes here the aforementioned *converse of Theorem 2.1,* concerning, namely, the *existence of an* $\mathcal{A}$*-connection,* in the particular case of a *line sheaf.* In fact, as already said, the analogous conclusion holds true, *for every vector sheaf,* in general.

Scholium 2.2.– Some final comments, in conjunction with our previous remarks in (2.25), pertaining to the notion of an $\mathcal{A}$-connection, are still here in order: Thus, as follows from the preceding, given a *vector sheaf* $\mathcal{E}$ and a *local frame* $\mathcal{U} = (U_\alpha)_{\alpha \in I}$ *of* $\mathcal{E}$, an $\mathcal{A}$*-connection* D of $\mathcal{E}$ gives rise to an *equivalence class* (cf. (2.26))

$$[\omega^{(\alpha)}], \tag{2.61}$$

where $\omega^{(\alpha)} \in M_n(\Omega)(U_\alpha)$, $\alpha \in I$, is given by (2.35) (cf. also (2.39)). Indeed, *such an equivalence class entails an* $\mathcal{A}$*-connection* too, as we shall see in the sequel; of course, *we have seen it* already *in the case of a line sheaf* (see (2.60) above, along with the comments following it).

Thus, according to the preceding and in anticipation also of Theorem 3.1

in the sequel, we can say, *roughly speaking,* that:

(2.62) an $\mathcal{A}$-*connection* of a *vector sheaf* is an *equivalence class of local* $\mathcal{A}$-*connection matrices* (viz. *matrix-valued "1-forms"*), which is *supplied by any* given *local frame of the vector sheaf* involved.

3. Transformation law of potentials. Existence of $\mathcal{A}$-connections

As already mentioned, our purpose by the ensuing discussion is to examine whether (2.51) yields a *sufficient condition for the existence of an* $\mathcal{A}$-*connection.* Indeed, our main objective hereby is to prove the following *basic result,* justifying thus completely our previous comments.

Theorem 3.1. *Let* $(\mathcal{A}, \partial, \Omega)$ *be a given differential triad on a topological space* X, $\mathcal{E}$ *a vector sheaf on* X, *with* $\operatorname{rk}\mathcal{E} = n \in \mathbf{N}$, *and*

$$(g_{\alpha\beta}) \in Z^1(\mathcal{U}, \mathcal{GL}(n, \mathcal{A})) \tag{3.1}$$

a coordinate 1-cocycle of $\mathcal{E}$, *with respect to a given local frame* $\mathcal{U} = (U_\alpha)_{\alpha \in I}$ *of* $\mathcal{E}$. *Moreover, let*

$$\omega \equiv (\omega^{(\alpha)}) \in C^0(\mathcal{U}, M_n(\Omega)) \tag{3.2}$$

be a 0-cochain of $\mathcal{U}$ *with values in the sheaf of* $\mathcal{A}$-*connection coefficients of* $\mathcal{E}$, *such that the following "transformation law of potentials" holds true;*

$$\omega^{(\beta)} = \operatorname{Ad}(g_{\alpha\beta}^{-1})\omega^{(\alpha)} + \tilde{\partial}(g_{\alpha\beta}), \tag{3.3}$$

for any $\alpha, \beta \in I$, *such that*

$$U_{\alpha\beta} \equiv U_\alpha \cap U_\beta \neq \emptyset. \tag{3.4}$$

Then, there exists a uniquely defined $\mathcal{A}$-*connection* D *of* $\mathcal{E}$, *whose local* $\mathcal{A}$-*connection matrices, relative to* $\mathcal{U}$, *are given by* (the components of) *(3.2).*

Proof. Based on Theorem 1.1, one can define, by employing tha *α-th component of* the 0-cochain ω in (3.2), i.e. the *section-matrix*

$$\omega^{(\alpha)} \equiv (\omega_{ij}^{(\alpha)}) \in M_n(\Omega)(U_\alpha) = M_n(\Omega(U_\alpha)) \tag{3.5}$$

(see also (2.35)), an $\mathcal{A}|_{U_\alpha}$*-connection of* $\mathcal{E}|_{U_\alpha}$*, for every* $\alpha \in I$; namely, *the* (local) *connection,* we are looking for, *is* simply *given by the corresponding relation to (1.15).* Therefore, if

$$e^{(\alpha)} \equiv \{U_\alpha; (e_i^{(\alpha)})_{1 \leqq i \leqq n}\} \tag{3.6}$$

is the respective *local gauge of* $\mathcal{E}$ *over* U_α (cf. (2.31), (2.44)), then the values of the latter connection, denoting it by D_α, on the *section-basis* $(e_i^{(\alpha)})_{1 \leqq i \leqq n}$ *of* $\mathcal{E}(U_\alpha)$ (see (2.32), (2.33)) are given by the relation;

$$D_\alpha(e_j^{(\alpha)}) = \sum_{i=1}^{n} e_i^{(\alpha)} \otimes \omega_{ij}^{(\alpha)} \in \Omega(\mathcal{E})(U_\alpha), \ 1 \leqq j \leqq n \tag{3.7}$$

(cf. also (1.10)). So *our claim* now *is that;*

$$D_\alpha(e_j^{(\beta)}) = D_\beta(e_j^{(\beta)}), \ \ 1 \leqq j \leqq n, \tag{3.8}$$

for every $\alpha, \beta \in I$*, satisfying (3.4);* of course, an obvious *restriction to* $U_{\alpha\beta}$ for both *of* D_α *and* D_β*,* along with a *similar one of* $e^{(\alpha)}$, are here in sight (see also e.g. (2.12)).

Thus, under the validity of (3.8), the two $\mathbf{C}$*-linear morphisms* D_α, D_β, in effect, *their restrictions,* as before, being equal on the (similarly restricted) *section-basis* $(e_i^{(\beta)})|_{U_{\alpha\beta}}$ *of* $\mathcal{E}(U_{\alpha\beta})$*, would be the same, for any* $\alpha, \beta \in I$*, as in (3.4).* This, in turn, would imply now that the *0-cochain*

$$(D_\alpha) \in C^0(\mathcal{U}, \mathcal{H}om_{\mathbf{C}}(\mathcal{E}, \Omega(\mathcal{E}))), \tag{3.9}$$

as defined by the preceding (cf. (1.15), (3.5)), would be actually a *0-cocycle;* viz. we would have then (cf. also e.g. Chapt. II; (6.4))

$$(D_\alpha) \in Z^0(\mathcal{U}, \mathcal{H}om_{\mathbf{C}}(\mathcal{E}, \Omega(\mathcal{E}))) = \mathcal{H}om_{\mathbf{C}}(\mathcal{E}, \Omega(\mathcal{E}))(X) = \mathrm{Hom}_{\mathbf{C}}(\mathcal{E}, \Omega(\mathcal{E})). \tag{3.10}$$

That is, one thus obtains a $\mathbf{C}$*-linear morphism*

$$D \equiv (D_\alpha) \in \mathrm{Hom}_{\mathbf{C}}(\mathcal{E}, \Omega(\mathcal{E})), \tag{3.11}$$

as above, satisfying also the corresponding *Leibniz condition,* this being actually true, according to the preceding, for each one of its components D_α, $\alpha \in I$. In other words, (3.11) *would give* us then *an* $\mathcal{A}$*-connection of* $\mathcal{E}$, which, in addition, would be *uniquely defined,* as well, *given the 0-cochain*

(3.2). Finally, the corresponding *local $\mathcal{A}$-connection matrices of D*, when the latter is given by (3.11), will be, by virtue of (1.11) and (3.7), those provided by (3.5). Thus, to complete the proof of the theorem, all that we still have to do, is just to give the *proof of the relation (3.8):*

Indeed, based on (3.7) and (2.48), one first obtains;

$$\begin{aligned} D_\beta(e_j^{(\beta)}) &= \textstyle\sum_{i=1}^n e_i^{(\beta)} \otimes \omega_{ij}^{(\beta)} = \sum_i(\sum_{k=1}^n g_{ki}^{(\alpha\beta)} e_k^{(\alpha)}) \otimes \omega_{ij}^{(\beta)} \\ &= \textstyle\sum_k e_k^{(\alpha)} \otimes (\sum_i g_{ki}^{(\alpha\beta)} \omega_{ij}^{(\beta)}), \end{aligned} \tag{3.12}$$

for any $1 \leqq j \leqq n$ *and* $\beta \in I$. On the other hand, by employing the notation of (1.20), one gets from (3.12);

$$D_\beta(e_j^{(\beta)})_k := \sum_{i=1}^n g_{ki}^{(\alpha\beta)} \cdot \omega_{ij}^{(\beta)} \equiv (g_{\alpha\beta} \cdot \omega^{(\beta)})_{kj}, \quad 1 \leqq k, j \leqq n, \tag{3.13}$$

where the *matrix-notation* in the last item above is, of course, clear (see, for instance, (2.35) and (2.49)). Now, the same matrix-notation facilitates, in effect, the application here, according to our hypothesis, of the *transformation law of potentials* in the form of (2.15). Thus, for the case at hand, one now obtains;

$$g_{\alpha\beta} \cdot \omega^{(\beta)} = \omega^{(\alpha)} \cdot g_{\alpha\beta} + \partial(g_{\alpha\beta}), \tag{3.14}$$

true, *for any* $\alpha, \beta \in I$, *satisfying (3.4)*. Indeed, we have here an *equivalent formulation,* of course, of (3.3). Yet, we should also recall at this place that in the last relation ∂ stands, in effect, for its *extension to* $M_n(\mathcal{A})$; (cf. e.g. (1.28)). Consequently, one now obtains, by (3.14),

$$\begin{aligned} (g_{\alpha\beta} \cdot \omega^{(\beta)})_{kj} &= (\omega^{(\alpha)} \cdot g_{\alpha\beta} + \partial(g_{\alpha\beta}))_{kj} = \partial(g_{\alpha\beta})_{kj} + (\omega^{(\alpha)} \cdot g_{\alpha\beta})_{kj} \\ &= \partial(g_{kj}^{(\alpha\beta)}) + \textstyle\sum_{i=1}^n \omega_{ki}^{(\alpha)} \cdot g_{ij}^{(\alpha\beta)} = D_\alpha(\sum_{i=1}^n g_{ij}^{(\alpha\beta)} e_i^{(\alpha)})_k = D_\alpha(e_j^{(\beta)})_k, \end{aligned} \tag{3.15}$$

for any $1 \leqq k, j \leqq n$. (In this regard, cf. also (2.48), (1.15), as well as, for instance, (1.29), or even (2.11)). Thus, in view of the *uniqueness of an expression, like (1.11), in terms of the* ω_{ij}'s, the above relations (3.13) and (3.15) *prove* already (3.8), which, thus, according to the preceding, terminates the proof of the theorem. ▮

Now, in view of its significance, we do reformulate our information, gathered so far by the previous two Theorems 2.1 and 3.1, into the form of the following.

Theorem 3.2. Suppose that we are given a *differential triad* $(\mathcal{A}, \partial, \Omega)$ on a *topological space* X and $\mathcal{E}$ a *vector sheaf* on X, of *rank* $n \in \mathbf{N}$. Moreover, let $\mathcal{U} = (U_\alpha)_{\alpha \in I}$ be a *local frame of* $\mathcal{E}$, along with the following *0-cochain* of $\mathcal{U}$;

$$\omega \equiv (\omega^{(\alpha)}) \in C^0(\mathcal{U}, M_n(\Omega)). \tag{3.16}$$

Then,

(3.17) *the 0-cochain (3.16) yields an $\mathcal{A}$-connection of the vector sheaf* $\mathcal{E}$ (cf. (3.11), (3.7)) *if, and only if, the* corresponding *transformation law of potentials holds true,*

the latter law (cf. (3.3), or even (3.14)) being associated with ω, as in (3.16), and with a given *coordinate 1-cocycle of* $\mathcal{E}$, relative to $\mathcal{U}$ (cf. (3.1)). ∎

Concerning the classical case of differential geometry, see, for instance, E. Binz - J. Śniatycki - H. Fischer [1: p. 172, Remarks, (i)], or even M. Nakahara [1: p. 335, (10.9)]; yet, see M. Göckeler - T. Schüker [1: p. 50, (4.42)], as well as, Y. Choquet-Bruhat, C. DeWitt-Morette and M. Dillard-Bleick [1: p. 364, Theorem].

4. Dual $\mathcal{A}$-connections. Local form

We have already considered in the foregoing (Chapt. VI; Subsection 5.(d)) *dual $\mathcal{A}$-connections* for *vector sheaves.* As the heading indicates, we further examine in this Section the local aspect of the same notion, yet, for *$\mathcal{A}$-modules,* in general, by employing the notion of a *local gauge* (cf., for example, (1.2)). We shall find applications of the present material later on (see thus the next Chapter VIII).

Now, to start with, suppose, as usual, that we are given a *differential triad* $(\mathcal{A}, \partial, \Omega)$ on a *topological space* X (see Chapt. VI; (3.1)), and let $\mathcal{E}$ be an *$\mathcal{A}$-module* on X. Moreover, assume that

$$e^U \equiv \{U; (e_i)_{1 \leqq i \leqq n}\} \tag{4.1}$$

is a *local gauge of* $\mathcal{E}$ over an *open* set $U \subseteq X$, of *rank* $n \in \mathbf{N}$ (cf. (1.2) - (1.4)). On the other hand, consider the *dual* $\mathcal{A}$*-module of* $\mathcal{E}$,

$$\mathcal{E}^* := \mathcal{H}om_{\mathcal{A}}(\mathcal{E}, \mathcal{A}) \tag{4.2}$$

(see Chapt. IV; Definition 5.1), and let

$$(e^U)^* \equiv \{U; (e_i^*)_{1 \leqq i \leqq n}\} \tag{4.3}$$

the *dual local gauge of* e^U, being thus a *local gauge of* $\mathcal{E}^*$, of the *same sort, as* e^U: That is, first, due to our hypothesis for e^U (cf. (1.2)), one obtains (see Chapt. II; (6.23), (6.22));

$$\begin{aligned} \mathcal{E}^*|_U = \mathcal{H}om_{\mathcal{A}}(\mathcal{E}, \mathcal{A})|_U &= \mathcal{H}om_{\mathcal{A}|_U}(\mathcal{E}|_U, \mathcal{A}|_U) = \mathcal{H}om_{\mathcal{A}|_U}(\mathcal{A}^n|_U, \mathcal{A}|_U) \\ &= \mathcal{H}om_{\mathcal{A}}(\mathcal{A}^n, \mathcal{A})|_U = \mathcal{A}^n|_U, \end{aligned} \tag{4.4}$$

within $\mathcal{A}|_U$*-isomorphisms*. Therefore, one also obtains

$$\mathcal{E}^*(U) = \mathcal{A}^n(U) = \mathcal{A}(U)^n, \tag{4.5}$$

up to $\mathcal{A}(U)$*-isomorphisms* (cf. (1.4)); viz. again a *free* $\mathcal{A}(U)$*-module,* of *rank* $n \in \mathbf{N}$. Furthermore, one has, by definition (see e.g. Chapt. IV; (5.18)),

$$e_i^*(e_j) := \delta_{ij} \in \mathcal{A}(U), \quad 1 \leqq i, j \leqq n \tag{4.6}$$

(Kronecker's sections), so that

$$(e_i^*)_{1 \leqq i \leqq n} \subseteq \mathcal{E}^*(U) \tag{4.7}$$

becomes the *dual* (section) *basis* of the given one of $\mathcal{E}(U)$, as in (4.1). The above explain also the meaning of (4.3).

Now, suppose further that we are given an $\mathcal{A}$*-connection* D *of* $\mathcal{E}$. Thus, our purpose now is to define an $\mathcal{A}|_U$*-connection of* $\mathcal{E}^*|_U$, which, by *abusing notation,* we denote by $D^*|_U$, such that one has;

$$D^*|_U : \mathcal{E}^*|_U \longrightarrow \mathcal{E}^*|_U \otimes_{\mathcal{A}|_U} \Omega|_U \tag{4.8}$$

(see Scholium 4.1 below for an explanation too of the notation applied in (4.8); cf. thus (4.15)). Yet, concerning the range of the last map, one has, by virtue of (4.4) (see also Chapt. II; (5.25) and (5.18)),

$$\mathcal{E}^*|_U \otimes_{\mathcal{A}|_U} \Omega|_U = \mathcal{A}^n|_U \otimes_{\mathcal{A}|_U} \Omega|_U = \mathcal{A}^n \otimes_{\mathcal{A}} \Omega|_U = \Omega^n|_U, \tag{4.9}$$

so that one finally obtains for (4.8);

$$D^*|_U : \mathcal{E}^*|_U = \mathcal{A}^n|_U \longrightarrow \Omega(\mathcal{E}^*)|_U = \Omega^n|_U. \tag{4.10}$$

Thus, consider *any (local) section* (cf. also (1.2))

$$\begin{aligned} u \in \mathcal{E}^*(U) = \mathcal{H}om_{\mathcal{A}}(\mathcal{E}, \mathcal{A})(U) = \mathrm{Hom}_{\mathcal{A}|_U}(\mathcal{E}|_U, \mathcal{A}|_U) \\ \mathrm{Hom}_{\mathcal{A}|_U}(\mathcal{A}^n|_U, \mathcal{A}|_U) = \mathcal{H}om_{\mathcal{A}}(\mathcal{A}^n, \mathcal{A})(U) = \mathcal{A}^n(U) = \mathcal{A}(U)^n. \end{aligned} \tag{4.11}$$

We thus arrived at (4.5) again; however, the first row above is of importance for our next definition in (4.12). Thus, *for* $u \in \mathcal{E}^*(U)$ *and* $s \in \mathcal{E}(U)$, *one has* $u(s) \in \mathcal{A}(U)$, so that, *for every* $s \in \mathcal{E}(U)$, one now defines;

$$(D^*|_U)(u)(s) \equiv D^*(u)(s) := \partial(u(s)) - (u \otimes 1)(D(s)), \tag{4.12}$$

where we set, as usual, $1 \equiv 1_{\Omega|_U} \in \mathrm{Aut}(\Omega|_U) = (\mathcal{A}ut\Omega)(U)$. (Of course, similar *restrictions to* U *of* the operators ∂ *and* D, as appeared in (4.2), are, in fact, implicitly understood, as well, cf. also Chapt. VI; (6.21)). Thus, we still set;

$$(D^*|_U)(u) \equiv D^*(u) := \partial \circ u - (u \otimes 1) \circ D, \tag{4.13}$$

for every $u \in \mathcal{E}^*(U)$ (cf. (4.11)).

Now, our claim is that (4.12), or even (4.13), provides an $\mathcal{A}|_U$*-connection of* $\mathcal{E}^*|_U$: Indeed, this follows easily from Chapt. VI; (5.13), when we *localize things* on the given by (4.1) *local gauge* e^U of the $\mathcal{A}$*-module* $\mathcal{E}$ considered (cf. e.g. (4.9), (4.10) above). ∎ In this context, we further remark that (see also Chapt. II; (6.23));

$$\mathcal{E}^*|_U = \mathcal{H}om_{\mathcal{A}}(\mathcal{E}, \mathcal{A})|_U = \mathcal{H}om_{\mathcal{A}|_U}(\mathcal{E}|_U, \mathcal{A}|_U) \equiv (\mathcal{E}|_U)^*, \tag{4.14}$$

for any given $\mathcal{A}$*-module* $\mathcal{E}$ on X, and *open* $U \subseteq X$.

Scholium 4.1.– What we have actually defined above, by (4.13), is $(D|_U)^*$; that is, the *dual* $\mathcal{A}|_U$*-connection of* $D|_U$, which is *defined on* $(\mathcal{E}|_U)^* = \mathcal{E}^*|_U$ (see (4.14)). Of course, when $D^* \equiv D_{\mathcal{E}^*}$, viz. the *dual* $\mathcal{A}$*-connection of* D, there does exist (take, for instance, $\mathcal{E}$ a vector sheaf on X, as in Chapt. VI; (5.31)), one obtains

$$(D|_U)^* = D^*|_U. \tag{4.15}$$

Thus, this also clarifies the *abused notation,* as already remarked, that was employed in (4.8) (and in (4.12) too), however, simply, for convenience. Yet, it also justifies better the heading of the running Section.

On the other hand, concerning (4.12), one also obtains (cf. (1.10), as well as, for instance, (4.18) below)

$$D(s) \in \Omega(\mathcal{E})(U) \equiv (\mathcal{E} \otimes_{\mathcal{A}} \Omega)(U) = \mathcal{E}(U) \otimes_{\mathcal{A}(U)} \Omega(U) = \Omega^n(U) = \Omega(U)^n, \tag{4.16}$$

for every $s \in \mathcal{E}(U)$ and $U \subseteq X$, as in (4.1). Moreover, one has (cf. (4.13), (4.10) and (4.5))

$$\begin{aligned} D^*(u) \in \Omega(\mathcal{E}^*)(U) &\equiv (\mathcal{E}^* \otimes_{\mathcal{A}} \Omega)(U) = \mathcal{E}^*(U) \otimes_{\mathcal{A}(U)} \Omega(U) \\ &= \Omega^n(U) = \Omega(U)^n, \end{aligned} \tag{4.17}$$

for every $u \in \mathcal{E}^*(U)$ (see (4.11), the *open* $U \subseteq X$ being, by hypothesis, as before, (the domain of) a *local gauge* of the $\mathcal{A}$*-module* $\mathcal{E}$ involved.

Accordingly, by considering now the *local gauges (4.1) and (4.3)* of the $\mathcal{A}$*-module* $\mathcal{E}$ and its *dual* $\mathcal{A}$*-module* $\mathcal{E}^*$, respectively, one obtains from (4.12) (cf. also (1.11) and (4.6), as well as, Chapt. VI; (1.7));

$$\begin{aligned} D^*(e_j^*)(e_i) &= \partial(e_j^*(e_i)) - (e_j^* \otimes 1)(D(e_i)) \\ &= \partial(\delta_{ij}) - (e_j^* \otimes 1)(\textstyle\sum_{k=1}^n e_k \otimes \omega_{ki}) = -\sum_k e_j^*(e_k) \otimes \omega_{ki} \\ &\equiv -\textstyle\sum_k e_j^*(e_k) \cdot \omega_{ki} = -\omega_{ji}, \end{aligned} \tag{4.18}$$

for any $1 \leqq i, j \leqq n$. Consequently, one now sets;

$$\omega_{ij}^* := D^*(e_j^*)(e_i) = -\omega_{ji}, \quad 1 \leqq i, j \leqq n. \tag{4.19}$$

Thus, by extension of the previous terminology, connected with (1.11), we can say that the (local) $\mathcal{A}|_U$*-connection matrix* of the $\mathcal{A}|_U$*-connection* $(D|_U)^*$ of $\mathcal{E}^*|_U$ is given by (4.19). So *one* finally *sets;*

$$\omega^* \equiv (\omega_{ij}^*) = -{}^t\omega \equiv (-\omega_{ji}) \in M_n(\Omega(U)) = M_n(\Omega)(U), \tag{4.20}$$

with U *given by (4.1).*

In particular, one then concludes that;

(4.21) given a *vector sheaf* $\mathcal{E}$ on X, *with* $\operatorname{rk}\mathcal{E} = n \in \mathbf{N}$, e^U a *local gauge* of it, as in (4.1), and D an $\mathcal{A}$*-connection of* $\mathcal{E}$, then (4.15) *holds true,* so that ω^*, *as given by (4.20), is,* according to (1.11), *the local* $\mathcal{A}$*-connection matrix of* $D^* \equiv D_{\mathcal{E}^*}$, the *dual* $\mathcal{A}$*-connection of* D, with respect to the given local gauge e^U of $\mathcal{E}$.

Applications of the preceding will appear throughout the sequel; see thus the proof of Theorem 5.1 in the next Section 5, as well as, the subsequent Chapter VIII, where (4.21) will be employed in deriving the fundamental *Bianchi's identity,* within the present abstract context.

5. Christoffel sections

As an application of our considerations in Section 1 above, we reproduce here, within our abstract framework, the familiar situation from the classical theory of the *Christoffel functions ("symbols").* So, as in the standard differential geometry of smooth manifolds, one obtains another way of identifying *locally* (but, see also the next Section 6) a given $\mathcal{A}$-connection *("law of (covariant) differentiation").* Furthermore, we shall have the opportunity of looking at an $\mathcal{A}$-connection, as a type of the familiar, from the classical theory, *"∇-operator"* (yet, *locally*! See thus (5.34) below).

So assume now that we have our hitherto standard framework, viz. a *differential triad*

$$(\mathcal{A}, \partial, \Omega) \tag{5.1}$$

on a *topological space* X. Yet, for the first time, thus far, we are going to specialize the given $\mathcal{A}$-module Ω, right away (see thus (5.3) below), by further imitating the classical case (cf. Chapt. VI; (2.3), (2.4)).

On the other hand, suppose we have a *vector sheaf* $\mathcal{E}$ on X, with

$$\mathrm{rk}\,\mathcal{E} = n \in \mathbf{N}, \tag{5.2}$$

which is also endowed with an *$\mathcal{A}$-connection D.* Furthermore, as already said, we assume, in particular, that

$$\Omega = \mathcal{E}^* \equiv \mathcal{H}om_{\mathcal{A}}(\mathcal{E}, \mathcal{A}); \tag{5.3}$$

hence, Ω *is a vector sheaf* on X too, of the *same rank, as* $\mathcal{E}$, being by (5.3), the *dual* ($\mathcal{A}$-module) *sheaf* of the given vector sheaf $\mathcal{E}$ on X (cf. Chapt. IV; Definition 5.1 and (5.19), along with Chapt. II; Lemma 6.1). That is, we have (cf. also (5.2));

(5.4) $$\operatorname{rk} \Omega = \operatorname{rk} \mathcal{E}^* = \operatorname{rk} \mathcal{E} = n \in \mathbf{N}.$$

Scholium 5.1.– Given that *any vector sheaf* $\mathcal{E}$ on X *is reflexive* (cf. Chapt. IV; Theorem 5.1), viz. one has

(5.5) $$\mathcal{E} = \mathcal{E}^{**},$$

we further obtain from (5.3), *concerning the given vector sheaf* $\mathcal{E}$, the relation;

(5.6) $$\mathcal{E} = \Omega^*$$

(within, in effect, an *$\mathcal{A}$-isomorphism of* the *$\mathcal{A}$-modules* concerned). Moreover, given the *$\mathcal{A}$-connection D of $\mathcal{E}$*, one also defines, according to the general pattern (see Chapt. VI; Subsection 5.(d)), its *dual $\mathcal{A}$-connection* $D_{\mathcal{E}^*} \equiv D^*$ on the dual $\mathcal{A}$-module (in fact, vector sheaf) of $\mathcal{E}$, $\mathcal{E}^* = \Omega$ (cf. (5.3)), so that (the vector sheaf now) Ω *admits an $\mathcal{A}$-connection,* as well, *the dual of D.*

Consequently, it would be *equivalent to* say, concerning *the previous data,* that;

(5.7) we are given a *differential triad* $(\mathcal{A}, \partial, \Omega)$ on a *topological space* X, such that Ω is, in particular, a *vector sheaf* on X, of *rank* $n \in \mathbf{N}$, admitting also an *$\mathcal{A}$-connection* D^*.

Thus, the succeeding discussion refers, in effect, to the *dual (vector) sheaf* of Ω, $\Omega^* \equiv \mathcal{E}$ (cf. (5.6)). So we still obtain that;

(5.8) $$\operatorname{rk} \mathcal{E} \equiv \operatorname{rk} \Omega^* = \operatorname{rk} \Omega = n \in \mathbf{N},$$

while (the *vector sheaf*) $\mathcal{E} \equiv \Omega^*$ admits an *$\mathcal{A}$-connection,* the *dual of* D^*, viz. we have

(5.9) $$(D^*)^* \equiv D^{**} = D.$$

We wrote here (5.9) in this form, in purpose; that is, as if D^* was the dual of an $\mathcal{A}$-connection D. So we call then $D^{**} \equiv (D^*)^*$, the *bidual $\mathcal{A}$-connection* of D. Indeed, what we really have here, is that;

(5.10) *(5.9) holds true, for every $\mathcal{A}$-connection D of a given vector sheaf $\mathcal{E}$ on X.*

The assertion follows from Theorem 1.1 in the preceding, in conjunction with (4.19), or even (4.20) above, and Theorem 5.1 in the sequel. ∎ The latter also supplements our previous considerations in Section 1 above.

Theorem 5.1. *Suppose that we are given a differential triad $(\mathcal{A}, \partial, \Omega)$ on a topological space X, and a vector sheaf $\mathcal{E}$ on X, of rank $n \in \mathbf{N}$. Moreover, let D and $\widetilde{D}$ be two given $\mathcal{A}$-connections of $\mathcal{E}$. Then, one obtains*

$$D = \widetilde{D}, \tag{5.11}$$

if, and only if, one has;

$$\omega^{(\alpha)} = \tilde{\omega}^{(\alpha)} \in M_n(\Omega(U_\alpha)), \quad \alpha \in I, \tag{5.12}$$

for the respective local $\mathcal{A}$-connection matrices of D and $\widetilde{D}$, relative to some local frame

$$\mathcal{U} = (U_\alpha)_{\alpha \in I} \tag{5.13}$$

of $\mathcal{E}$ (cf. (1.13), along with (2.35)).

Proof. In view of what has been said above for (1.11), it is enough, of course, to prove the *"if"* part of the assertion: Thus, *assuming (5.12),* we shall prove that

$$D(z) = \widetilde{D}(z), \tag{5.14}$$

for every point $z \in \mathcal{E}$. So if

$$\pi : \mathcal{E} \longrightarrow X \tag{5.15}$$

stands for the given (vector) *sheaf* $\mathcal{E}$ on X (Chapt. I; (1.2)), such that $\pi(z) = x \in X$, for some given $z \in \mathcal{E}$, let $x \in U_\alpha \in \mathcal{U}$, *for some* $\alpha \in I$, according to the hypothesis for $\mathcal{U}$. Therefore (cf. Chapt. I; (2.9)), one may assume that there exists a *(local) section* $s \in \mathcal{E}(U)$, for a particular *open* $U \subseteq U_\alpha$, *such that*

$$z = s(x). \tag{5.16}$$

On the other hand, by employing the *local gauge*

$$(5.17)\qquad e^{(\alpha)} \equiv \left\{U_\alpha; (e_i^{(\alpha)})_{1 \leqq i \leqq n}\right\}$$

of $\mathcal{E}$, which is associated with $U_\alpha \in \mathcal{U}$, as above, we shall have (cf., for instance, (2.31) along with (1.6.1))

$$(5.18)\qquad s = \sum_{i=1}^{n} s_i \cdot e_i^{(\alpha)}.$$

(Note that one actually employs here a *restriction to* $U \subseteq U_\alpha$ *of* the local gauge $e^{(\alpha)}$, as in (5.17), getting thus a *new local gauge of* $\mathcal{E}$ *over* U). So one now obtains, by virtue of (5.18), (1.15) and (5.12);

$$(5.19)\qquad D(s) = \sum_i e_i^{(\alpha)} \otimes \left(\partial s_i + \sum_j s_j \omega_{ij}^{(\alpha)}\right) = \sum_i e_i^{(\alpha)} \otimes \left(\partial s_i + \sum_j s_j \tilde{\omega}_{ij}^{(\alpha)}\right) = \widetilde{D}(s).$$

Consequently, in view of (5.16), one finally has;

$$(5.20)\qquad D(z) = D(s(x)) = \widetilde{D}(s(x)) = \widetilde{D}(z)$$

(see also the succeeding Scholium 5.2 for the notation applied above), so we arrived at *the desired relation (5.14)*, which thus finishes the proof. ∎

Scholium 5.2.– Concerning the notation that was employed in (5.20) above, we should also make the following explanatory comments: Namely, when we consider a given $\mathcal{A}$-*connection*

$$(5.21)\qquad D : \mathcal{E} \longrightarrow \Omega(\mathcal{E})$$

of an $\mathcal{A}$-*module* $\mathcal{E}$ as a (**C**-linear) *morphism of the sheaves* (of $\mathcal{A}$-modules) involved, then (cf. Chapt. I; (12.49), (12.50)), we also obtain a similar *morphism of* the corresponding *(complete) presheaves of sections,* viz.

$$(5.22)\qquad \overline{D} : \Gamma(\mathcal{E}) \longrightarrow \Gamma(\Omega(\mathcal{E})).$$

It is still *worth noting* on this occasion that *it is the last (presheaf) morphism,* which is *actually employed, hitherto,* in all the preceding calculations, *when D is applied on sections;* thus, (5.14) is, in fact, the first genuine evaluation of D, so far(!), *viewed,* by its very definition (cf. (5.21)), *as a sheaf morphism.* (Cf. also e.g. the relevant comments on ∂ (cf. (5.1)) in Chapt. VI; (1.3)).

Thus (the same Chapt. I; (12.50)), for every *(local) section* $s \in \mathcal{E}(U)$, and any *open* $U \subseteq X$, one sets;

$$\overline{D}(s) := D \circ s \equiv D(s), \tag{5.23}$$

employing thus, for simplicity, an obvious *abuse of notation,* as in the last term above. (In this context, see also Chapt. I; Proposition 2.1). Hence, evaluating now (5.23) at *any point* $x \in U \equiv \mathrm{Dom}(s)$, as above, *one has;*

$$\overline{D}(s)(x) := (D \circ s)(x) \equiv D(s(x)) \equiv D(s)(x), \tag{5.24}$$

where we still *abuse notation by the last term* of (5.24).

Therefore, more precisely, by applying the notation of (5.22) to (5.19), one essentially obtains

$$\overline{D}(s) = \overline{\overline{\widetilde{D}}}(s), \tag{5.25}$$

so that, *for every* $x \in \mathrm{Dom}(s) \equiv U$, one finally has, by virtue of (5.16), (5.24) and (5.25);

$$\begin{aligned} D(z) &= D(s(x)) = (D \circ s)(x) = \overline{D}(s)(x) = \overline{\overline{\widetilde{D}}}(s)(x) \\ &= (\widetilde{D} \circ s)(x) = \widetilde{D}(s(x)) = \widetilde{D}(z). \end{aligned} \tag{5.26}$$

So this unambiguously explains now (5.20).

Thus, by returning now to the given $\mathcal{A}$-*connection* D of the *vector sheaf* $\mathcal{E}$ condidered and looking further at the *range of* D, one has, by our hypothesis for Ω (cf. (5.3));

$$D : \mathcal{E} \longrightarrow \Omega(\mathcal{E}) \equiv \mathcal{E} \otimes_{\mathcal{A}} \Omega = \mathcal{E} \otimes_{\mathcal{A}} \mathcal{E}^* = \mathcal{H}om_{\mathcal{A}}(\mathcal{E}, \mathcal{E}) \equiv \mathcal{E}nd\, \mathcal{E} \tag{5.27}$$

(see also Chapt. IV; (6.1) and Chapt. II; (6.28)). In other words, one obtains a **C**-*linear morphism*

$$D \in \mathrm{Hom}_{\mathbf{C}}(\mathcal{E}, \mathcal{E}nd\, \mathcal{E}), \tag{5.28}$$

which further satisfies the *Leibniz condition,* as well (cf. Chapt. VI; (3.6)).

Therefore, according to the general theory (see Chapt. I; (13.19)), D, as in (5.28), entails a corresponding *morphism of* (complete) *presheaves*, still denoted by D (cf. (5.22), (5.23) above); that is (cf. Chapt. I; (6.7)), D provides, in effect, a *family of local* (presheaf) *morphisms*, indexed by the open subsets of X; viz. one has

$$D = (D^U \equiv D|_U)_{U \subseteq X,\, open}, \tag{5.29}$$

such that

$$D^U \in \mathrm{Hom}_{\mathbf{C}}(\mathcal{E}(U), (\mathcal{E}nd\,\mathcal{E})(U)), \tag{5.30}$$

for every open set $U \subseteq X$. Consequently, *for every (local) section* $s \in \mathcal{E}(U)$, one gets, by (5.30),

$$\begin{aligned}(D|_U)(s) \equiv D^U(s) \equiv D(s) \in (\mathcal{E}nd\,\mathcal{E})(U) &= \mathrm{Hom}_{\mathcal{A}|_U}(\mathcal{E}|_U, \mathcal{E}|_U) \\ &\equiv \mathrm{End}(\mathcal{E}|_U).\end{aligned} \tag{5.31}$$

Thus, by further considering *any other (local) section*

$$t \in \mathcal{E}(U) = (\mathcal{E}|_U)(U) \tag{5.32}$$

(cf. Chapt. I; (11.40)), one has, by (5.31);

$$D(s)(t) \in \mathcal{E}(U). \tag{5.33}$$

We summarize the preceding by the following statement;

> under the *assumption (5.3)*, given an *$\mathcal{A}$-connection D of a vector sheaf $\mathcal{E}$* on X and an *open* $U \subseteq X$, one obtains a map
>
> $$\nabla : \mathcal{E}(U) \times \mathcal{E}(U) \longrightarrow \mathcal{E}(U), \tag{5.34}$$
>
> defined by the relation (see (5.31), (5.33))
>
> $$\nabla(t,s) \equiv \nabla_t(s) := [D(s)](t), \tag{5.35}$$
>
> *for any* $s, t \in \mathcal{E}(U)$.

In this context, one further concludes that:

the operator (5.34), as given by (5.35), *displays the properties of a* classical *"linear connection"; viz., for any* $\alpha \in \mathcal{A}(U)$ *and* $s, t \in \mathcal{E}(U)$, with U *open* in X, *one has*

$$\nabla(\alpha \cdot t, s) = \alpha \cdot \nabla(t, s), \tag{5.36}$$

as well as (Leibniz condition)

$$\nabla(t, \alpha \cdot s) = \alpha \cdot \nabla(t, s) + (\partial\alpha)(t) \cdot s. \tag{5.37}$$

Yet, as a consequence of (5.36) and (5.37), one also infers that;

(5.38) the operator (5.34) *is bilinear.*

(see Chapt. VI; (1.5), along with Chapt. II; (1.38), (1.67)).

Proof of (5.36) and (5.37). First, concerning (5.36), one obtains, by (5.35);

$$\nabla(\alpha \cdot t, s) = D(s)(\alpha \cdot t) = \alpha \cdot D(s)(t) = \alpha \cdot \nabla(t, s), \tag{5.39}$$

which thus *proves* (5.36).

On the other hand, one obtains,

$$\begin{aligned} \nabla(t, \alpha s) &= D(\alpha s)(t) = (\alpha \cdot Ds + s \otimes \partial\alpha)(t) = \alpha \cdot D(s)(t) + (s \otimes \partial\alpha)(t) \\ &\equiv \alpha \cdot D(s)(t) + (\partial\alpha)(t) \cdot s = \alpha \cdot \nabla(t, s) + (\partial\alpha)(t) \cdot s, \end{aligned} \tag{5.40}$$

which, of course, *proves (5.37), proviso* the following relation, that was actually employed in the last calculations in (5.40); thus, *we have set*

$$(s \otimes \partial\alpha)(t) = (\partial\alpha)(t) \cdot s, \tag{5.41}$$

for any $\alpha \in \mathcal{A}(U)$ *and* $s, t \in \mathcal{E}(U)$, as in (5.37): Indeed, one first obtains the next *canonical* $\mathcal{A}(U)$*-morphism,* by virtue of (5.3),

$$\begin{aligned} \mathcal{E}(U) \otimes_{\mathcal{A}(U)} \Omega(U) = \mathcal{E}(U) \otimes_{\mathcal{A}(U)} \mathcal{E}^*(U) &= \mathcal{E}(U) \otimes_{\mathcal{A}(U)} \mathrm{Hom}_{\mathcal{A}|_U}(\mathcal{E}|_U, \mathcal{A}|_U) \\ &\longrightarrow \mathrm{Hom}_{\mathcal{A}|_U}(\mathcal{E}|_U, \mathcal{E}|_U) = (\mathcal{E}nd\, \mathcal{E})(U), \end{aligned} \tag{5.42}$$

so that *one* actually *defines;*

$$(s \otimes u)(t) := u(t) \cdot s, \tag{5.43}$$

for any $s, t \in \mathcal{E}(U)$ *and* $u \in \mathcal{E}^*(U) \equiv \Omega(U)$, with U *open* in X (see also Chapt. IV; (6.4), (6.5)). Thus, (5.41) is just a particular instance of (5.43), given that, in view of (5.3), *one has;*

$$\partial(\alpha) \in \Omega(U) = \mathcal{E}^*(U) = \mathcal{H}om_{\mathcal{A}}(\mathcal{E}, \mathcal{A})(U) = \mathrm{Hom}_{\mathcal{A}|_U}(\mathcal{E}|_U, \mathcal{A}|_U), \tag{5.44}$$

for every $\alpha \in \mathcal{A}(U)$. Hence, *for every* $t \in \mathcal{E}(U)$ (cf. also (5.32)), one obtains

$$[\partial(\alpha)](t) \equiv (\partial\alpha)(t) \equiv \overline{\partial(\alpha)}_U(t) \in \mathcal{A}(U), \tag{5.45}$$

since one has (cf. e.g. (5.29), (5.30))

$$\overline{\partial(\alpha)}_U(t) \in \mathrm{Hom}_{\mathcal{A}(U)}(\mathcal{E}(U), \mathcal{A}(U)) \equiv \mathcal{E}(U)^*, \tag{5.46}$$

by referring, in particular, to the given *open* $U \subseteq X$, when $\partial(\alpha)$ is viewed as an element of the last set in (5.44); thus, an analogous relation to (5.46) is certainly valid, *for every open* $V \subseteq U$. The preceding supply thus already the *full justification of (5.37),* and we are done. ∎

We are thus ready now to accomplish our program, enunciated at the beginning of this Section; we turn to this task straight away, through the succeeding discussion.

So consider first a *local gauge,* say

$$e^U \equiv \{U; (e_1, \ldots, e_n)\}, \tag{5.47}$$

of the given *vector sheaf* $\mathcal{E}$ over an *open* set $U \subseteq X$. (Notice that, due to the hypothesis for $\mathcal{E}$, such objects are come across in abundance; in fact, every point $x \in U$ is contained in a set U, as above. See, for instance, (2.30) – (2.33) in the preceding and (5.2)). Thus, by virtue of our hypothesis (5.3), the *dual local gauge of* e^U

$$(e^U)^* \equiv \{U; (e_1^*, \ldots, e_n^*)\}, \tag{5.48}$$

is a *local gauge of the vector sheaf* $\Omega = \mathcal{E}^*$ (cf. (4.3), (4.6)).

Thus, we come next to define, through (5.35) and (5.47), the following *(local) sections of* (the *structure sheaf*) $\mathcal{A}$ *over* U; viz. we set (see also (5.33), along with e.g. (2.33)):

$$\begin{aligned} \Gamma_{ij}^k := D(e_j)(e_i)_k &\equiv \nabla(e_i, e_j)_k \equiv \nabla_{e_i}(e_j)_k \\ &= e_k^*(D(e_j)(e_i)) \in \mathcal{A}(U), \end{aligned} \tag{5.49}$$

for any $1 \leq i, j, k \leq n = \mathrm{rk}\,\mathcal{E}$. We call (5.49) the *Christoffel sections* of the

$\mathcal{A}$-connection D of $\mathcal{E}$, which are associated with the given local gauge e^U of the vector sheaf $\mathcal{E}$, as in (5.47).

Therefore, one obtains from (5.49);

$$\nabla_{e_i}(e_j) = \nabla(e_i, e_j) = \sum_{k=1}^{n} \Gamma_{ij}^k \cdot e_k, \quad 1 \leqq i, j \leqq n. \tag{5.50}$$

As a result, our next observation is now that;

(5.51) *the* given *$\mathcal{A}$-connection D of the vector sheaf $\mathcal{E}$,* when *restricted to U* (domain of definition of a *local gauge e^U of $\mathcal{E}$*, cf. (5.47)), *is uniquely determined through the* corresponding to U, n^3 ($n = \operatorname{rk}\mathcal{E}$) *Christoffel sections* of D (cf. (5.49)). Of course, this always, *under the* particular *assumption (5.3).*

Indeed, *one proves that* (cf. (5.35)):

$$\nabla(t, s) = \sum_{k=1}^{n} \left(\partial(s_k)(t) + \sum_{i,j} t_i s_j \Gamma_{ij}^k \right) e_k, \tag{5.52}$$

for any (local) sections

$$t = \sum_{i=1}^{n} t_i e_i \quad \textit{and} \quad s = \sum_{j=1}^{n} s_j e_j \quad \textit{in} \quad \mathcal{E}(U) = \mathcal{A}(U)^n \tag{5.53}$$

(see also e.g. (1.6.1) and (1.4)). Therefore, based on (5.52), the *coordinates of* $\nabla(t, s) \in \mathcal{E}(U) = \mathcal{A}(U)^n$, *with respect to* e^U, as in (5.47), are given by the relation (cf. also (4.6));

$$\nabla(t, s)_k = e_k^*(\nabla(t, s)) = \partial(s_k)(t) + \sum_{i,j} t_i s_j \Gamma_{ij}^k, \quad 1 \leqq k \leqq n. \tag{5.54}$$

Proof of (5.52): Our claim on (5.52) is a routine consequence of the properties (5.36), (5.37) and (5.38) of the operator ∇, applied to the elements of (5.53); thus, one has

$$\begin{aligned} \nabla(t, s) &= \nabla(\textstyle\sum_i t_i e_i, \sum_j s_j e_j) = \sum_{i,j} t_i \nabla(e_i, s_j e_j) \\ &= \textstyle\sum_{i,j} t_i (s_j \nabla(e_i, e_j) + \partial(s_j)(e_i) e_j) \\ &= \textstyle\sum_{i,j} t_i s_j (\sum_k \Gamma_{ij}^k e_k) + \sum_{i,j} t_i \partial(s_j)(e_i) e_j \\ &= \textstyle\sum_k (\sum_{i,j} t_i s_j \Gamma_{ij}^k) e_k + \sum_j \partial(s_j)(\sum_i t_i e_i) e_j \\ &= \textstyle\sum_k (\partial(s_k)(t) + \sum_{i,j} t_i s_j \Gamma_{ij}^k) e_k, \end{aligned} \tag{5.55}$$

that is, *the desired relation (5.52)*. (In the last calculations above we have also made use of the analogous facts to (5.44) – (5.46), as before, for the particular case at hand). ∎

Now, for convenience, we gather the preceding together into the following.

Theorem 5.2. *Suppose that we are given a differential triad $(\mathcal{A}, \partial, \Omega)$ on a topological space X and a vector sheaf $\mathcal{E}$ on X, of rank $n \in \mathbf{N}$, such that one has;*

$$\Omega = \mathcal{E}^*. \tag{5.56}$$

Moreover, assume that $\mathcal{E}$ admits an $\mathcal{A}$-connection, while let also

$$e^U \equiv \{U; (e_1, \dots, e_n)\} \tag{5.57}$$

be a local gauge of $\mathcal{E}$.

Then, D restricted to U is uniquely defined through the Christoffel sections of D, associated with (5.57) (see (5.49)).

Proof. The assertion follows already, as explained before, from (5.52), in conjunction with (5.35). (See also (5.31), (5.32)). ∎

On the other hand, the standard familiar situation, pertaining to the *Christoffel* functions (*sections* in our case, cf. (5.49)), is further extended, within the present framework, by realizing here too that one can actually obtain the converse of the information supplied by the previous Theorem 5.2; namely, *the aforementioned sections define* (*locally,* always(!), but, see also the next Section 6) *an $\mathcal{A}$-connection of $\Omega^* \equiv \mathcal{E}$* (see (5.3), (5.6)). Thus, precisely speaking, one has the following.

Theorem 5.3. *Let $(\mathcal{A}, \partial, \Omega)$ be a given differential triad on a topological space X, with Ω being a vector sheaf on X, of rank $n \in \mathbf{N}$, and let*

$$e^U \equiv \{U; (e_1, \dots, e_n)\} \tag{5.58}$$

be a local gauge of $\Omega^* \equiv \mathcal{E}$ (the *dual of* Ω). *Moreover, suppose that we are given the following n^3 (local) sections of $\mathcal{A}$ over the open $U \subseteq X$;*

$$\Gamma^k_{ij} \in \mathcal{A}(U), \quad 1 \leqq i, j, k \leqq n = \operatorname{rk} \Omega. \tag{5.59}$$

Then, there exists a uniquely defined $\mathcal{A}|_U$-connection of $\Omega^*|_U \equiv \mathcal{E}|_U$, having the given sections (5.59) as the corresponding Christoffel sections, which are associated with (5.58).

Proof. By considering the given *(local) sections* (5.59) and the *local gauge* e^U (cf. (5.58)) of the *dual of* Ω,

$$\mathcal{E} \equiv \Omega^*, \tag{5.60}$$

for any *local sections* of $\mathcal{E}$ over the *open* set $U \subseteq X$,

$$t = \sum_{i=1}^{n} t_i e_i \quad \textit{and} \quad s = \sum_{j=1}^{n} s_j e_j, \tag{5.61}$$

where $n = \operatorname{rk} \mathcal{E} \equiv \operatorname{rk} \Omega^* = \operatorname{rk} \Omega$ (cf., for instance, (5.4) or even (5.8)), one sets;

$$\nabla(t, s) := \sum_{k=1}^{n} \alpha_k \cdot e_k, \tag{5.62}$$

such that

$$\alpha_k := \partial(s_k)(t) + \sum_{i=1}^{n} \sum_{j=1}^{n} t_i s_j \Gamma_{ij}^k \in \mathcal{A}(U), \ 1 \leqq k \leqq n. \tag{5.63}$$

Thus, one actually obtains a map

$$\nabla^U \equiv \nabla : \mathcal{E}(U) \times \mathcal{E}(U) \longrightarrow \mathcal{E}(U), \tag{5.64}$$

having the formal properties of a *linear* $\mathcal{A}(U)$*-connection on* $\mathcal{E}(U)$; namely, this is:

i) *bilinear,* relative to the **C**-vector spaces concerned (cf. Chapt. II; (1.38), (1.51), or even (1.67)), as easily follows from (5.62) and (5.63).

ii) $\mathcal{A}(U)$*-linear,* with respect to the *first variable;* this is still obvious by (5.61), (5.63), along with the fact that, *for any* $\alpha \in \mathcal{A}(U)$, *one obtains* (cf. also (5.44) – (5.46), as well as, (5.5) and (5.60));

$$\partial(\alpha) \in \Omega(U) = \mathcal{E}^*(U) = \mathcal{H}om_{\mathcal{A}}(\mathcal{E}, \mathcal{A})(U) = \operatorname{Hom}_{\mathcal{A}|_U}(\mathcal{E}|_U, \mathcal{A}|_U). \tag{5.65}$$

Finally, one concludes that;

iii) *the same map* $\nabla^U \equiv \nabla$, as given by (5.64) and (5.62), *satisfies the Leibniz condition* (5.37): Indeed, for any $s, t \in \mathcal{E}(U)$, as in (5.61), and $\alpha \in \mathcal{A}(U)$, one obtains (see also (5.63), (5.41), along with Chapt. VI; (1.2));

$$
(5.66)\qquad
\begin{aligned}
\nabla(t,\alpha s) &= \textstyle\sum_k \left(\partial(\alpha s_k)(t) + \sum_{i,j} t_i(\alpha s_j)\Gamma^k_{ij}\right) e_k \\
&= \textstyle\sum_k \left(\alpha\cdot\partial(s_k)(t) + (s_k \otimes \partial\alpha)(t) + \alpha\cdot\sum_{i,j} t_i s_j \Gamma^k_{ij}\right) e_k \\
&= \alpha\cdot\textstyle\sum_k(\partial(s_k)(t) + \sum_{i,j} t_i s_j\Gamma^k_{ij})e_k + \partial(\alpha)(t)\cdot\sum_k s_k e_k \\
&= \alpha\cdot\nabla(t,s) + \partial(\alpha)(t)\cdot s,
\end{aligned}
$$

that is, the assertion.

Now, by setting

$$
(5.67)\qquad D(s)(t) := \nabla(t,s),
$$

for any $s, t \in \mathcal{E}(U)$ (note that, by virtue of our hypothesis for e^U, the latter sections can always be put in the form (5.61)), one realizes that

$$
(5.68)\qquad D(s) \in (\mathcal{E}nd\,\mathcal{E})(U) = \mathrm{End}(\mathcal{E}|_U) \equiv \mathrm{Hom}_{\mathcal{A}|_U}(\mathcal{E}|_U, \mathcal{E}|_U),
$$

for every $s \in \mathcal{E}(U)$, by the very properties of (5.64), as proved above; in particular, by still employing here an obvious *abuse of notation* for (5.68) (applying thus the same notation too for its *"U-component"*, see e.g. (5.29)), *one actually has*

$$
(5.69)\qquad D(s) \in \mathrm{Hom}_{\mathcal{A}(U)}(\mathcal{E}(U), \mathcal{E}(U)),
$$

for every $s \in \mathcal{E}(U)$. Furthermore, *one obtains* (caution! See also e.g. (5.30))

$$
(5.70)\qquad D(\alpha\cdot s) = \alpha\cdot D(s) + s\otimes\partial(\alpha),
$$

for any $\alpha \in \mathcal{A}(U)$ *and* $d \in \mathcal{E}(U)$: Indeed, one has (cf. also (5.67), in conjunction with (5.66) and (5.41));

$$
(5.71)\qquad
\begin{aligned}
D(\alpha s)(t) &= \nabla(t,\alpha s) = \alpha\cdot\nabla(t,s) + (\partial\alpha)(t)\cdot s \\
&= \alpha\cdot D(s)(t) + (s\otimes\partial\alpha)(t) = (\alpha\cdot D(s) + s\otimes\partial\alpha)(t),
\end{aligned}
$$

for every $t \in \mathcal{E}(U)$. Therefore, one gets (5.70), given that, both summands in the second member of (5.70), viz. $\alpha\cdot D(s)$ *and,* by virtue of (5.42), (5.43) (see also (5.44)), $s\otimes\partial(\alpha)$ *belong to the same* $\mathcal{A}(U)$*-module, as* $D(s)$ *does* (cf. (5.69)).

Consequently, *the map*

$$
(5.72)\qquad D^U \equiv D : \mathcal{E}|_U \longrightarrow \mathcal{E}\otimes_{\mathcal{A}}\Omega|_U = \mathcal{E}\otimes_{\mathcal{A}}\mathcal{E}^*|_U = (\mathcal{E}nd\,\mathcal{E})|_U
$$

(cf also (5.60), (5.27)), which is defined, *section-wise*, by (5.67), *yields an* $\mathcal{A}|_U$*-connection of* $\mathcal{E}|_U$ ($\equiv \Omega^*|_U$). Furthermore, *the same connection has*

(5.59) as Christoffel sections, associated with the given local gauge e^U, as this follows from the relation;

$$D(e_j)(e_i)_k = \nabla(e_i, e_j)_k = \Gamma^k_{ij}, \quad 1 \leqq i, j, k \leqq n \tag{5.73}$$

(see also (5.47), (5.49), (5.67) and (5.63), along with Chapt. VI; (1.5)). Finally, such an $\mathcal{A}|_U$-connection on $\mathcal{E}|_U$ (viz. with given Christoffel sections on U), is *uniquely defined,* as this follows straightforwardly from a relevant localization to U of Theorem 5.2 above, and this terminates the proof. ∎

On the other hand, by pursuing further the classical pattern, one also obtains *another expression for the Christoffel sections,* as in (5.49), of the given $\mathcal{A}$-connection D, *by means of the local $\mathcal{A}$-connection matrices* of D, as they were defined in the preceding by (1.13). (See thus (5.77) below).

Thus, suppose, in particular, that *we have the hypotheses of Theorem 5.2* above. Moreover, let

$$\omega \equiv (\omega_{ij}) \in M_n(\Omega(U)) = M_n(\Omega)(U) \tag{5.74}$$

be the *local $\mathcal{A}$-connection matrix* of the given *$\mathcal{A}$-connection D of $\mathcal{E}$*, which can be associated with a *local gauge e^U of $\mathcal{E}$* (cf. (5.57) and (1.13)), such that one has (cf. (5.56))

$$\omega_{ij} \in \Omega(U) = \mathcal{E}^*(U), \quad 1 \leqq i, j \leqq n = \operatorname{rk} \mathcal{E}. \tag{5.75}$$

Accordingly, using the *dual local gauge* $(e^U)^*$ of e^U (cf. (5.48)), as well as, the corresponding to these data *Christoffel sections* (see (5.49)), *one can define* the previous sections (5.75), hence, *the matrix* ω, through the following relation;

$$\omega_{ij} := \sum_{k=1}^{n} \Gamma^k_{ij} \cdot e^*_k \in \mathcal{E}^*(U) = \Omega(U), \quad 1 \leqq i, j \leqq n. \tag{5.76}$$

Thus, *equivalently,* one obtains that (see (4.6));

$$\omega_{ij}(e_k) = \Gamma^k_{ij}, \quad 1 \leqq i, j, k \leqq n. \tag{5.77}$$

Now, as a result of the preceding discussion (cf. (5.76) above and Theorem 5.2), one concludes that:

(5.78) under the *hypothesis of Theorem 5.2,* an *$\mathcal{A}$-connection D* of *$\mathcal{E}$ is, locally, uniquely defined either through the local $\mathcal{A}$-connection matrix ω, or the* corresponding *Christoffel sections* $\{\Gamma^k_{ij}\}$ *of D,* where both objects are *referred to a* given *local gauge e^U of $\mathcal{E}$.*

In this context, we also note that, by further following the classical pattern, one can give a *relation connecting the Christoffel sections, when a gauge transformation is under consideration.* We do not give it, however, due to its somewhat elaborate (routine, anyway!) technical nature. Yet, we shall be concerned with the Christoffel sections of a given $\mathcal{A}$-connection, several times in the sequel.

6. Complete parallelism

The concept in the title of this Section belongs, in effect, to the next Chapter VIII, that is dealt with the *"curvature"* of a given $\mathcal{A}$-connection. However, we have already anticipated in the preceding that notion, by considering *flat vector sheaves* (Chapt. V; Section 5), as well as, *integrable $\mathcal{A}$-connections* (Chapt. VI; Subsection 11.(a)), to which the present notion is intimately related; in a sense, all these concepts are (in concrete instances, including the classical case, of course), equivalent, as we are going to see (partly, the whole story will be told in Chapter VIII), straight away, by the ensuing discussion.

Thus, suppose that we are given a *differential triad* $(\mathcal{A}, \partial, \Omega)$ on a *topological space* X, along with an *$\mathcal{A}$-module* $\mathcal{E}$ on X and an *$\mathcal{A}$-connection D of $\mathcal{E}$*. Moreover, let

$$e^U \equiv \{U; (e_1, \ldots, e_n)\} \tag{6.1}$$

be a *local gauge of* $\mathcal{E}$, of *rank* $n \in \mathbf{N}$, over the *open* set $U \subseteq X$ (see (1.5)).

Now, we say that the $\mathcal{A}$-connection D, as before, defines a *complete parallelism of $\mathcal{E}$ over U*, whenever one has

$$D(e_i) = 0, \quad 1 \leqq i \leqq n. \tag{6.2}$$

In this context, we also write, for short, instead of (6.2), just

(6.3) $$D(e^U) = 0.$$

On the other hand, a given *(local) section* $s \in \mathcal{E}(V)$ over the *open* set $V \subseteq X$ is said to be *parallel* (or even *horizontal*), with respect to the $\mathcal{A}$-connection D of $\mathcal{E}$, as above, if one has

(6.4) $$D(s) = 0.$$

Thus, when (6.2) is satisfied (cf. also (1.3)), one also speaks for e^U, as a *parallel* (or even *horizontal*) *local gauge* of $\mathcal{E}$.

The preceding nomenclature is further specialized by the following.

Definition 6.1. Suppose we have a vector sheaf $\mathcal{E}$ on X, as before, of rank $n \in \mathbf{N}$, which is also endowed with an $\mathcal{A}$-connection D. Then, we say that $\mathcal{E}$ has a *complete parallelism,* with respect to D (or even that D *provides a complete parallelism* of $\mathcal{E}$), whenever $\mathcal{E}$ has a *parallel* (or *horizontal*) *local frame,* with respect to D.

Namely, we assume that there exists a *local frame* of $\mathcal{E}$,

(6.5) $$\mathcal{U} = (U_\alpha)_{\alpha \in I},$$

such that *the* corresponding *local gauges* (see (2.31), (2.32))

(6.6) $$e^{(\alpha)} \equiv \{U_\alpha; (e_i^{(\alpha)})_{1 \leqq i \leqq n}\}$$

are parallel; that is, one has (cf. (6.2), (6.3))

(6.7) $$D(e^{(\alpha)}) = 0, \quad \alpha \in I.$$

Now, in order also to respond to the comments at the beginning of this Section, it is instructive to consider at this place a *generalization* of the notion *of flat vector sheaf* (cf. thus (6.12) in the sequel). The same extended notion of the latter will be of importance later on too. So we further set the following.

Definition 6.2. Suppose that we are given a differential triad $(\mathcal{A}, \partial, \Omega)$ on a topological space X, and let $\mathcal{E}$ be a vector sheaf on X, with $\operatorname{rk} \mathcal{E} = n \in \mathbf{N}$. Now, we say that $\mathcal{E}$ is *∂-flat,* whenever there exists a coordinate 1-cocycle of $\mathcal{E}$,

(6.8) $$(g_{\alpha\beta}) \in Z^1(\mathcal{U}, \mathcal{GL}(n, \mathcal{A})),$$

with respect to a local frame $\mathcal{U}$ of $\mathcal{E}$ (cf. Chapt. V; (1.8)), such that

$$\partial(g_{\alpha\beta}) = 0. \tag{6.9}$$

In the last relation ∂ stands, in effect, for its *extension to* $M_n(\mathcal{A})$ (*Whitney sum,* cf. Chapt. VI; (1.15)). So by extending the terminology of (6.4) (to the *$\mathcal{A}$-connection ∂ of $\mathcal{A}$*, in fact, to its *extension to* $M_n(\mathcal{A})$), we can say that;

(6.10) a given *vector sheaf* $\mathcal{E}$ on X is *∂-flat,* whenever it possesses a coordinate 1-cocycle, consisting of *∂-parallel,* or even *∂-horizontal, (coordinate) matrices.*

Notice that, by definition (see (6.8)), one has;

$$g_{\alpha\beta} \in \mathcal{GL}(n, \mathcal{A})(U_{\alpha\beta}) = \mathrm{GL}(n, \mathcal{A}(U_{\alpha\beta})), \tag{6.11}$$

for any $\alpha, \beta \in I$, *with* $U_{\alpha\beta} \equiv U_\alpha \cap U_\beta \neq \emptyset$.

So, as already mentioned, one concludes, in fact, according to the very definitions, that;

(6.12) *every flat vector sheaf* on X *is ∂-flat.* (See also Chapt. V; Definition 5.1).

Thus, see Chapt. V; (5.7), Theorem 5.1; in particular, (5.18) and (5.24), along with Chapt. VI; Lemma 1.1. ∎

We are ready now to come to our first result, pertaining to the preceding notions. So, as we shall presently see, *every ∂-flat vector sheaf possesses always* a certain special type of *$\mathcal{A}$-connection,* we already met before (viz. an *integrable $\mathcal{A}$-connection*). Thus, more precisely, one has the following.

Theorem 6.1. *Suppose that we have a differential triad $(\mathcal{A}, \partial, \Omega)$ on a topological space X, and let $\mathcal{E}$ be a vector sheaf on X, of rank $n \in \mathbf{N}$. Then, the three following assertions are equivalent:*

1) $\mathcal{E}$ is ∂-flat.

2) $\mathcal{E}$ has an integrable $\mathcal{A}$-connection (cf. Chapt. VI; Definition 11.2).

3) $\mathcal{E}$ admits an $\mathcal{A}$-connection D, providing a complete parallelism.

Proof. We prove first that *1) implies 2):* So let

$$\mathcal{U} = (U_\alpha)_{\alpha \in I} \tag{6.13}$$

be a *local frame of* $\mathcal{E}$, with respect to which we have a *coordinate 1-cocycle* (6.8), for which (6.9) *holds true.* Thus, by considering the corresponding to (6.13) *Levi-Civita ($\mathcal{A}$-connection) 1-cocycle* of $\mathcal{E}$, one obtains in view of (6.9) (see also Chapt. VI; (8.34) and Theorem 10.1, in particular (10.22), (10.23)),

$$\delta(D_\alpha) = (D_\beta - D_\alpha) \underset{\alpha}{\sim} \tilde{\partial}(g_{\alpha\beta}) = g_{\alpha\beta}^{-1}\partial(g_{\alpha\beta}) = 0. \tag{6.14}$$

That is, *the Levi-Civita 0-cochain* (D_α), as above (cf. also Chapt. VI; (8.32.1)), *is,* in fact, *a 0-cocycle,* so that it defines an $\mathcal{A}$-connection of $\mathcal{E}$, which is, of course, integrable (see Chapt. VI; (11.24)), viz. the assertion. *Convesely,* assume that $\mathcal{E}$ admits an *integrable $\mathcal{A}$-connection* $\widetilde{D}$, with respect to (6.13) (cf. Chapt. VI; Definition 11.2). Therefore (see Chapt. VI; (6.23)), one has

$$0 = \delta(\widetilde{D}_\alpha) = \delta(D_\alpha) \underset{\alpha}{\sim} \tilde{\partial}(g_{\alpha\beta}) = g_{\alpha\beta}^{-1}\partial(g_{\alpha\beta}), \tag{6.15}$$

hence, $\partial(g_{\alpha\beta}) = 0$, so that *2) implies 1),* as well.

Furthermore, we prove that *1) and 3) are equivalent:* Indeed, *if 1) is valid,* and we are given (6.13), one can further define a *0-cochain of local $\mathcal{A}$-connection matrices,* associated with it, viz.

$$\omega \equiv (\omega^{(\alpha)}) \in C^0(\mathcal{U}, M_n(\Omega)), \tag{6.16}$$

by simply setting

$$\omega^{(\alpha)} \equiv (\omega_{ij}^{(\alpha)}) := 0 \in M_n(\Omega)(U_\alpha) = M_n(\Omega(U_\alpha)), \tag{6.17}$$

for every $\alpha \in I$. However, *by virtue of (6.9), the same 0-cochain satisfies* now *the transformation law of potentials* (cf. (3.3)), so that (Theorem 3.1) $\mathcal{E}$ *admits (uniquely) an $\mathcal{A}$-connection D,* having the above $\omega^{(\alpha)}$'s. In particular, *1) implies 3),* since now, by (6.17) and (2.46), one has;

$$D(e_j^{(\alpha)}) = \sum_i e_i^{(\alpha)} \otimes \omega_{ij}^{(\alpha)} = 0, \tag{6.18}$$

for any $1 \leqq i, j \leqq n$, *and* $\alpha \in I$, that is, (6.7) *is fulfilled.* Conversely, *if (6.7) is valid,* then by the same relation (6.18), one concludes that

$$\omega^{(\alpha)} = 0, \quad \alpha \in I, \tag{6.19}$$

so that by still employing the transformation law of potentials, one further obtains that

$$\tilde{\partial}(g_{\alpha\beta}) = g_{\alpha\beta}^{-1}\partial(g_{\alpha\beta}) = 0, \tag{6.20}$$

hence, $\partial(g_{\alpha\beta}) = 0$. Thus, *3) implies 1), as well,* and this completes the proof of the theorem. ∎

Scholium 6.1.- The aim of the following lines is to offer *another proof of prop. 1) implies 2) of* the previous *Theorem 6.1.* (Although the present proof is rather longer, than the one given above, it has, however, an interest **per se**; cf. thus (6.27), (6.28) and (6.29) below). So, following the argument in the first step of the proof given before, one defines an *$\mathcal{A}$-connection* $\widetilde{D}$ of $\mathcal{E}$, *whose local $\mathcal{A}$-connection matrices were defined by (6.17).* We prove here that *this same* $\widetilde{D}$ (not only the Levi-Civita one, as above) *is an integrable $\mathcal{A}$-connection of $\mathcal{E}$ too;* viz. one has

$$\widetilde{D}_\alpha \equiv \widetilde{D}|_{U_\alpha} = D_\alpha, \quad \alpha \in I, \tag{6.21}$$

where the last item of (6.21), an $\mathcal{A}|_{U_\alpha}$*-connection of* $\mathcal{E}|_{U_\alpha}$, is given by Chapt. VI; (8.29): Indeed, *for any*

$$s = \sum_{i=1}^{n} s_i e_i^{(\alpha)} \in \mathcal{E}(U_\alpha), \quad \alpha \in I, \tag{6.22}$$

(see (1.6)), *one obtains* (cf. (1.15))

$$\widetilde{D}_\alpha(s) = \sum_{i=1}^{n} e_i^{(\alpha)} \otimes \left(\partial(s_i) + \sum_{j=1}^{n} s_j \omega_{ij}^{(\alpha)} \right), \tag{6.23}$$

so that, in view of (6.17), *one has;*

$$\widetilde{D}_\alpha(s) = \sum_{i=1}^{n} e_i^{(\alpha)} \otimes \partial(s_i). \tag{6.24}$$

On the other hand, based on Chapt. VI; (8.29), one sets

$$D_\alpha := (\eta_\alpha^{-1} \otimes 1) \circ \partial_\alpha \circ \eta_\alpha \quad \alpha \in I. \tag{6.25}$$

Therefore, since, by definition (cf. Scholium 1.1), one has

$$e_i^{(\alpha)} := \eta_\alpha^{-1} \circ \varepsilon_i^{(\alpha)} \quad 1 \leqq i \leqq n, \tag{6.26}$$

for every $\alpha \in I$, with $\varepsilon_i^{(\alpha)} = (\delta_{ij}) \in \mathcal{A}^n(U_\alpha) = \mathcal{A}(U_\alpha)^n$ (canonical *Kronecker section-basis of $\mathcal{A}^n$* over U_α), *one obtains,* by virtue of (6.25) (see also (5.5), (1.7));

(6.27) $$D_\alpha(e_i^{(\alpha)}) = 0, \quad 1 \leqq i \leqq n, \quad \textit{and} \quad \alpha \in I.$$

(As a *byproduct,* one thus concludes that *when the Levi-Civita $\mathcal{A}$-connection $D = (D_\alpha)$ of $\mathcal{E}$ exists,* then, according to (6.27), one concludes (directly!) that *the latter* still *supplies a complete parallelism;* a consequence, otherwise, of Chapt. VI; (11.24), in conjunction with Theorem 6.1 above). Hence (see, for instance, (6.18)), if $\tilde{\omega}^{(\alpha)} \in M_n(\Omega)(U_\alpha)$, $\alpha \in I$, denote the local $\mathcal{A}$-connection matrices of D, one has, by (6.27),

(6.28) $$\tilde{\omega}^{(\alpha)} \equiv (\omega_{ij}^{(\alpha)}) = 0,$$

so that one still concludes from (1.15) that

(6.29) $$D_\alpha(s) = \sum_{i=1}^{n} e_i^{(\alpha)} \otimes \partial(s_i),$$

for every $s \in \mathcal{E}(U_\alpha)$, and $\alpha \in I$, which thus proves (6.21), in view of (6.24). ∎

On the other hand, by supplementing our argument in Chapt. VI; (11.24), we further remark that;

(6.31) *if $\widetilde{D}$ is an integrable $\mathcal{A}$-connection of* (a vector sheaf) $\mathcal{E}$, then *the Levi-Civita 0-cochain*

(6.31.1) $$(D_\alpha) \in C^0(\mathcal{U}, M_n(\Omega))$$

(see Chapt. VI; (11.18)) *is,* in fact, *a 0-cocycle, yielding thus the integrable* (always! whenever it exists) *Levi-Civita $\mathcal{A}$-connection $D = (D_\alpha)$ of $\mathcal{E}$.* (Cf. also Chapt. VI; Definition 11.1).

Furthermore, as a consequence of (6.23) and (6.17), in conjunction with Theorem 6.1, we also conclude that;

(6.32) *a given $\mathcal{A}$-connection $\widetilde{D}$ of* (a vector sheaf) *$\mathcal{E}$ is integrable, if, and only if, one has*

(6.32.1) $$\widetilde{D}_\alpha(s) = \sum_{i=1}^{n} e_i^{(\alpha)} \otimes \partial(s_i),$$

for any (local) section $s = \sum_{i=1}^{n} s_i e_i^{(\alpha)} \in \mathcal{E}(U_\alpha)$, and for every $\alpha \in I$ (cf. (6.24) for the notation applied).

For the *"if"* part of our claim apply (6.32.1) to $e_i^{(\alpha)}$ to get (6.7) (cf. also

(6.26), (6.27), so that the assertion follows then from Theorem 6.1. The rest is already provided by the preceding. ∎

Finally, it is worth noticing at this place that, as a result of (6.21), along with (6.24) and (6.29), one also obtains for the corresponding (local) *components* of the *$\mathcal{A}$-connections $\widetilde{D} = (\widetilde{D}_\alpha)$ and $D = (D_\alpha)$ of $\mathcal{E}$*, as above, the relations;

$$\widetilde{D}_\alpha(s)_i = D_\alpha(s)_i = \partial(s_i), \quad 1 \leqq i \leqq n, \tag{6.33}$$

for any $s \in \mathcal{E}(U_\alpha)$, as in (6.22), and $\alpha \in I$. (This should also be compared with our previous remark in Chapt. VI; (8.32.3)). Thus, stated otherwise, (6.33) means that:

(6.34) *an integrable $\mathcal{A}$-connection* of a given vector sheaf $\mathcal{E}$, *hence also* (cf. Theorem 6.1), *an $\mathcal{A}$-connection of $\mathcal{E}$, supplying a complete parallelism, are locally the* given *standard flat $\mathcal{A}$-connection ∂ of $\mathcal{A}$* (indeed, *its extension to $M_n(\mathcal{A})$, with $n =$* $\operatorname{rk}\mathcal{E} \in \mathbf{N}$).

Scholium 6.2.– The previous notion of *∂-flatness* is proved to be, in effect, the *appropriate generalization,* within the present abstract setting of the classical notion *of flatness* (viz. *zero curvature,* cf. the next Chapter VIII), which in standard terms reduces to the fact that

$$\ker \partial = \mathbf{C} \tag{6.35}$$

(here, of course, its *extension to $M_n(\mathcal{A})$,* see (6.9), (6.11) above, in conjunction with Chapt. VI; (1.22), (1.23), and Chapt. V; (5.25)).

Now, the last relation, as above, *is likely not to hold true, in general.* Yet, as we shall see in the subsequent Chapter *∂-flatness implies curvature zero* for the respective (integrable) $\mathcal{A}$-connection (Theorem 6.1) and conversely, as well, while the same conditions are implied, in turn, by a *flat vector sheaf*, in the sense of (6.12) above. However, to all these things we shall return, with more details, at the pertinent places in the sequel.

In this context, by further looking at the local expressions of a given $\mathcal{A}$-connection D, in terms of the "ω's", as above, as well as, of the corresponding Christoffel sections, one concludes the following:

Namely, assume that $\mathcal{E}$ is a *vector sheaf* on X, of *rank* $n \in \mathbf{N}$, endowed with an $\mathcal{A}$-*connection* D, and let $\mathcal{U} = (U_\alpha)_{\alpha \in I}$ be a *local frame* of $\mathcal{E}$ (cf. (6.5), (6.6)). Moreover, let

$$\omega \equiv (\omega^{(\alpha)}) \in C^0(\mathcal{U}, M_n(\Omega)) \tag{6.36}$$

be the corresponding *0-cochain of local (A-connection) matrices* of D, which are associated with $\mathcal{U}$ (cf., for instance, (2.39.1) in the foregoing). Then

(6.37) *D yields a complete parallelism* (cf. Definition 6.1), *if, and only if, one has*

$$\omega^{(\alpha)} = 0, \tag{6.37.1}$$

for every $\alpha \in I$.

On the other hand, if, in addition, (5.3) is satisfied, then one concludes that;

(6.37) is still equivalent with the condition

$$^{(\alpha)}\Gamma^k_{ij} = 0, \tag{6.38}$$

for any $1 \leqq i, j, k \leqq n = \operatorname{rk} \mathcal{E}$, *and* $\alpha \in I$.

(The notation applied in (6.38) indicates obviously the dependence of the respective Christoffel sections on the various indices α in I). So (6.38) is, in fact, an immediate consequence of (6.7) and (2.46), while (6.38) follows from (6.7), in conjunction with (5.77) (cf. also (5.76)). ∎ Thus, putting now (6.37) and (6.38) together, we certainly infer that:

(6.39) by *assuming (5.3), the following conditions are mutually equivalent:*

i) $D(e^{(\alpha)}) = 0$, ii) $\omega^{(\alpha)} = 0$, *for every* $\alpha \in I$, *and* iii) $^{(\alpha)}\Gamma^k_{ij} = 0$, *for* any $1 \leqq i, j \leqq n$, *and* $\alpha \in I$.

So, *if any one of the above three equivalent conditions is satisfied,* then *the* given $\mathcal{A}$-*connection* D, as before, *provides a complete parallelism* and conversely, of course.

The intimate relation of all the preceding notions with that one of a *"flat $\mathcal{A}$-connection"* (cf. Scholium 6.2 above) will be examined, in detail, as already mentioned, in the next Chapter VIII.

7. ∂-flat line sheaves. Generalized Selesnick - Chern class of a line sheaf

We consider by the ensuing discussion, a certain particular 2-dimensional cohomology class, which as we shall presently see (cf. Theorem 7.1 below), by analogy with its complex counterpart, the (complex) Chern class of a line sheaf (cf. Chapt. V; Theorem 6.1), characterizes ∂-flatness; in this context, we have already seen (cf. (6.12), along with Scholium 6.2) that the latter notion generalizes the classical one of flatness. Yet, to keep track of the evolution of the idea around the aforementioned cohomology class, we still note that this has first been applied by S.S. Selesnick [2] to the particular case of a *line bundle* over the *spectrum* of a suitable *Banach algebra.* (In this regard, see also Chapt. VI; Note 10.1, as well as, Chapter XI in the sequel).

Throughout the succeeding discussion we shall make repeated use of our previous terminology and results in Chapt. V; Section 6. Thus, to start with, suppose that we are given an *exponential algebraized space* $(X, \mathcal{A})$ (cf. Chapt. V; Definition 6.1), which we further assume to be endowed with a *differential triad* $(\mathcal{A}, \partial, \Omega)$, in such a manner that the following diagram is *commutative;*

$$\begin{array}{ccccccccc} 0 & \longrightarrow & \mathbf{Z} & \longrightarrow & \mathcal{A} & \xrightarrow{e} & \mathcal{A}^{\cdot} & \longrightarrow & 1 \\ & & & & & \partial\searrow & \swarrow\frac{1}{2\pi i}\tilde{\partial} & & \\ & & & & & \partial\mathcal{A}. & & & \end{array} \tag{7.1}$$

We say that (7.1) defines an *exponential sheaf diagram.* Yet, for brevity's sake we shall refer to the previous data, as a *topological space X with an exponential sheaf diagram.* The even shorter term, that, namely, X is a *sheaf exponential space,* will often be of use below.

Now, as a result of the *short exact sequence,* that appears in (7.1) and of Chapt. VI; Lemma 1.1, one has the following *morphism of short exact sequences;*

$$(7.2)\qquad \begin{array}{ccccccccc} 0 & \longrightarrow & \mathbf{Z} & \longrightarrow & \mathcal{A} & \xrightarrow{e} & \mathcal{A}^{\cdot} & \longrightarrow & 1 \\ & & \Big\downarrow \zeta & & \Big\| & & \Big\downarrow \frac{1}{2\pi i}\tilde{\partial} & & \\ 0 & \longrightarrow & \ker\partial & \longrightarrow & \mathcal{A} & \longrightarrow & \partial\mathcal{A} & \longrightarrow & 0. \end{array}$$

Therefore (see e.g. Chapt. III; Theorem 3.1, prop. 4)), one gets the following *commutative diagram in (sheaf) cohomology;*

$$(7.3)\qquad \begin{array}{ccc} H^1(X,\mathcal{A}^{\cdot}) & \xrightarrow{\quad\delta\quad} & H^2(X,\mathbf{Z}) \\ \frac{1}{2\pi i}\tilde{\partial}\Big\downarrow & c_{\mathcal{A}} & \Big\downarrow\zeta \\ H^1(X,\partial\mathcal{A}) & \xrightarrow{\quad\gamma\quad} & H^2(X,\ker\partial). \end{array}$$

Thus we can now set the following.

Definition 7.1. Suppose we have a sheaf exponential space X (see (7.1)), and let $\mathcal{L}$ be a given line sheaf on X. Moreover, consider the Picard group of X (cf. Chapt. V; Theorem 4.1), so that one has;

$$(7.4)\qquad \mathcal{L} \equiv [\mathcal{L}] \in \operatorname{Pic}(X) = H^1(X,\mathcal{A}^{\cdot}).$$

Thus, one defines the *Chern class of $\mathcal{L}$, relative to $\mathcal{A}$*, or just the *Selesnick-Chern class of $\mathcal{L}$*, as the 2-dimensional cohomology class $c_{\mathcal{A}}(\mathcal{L})$, which is given by the relation (cf. (7.3));

$$(7.5)\qquad c_{\mathcal{A}}(\mathcal{L}) := \gamma\left(\frac{1}{2\pi i}\tilde{\partial}([(g_{\alpha\beta})])\right) \in H^2(X,\ker\partial),$$

where $(g_{\alpha\beta}) \in Z^1(\mathcal{U},\mathcal{A}^{\cdot})$ stands for a coordinate 1-cocycle of $\mathcal{L}$ (cf. Chapt. V; (1.8), for $n = 1$).

Now, as already mentioned at the beginning of this Section, a basic application of the previous cohomology class is that its vanishing yields a criterion for the *∂-flatness* of the given line sheaf $\mathcal{L}$ (which, in effect, means (Theorem 6.1) that $\mathcal{L}$ admits an integrable $\mathcal{A}$-connection); so cf. Theorem 7.1, or even Corollary 7.1 below. However, before we come to the cited results, it is constructive first to give the following key-lemmas for the proof of the former. (In this context, see also S.A. Selesnick [2: p. 66, Theorem 3.6]). So we first have the following result, referring actually to the *kernel* and *image of* the *logarithmic derivation* $\tilde{\partial}$, with respect to those of ∂; viz. one obtains.

Lemma 7.1. *Suppose we have a differential triad* $(\mathcal{A}, \partial, \Omega)$ *on a topological space* X. *Then,* applying the terminology of Chapt. VI; Definition 1.2, *one has;*

$$\ker \tilde{\partial} = (\ker \partial) \cap \mathcal{A}^{\cdot}. \tag{7.6}$$

In particular, suppose that we are given, in addition, the following commutative triangle of morphisms of sheaves (of abelian) groups, see also (7.1), for the notation applied),

$$\begin{array}{ccc} \mathcal{A} & \xrightarrow{\ e\ } & \mathcal{A}^{\cdot} \\ & \partial \searrow \quad \swarrow \frac{1}{2\pi i}\tilde{\partial} & \\ & \Omega, & \end{array} \tag{7.7}$$

with the morphism e *being also surjective. Then, one has;*

$$\tilde{\partial}(\mathcal{A}^{\cdot}) = \partial(\mathcal{A}) = 2\pi i \cdot \partial(\mathcal{A}). \tag{7.8}$$

Proof. Our claim for (7.6) is, of course, an immediate consequence of the very definitions (cf. thus Chapt. VI; (1.24), (1.25)).

Now, concerning (7.8), based on our supplementary assumption for (7.1) and the map e, as well as, on the **C**-*linearity of* ∂, one obtains;

$$\tilde{\partial}(\mathcal{A}^{\cdot}) = \tilde{\partial}(e(\mathcal{A})) = 2\pi i \cdot \partial(\mathcal{A}) = \partial(2\pi i \cdot \mathcal{A}) = \partial(\mathcal{A}), \tag{7.9}$$

which thus proves (7.8), and this completes the proof. ∎

Now, as an application of the previous two elementary, however, useful formulae (7.6) and (7.8), one further obtains the following key-result for our subsequent considerations. Thus, we have the next.

Lemma 7.2. *(S.A. Selesnick). Assume that we are given a differential triad* $(\mathcal{A}, \partial, \Omega)$ *on a topological space* X, *along with a commutative triangle, as in (7.7), the morphism* e *being surjective. Moreover, let*

$$(g_{\alpha\beta}) \in Z^1(\mathcal{U}, \mathcal{A}^{\cdot}) \tag{7.10}$$

be an $\mathcal{A}^{\cdot}$-valued 1-cocycle of an open covering $\mathcal{U} = (U_\alpha)_{\alpha \in I}$ of X. Then, the two following assertions are equivalent:

i) $\partial(g_{\alpha\beta}) = 0$.

ii) $[\tilde{\partial}(g_{\alpha\beta})] = 0 \in H^1(X, \Omega)$.

Proof. By definition (see Chapt. VI; (1.24), (1.25)), one has for the given 1-cocycle (7.10);

$$\tilde{\partial}(g_{\alpha\beta}) = g_{\alpha\beta}^{-1}\partial(g_{\alpha\beta}). \tag{7.11}$$

Therefore, it suffices only to prove that *ii) implies i), as well:* Thus, consider the following *short exact sequence of* (abelian) *group sheaves;*

$$1 \longrightarrow \ker\tilde{\partial} \longrightarrow \mathcal{A}^{\cdot} \longrightarrow \tilde{\partial}(\mathcal{A}^{\cdot}) \longrightarrow 0. \tag{7.12}$$

However, according to our hypothesis, the above *formulae (7.6) and (7.8) hold true,* so that (7.12) yields now the *short exact sequence;*

$$1 \longrightarrow (\ker\partial) \cap \mathcal{A}^{\cdot} \longrightarrow \mathcal{A}^{\cdot} \longrightarrow \partial(\mathcal{A}) \longrightarrow 0. \tag{7.13}$$

Hence, by looking now at the corresponding *long exact sequence in cohomology* (see e.g. Chapt. III; Theorem 2.1, prop. 3)), one has

$$\cdots \longrightarrow H^1(X, \ker\partial \cap \mathcal{A}^{\cdot}) \longrightarrow H^1(X, \mathcal{A}^{\cdot}) \xrightarrow{(1/2\pi i)\tilde{\partial}} H^1(X, \partial\mathcal{A}) \longrightarrow \cdots \tag{7.14}$$

Therefore, *if cond. ii) is valid,* then one has;

$$\left(\frac{1}{2\pi i}\tilde{\partial}\right)([(g_{\alpha\beta})]) = \frac{1}{2\pi i}\left[\tilde{\partial}(g_{\alpha\beta})\right] = 0, \tag{7.15}$$

that is, one obtains, in view of (7.14),

$$\begin{aligned}[(g_{\alpha\beta})] &\in \ker(H^1(X, \mathcal{A}^{\cdot}) \xrightarrow{(1/2\pi i)\tilde{\partial}} H^1(X, \partial\mathcal{A})) \\ &= \operatorname{im}(H^1(X, \ker\partial \cap \mathcal{A}^{\cdot}) \longrightarrow H^1(X, \mathcal{A}^{\cdot})).\end{aligned} \tag{7.16}$$

In other words, *the* given *1-cocycle (7.10) should take values from* $\ker\partial \cap \mathcal{A}^{\cdot}$; viz. one has

$$(g_{\alpha\beta}) \in Z^1(\mathcal{U}, \ker\partial \cap \mathcal{A}^{\cdot}) \hookrightarrow Z^1(\mathcal{U}, \ker\partial), \tag{7.17}$$

so that $\partial(g_{\alpha\beta}) = 0$, as desired, and this terminates the proof. ∎

Stated otherwise, Selesnick (ibid.) says that;

(7.17′) *the logarithmic derivative of a coordinate 1-cocycle of a line sheaf* (on a space X, as in Lemma 7.2) *can be a coboundary* (if, and) *only if, the line sheaf in question is ∂-flat.*

We now come to the promised result, *relating $c_{\mathcal{A}}(\mathcal{L})$ with ∂-flatness.* That is, one has the following.

Theorem 7.1. *Suppose that we have a sheaf exponential space X* (cf. (7.1)), *and let $\mathcal{L}$ be a given line sheaf on X. Moreover, consider the following assertions:*

1) $\mathcal{L}$ is ∂-flat; i.e., one has

$$\partial(g_{\alpha\beta}) = 0, \tag{7.18}$$

for some coordinate 1-cocycle of $\mathcal{L}$, $(g_{\alpha\beta}) \in Z^1(\mathcal{U}, \mathcal{A}^{\cdot})$.

2) One has;

$$c_{\mathcal{A}}(\mathcal{L}) = 0. \tag{7.19}$$

Then, 1) implies 2).

In particular, assume, in addition, that the following condition holds true;

$$H^1(X, \mathcal{A}) = H^2(X, \mathcal{A}) = 0. \tag{7.20}$$

Then, 2) implies 1), as well, so that the two above conditions are then equivalent.

Note 7.1.– The above condition (7.20) has already been considered in the preceding (see Chapt. V; (6.33)), to assure, for instance, within the present abstract framework, the so-called *Chern isomorphism* δ (ibid.; (6.34), (6.35)). Thus, we refer to the relevant comments there, for particular instances, where the said condition is in force.

Proof of Theorem 7.1. The first part of the assertion is certainly clear, in view of the very definition of $c_{\mathcal{A}}(\mathcal{L})$ (see (7.5)) and of the above Lemma 7.2 (i) implies ii); note that γ in (7.3) is a **C**-*linear map* between the **C**-*vector spaces* concerned). That is, one has

$$\begin{aligned} c_{\mathcal{A}}(\mathcal{L}) &:= \gamma\left(\tfrac{1}{2\pi i}\tilde{\partial}([(g_{\alpha\beta})])\right) = \tfrac{1}{2\pi i}\gamma\left(\tilde{\partial}([(g_{\alpha\beta})])\right) \\ &= \tfrac{1}{2\pi i}\gamma([\tilde{\partial}(g_{\alpha\beta})]) \in H^2(X, \ker\partial), \end{aligned} \tag{7.21}$$

which thus justifies our argument.

On the other hand, based on the *short exact sequence* at the bottom row of (7.2), one has the following *long exact sequence in (sheaf) cohomology* (cf. e.g. Chapt. III; Theorem 3.1, prop. 3))

$$(7.22)\quad \cdots \longrightarrow H^1(X,\mathcal{A}) \longrightarrow H^1(X,\partial\mathcal{A}) \longrightarrow H^2(X,\ker\partial) \longrightarrow H^2(X,\mathcal{A}) \longrightarrow \cdots$$

Consequently, *if* the supplementary *cond. (7.20) is valid,* then one gets from (7.22) the next *exact sequence* (see also (7.3));

$$(7.23)\qquad 0 \longrightarrow H^1(X,\partial\mathcal{A}) \xrightarrow{\gamma} H^2(X,\ker\partial) \longrightarrow 0.$$

Thus, the **C**-linear map γ *becomes,* in fact, *a* **C**-*vector space isomorphism; therefore, if (7.19) is valid,* then *one has* (cf. also (7.21))

$$(7.24)\qquad c_{\mathcal{A}}(\mathcal{L}) = \frac{1}{2\pi i}\gamma\left(\left[\tilde{\partial}(g_{\alpha\beta})\right]\right) = 0,$$

so that, in view of the *isomorphism* γ, one obtains

$$(7.25)\qquad \left[\tilde{\partial}(g_{\alpha\beta})\right] = 0.$$

Hence (Lemma 7.2), *one has* $\partial(g_{\alpha\beta}) = 0$ *too,* viz. *2) implies 1), as well* (under (7.20)!), and the proof is complete. ∎

Now, due to its significance for the sequel, we reset the second part of the previous theorem into the form of the following corollary. (We come to it again in the next Chapter VIII). So we have.

Corollary 7.1. *Let X be a given sheaf exponential space, in such a manner that the following condition holds true;*

$$(7.26)\qquad H^1(X,\mathcal{A}) = H^2(X,\mathcal{A}) = 0.$$

Moreover, let $\mathcal{L}$ be a line sheaf on X. Then, $\mathcal{L}$ is ∂-flat if, and only if, one has $c_{\mathcal{A}}(\mathcal{L}) = 0$. ∎

The previous data will often be applied in the sequel too. So, for convenience, we still set the following.

Definition 7.2. A topological space satisfying the hypothesis of the previous Corollary 7.1, is called a *strict sheaf exponential space.*

Yet, as we shall see later on, under further natural assumptions for the space X, the preceding conditions, as given e.g. by Corollary 7.1, are still equivalent with the line sheaf $\mathcal{L}$ having a *flat $\mathcal{A}$-connection* (see thus the next Chapter VIII).

7.(a). Complex Chern class of a line sheaf (contn'd).– By anticipating later applications, it is still appropriate to give here the relation of the above cohomology class $c_{\mathcal{A}}(\mathcal{L})$, that is associated with any given line sheaf $\mathcal{L}$ on X, with the usual *complex Chern class of* $\mathcal{L}$; the latter class has been already considered in the preceding (cf. Chapt. V; (6.17)).

Thus, suppose that we are given a *line sheaf*

$$\mathcal{L} \equiv [\mathcal{L}] \equiv [(g_{\alpha\beta})] \in H^1(X, \mathcal{A}^{\cdot}) = \mathbf{\Phi}^1_{\mathcal{A}}(X), \tag{7.27}$$

as before, such that

$$(g_{\alpha\beta}) \in Z^1(\mathcal{U}, \mathcal{A}^{\cdot}) \tag{7.28}$$

is a *coordinate 1-cocycle of* $\mathcal{L}$ (cf. Chapt. V; (6.11), (2.59)). Furthermore, consider the following *(canonical) commutative triangle;*

$$\begin{array}{ccc} \mathbf{Z} & \xrightarrow{\;j\;} & \mathbf{C} \\ & {\scriptstyle\zeta}\searrow \quad \swarrow{\scriptstyle\varepsilon} & \\ & \ker\partial, & \end{array} \tag{7.29}$$

where $\mathbf{Z}$ and $\mathbf{C}$ are viewed, of course, as the corresponding *constant sheaves* (of *abelian groups*) on X, the latter space being endowed, as usual, with a *differential triad* $(\mathcal{A}, \partial, \Omega)$. Notice that (7.29) is a consequence of the fact that

$$\mathbf{C} \underset{\varepsilon}{\overset{\subseteq}{\rightarrow}} \ker\partial \subseteq \mathcal{A} \tag{7.30}$$

(cf. Chapt. VI; Lemma 1.1).

In particular, suppose now that X is a given *sheaf exponential space* (see (7.1), so that the commutative diagram (7.3) *is in force.* Therefore, based further on Chapt. III; (3.22), (3.23), along with (7.29) above, one obtains the following *commutative diagram,* supplementing (7.3);

(7.31)

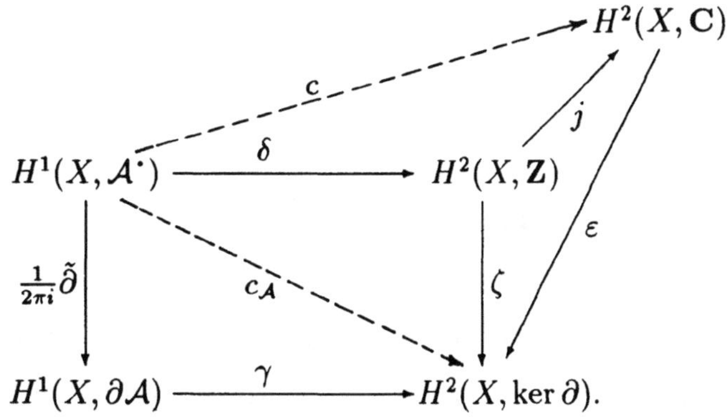

Thus, one now defines the *complex Chern class* of the given line sheaf $\mathcal{L}$ (see also (7.27)), by the relation (cf. (7.31));

(7.32) $$c(\mathcal{L}) := j(\delta(\mathcal{L})).$$

That is, one thus considers the *"complexification" of the integral Chern class of* $\mathcal{L}$, $\delta(\mathcal{L})$ (cf. (7.31) and Chapt. V; (6.12)). So we still call, occasionally,

(7.33) $$z(\mathcal{L}) := \delta(\mathcal{L}) \in H^2(X, \mathbf{Z}),$$

the *(integral) Chern class of* $\mathcal{L}$. (In this regard, see also Chapter IX in the sequel). Yet, based on (7.31) and (7.5), *one obtains;*

(7.34) $$c_{\mathcal{A}}(\mathcal{L}) = \varepsilon(c(\mathcal{L})) = \frac{1}{2\pi i}\gamma([\tilde{\partial}(g_{\alpha\beta})]).$$

In particular, *if (7.20) holds true,* then (7.34) yields;

(7.35) $$2\pi i \cdot c_{\mathcal{A}}(\mathcal{L}) = [\tilde{\partial}(g_{\alpha\beta})].$$

In this context, our main conclusion is displayed, however, by the following.

Theorem 7.2. *Suppose that X is a sheaf exponential space, and let $\mathcal{L}$ be a given line sheaf on it. Then, the condition*

(7.36) $$c(\mathcal{L}) = 0$$

implies that $\mathcal{L}$ is a flat line sheaf on X, so that $\mathcal{L}$ is ∂-flat, as well (see also (6.12)).

Proof. Consider the standard *short exact exponential sequence*

(7.37) $$0 \longrightarrow \mathbf{Z} \xrightarrow{j} \mathbf{C} \xrightarrow{e} \mathbf{C}^{\cdot} \longrightarrow 1,$$

such that

(7.38) $$e(t) := \exp 2\pi i t, \quad t \in \mathbf{C}$$

(see also Chapt. V; (6.18), (6.19)). Hence, by considering further the corresponding to (7.37) *long exact sequence in cohomology,* one obtains

(7.39) $$\cdots \longrightarrow H^1(X, \mathbf{C}^{\cdot}) \longrightarrow H^2(X, \mathbf{Z}) \xrightarrow{j} H^2(X, \mathbf{C}) \longrightarrow \cdots$$

Therefore, supposing that $c(\mathcal{L}) = 0$, one also gets (cf. (7.32))

(7.40) $$c(\mathcal{L}) = j(\delta(\mathcal{L})) = 0,$$

that is (see also (7.39)),

(7.41) $$\begin{aligned} \delta(\mathcal{L}) &\in \ker(H^2(X, \mathbf{Z}) \xrightarrow{j} H^2(X, \mathbf{C})) \\ &= \operatorname{im}(H^1(X, \mathbf{C}^{\cdot}) \longrightarrow H^2(X, \mathbf{Z})). \end{aligned}$$

Consequently, by considering a *coordinate 1-cocycle of $\mathcal{L}$, as in (7.28),* representing $\mathcal{L}$ in $H^1(X, \mathcal{A}^{\cdot})$ (see (7.27)), hence, in turn, in $H^2(X, \mathbf{Z})$, via the *Chern morphism* δ (cf. (7.31)), one infers from (7.41) that *this 1-cocycle should take values from* $\mathbf{C}^{\cdot}$; that is, one actually has that

(7.42) $$(g_{\alpha\beta}) \in Z^1(\mathcal{U}, \mathbf{C}^{\cdot}) \underset{\varepsilon}{\subsetneq} Z^1(\mathcal{U}, \mathcal{A}^{\cdot}).$$

However, *this proves,* in effect, *our assertion,* according to the very definition of a *flat line sheaf* (cf. Chapt. V; (5.24), for $n = 1$), and the proof is complete. ∎

Therefore, based on (6.12) and on Theorem 7.1 (viz. 1) implies 2)), or even just (7.34), one now concludes that;

(7.43) $$c(\mathcal{L}) = 0 \ \textit{implies} \ c_{\mathcal{A}}(\mathcal{L}) = 0.$$

On the other hand, based on the very definitions, it is not hard to see that;

(7.44) *in the* particular *case* that the *cond. (6.35) holds true, all the preceding notions coincide.*

Of course, this happens already in the classical theory; however, see also Chapters X and XI in the sequel. Yet, on all these we shall return, when the notion of *"curvature"* of a given $\mathcal{A}$-connection will be available. So we still defer the relevant discussion for the next Chapter VIII.

8. Symmetric $\mathcal{A}$-connections

We apply below our earlier considerations in Chapt. IV; Section 8, pertaining to *Riemannian $\mathcal{A}$-modules.* Of course, as is the case with our classical model, viz. the *differential geometry of $\mathcal{C}^\infty$-manifolds,* the preponderant notion hereby is that one of a *"compatible $\mathcal{A}$-connection",* in a sense, that will be made precise presently below.

Thus, suppose that we are given an *ordered algebraized space* $(X, \mathcal{A})$ (cf. Chapt. IV; Definition 8.1, together with the comments after it). Moreover, we assume that

$$(\mathcal{A}, \rho) \tag{8.1}$$

is a *Riemannian $\mathcal{A}$-module;* we recall (ibid.; Definition 8.3) that this means that

$$\rho : \mathcal{A} \oplus \mathcal{A} \longrightarrow \mathcal{A} \tag{8.2}$$

is an *$\mathcal{A}$-valued strongly non-degenerate inner product* on $\mathcal{A}$, that is, a *Riemannian $\mathcal{A}$-metric on $\mathcal{A}$*, which thus implies a *similar one on $\mathcal{A}^n$*, denoted by $\overline{\rho}$ (viz. *Whitney sum of ρ, n-times:* ibid.; (8.37), (8.37′), or even (8.40)).

On the other hand, we know that, under further suitable restrictions on X and $\mathcal{A}$, one can prove the existence of a Riemannian $\mathcal{A}$-metric on every vector sheaf $\mathcal{E}$ on X, provided we are given (8.1), as above (ibid.; Theorem 8.3), something that we are going to apply in the sequel. However, for generality's, as well as, for simplicity's sake, we just assume now that

$$(\mathcal{E}, \rho) \tag{8.3}$$

is a given *Riemannian vector sheaf* on X (ibid.; Definition 8.3). Then, as already known (ibid.; Theorem 8.1), given a *local frame of $\mathcal{E}$*,

$$\mathcal{U} = (U_\alpha)_{\alpha \in I}, \tag{8.4}$$

the *(Riemannian)* $\mathcal{A}$*-metric* ρ on $\mathcal{E}$ is (uniquely) determined by a *0-cocycle*

$$\rho_{\mathcal{U}} \equiv \rho \equiv (\rho_\alpha) \in \prod_\alpha M_n(\mathcal{A})(U_\alpha) = \prod_\alpha M_n(\mathcal{A}(U_\alpha)), \tag{8.5}$$

such that one actually has

$$\rho \equiv (\rho_\alpha) \in Z^0(\mathcal{U}, \mathcal{GL}(n, \mathcal{A})) = \mathrm{GL}(n, \mathcal{A}(X)), \tag{8.6}$$

with $n = \operatorname{rk} \mathcal{E} \in \mathbf{N}$; viz., as we say, one considers the *Gram cocycle of the* $\mathcal{A}$*-metric* ρ, associated with $\mathcal{U}$ (Chapt. IV; (8.29), along with the comments, following the end of the proof of Theorem 8.1).

Furthermore, suppose now that we are also given on the *topological space* X, as a bove, a *differential triad* $(\mathcal{A}, \partial, \Omega)$, while we still assume the relation

$$\mathcal{E}^* = \Omega \tag{8.7}$$

(see, for instance, (5.3) together with Scholium 5.1 in the preceding for a similar situation). Thus, Ω ia s *vector sheaf* on X too, of the *same rank, as* $\mathcal{E}$ (cf. (5.4)).

Furthermore, by considering a *local gauge of* $\mathcal{E}$,

$$e^U \equiv \{U; (e_1, \ldots, e_n)\} \tag{8.8}$$

(cf., for example (4.1) in the preceding), one defines the corresponding *"nabla" operator,*

$$\nabla : \mathcal{E}(U) \times \mathcal{E}(U) \longrightarrow \mathcal{E}(U), \tag{8.9}$$

according to (5.35), given the $\mathcal{A}$*-connection* D *of* $\mathcal{E}$, along with the *validity of (8.7).* Now, *for any local sections*

$$s = \sum_{i=1}^n s_i e_i \ \ and \ \ t = \sum_{j=1}^n t_j e_j \ \ in\ \mathcal{E}(U), \tag{8.10}$$

with U*, as in (8.8),* one defines the following *Lie product of (local) sections;*

$$[s, t] := \sum_{i=1}^n \left(\sum_{j=1}^n (t_j \partial(s_i) - s_j \partial(t_i))(e_j) \right) e_i \in \mathcal{E}(U). \tag{8.11}$$

Thus, *equivalently,* by looking at the i*-th component of* $[s, t] \in \mathcal{E}(U)$*, relative to* e^U (see (8.8)), *one sets;*

$$[s,t]_i := \sum_{j=1}^{n}(t_j\partial(s_i) - s_j\partial(t_i))(e_j), \quad 1 \leqq i \leqq n = \operatorname{rk}\mathcal{E}. \tag{8.12}$$

So, as a result of (8.12), by further taking the corresponding *(local) section basis of* $\mathcal{E}(U)$ (cf. (1.3), (1.4)),

$$(e_i)_{1\leqq i\leqq n} \subseteq \mathcal{E}(U) \subseteq \mathcal{A}(U)^n, \tag{8.13}$$

one obtains (cf. also Chapt. VI; (1.6), as well as, (6.26) above);

$$[e_i, e_j] = 0, \quad 1 \leqq i, j \leqq n = \operatorname{rk}\mathcal{E}. \tag{8.14}$$

Thus, we come now to the following basic notion for the ensuing discussion, as given by the next.

Definition 8.1. Suppose we have a Riemannian vector sheaf $(\mathcal{E}, \rho)$ on a topological space X, the latter being endowed with a differential triad $(\mathcal{A}, \partial, \Omega)$ such that (8.7) be satisfied. Then, a given *$\mathcal{A}$-connection D of $\mathcal{E}$* is said to be *compatible with the Riemannian $\mathcal{A}$-metric ρ of $\mathcal{E}$*, or even, for short, a *Riemannian $\mathcal{A}$-connection* of $\mathcal{E}$, whenever the two following conditions are fulfilled:

1) For any (local) sections $s, t \in \mathcal{E}(U)$, with U, as in (8.8), one has (*symmetry of D*);

$$\nabla(s,t) - \nabla(t,s) = [s,t]. \tag{8.15}$$

2) For any $s, t, z \in \mathcal{E}(U)$, as before, the following relation holds true (*Ricci identity*);

$$\partial(\rho(s,t))(z) = \rho(\nabla(z,s), t) + \rho(s, \nabla(z,t)). \tag{8.16}$$

Now, by applying the local section basis (8.13) of $\mathcal{E}(U)$, one first obtains, in view of (8.15) and (8.14), the following *symmetry condition of a Riemannian $\mathcal{A}$-connection D* (*equivalent,* of course, *to (8.14)*);

$$\nabla(e_i, e_j) = \nabla(e_j, e_i), \quad 1 \leqq i, j \leqq n. \tag{8.17}$$

Furthermore, one has the following *equivalent formulation of Ricci's identity;*

$$\partial(\rho_{ij})(e_k) = \rho(\nabla(e_k, e_i), e_j) + \rho(e_i, \nabla(e_k, e_j)), \tag{8.18}$$

with $1 \leqq i, j, k \leqq n$. Here

(8.19) $$\rho_U \equiv (\rho_{ij}) \in M_n(\mathcal{A})(U) = M_n(\mathcal{A}(U))$$

stands for the local form of (the restriction of) ρ on $U \subseteq X$, such that

(8.20) $$\rho_{ij} := \rho(e_i, e_j) \in \mathcal{A}(U), \quad 1 \leqq i, j \leqq n,$$

(see also Chapt. IV; (8.34), (8.35), for $U \equiv U_\alpha$ $\alpha \in I$, as in (8.4)). Yet, in view of (8.6) (cf. also e.g. Chapt. IV; (8.26.4)), one has

(8.21) $$\rho_U \equiv (\rho_{ij}) \in \mathrm{GL}(n, \mathcal{A}(U)) = \mathcal{GL}(n, \mathcal{A})(U).$$

Now, our assertion about (8.18) follows from the very properties of ∇ (see (5.34) – (5.38)), from those of ∂ (cf. Chapt. VI; Definition 1.1) and the *$\mathcal{A}$-bilinearity of ρ* (Chapt. IV; Definition 8.2), as well as, from our hypothesis for (8.13) (see (1.6.1)). ∎

On the other hand, as already seen in Section 5 in the preceding, the previous data supply also the appropriate framework for the definition of the *(local) Christoffel sections* of the given *$\mathcal{A}$-connection D* of $\mathcal{E}$, which are associated with the *local gauge e^U of $\mathcal{E}$*, as in (8.8). Thus, for convenience, we recall that one can define the following local sections of $\mathcal{A}$ over U (cf. (5.49));

(8.22) $$\Gamma^k_{ij} := \nabla(e_i, e_j)_k \in \mathcal{A}(U), \quad 1 \leqq i, j, k \leqq n,$$

that is, *equivalently*, one has (see also (1.6))

(8.23) $$\nabla(e_i, e_j) = \sum_{k=1}^{n} \Gamma^k_{ij} e_k, \quad 1 \leqq i, j \leqq n.$$

Therefore, by virtue of (8.17), one has, here too, the classical *symmetry condition* of a Riemannian $\mathcal{A}$-connection, *in terms of the* corresponding *Christoffel sections;* viz. one obtains

(8.24) $$\Gamma^k_{ij} = \Gamma^k_{ji}, \quad 1 \leqq i, j, k \leqq n,$$

which thus, in view of (8.22), is an *equivalent expression of (8.17)*. Furthermore, as a consequence of (8.23) (see also (8.20)), one first obtains;

(8.25) $$\begin{aligned} \rho(\nabla(e_i, e_j), e_k) &= \rho(\textstyle\sum_{l=1}^{n} \Gamma^l_{ij} e_l, e_k) = \sum_l \Gamma^l_{ij} \rho(e_l, e_k) \\ &= \textstyle\sum_l \Gamma^l_{ij} \rho_{lk}. \end{aligned}$$

Thus, based further on (8.18), one can now compute the *Christoffel sections, in terms of the Riemannian $\mathcal{A}$-metric ρ:*

Namely, by symmetry, one first obtains the two other analogous relations to (8.18), so that by further applying the standard procedure, and due also to (8.17) and the *symmetry of* ρ (cf. Chapt. IV; (8.13) and (8.22.2)), one finally gets, for the case under consideration, the *1st Christoffel identity;*

$$\rho(\nabla(e_i, e_j), e_k) = \frac{1}{2}(\partial(\rho_{jk})(e_i) + \partial(\rho_{ik})(e_j) - \partial(\rho_{ij})(e_k)). \tag{8.26}$$

Consequently, based on (8.25) and (8.26), one now obtains

$$\sum_{l=1}^{n} \Gamma_{ij}^{l} \rho_{lk} = \frac{1}{2} A, \tag{8.27}$$

with the obvious meaning of "A", so that, by still applying the standard procedure, one finds, within the present framework the *2nd Christoffel identity,* as actually promised at the beginning; that is, one has

$$\Gamma_{ij}^{l} = \frac{1}{2} \sum_{k=1}^{n} (\partial(\rho_{jk})(e_i) + \partial(\rho_{ik})(e_j) - \partial(\rho_{ij})(e_k))\rho^{kl}, \tag{8.28}$$

for any $1 \leqq i, j, l \leqq n$, where, as usual (from the classical theory!), we set

$$(\rho^{kl}) := (\rho_{lk})^{-1} \in \mathrm{GL}(n, \mathcal{A}(U)) \tag{8.29}$$

(see e.g. (8.21)).

Now, as a consequence of the previous discussion, we thus arrived at the following basic conclusion, as given by the next.

Lemma 8.1. *Consider a topological space* X, *endowed with a differential triad* $(\mathcal{A}, \partial, \Omega)$, *such that* $(X, \mathcal{A})$ *is an ordered algebraized space. Moreover, let* $(\mathcal{E}, \rho)$ *be a Riemannian vector sheaf on* X, *with* $\operatorname{rk} \mathcal{E} = n \in \mathbf{N}$, *and* $e^U \equiv \{U; (e_1, \ldots, e_n)\}$ *a local gauge of* $\mathcal{E}$ *over an open set* $U \subseteq X$, *while we also assume that*

$$\mathcal{E}^* = \Omega. \tag{8.30}$$

Then, the n^3 *(local) sections of* $\mathcal{A}$ *over* U,

$$\Gamma_{ij}^{l} \in \mathcal{A}(U), \quad 1 \leqq i, j, l \leqq n, \tag{8.31}$$

as they are given by (8.28), may be considered as the (local) Christoffel sections of a unique Riemannian $\mathcal{A}|_U$*-connection of* $\mathcal{E}|_U$.

Proof. Indeed, according to Theorem 5.3 (see also (5.78)), *such* n^3 *local* *sections of $\mathcal{A}$ over U provide a uniquely defined $\mathcal{A}|_U$-connection of $\mathcal{E}|_U$.* Thus, we further prove that *the latter is, in effect, a Riemannian $\mathcal{A}|_U$-connection.*

So, concerning the *symmetry of the $\mathcal{A}|_U$-connection* at issue, we remark that, since the given Riemannian $\mathcal{A}$-metric ρ of $\mathcal{E}$ is, by hypothesis, symmetric, one has, by definition (cf. (8.20))

$$\rho_{ij} = \rho_{ji}, \quad 1 \leqq i, j \leqq n = \operatorname{rk} \mathcal{E}, \tag{8.32}$$

or even in *matrix form* (cf. (8.21)) the relation

$$\rho = {}^t\rho \tag{8.33}$$

(see Chapt. IV; (8.26.1), (8.26.3)). Therefore, due to (8.28), one further obtains

$$\Gamma^l_{ij} = \Gamma^l_{ji}, \quad 1 \leqq i, j, l \leqq n. \tag{8.34}$$

However, the last relations are, in fact, *equivalent with the symmetry of the* (local) *$\mathcal{A}|_U$-connection of $\mathcal{E}|_U$*, defined by the given Christoffel sections (8.31), as this has already been explained above, in the case of the relation (8.24).

Thus, one has still to prove the *Ricci identity,* in its equivalent formulation, for instance, *via (8.18);* indeed, this will be the result of the subsequent complementary discussion:

So to start with, assume that we are given, of course, the *framework of Lemma 8.1,* and consider the *Ricci identity* (cf. (8.16)), in its *equivalent version, as given by (8.18).* Thus, by applying now (8.23) to (8.18), and by taking the *$\mathcal{A}$-bilinearity of the $\mathcal{A}$-metric ρ* into account (see Chapt. IV; (8.26.2)), one gets

$$\begin{aligned} \partial(\rho_{ij})(e_k) &= \rho(\nabla(e_k, e_i), e_j) + \rho(e_i, \nabla(e_k, e_j)) \\ &= \rho(\textstyle\sum_l \Gamma^l_{ki} e_l, e_j) + \rho(e_i, \sum_l \Gamma^l_{kj} e_l) \\ &= \textstyle\sum_l (\Gamma^l_{ki}\rho(e_l, e_j) + \Gamma^l_{kj}\rho(e_i, e_l)). \end{aligned} \tag{8.35}$$

Therefore, *one* finally *obtains;*

$$\partial(\rho_{ij})(e_k) = \sum_{l=1}^{n} (\Gamma^l_{ki}\rho_{lj} + \Gamma^l_{kj}\rho_{li}), \quad 1 \leqq i, j, k \leqq n. \tag{8.36}$$

(In the last term of (8.36) we made use of (8.32)). Now, it is easy to check that *the above argument can* actually *be reversed,* so that one concludes, in fact, that:

(8.37) *the Ricci identity,* as given by (8.18), *is equivalent with the rel. (8.36).*

Yet, in view of (8.28) and (8.36), one further obtains, as in the classical case, that;

(8.37′) $\partial(\rho_{ij}) = 0$, *if, and only if,* $\Gamma_{ij}^k = 0$,

with $1 \leqq k \leqq n$, and $1 \leqq i, j \leqq n$.

Thus, based further on the last conclusion, we can now proceed, as already promised, to the

End of the proof of Lemma 8.1. Based on (8.28) and (8.29), and by making the obvious substitutions in (8.36), one can prove, following thus the analogous to our case classical argument, that *the 2nd Christoffel identity, viz. (8.28), entails (8.36).* However, according to (8.37), the rel. (8.36) is equivalent to Ricci's identity, as the latter is given, equivalently, by (8.18), which thus terminates the proof of the lemma. (The last argument, in conjunction with our considerations in the foregoing, leading to (8.28), prove also, incidentally, that, within the context e.g. of Definition 8.1, the rels. *(8.28) and (8.18) are* still *equivalent*). ∎

We proceed now to give further *equivalent expressions of the Ricci identity;* in particular, we wish to have (8.18), in terms of the *local $\mathcal{A}$-connection matrix-valued forms* of a given *Riemannian $\mathcal{A}$-connection,* as before. As we shall see these alternatives of (8.18) will be of much use right in the sequel of our discussion:

Thus, assuming again that we have the *context of Lemma 8.1,* we first recall, for convenience, that, by virtue of the given *Christoffel sections* (8.28), one can further define the following (local) sections of $\Omega = \mathcal{E}^*$ (see (8.30)) over U, by the relations;

(8.38) $$\omega_{ij}(e_k) := \Gamma_{ij}^k, \quad 1 \leqq i, j, k \leqq n.$$

That is, *equivalently,* one has

(8.39) $$\omega_{ij} = \sum_{k=1}^{n} \Gamma_{ij}^k e_k^* \in \Omega(U), \quad 1 \leqq i, j \leqq n,$$

such that

$$(e^U)^* \equiv \{U; (e_i^*)_{1 \leqq i \leqq n}\} \tag{8.40}$$

stands for the *dual local gauge of* e^U, hence, a *local gauge of* (the *vector sheaf*) $\mathcal{E}^* = \Omega$ over the *open* $U \subseteq X$. (In this context, cf. (5.75) and (5.76) in the foregoing).

Accordingly, based now on (8.34) and (8.38), *one obtains;*

$$\omega_{ij} = \omega_{ji} \in \Omega(U), \quad 1 \leqq i, j \leqq n, \tag{8.41}$$

or in *matrix notation,*

$$\omega = {}^t\omega \in M_n(\Omega(U)) = M_n(\Omega)(U). \tag{8.42}$$

On the other hand, as already noticed (cf. (8.37)), *the Ricci identity* (cf. (8.16)), as (equivalently) given by (8.18), *is* further *equivalent with* (8.36). Thus, applying now (8.38) to (8.36), one obtains (cf. also (8.32));

$$\begin{aligned} \partial(\rho_{ij})(e_k) &= \textstyle\sum_l (\Gamma^l_{ki}\rho_{lj} + \Gamma^l_{kj}\rho_{li}) = \sum_l (\omega_{ki}(e_l)\rho_{lj} + \omega_{kj}(e_l)\rho_{li}) \\ &= \textstyle\sum_l (\rho_{il}\omega_{kj} + \rho_{jl}\omega_{ki})(e_l), \end{aligned} \tag{8.43}$$

with $1 \leqq k \leqq n$, so that we have

$$\partial(\rho_{ij}) = \sum_{k=1}^{n} (\rho_{ik}\omega_{kj} + \rho_{jk}\omega_{ki}), \quad 1 \leqq i, j \leqq n. \tag{8.44}$$

Therefore, by employing *matrix notation,* and in view also of (8.32) and (8.42), one has;

$$\partial(\rho) = \rho \cdot \omega + {}^t\rho \cdot {}^t\omega = \rho\omega + \rho\omega = (\rho + \rho) \cdot \omega = 2\rho\omega, \tag{8.45}$$

so that *one* finally *obtains* (cf. also (8.21))

$$\omega = \frac{1}{2}\rho^{-1}\partial(\rho) \equiv \frac{1}{2}\tilde{\partial}(\rho). \tag{8.46}$$

By extending the classical terminology, we call (8.46) the *Levi-Civita identity.*

In this context, it is now easy to see that the previous argument, that led to (8.46) *can,* in fact, *be reversed,* under the further assumption, of course, that (8.41), or *equivalently, (8.42), holds true.* Therefore, one thus concludes that;

(8.47) supposing that (8.41), or equivalently (8.42), holds true, *the Ricci and the Levi-Civita identities,* (8.18) and (8.46), respectively, *are,* in effect, *equivalent.* Indeed, under the said assumption (but, see also (8.48) below), *the rels. (8.18), (8.44) and (8.46) are equivalent.*

The above conclusion will be applied just below, however, in a strengthened form. Thus, we still remark that:

(8.48) as already noted by (8.47), *the Levi-Civita identity (8.46) implies the Ricci identity,* under the assumption, however, of (8.41). Now, the point here is that, as easily seen, *this* hypothesis about (8.41), *is* actually *redundant,* due to (8.46) itself.

As we said, it is in the form of (8.48) that we are going to apply the preceding. Thus, although a more general case will be treated right in the next Section 9, we do give here an application of the above discussion, pertaining to the *Fundamental Lemma* of Riemannian geometry, as this classical result appears within the present abstract setting. So one has the next.

Theorem 8.1. *(Fundamental Lemma of Riemannian* $\mathcal{A}$*-connections). Suppose we are given an ordered algebraized space* $(X, \mathcal{A})$*, with* X *a paracompact (Hausdorff) space, such that* $(\mathcal{A}, \overline{\rho})$ *is a strictly positive fine* $\mathcal{A}$*-module, with* $\overline{\rho}$ *a Riemannian* $\mathcal{A}$*-metric. Furthermore, assume that* X *carries a differential triad* $(\mathcal{A}, \partial, \Omega)$*, while we still adopt that* Ω *is a vector sheaf on* X*, with* $\operatorname{rk}\Omega = n \in \mathbf{N}$*. Then, the dual (vector sheaf) of* Ω*, say,*

$$\mathcal{E} \equiv \Omega^* := \mathcal{H}om_{\mathcal{A}}(\Omega, \mathcal{A}), \tag{8.49}$$

is (canonically) endowed with a unique Riemannian $\mathcal{A}$*-connection.*

Proof. Based on our hypothesis, we first remark (cf. Chapt. IV; Theorem 8.3) that the *vector sheaf* $\mathcal{E} = \Omega^*$ (see (8.49)) can be endowed with a *Riemannian* $\mathcal{A}$*-metric* ρ, so that the pair

$$(\mathcal{E}, \rho) \tag{8.50}$$

constitutes a *Riemannian vector sheaf* on X, such that

$$\mathcal{E}^* = \Omega \tag{8.51}$$

(cf. (8.49) above and Chapt. IV; Theorem 5.1), that is, (8.30) *is* thus *fulfilled.*

Now, suppose that

$$\mathcal{U} = (U_\alpha)_{\alpha \in I} \tag{8.52}$$

is a *local frame* of our vector sheaf $\mathcal{E}$, as above; therefore, for each *local gauge* $e^{(\alpha)}$ *of* $\mathcal{E}$ *over* $U_\alpha \in \mathcal{U}$, $\alpha \in I$ (see, for instance, (2.31), (2.32)), one considers the respective *Riemannian* $\mathcal{A}|_{U_\alpha}$*-metric of* $\mathcal{E}|_{U_\alpha}$, *restriction of* ρ in (8.50) *to* U_α,

$$\rho_\alpha := \rho|_{U_\alpha} \equiv (\rho_{ij}^{(\alpha)}) \in \mathrm{GL}(n, \mathcal{A}(U_\alpha)) = \mathcal{GL}(n, \mathcal{A})(U_\alpha) \tag{8.53}$$

(cf. (8.6), (8.20) and (8.21)), hence, the corresponding *Christoffel sections in* $\mathcal{A}(U_\alpha)$, as they are given now, *for each index* $\alpha \in I$, by an analogous formula to (8.28). Consequently, in view of Lemma 8.1, one infers that *each one of* the vector sheaves $\mathcal{E}|_{U_\alpha}$ over U_α, $\alpha \in I$, *is endowed with a uniquely defined Riemannian* $\mathcal{A}|_{U_\alpha}$*-connection,* say $\overline{D}_\alpha$, whose *local* $\mathcal{A}|_{U_\alpha}$*-connection matrix* is given (cf. (8.46)) by the relation

$$\overline{\omega}^{(\alpha)} = \frac{1}{2}\tilde{\partial}(\rho_\alpha), \quad \alpha \in I. \tag{8.54}$$

On the other hand, by hypothesis, $\mathcal{A}$ is a *fine* $\mathbf{C}$*-algebra sheaf* over the *paracompact* (Hausdorff) *space* X, so that the given $\mathcal{A}$*-module* $\mathcal{E} = \Omega^*$ *is a fine sheaf* on X too (cf., for example, Chapt. III; (8.56)). Therefore, the standard *Levi-Civita 0-cochain,* which is (canonically) associated with the given local frame $\mathcal{U}$ of our vector sheaf $\mathcal{E}$, as before (cf. (8.52) and Chapt. VI; (8.32.1)),

$$(D_\alpha) \in C^0(\mathcal{U}, \mathcal{H}om_{\mathbf{C}}(\mathcal{E}, \Omega(\mathcal{E}))), \tag{8.55}$$

yields now an $\mathcal{A}$*-connection* D *of* $\mathcal{E}$ (see Chapt. VI; Lemma 16.1. In this context, it is certainly clear that (8.52) may be taken to be a *locally finite local frame of* $\mathcal{E}$, in view of our hypothesis for X; see e.g. Chapt. VI; (16.16)). Thus (see also Chapt. VI; Theorem 7.1), one can define now, *for every* $\alpha \in I$, a (local) $\mathcal{A}|_{U_\alpha}$*-connection of* $\mathcal{E}|_{U_\alpha}$, by the relation,

$$\widetilde{D}_\alpha := D_\alpha + \overline{\omega}^{(\alpha)}; \tag{8.56}$$

that is, *equivalently,* one sets (cf. also (8.54))

$$\widetilde{D} := D + \overline{\omega} = D + \frac{1}{2}\tilde{\partial}(\rho), \tag{8.57}$$

such that *one actually has*

$$\overline{\omega} \equiv \frac{1}{2}\tilde{\partial}(\rho) \in \Omega(\mathcal{E}nd\,\mathcal{E})(X), \tag{8.58}$$

which thus, in view of the cited result of Chapter VI, yields that $\widetilde{D}$, as given by (8.57), is a well-defined *$\mathcal{A}$-connection of $\mathcal{E}$*. Now, concerning our last claim for $\overline{\omega}$, one obtains (see also (8.6) above, together with Chapt. III; (4.55), (4.28) and Chapt. VI; (9.4));

$$\begin{aligned}\tilde{\partial}(\rho) &\equiv \tilde{\partial}((\rho_{ij})) := (\tilde{\partial}(\rho_{ij})) \in M_n(\Omega(X)) = M_n(\Omega)(X)\\ &= Z^0(\mathcal{U}, M_n(\Omega)) \subseteq \textstyle\prod_\alpha M_n(\Omega)(U_\alpha) = \prod_\alpha \Omega(\mathcal{E}nd\,\mathcal{E})(U_\alpha),\end{aligned} \tag{8.59}$$

so that, in particular, *one has*

$$\tilde{\partial}(\rho) \in Z^0(\mathcal{U}, \Omega(\mathcal{E}nd\,\mathcal{E})) = \Omega(\mathcal{E}nd\,\mathcal{E})(X), \tag{8.60}$$

which thus justifies (8.58). ∎ Now, we prove further that $\widetilde{D}$, as defined by (8.57), *is,* in fact, *a Riemannian $\mathcal{A}$-connection of $\mathcal{E}$*:

Indeed, as already known, *the Levi-Civita $\mathcal{A}$-connection D*, as defined above, *is integrable* (see Chapt. VI; (11.24)); hence, one still obtains that (cf. (6.28));

$$\omega^{(\alpha)} = 0, \quad \alpha \in I. \tag{8.61}$$

Accordingly, in view of (8.56) and (8.54), one has

$$\tilde{\omega}^{(\alpha)} = \overline{\omega}^{(\alpha)} = \frac{1}{2}\tilde{\partial}(\rho_\alpha), \quad \alpha \in I, \tag{8.62}$$

the first item of (8.62) standing, of course, for the *local $\mathcal{A}$-connection matrix of the $\mathcal{A}$-connection $\widetilde{D}$*, as the latter is given by (8.57). However, as already remarked (cf. (8.48) above), the previous rel. (8.62) *provides the symmetry of $\widetilde{D}$, as well as, the Ricci identity* (cf. (8.47)). Yet, by virtue of (8.51), the corresponding *Christoffel sections of $\widetilde{D}$* are given (locally) by the analogous relations to (8.28) above (see also (8.38)), yielding thus the *uniqueness of $\widetilde{D}$* (cf. also Lemma 8.1), which now completes the proof of the theorem. ∎

Scholium 8.1.– It is profitable, in anticipation of a later application, to notice at this point the use, we made above, of the *extra hypothesis* (8.49) or, *equivalently,* (8.51) (see Scholium 5.1 in the foregoing); so the latter condition has been employed above, in order *to get* the form (8.28) of the Christoffel sections, hence (Lemma 8.1), *the uniqueness of the* (Riemannian) *$\mathcal{A}$-connection $\widetilde{D}$*, as in (8.57). Indeed, as we shall see in the next Section 9, the same proof, as above, works for the general case too, viz. if *no extra assumption, like (8.51), is made* (cf. Theorem 9.1 in the sequel). However, this *at the cost of the uniqueness of* $\widetilde{D}$, as before, while we also have to, appropriately, reformulate Definition 8.1 (see thus Definition 9.1 below).

Now, before we end this Section, we *still* consider *another equivalent formulation of the Levi-Civita identity,* viz. of (8.46):

Thus, suppose that we are given, as usual, a *differential triad*

$$(\mathcal{A}, \partial, \Omega) \tag{8.63}$$

on a *topological space* X, in such a manner that

$$(X, \mathcal{A}) \tag{8.64}$$

is, in particular, an *ordered algebraized space* (Chapt. IV; Definition 8.1). Furthermore, assume that

$$(\mathcal{E}, \rho) \tag{8.65}$$

is a *Riemannian vector sheaf* on X (see Chapt. IV; Definition 8.3). Thus (ibid.; (8.19), (8.17)), the given *Riemannian $\mathcal{A}$-metric* ρ on $\mathcal{E}$, as in (8.65), induces, by definition, an *$\mathcal{A}$-isomorphism* of $\mathcal{E}$ onto its *dual* (vector sheaf) $\mathcal{E}^* := \mathcal{H}om_{\mathcal{A}}(\mathcal{E}, \mathcal{A})$,

$$\mathcal{E} \underset{\tilde{\rho}}{\cong} \mathcal{E}^*, \tag{8.66}$$

such that one sets;

$$\tilde{\rho}(s)(t) := \rho(s, t), \tag{8.67}$$

for any (local) sections s, t *in* $\mathcal{E}(U)$, and *open* $U \subseteq X$. So our next objective is to show now the following.

Theorem 8.2.. *Suppose we are given an ordered algebraized space* $(X, \mathcal{A})$, *endowed with a differential triad* $(\mathcal{A}, \partial, \Omega)$. *Moreover, let*

$$(\mathcal{E}, \rho) \tag{8.68}$$

be a Riemannian vector sheaf on X, *with* $\operatorname{rk} \mathcal{E} = n \in \mathbf{N}$, *admitting also an $\mathcal{A}$-connection* D. *Then, D satisfies the Levi-Civita identity* (cf. (8.46)) *if, and only if, it is gauge equivalent to* D^*, *the dual $\mathcal{A}$-connection of D* (cf. Chapt. VI; (5.31)), *via the $\mathcal{A}$-isomorphism*

$$\tilde{\rho} : \mathcal{E} \longrightarrow \mathcal{E}^* \equiv \mathcal{H}om_{\mathcal{A}}(\mathcal{E}, \mathcal{A}), \tag{8.69}$$

provided by ρ (see (8.66), (8.67)).

Equivalently, (8.46) holds true, if, and only if, $\tilde{\rho}$ *is horizontal, with respect to the* $\mathcal{A}$*-connection of the vector sheaf* $\mathcal{H}om_{\mathcal{A}}(\mathcal{E}, \mathcal{E}^*) = (\mathcal{E} \otimes_{\mathcal{A}} \mathcal{E})^*$ (cf. Chapt. IV; (6.16), along with Chapt. VI; (5.13) and (5.25)); *viz. if, and only if, one has*

$$D_{\mathcal{H}om_{\mathcal{A}}(\mathcal{E},\mathcal{E}^*)}(\tilde{\rho}) = 0. \tag{8.70}$$

Proof. First, we remark that *(8.70) is,* indeed, by the vey definitions, *tantamount to* the $\mathcal{A}$-connection D *being gauge equivalent with* D^**, via the* $\mathcal{A}$*-isomorphism* $\tilde{\rho}$; viz. to the relation

$$D^* = (\tilde{\rho} \otimes 1) \circ D \circ \tilde{\rho}^{-1}, \tag{8.71}$$

or even to the *commutativity* of the following *diagram,*

(8.72) 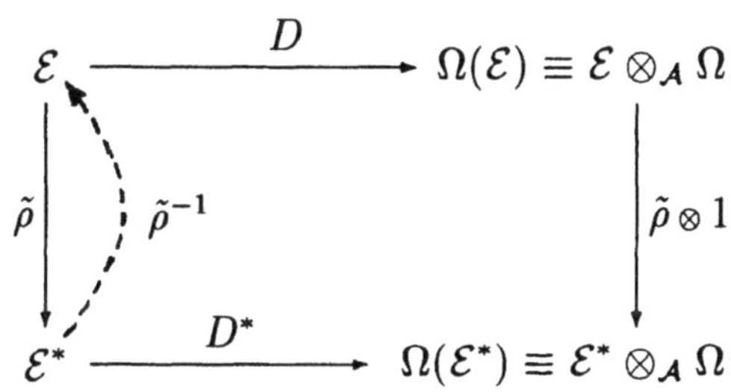

(see thus Chapt. VI; (5.41), (5.43) and (5.13), along with (8.25)).

Thus, *assume* first *that (8.70), or,* as already explained, *equivalently, (8.71) holds true;* yet, *suppose,* namely, *that we have*

$$D^* \circ \tilde{\rho} = (\tilde{\rho} \otimes 1) \circ D. \tag{8.73}$$

Then, we assert that *(8.46) is in force:* In this context, we recall first that ω (cf. also (8.58)) is determined, *locally,* by a relation, like (1.11). Hence, assuming (8.73), we are going to verify (8.46), locally, in terms of a *local gauge of* $\mathcal{E}$, say, e^U (see e.g. (8.8)), and the corresponding *(local) section basis of* $\mathcal{E}(U)$, as in (8.13). So, applying now (8.73) to the same basis, one obtains;

$$D^*(\tilde{\rho}(e_i))(e_j) = (\tilde{\rho} \otimes 1)(D(e_i))(e_j), \quad 1 \leqq i, j \leqq n = \operatorname{rk} \mathcal{E}. \tag{8.74}$$

Accordingly, one has from the 1st member of the last relation (cf. Chapt. VI; (5.33), together with (8.67), (8.20) and (1.11) in the preceding);

$$
\begin{aligned}
(8.75)\qquad D^*(\tilde{\rho}(e_i))(e_j) &= \partial(\tilde{\rho}(e_i)(e_j)) - (\tilde{\rho}(e_i)\otimes 1)(D(e_j)) \\
&= \partial(\rho(e_i,e_j)) - (\tilde{\rho}(e_i)\otimes 1)(\textstyle\sum_{k=1}^{n} e_k\otimes\omega_{kj}) \\
&= \partial(\rho_{ij}) - \textstyle\sum_k \tilde{\rho}(e_i)(e_k)\otimes\omega_{kj} = \partial(\rho_{ij}) - \sum_k \rho(e_i,e_k)\cdot\omega_{kj} \\
&= \partial(\rho_{ij}) - \textstyle\sum_k \rho_{ik}\omega_{kj}.
\end{aligned}
$$

On the other hand, the 2nd member of (8.74) yields (cf. also Chapt. IV; (6.5)),

$$
\begin{aligned}
(8.76)\qquad (\tilde{\rho}\otimes 1)(D(e_i))(e_j) &= (\tilde{\rho}\otimes 1)(\textstyle\sum_k e_k\otimes\omega_{ki})(e_j) \\
&= (\textstyle\sum_k \tilde{\rho}(e_k)\otimes\omega_{ki})(e_j) = \sum_k \tilde{\rho}(e_k)(e_j)\otimes\omega_{ki} \\
&= \textstyle\sum_k \rho(e_k,e_j)\cdot\omega_{ki} = \sum_k \rho_{kj}\omega_{ki}.
\end{aligned}
$$

Consequently, in view of (8.74), (8.75) and (8.76), one finally obtains;

$$
(8.77)\qquad \partial(\rho_{ij}) - \sum_k \rho_{ik}\omega_{kj} = \sum_k \rho_{kj}\omega_{ki} = \sum_k \rho_{jk}\omega_{ki}
$$

(the last equality in (8.77) is due, of course, to (8.32). Now, (8.77) is written in *matrix form;*

$$
(8.78)\qquad \partial(\rho) - \rho\omega = \rho\omega.
$$

Therefore, one has

$$
(8.79)\qquad \partial(\rho) = \rho\omega + \rho\omega = (\rho+\rho)\cdot\omega = 2\rho\omega,
$$

so that one finally obtains the *Levi-Civita identity* (cf. (8.46));

$$
(8.80)\qquad \omega = \frac{1}{2}\tilde{\partial}(\rho) \in M_n(\Omega(U)).
$$

Of course, we have arrived at (8.80), by employing a *local argument,* with respect, namely, to a *local gauge* e^U of $\mathcal{E}$, as e.g. in (8.8). However, by further considering a *local frame of* $\mathcal{E}$ (see, for instance, (8.52)), and by still following a similar procedure to that one for (8.58) (see also (8.59), (8.60)), one can think of (8.80) *globally;* viz. one has

$$
(8.81)\qquad \omega = \frac{1}{2}\tilde{\partial}(\rho) \in \Omega(\mathcal{E}nd\,\mathcal{E})(X).
$$

On the other hand, it is quite obvious that *the above argument can be reversed;* thus, (8.80) entails (8.87), so that, in turn, (8.74) or, equivalently, (8.73), hence, (8.70) too, which thus terminates the proof of the theorem. ∎

Now, as already remarked (see (8.48)), *the Levi-Civita identity (8.46) implies the symmetry of* ω, hence, in the case that the Christoffel sections can be defined (cf. (8.51) and (8.22)), *cond. (8.24) too.* Accordingly, taking now (8.47) into account, one concludes, on the basis of the previous Theorem 8.2, that:

(8.82) Given an $\mathcal{A}$-*connection* D of a *Riemannian vector sheaf* $(\mathcal{E}, \rho)$, as before, *the following* assertions *are equivalent:*

1) *The Levi-Civita identity is valid* (see (8.80) or, *equivalently,* (8.81)).

2) D *and its dual* $\mathcal{A}$-*connection* D^* *are gauge equivalent, through* $\tilde{\rho}$ (cf. (8.69)).

3) The *cond. (8.70) holds true; viz.* $\tilde{\rho}$, as in 2), *is horizontal, with respect to the* $\mathcal{A}$-*connection of* (the vector sheaf) $\mathcal{H}om_{\mathcal{A}}(\mathcal{E}, \mathcal{E}^*) = \mathcal{E}^* \otimes_{\mathcal{A}} \mathcal{E}^* = (\mathcal{E} \otimes_{\mathcal{A}} \mathcal{E})^*$ (cf. also Chapt. IV; (6.16)).

4) The *cond. (8.42) and the Ricci identity,* as in (8.44), *are in force.* ∎

9. Riemannian vector sheaves. Fundamental lemma (general case)

As the heading indicates, we give in this Section the analogue in our case of the *"Fundamental lemma of Riemannian geometry"* of the standard theory. So, under suitable further restrictions on X and $\mathcal{A}$ of a given pair $(X, \mathcal{A})$, as in the preceding Section 8, one succeeds, by analogy with the classical situation, to glue together local informations about Riemannian $\mathcal{A}$-connections, to get such a one globally. However, before we embark on the statement of our main result, we have first to appropriately modify, as already mentioned, our previous Definition 8.1, to fit the general case, that is to be considered in the sequel. In this context, we further remark that *the two definitions,* thus resulted, *are,* in effect, *equivalent* in the special case one has *cond. (8.7) fulfilled,* as follows from (8.47) and (8.48), in conjunction also with Theorem 8.2 above. Thus, we now have the next.

Definition 9.1. Suppose we have an ordered algebraized space $(X, \mathcal{A})$, such that $(\mathcal{A}, \partial, \Omega)$ is a differential triad on X. Moreover, let $(\mathcal{E}, \rho)$ be a Riemannian vector sheaf on X. Then, a given *$\mathcal{A}$-connection D of $\mathcal{E}$* is said to be *compatible with the Riemannian $\mathcal{A}$-metric ρ*, whenever one has;

$$(9.1) \qquad D_{\mathcal{H}om_{\mathcal{A}}(\mathcal{E}, \mathcal{E}^*)}(\tilde{\rho}) = 0$$

(see (8.70) for the notation applied above).

Now, we continue by formulating straight away our main object of the succeeding discussion. For details, concerning the terminology applied, we still refer to Chapt. IV; Section 8. Thus, one has the following basic result.

Theorem 9.1. (Fundamental lemma of Riemannian vector sheaves). *Suppose we are given an ordered algebraized space $(X, \mathcal{A})$, with X (Hausdorff) paracompact and $\mathcal{A}$ a strictly positive fine $\mathcal{A}$-module, endowed with a Riemannian $\mathcal{A}$-metric $\overline{\rho}$. Moreover, assume that X is equipped with a differential triad $(\mathcal{A}, \partial, \Omega)$. Then, every vector sheaf $\mathcal{E}$ on X becomes a Riemannian vector sheaf $(\mathcal{E}, \rho)$, admitting further an $\mathcal{A}$-connection compatible with the Riemannian $\mathcal{A}$-metric ρ.*

Proof. First, based on our hypothesis for X and $\mathcal{A}$, we know that *the* given *vector sheaf $\mathcal{E}$ admits a Riemannian $\mathcal{A}$-metric ρ* (see Chapt. IV; Theorem 8.3), so that

$$(9.2) \qquad (\mathcal{E}, \rho)$$

is a *Riemannian vector sheaf* on X (ibid.; Definition 8.3).

Now, consider a *local frame of $\mathcal{E}$*,

$$(9.3) \qquad \mathcal{U} = (U_\alpha)_{\alpha \in I},$$

which further might be taken to be (as an open covering of X) *locally finite* (see e.g. Chapt. IV; (8.42)). Moreover, since by hypothesis $\mathcal{A}$ is a *fine* **C**-*algebra sheaf* over the *paracompact* (Hausdorff) space X, *the* given *$\mathcal{A}$-module* (in fact, vector sheaf) *$\mathcal{E}$ is a fine sheaf* too (cf. Chapt. III; (8.56)), so that the above *locally finite local frame of* $\mathcal{E}$ has a *subordinate partition of unity* for $\mathcal{E}$ (see Chapt. III; Definition 8.1). Therefore, by further considering the *Levi-Civita 0-cochain,* which is associated eith $\mathcal{U}$,

$$(D_\alpha) \in C^0(\mathcal{U}, \mathcal{H}om_{\mathbf{C}}(\mathcal{E}, \Omega(\mathcal{E}))) \tag{9.4}$$

(cf. Chapt. VI; (8.32.1)), one gets an *$\mathcal{A}$-connection D of $\mathcal{E}$*, which is given by the relation;

$$D := \sum_\alpha (\phi_\alpha \otimes 1) \circ D_\alpha \tag{9.5}$$

with

$$(\phi_\alpha)_{\alpha \in I} \subseteq \operatorname{End} \mathcal{E} \tag{9.6}$$

the *partition of unity* for $\mathcal{E}$, subordinate to $\mathcal{U}$, as above (see Chapt. VI; (16.7)). In this context, it is still useful to note for use right below (anyway, we have done it again in the preceding, in the proof of Theorem 8.1) that the *Levi-Civita $\mathcal{A}$-connection D*, as given by (9.5), is integrable, so that it has *trivial local $\mathcal{A}$-connection matrices,* still associated with $\mathcal{U}$; viz. we have

$$\omega^{(\alpha)} \equiv (\omega_{ij}^{(\alpha)}) = 0 \in M_n(\Omega(U_\alpha)), \quad \alpha \in I \tag{9.7}$$

(see Chapt. VI; (11.22), along with (6.28) above). So consider now the following *$\mathcal{A}$-connection of $\mathcal{E}$*;

$$\widetilde{D} := D + \frac{1}{2}\tilde{\partial}(\rho). \tag{9.8}$$

Now, as already remarked by (8.58), one has

$$\frac{1}{2}\tilde{\partial}(\rho) \in \Omega(\mathcal{E}nd\,\mathcal{E})(X), \tag{9.9}$$

so that the above *rel. (9.8)* certainly *provides an $\mathcal{A}$-connection of $\mathcal{E}$* (cf. Chapt. VI; Theorem 7.1). On the other hand, as already pointed out, by the proof of Theorem 8.1 (see thus (8.57)), the previous *$\mathcal{A}$-connection (9.8) satisfies prop. 4) of (8.82).* Accordingly, still by (8.82), *the $\mathcal{A}$-connection at issue fulfills (8.70), as well,* which thus terminates the proof (see also Definition 9.1). ∎

Remarks 9.1.– i) Looking more closely at the previous proof, one ascertains that what we really need is our *vector sheaf* $\mathcal{E}$ to satisfy the following condition:

(9.10) $\mathcal{E}$ has a *locally finite local frame* $\mathcal{U}$, such that there exists a *strictly positive partition of unity subordinate to* $\mathcal{U}$. We then call $\mathcal{E}$ a *numerable vector sheaf* on X.

So the form, which the above Theorem 9.1 can take, is actually the following one:

(9.11) We have an *ordered algebraized space* $(X, \mathcal{A})$, with X any *topological space,* which is further endowed with a *differential triad* $(\mathcal{A}, \partial, \Omega)$. Then, *every numerable Riemannian vector sheaf* $(\mathcal{E}, \rho)$ on X, *admits an $\mathcal{A}$-connection, compatible with the* given *Riemannian $\mathcal{A}$-metric ρ.* ∎

Furthermore, we can suppose *instead,* that $(\mathcal{A}, \overline{\rho})$ is a *Riemannian $\mathcal{A}$-module* and $\mathcal{E}$ just a *numerable vector sheaf* on X (cf. (9.10)). Then, we have the *same conclusion, as in (9.11).* ∎ Thus, guided by the preceding, we might also consider, in the spirit of (9.10) above, an *analogous strengthening of* various similar results, as, for instance, of *Chapt. IV; Theorem 8.3.*

ii) Concerning the same notion above, as given by (9.10), we still note the following:

(9.12) Suppose that we are given a *continuous map* $f : X \longrightarrow Y$ and a *numerable vector sheaf* $\mathcal{E}$ *on* Y. Then, the *pull-back of* $\mathcal{E}$, *via* f, *over* X, $f^*(\mathcal{E})$ (see Chapt. I; (14.15)), *is a numerable vector sheaf on* X, as well.

First, if $(Y, \mathcal{B})$ is an *ordered algebraized space* over Y, *its pull-back, via* f, $(X, f^*(\mathcal{B}) \equiv \mathcal{A})$, *is an ordered algebraized space* on X, as one easily verifies, based on the very definitions (cf. Chapt. I; (14.23) and Chapt. IV; Definition 8.1). Thus, if now $\mathcal{V} = (V_\alpha)_{\alpha \in I}$ is a *locally finite local frame of* $\mathcal{E}$, then the *open covering of* $X, \mathcal{U} \equiv f^{-1}(\mathcal{V}) := (f^{-1}(V_\alpha))_{\alpha \in I}$ is a *similar one for* $f^*(\mathcal{E})$. Yet, if $(\phi_\alpha) \subseteq \operatorname{End} \mathcal{E}$ is a *strictly positive partition of unity for* $\mathcal{E}$, *subordinate to* $\mathcal{V}$, then

$$(f^*(\phi_\alpha))_{\alpha \in I} \subseteq \operatorname{End} f^*(\mathcal{E}) \tag{9.13}$$

is an *analogous family for* $f^*(\mathcal{E})$, *subordinate to* $\mathcal{U}$, as above (see Chapt. I; (14.43) and (14.26), in conjunction with Chapt. IV; Definition 8.4).

The previous considerations suggest now, of course, the case that, by suitably choosing Y and $\mathcal{E}$, as before, one can look at a *numerable vector sheaf*, like $f^*(\mathcal{E})$, as considered here, however, *over a not necessarily paracompact* (Hausdorff) *space* X. (For the classical case of *vector bundles*, see e.g. D. Husemoller [1: p.48, Proposition 9.3]).

10. Hermitian $\mathcal{A}$-connections

We have already considered in the foregoing *hermitian* $\mathcal{A}$*-modules* over a topological space X, base space of a given *involutive* $\mathbf{C}$*-algebraized space*

$$(X, \mathcal{A}, -) \tag{10.1}$$

(see Chapt. IV; (9.5)); here we still assume, tacitly, that

$$(X, \mathcal{A}) \tag{10.2}$$

is, in particular, an *ordered algebraized space* (ibid.; Definition 8.1), when one has to consider *"positivity"* questions. Furthermore, suppose that X is also endowed with a *differential triad*

$$(\mathcal{A}, \partial, \Omega), \tag{10.3}$$

and let

$$(\mathcal{E}, \rho) \tag{10.4}$$

be a *hermitian* $\mathcal{A}$*-module* on X (cf. Chapt. IV; Definition 9.1). So one can first obtain an *extension of* ρ (for simplicity, we retain the previous symbol for the extended ρ too),

$$\rho : \Omega(\mathcal{E}) \oplus \mathcal{E} \longrightarrow \Omega, \tag{10.5}$$

such that, by taking the *decomposable tensors* of $\Omega(\mathcal{E})_{\mathbf{C}} \equiv \mathcal{E} \otimes \Omega$, one sets;

$$\rho(s \otimes t, s') := \rho(s, s') \cdot t, \tag{10.6}$$

for any (local) sections s, s' *in* $\mathcal{E}(U)$ *and* $t \in \Omega(U)$, with U *open* in X.

Now, since the given *hermitian* $\mathcal{A}$*-metric* ρ on $\mathcal{E}$, as in (10.4), is $\mathcal{A}$*-linear* in the *first variable* (see Chapt. IV; (9.7)), the previous relation (10.6) can be *extended by $\mathcal{A}$-linearity to* all of $\Omega(\mathcal{E}) \equiv \mathcal{E} \otimes_{\mathcal{A}} \Omega$, so that (10.5) *is well-defined.* Yet, based on (10.6) and on Chapt. IV; (9.8) *one* still *sets;*

$$
\begin{aligned}
\rho(s, s' \otimes t) &:= \overline{\rho(s' \otimes t, s)} = \overline{\rho(s', s) \cdot t} = \overline{\rho(s', s)} \cdot \bar{t} \\
&= \rho(s, s') \cdot \bar{t},
\end{aligned}
\tag{10.7}
$$

for any $s, s' \in \mathcal{E}(U)$ and $t \in \Omega(U)$. Here the "–" over $t \in \Omega(U)$ refers, *by definition,* to the *(coordinates) components of* t, relative to a given *local gauge of* Ω over the *open* $U \subseteq X$, *the former being* thus (local) sections of $\mathcal{A}$ over U, viz. *elements of* $\mathcal{A}(U)$ (cf., for instance, (1.6.1)). So we can set now the following.

Definition 10.1. Suppose we have the framework of (10.1), (10.3) and let $(\mathcal{E}, \rho)$ be a hermitian $\mathcal{A}$-module on X. Then, a given *$\mathcal{A}$-connection D of* $\mathcal{E}$ is said to be *compatible with the hermitian $\mathcal{A}$-metric,* or just a *hermitian $\mathcal{A}$-connection,* whenever the following relation holds true (cf. also (10.5), (10.6), (10.7));

$$\partial(\rho(s, t)) = \rho(Ds, t) + \rho(s, Dt), \tag{10.8}$$

for any $s, t \in \mathcal{E}(U)$, with U *open* in X.

Thus, suppose now that

$$e^U \equiv \{U; (e_1, \ldots, e_n)\} \tag{10.9}$$

is a *local gauge,* of *rank* $n \in \mathbf{N}$, over the *open* set $U \subseteq X$ of the given *$\mathcal{A}$-module* $\mathcal{E}$, with

$$(e_i)_{1 \leqq i \leqq n} \subseteq \mathcal{E}(U) \tag{10.10}$$

the corresponding *section-basis of* $\mathcal{E}(U)$ (see e.g. (1.2) – (1.5) in the preceding).

So, by considering again a *hermitian $\mathcal{A}$-connection* D of $\mathcal{E}$ and *translating* further (10.8), *in terms of the* above *section-basis* of $\mathcal{E}(U)$, one has

$$\partial(\rho(e_i, e_j)) = \rho(De_i, e_j) + \rho(e_i, De_j), \quad 1 \leqq i, j \leqq n. \tag{10.11}$$

Furthermore, setting, as before,

$$\rho_{ij} := \rho(e_i, e_j) \in \mathcal{A}(U), \quad 1 \leqq i, j \leqq n, \tag{10.12}$$

and using also the *local $\mathcal{A}$-connection matrix of D*, associated with the given local gauge e^U,

$$\omega \equiv (\omega_{ij}) \in M_n(\Omega(U)) = M_n(\Omega)(U), \tag{10.13}$$

(see (1.13)), one has from (10.11) (cf. also (1.11));

$$\begin{aligned} \partial(\rho_{ij}) &= \rho(\textstyle\sum_k e_k \otimes \omega_{ki}, e_j) + \rho(e_i, \textstyle\sum_k e_k \otimes \omega_{kj}) \\ &= \textstyle\sum_k \rho(e_k \otimes \omega_{ki}, e_j) + \textstyle\sum_k \rho(e_i, e_k \otimes \omega_{kj}). \end{aligned} \tag{10.14}$$

Thus, by further applying (10.6) and (10.7), along with Chapt. IV; (9.8), one obtains, by (10.14),

$$\partial(\rho_{ij}) = \sum_k \rho_{kj} \cdot \omega_{ki} + \sum_k \rho_{ik} \cdot \overline{\omega}_{kj}; \tag{10.15}$$

hence, in view again of the *hermitian symmetry of* ρ (see Chapt. IV; (9.8)), one gets, in *matrix form*

$$\partial(\rho) = \rho \cdot \omega + \rho \cdot {}^t\overline{\omega} = \rho(\omega + {}^t\overline{\omega}), \tag{10.16}$$

so that *one* finally *obtains*

$$\omega + {}^t\overline{\omega} = \tilde{\partial}(\rho). \tag{10.17}$$

So we arrived at a *necessary condition for being the* given $\mathcal{A}$*-connection* D *of* $\mathcal{E}$ *hermitian.*

On the other hand, using the expression (1.6.1), for any $s \in \mathcal{E}(U)$, as before, and following further the classical technique (cf., for instance, R.O. Wells [1: p. 76]), one easily sees that *the above condition (10.17) is a sufficient one, as well.* Thus, we can now summarize the preceding, as follows:

(10.18) Suppose we have the *framework of (10.1), (10.3)* and let $(\mathcal{E}, \rho)$ ba a *hermitian* $\mathcal{A}$*-module* on X. Then, a given $\mathcal{A}$*-connection* D *of* $\mathcal{E}$ *is hermitian if, and only if,* one has the relation;

$$\tilde{\partial}(\rho) = \omega + {}^t\overline{\omega}, \tag{10.18.1}$$

with respect to a *local gauge* e^U *of* $\mathcal{E}$ (cf. (10.9)).

Now, another fact that will be of use here, is that;

(10.19) by imitating the *Gram-Schmidt process,* given in Chapt. IV; Section 10, in the case of a Riemannian $\mathcal{A}$-module, we can repeat the relevant argument for a *hermitian $\mathcal{A}$-module*, as well, obtaining thus *orthonormal local frames* of a given *hermitian vector sheaf.* (See also, in particular, Chapt. IV; Theorem 10.1).

So, based on the preceding and by further mimiking the classical pattern, we have in the case, under consideration, the following analogue of our previous Theorem 9.1.

Theorem 10.1. *Consider an enriched ordered and involutive algebraized space* $(X, \mathcal{A})$ (cf. also Chapt. IV; Definition 10.1), *with X a (Hausdorff) paracompact space and $\mathcal{A}$ a fine $\mathcal{A}$-module such that $(\mathcal{A}, \rho)$ is a hermitian $\mathcal{A}$-module. Moreover, suppose that X is further endowed with a differential triad $(\mathcal{A}, \partial, \Omega)$, with Ω being a vector sheaf on X, while $\mathcal{E}$ is also a given vector sheaf on X. Then, $\mathcal{E}$ admits a hermitian $\mathcal{A}$-connection. Thus, in particular, this holds true for (the vector sheaf) Ω, as well.* ∎

Concerning the standard case, see, for instance, R.O. Wells [1: p. 76, Proposition 1.11]. Finally, analogous comments to those in Remark 9.1; i) above, are here also in force, as it regards our previous theorem.

11. Cohomological classification of Maxwell fields (contn'd)

Our aim by the following discussion is to prove Theorem 18.2 of Chapter VI in the foregoing. It refers to a *cohomological characterization of line sheaves, having $\mathcal{A}$-connections,* what actually we have called (Chapt. VI; (18.11)) *Maxwell fields* on X.

So, to begin with, suppose, as usual, that we are given a *differential triad* $(\mathcal{A}, \partial, \Omega)$ on a *topological space* X. Moreover, let

(11.1) $$(\mathcal{L}, D)$$

be a pair, consisting of a *line sheaf* $\mathcal{L}$ on X, along with an *$\mathcal{A}$-connection D of $\mathcal{L}$,* that is, as already said, a *Maxwell field* on X.

Thus, we further denote by

$$\Phi^1_{\mathcal{A}}(X)^{\nabla} \tag{11.2}$$

the *set of* (isomorphism classes of) *Maxwell fields on* X (see Chapt. VI; (18.12)). Now, the latter set is, in effect, an *abelian group,* under "$\otimes_{\mathcal{A}}$" (ibid.; Theorem 18.1). So we come next, as promised, to the

Proof of Theorem 18.2 of Chapt. VI. Indeed, what we are actually going to prove, is the following *isomorphism of abelian groups;*

$$\Phi^1_{\mathcal{A}}(X)^{\nabla} = \check{\mathbf{H}}^1(X, \mathcal{A}^{\cdot} \xrightarrow{\tilde{\partial}} \Omega). \tag{11.3}$$

Now, the second member of (11.3) stands for the *1st hypercohomology group of X, with coefficients in the 2-term $\mathcal{A}$-complex* of X,

$$\mathcal{E}^{\cdot} : \mathcal{A}^{\cdot} \xrightarrow{\tilde{\partial}} \Omega \longrightarrow 0 \cdots, \tag{11.4}$$

which is defined by the *logarithmic derivation* of ∂,

$$\tilde{\partial} : \mathcal{A}^{\cdot} \xrightarrow{\tilde{\partial}} \Omega \tag{11.5}$$

(see Chapt. VI; (1.24), (1.25), for $\mathcal{E} \equiv \Omega$, along with (3.1)). Thus, we can further employ our previous argument in Chapt. III; Example 7.1; indeed, the latter example has been manufactured with the present application in mind:

So, by considering a *Maxwell field* $(\mathcal{L}, D)$ on X, as in (11.1), let

$$g \equiv (g_{\alpha\beta}) \in Z^1(\mathcal{U}, \mathcal{A}^{\cdot}) \tag{11.6}$$

be a *coordinate 1-cocycle of* $\mathcal{L}$, relative to a given *local frame* of $\mathcal{L}$,

$$\mathcal{U} = (U_\alpha)_{\alpha \in I}. \tag{11.7}$$

Furthemore, suppose that

$$\theta \equiv (\theta_\alpha) \in C^0(\mathcal{U}, \Omega) \equiv \prod_{\alpha \in I} \Omega(U_\alpha) \tag{11.8}$$

is the corresponding to $\mathcal{U}$, *0-cochain of local $\mathcal{A}$-connection "forms"* defining the given $\mathcal{A}$-connection D of $\mathcal{L}$ (cf. (2.39.1), for $n = 1$). Thus, as already known (cf. (2.59)), one obtains;

$$\delta(\theta_\alpha) = \tilde{\partial}(g_{\alpha\beta}). \tag{11.9}$$

However, according to Chapt. III; (7.55), the pair

$$z \equiv (g, \theta) \equiv ((g_{\alpha\beta}), (\theta_\alpha)), \tag{11.10}$$

as resulted by the previous data, determines a *1-dimensional (hyper)cohomology class,* as follows,

$$[z] \in \check{\mathbf{H}}^1(X, \mathcal{E}^{\cdot}). \tag{11.11}$$

Thus, we have already established a map

$$\chi : \mathbf{\Phi}^1_{\mathcal{A}}(X)^{\nabla} \longrightarrow \check{\mathbf{H}}^1(X, \mathcal{E}^{\cdot}), \tag{11.12}$$

given by (11.10) and (11.11): Indeed, suppose we have;

$$(\mathcal{L}, D) \sim (\mathcal{L}', D') \ \ in \ \ \mathbf{\Phi}^1_{\mathcal{A}}(X)^{\nabla}. \tag{11.13}$$

Thus (see Chapt. VI; (18.6) and Chapt. V; (2.24)), one concludes that; i) there exists an *$\mathcal{A}$-isomorphism*

$$h \in \mathrm{Isom}_{\mathcal{A}}(\mathcal{L}, \mathcal{L}') = Z^0(\mathcal{U}, \mathcal{I}som_{\mathcal{A}}(\mathcal{L}, \mathcal{L}')), \tag{11.14}$$

such that one has

$$\mathcal{L} \underset{h}{\sim} \mathcal{L}' \ \ in \ \ \mathbf{\Phi}^1_{\mathcal{A}}(X). \tag{11.15}$$

That is (cf. Chapt. V; Lemma 2.1), there exists a *0-cochain*

$$(s_\alpha) \in \check{C}^0(\mathcal{U}, \mathcal{A}^{\cdot}) = \prod_\alpha \mathcal{A}^{\cdot}(U_\alpha) = \prod_\alpha \mathcal{A}(U_\alpha)^{\cdot}, \tag{11.16}$$

such that *the following diagram commutes;*

$$\begin{array}{ccc} \mathcal{L}|_{U_\alpha} & \xrightarrow{\ h_\alpha\ } & \mathcal{L}'|_{U_\alpha} \\ \phi_\alpha \downarrow & & \downarrow \psi_\alpha \\ \mathcal{A}|_{U_\alpha} & \xrightarrow{\ s_\alpha\ } & \mathcal{A}|_{U_\alpha}, \end{array} \tag{11.17}$$

where one still has, *by definition,*

$$g_{\alpha\beta} = \phi_\alpha \circ \phi_\beta^{-1} \ \text{ and } \ g'_{\alpha\beta} = \psi_\alpha \circ \psi_\beta^{-1}, \tag{11.18}$$

for the *coordinate 1-cocycles of* the given line sheaves $\mathcal{L}$ *and* $\mathcal{L}'$, *respectively,* that are associated with $\mathcal{U}$, as in (11.7), the latter open covering of X, being now considered as a *common local frame of* $\mathcal{L}$ *and* $\mathcal{L}'$ (cf. also Chapt. V; (1.3.2)). Moreover, one gets, *equivalently to (11.15),* the relation;

$$g'_{\alpha\beta} = s_\alpha \cdot g_{\alpha\beta} \cdot s_\beta^{-1} = (s_\beta^{-1} s_\alpha) \cdot g_{\alpha\beta} = \delta(s_\alpha^{-1}) \cdot g_{\alpha\beta}. \tag{11.19}$$

Therefore, one has;

$$(g'_{\alpha\beta}) \cdot (g_{\alpha\beta})^{-1} \equiv g' \cdot g^{-1} = \delta(s_\alpha^{-1}) \in \delta(\check{C}^0(\mathcal{U}, \mathcal{A}^\cdot)) \equiv \operatorname{im}\delta \subseteq \check{C}^1(\mathcal{U}, \mathcal{A}^\cdot). \tag{11.20}$$

the multiplication in the first member of (11.20) being considered in the *abelian group* $Z^1(\mathcal{U}, \mathcal{A}^\cdot)$ (again Chapt. V; (2.11), for $n = 1$).

ii) *As a result of (11.13),* one further obtains, for the $\mathcal{A}$*-connections* D *and* D' involved, the relation (see Chapt. VI; (18.8));

$$D' = (h \otimes 1) \circ D \circ h^{-1}. \tag{11.21}$$

Now, the above global relation is *equivalently* expressed, in view of (11.7), through the *local* $\mathcal{A}$*-connection "forms"* of D and D' (see e.g. (11.8)) and the corresponding to the given h, as in (11.14) *"gauge transformation"*

$$(s_\alpha^{-1}) \in \check{C}^0(\mathcal{U}, \mathcal{A}^\cdot), \tag{11.22}$$

by the following relation (see also, for instance, (2.15) or even (1.41));

$$s_\alpha^{-1} \cdot \theta'_\alpha = \theta_\alpha \cdot s_\alpha^{-1} + \partial(s_\alpha^{-1}), \quad \alpha \in I, \tag{11.23}$$

so that *one has*

$$\theta'_\alpha = s_\alpha \cdot \theta_\alpha \cdot s_\alpha^{-1} + s_\alpha \cdot \partial(s_\alpha^{-1}) = \theta_\alpha + \tilde{\partial}(s_\alpha^{-1}), \quad \alpha \in I. \tag{11.24}$$

Therefore, one obtains (cf. also (11.22))

$$\begin{aligned} \theta' - \theta \equiv (\theta'_\alpha) - (\theta_\alpha) = (\theta'_\alpha - \theta_\alpha) = \tilde{\partial}(s_\alpha^{-1}) \in \tilde{\partial}(\check{C}^0(\mathcal{U}, \mathcal{A}^\cdot)) \\ \equiv \operatorname{im}\tilde{\partial} \subseteq \check{C}^0(\mathcal{U}, \Omega). \end{aligned} \tag{11.25}$$

Thus, *in view of (11.20) and (11.25), one has;*

$$\begin{aligned} (\mathcal{L}', D') - (\mathcal{L}, D) &\equiv (g', \theta') - (g, \theta) = (g' \cdot g^{-1}, \theta' - \theta) \\ &= (\delta(s_\alpha^{-1}), \tilde{\partial}(s_\alpha^{-1})) = (\delta, \tilde{\partial})(s_\alpha^{-1}) \equiv (\delta \oplus \tilde{\partial})(s_\alpha^{-1}) \\ &\in \operatorname{im}(\delta + \tilde{\partial}) \equiv \operatorname{im} D^0 \subseteq \check{C}^1(\mathcal{U}, \mathcal{A}^\cdot) \oplus \check{C}^0(\mathcal{U}, \Omega) \equiv \mathcal{S}^1. \end{aligned} \tag{11.26}$$

Here we have also set (cf. (11.4) and Chapt. III; (7.13), (7.16))

$$D^0 = \delta \oplus \tilde{\partial} : \check{C}^0(\mathcal{U}, \mathcal{A}^\cdot) \equiv \mathcal{S}^0 \longrightarrow \mathcal{S}^1 \equiv \check{C}^1(\mathcal{U}, \mathcal{A}^\cdot) \oplus \check{C}^0(\mathcal{U}, \Omega). \tag{11.27}$$

Similarly, one obtains (ibid., and (11.27))

$$D^1 : \mathcal{S}^1 \longrightarrow \mathcal{S}^2 \equiv \check{C}^2(\mathcal{U}, \mathcal{A}^\cdot) \oplus \check{C}^1(\mathcal{U}, \Omega), \tag{11.28}$$

such that one has

$$D^1 = (\delta - \tilde{\partial}) + \delta. \tag{11.29}$$

(*Warning!* For convenience, we abused notation here, by setting, $\delta^{0,1} \equiv \delta$, as well as, $\delta^{1,0} \equiv \delta$; (cf. Chapt. III; (7.16) and the following diagram (11.32)). Therefore, in view of (11.9) and (11.6), *one obtains*

$$(g, \theta) \in \ker D^1 = \ker(\delta - \tilde{\partial}) \oplus \ker \delta, \tag{11.30}$$

for any $(\mathcal{L}, D) \equiv (g, \theta)$, as above. Accordingly, *by virtue of (11.26) and (11.30), one* now *obtains* (cf. also (11.4) and Chapt. III; (7.45));

$$[(\mathcal{L}, D)] = [(\mathcal{L}', D')] \in \check{\mathbf{H}}^1(X, \mathcal{E}^\cdot) := \ker D^1 / \operatorname{im} D^0. \tag{11.31}$$

That is, *(11.13) entails (11.31),* which thus proves that *the map (11.12),* as given by (11.10) and (11.11) (cf. also (11.30)), *is well-defined,* as asserted.

For convenience, we still depict the previous situation, concerning the various *hypercohomology* spaces and operators, through the *following diagram (11.32),* which will also be of use presently below;

(11.32)

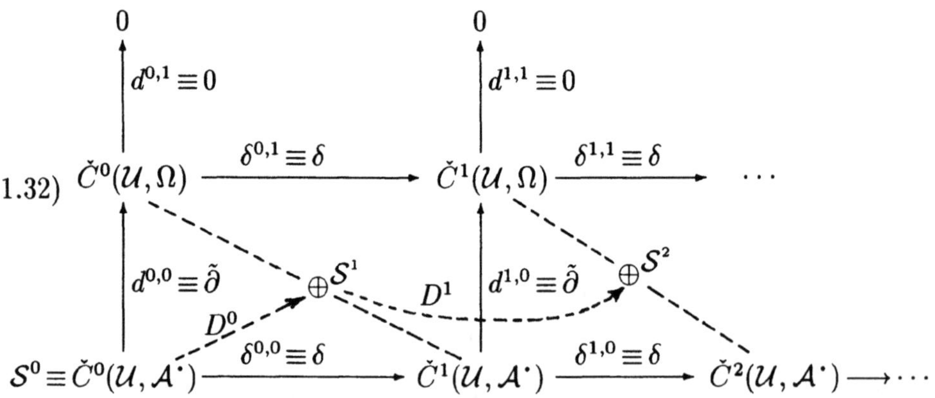

Now, we further prove that *the map (11.12) is one-to-one:* Thus, suppose that

$$[(\mathcal{L}, D)] = 0 \in \check{\mathbf{H}}^1(X, \mathcal{E}^{\cdot}), \tag{11.33}$$

for some $(\mathcal{L}, D) \equiv (g, \theta)$, as above (cf. (11.10)). Therefore, one concludes that (see (11.31) and (11.27))

$$(\mathcal{L}, D) \equiv (g, \theta) \in \operatorname{im} D^0 \equiv \operatorname{im}(\delta + \tilde{\partial}); \tag{11.34}$$

in other words, there exists a *0-cochain*

$$(s_\alpha) \in \check{C}^0(\mathcal{U}, \mathcal{A}^{\cdot}), \tag{11.35}$$

in such a manner that one has

$$(\mathcal{L}, D) \equiv (g, \theta) = (\delta + \tilde{\partial})(s_\alpha) \in \operatorname{im} D^0. \tag{11.36}$$

Consequently, one obtains (cf. also (11.6) and (11.8));

$$g \equiv (g_{\alpha\beta}) = \delta(s_\alpha) \equiv (\delta(s_\alpha)) = (s_\beta s_\alpha^{-1}), \tag{11.37}$$

as well as,

$$\theta \equiv (\theta_\alpha) = \tilde{\partial}(s_\alpha) \equiv (\tilde{\partial}(s_\alpha)) = (s_\alpha^{-1}\partial(s_\alpha)). \tag{11.38}$$

In this context, we still note, in passing, that, in view of (11.37) and (11.38), in conjunction with Chapt. VI; (1.27), one gets, of course,

$$\delta(\theta_\alpha) = \tilde{\partial}(g_{\alpha\beta}), \tag{11.39}$$

a fact, that was to be expected, for that matter, by virtue of (11.33) and (11.9).

Accordingly, setting now

$$g'_{\alpha\beta} := s_\alpha \cdot g_{\alpha\beta} \cdot s_\beta^{-1}, \tag{11.40}$$

one obtains, in view of (11.37), that

$$g'_{\alpha\beta} = 1 \; (\equiv 1_{\alpha\beta}), \tag{11.41}$$

while we still have, by (11.40), (11.35), in conjunction with Chapt. V; Lemma 2.1, that;

$$\mathcal{L} \equiv (g_{\alpha\beta}) \sim (g'_{\alpha\beta}) \equiv \mathcal{L}' \tag{11.42}$$

in $\mathbf{\Phi}^1_{\mathcal{A}}(X)$. That is (cf. (11.41) and Chapt. V; (4.3)),

(11.43) *the line sheaf* $\mathcal{L}$, as in (11.33), given further by (11.37), *is* $\mathcal{A}$*-isomorphic* to the *"trivial" line sheaf* (viz. the *free* $\mathcal{A}$*-module* of *rank* 1) $\mathcal{A}$. Thus, to the *neutral element* of the *multiplicative* (under $\otimes_{\mathcal{A}}$) *abelian group* $\boldsymbol{\Phi}^1_{\mathcal{A}}(X)$ (cf. Chapt. V; Theorem 4.1).

On the other hand, by further setting

$$\theta'_\alpha = \theta_\alpha + \tilde{\partial}(s_\alpha^{-1}) \in \Omega(U_\alpha), \quad \alpha \in I \tag{11.44}$$

(see also (11.35), or even (11.22)), one gets a *0-cochain*

$$(\theta'_\alpha) \in \check{C}^0(\mathcal{U}, \Omega) = \prod_\alpha \Omega(U_\alpha), \tag{11.45}$$

which, in effect, *supplies an* $\mathcal{A}$*-connection* D' *of* $\mathcal{L}'$; in other words, one verifies that *(11.45) satisfies the* corresponding *transformation law of potentials* (cf. Theorem 3.2, along with (2.59)): Indeed, one obtains;

$$\begin{aligned}\delta(\theta'_\alpha) &= \delta(\theta_\alpha) + \delta\tilde{\partial}(s_\alpha^{-1}) = \tilde{\partial}(g_{\alpha\beta}) + \tilde{\partial}(s_\beta^{-1}) - \tilde{\partial}(s_\alpha^{-1}) \\ &= \tilde{\partial}(s_\beta s_\alpha^{-1}) - \tilde{\partial}(s_\beta) + \tilde{\partial}(s_\alpha) = \tilde{\partial}(s_\beta) - \tilde{\partial}(s_\alpha) - \tilde{\partial}(s_\beta) + \tilde{\partial}(s_\alpha) \\ &= 0 = \tilde{\partial}(1) = \tilde{\partial}(g'_{\alpha\beta}).\end{aligned} \tag{11.46}$$

(In this regard, cf. Chapt. VI; (1.24) and (1.7), along with the previous rels. (11.44), (11.39), (11.37) and (11.41)). Thus, (11.46) justifies now our assertion. Furthermore, *the* two $\mathcal{A}$*-connections* D *and* D' *of* $\mathcal{L}'$, as given by (11.38) and (11.45), respectively, *satisfy the corresponding relation to (11.24),* by the very definition of (θ'_α) (see (11.44)). Thus, one actually obtains that;

$$(\mathcal{L}, D) \sim (\mathcal{L}', D') \equiv (1, D') \tag{11.47}$$

in $\boldsymbol{\Phi}^1_{\mathcal{A}}(X)^\nabla$ (cf. Chapt. VI; (18.6)). So we have proved that *(11.33) implies (11.47), that was,* in fact (cf. also (11.43)), *to be proved.*

Furthermore, we prove that *the map (11.12) is onto:* Thus, suppose we have an element

$$[z] \in \check{\mathbf{H}}(X, \mathcal{E}^\cdot). \tag{11.48}$$

Therefore (see e.g. (11.31), (11.30)), one has

$$z \in \ker D^1 = \ker \delta \oplus \ker(\delta - \tilde{\partial}), \tag{11.49}$$

so that one finally concludes that (cf. also, for instance, (11.32), or even

(11.27), (11.28));

$$z \equiv (g, \theta) \equiv ((g_{\alpha\beta}), (\theta_\alpha)) \in \check{C}^1(\mathcal{U}, \mathcal{A}^{\cdot}) \oplus \check{C}^0(\mathcal{U}, \Omega) \tag{11.50}$$

(for any *representative* z of the cohomology class considered in (11.48)). Consequently, in view of (11.49) and (11.50), one *first* obtains

$$\delta(g) \equiv \delta(g_{\alpha\beta}) = 0, \tag{11.51}$$

so that one thus concludes that

$$(g_{\alpha\beta}) \in Z^1(\mathcal{U}, \mathcal{A}^{\cdot}); \tag{11.52}$$

in other words, one thus obtains a *line sheaf* $\mathcal{L}$ on X, which is determined through its *coordinate 1-cocycle* (11.52) (see Chapt. V; Theorem 4.1, or even (2.60) therein). *Secondly,* according to the same rels. (11.49) and (11.50), one further obtains

$$\delta(\theta_\alpha) = \tilde{\partial}(g_{\alpha\beta}) \in Z^1(\mathcal{U}, \Omega) \tag{11.53}$$

(cf. also Chapt. VI; (10.25) above, for $n = 1$). Thus, *the 1-cocycle* $\tilde{\partial}(g_{\alpha\beta})$, as above, *is* actually *a coboundary,* so that one has

$$[\tilde{\partial}(g_{\alpha\beta})] = 0 \in \check{H}^1(X, \Omega). \tag{11.54}$$

However, the last condition implies already (*Atiyah's criterion,* cf. Chapt. VI; Theorem 11.1, along with (10.28.1)) that *our line sheaf* $\mathcal{L}$, as given by (11.52), *admits an $\mathcal{A}$-connection D,* whose *local $\mathcal{A}$-connection forms* are given by the *0-cochain*

$$(\theta_\alpha) \in \check{C}^0(\mathcal{U}, \Omega) \tag{11.55}$$

as appeared in (11.53) (see Chapt. VI; (11.7), (11.8) and (10.23); yet, see (1.24) in the preceding). Thus, we arrived at an element

$$[(\mathcal{L}, D)] \in \Phi^1_{\mathcal{A}}(X)^{\nabla}, \tag{11.56}$$

whose image, through the map (11.12), is the given element z, as in (11.48), according to (11.50) and our previous interpretation of (11.52) and (11.55); i.e., *we have*

$$\chi([(\mathcal{L}, D)]) = z, \tag{11.57}$$

as asserted.

So we have established, so far, through the map (11.12), a *bijection between* the two *(abelian) groups* involved. Now, our final claim is that *the same map is,* in effect, *a group morphism:* Namely, suppose that we have *two elements of (11.2)*

$$(\mathcal{L}, D) \equiv (g, \theta) \ \textit{and} \ (\mathcal{L}', D') \equiv (g', \theta') \tag{11.58}$$

(see e.g. (11.10). As usual so far, we do not distinguish, for convenience, the latter elements from their corresponding (isomorphism) classes in $\mathbf{\Phi}^1_{\mathcal{A}}(X)^{\nabla}$). Thus, one obtains (cf. Chapt. VI; (18.2), (18.15), (18.16), (18.17) and (18.21), along with Chapt. V; (4.3));

$$\begin{aligned}(\mathcal{L}', D') \otimes (\mathcal{L}, D)^{-1} &= (\mathcal{L}' \otimes_{\mathcal{A}} \mathcal{L}^{-1}, D' \otimes D^{-1}) \\ &= (\mathcal{L}' \otimes_{\mathcal{A}} \mathcal{L}^{*}, D' \otimes D^{*}) \equiv (h, \omega),\end{aligned} \tag{11.59}$$

such that one has(see also Chapt. V; (4.16), (4.17))

$$h = g' \cdot g^{-1} \in Z^1(\mathcal{U}, \mathcal{A}^{\cdot}), \tag{11.60}$$

while we still obtain that

$$\omega = \theta' - \theta \in \check{C}^0(\mathcal{U}, \Omega), \tag{11.61}$$

as an *$\mathcal{A}$-connection 0-cochain for the $\mathcal{A}$-connection $D' \otimes D^*$.* The assertion here follows from Chapt. VI; (5.10), in conjunction with (1.11) and (4.20) in the preceding. On the other hand, by looking at the pairs (11.58), as above, and based on (11.39), one further obtains from (11.61) and (11.60);

$$\begin{aligned}\delta(\omega) = \delta(\theta' - \theta) = \delta(\theta') - \delta(\theta) &= \tilde{\partial}(g') - \tilde{\partial}(g) = \tilde{\partial}(g') + \tilde{\partial}(g^{-1}) \\ &= \tilde{\partial}(g'g^{-1}) = \tilde{\partial}(h).\end{aligned} \tag{11.62}$$

That is, one has here again *the analogous relation to (11.39), concerning the pair* (h, ω), as in (11.59); hence, *the pair* in question,

$$(h, \omega) = (g'g^{-1}, \theta' - \theta) \tag{11.63}$$

determines thus, in view of (11.62), *an element of* $\check{\mathbf{H}}^1(X, \mathcal{E}^{\cdot})$. Indeed, as such, viz. *the image of (11.63)* (in fact, of the respective *isomorphism class*), *through the map (11.12), has as representative in the latter (additive) group the element*

$$(\mathcal{L}', D') - (\mathcal{L}, D) \equiv (g', \theta') - (g, \theta) = (g'g^{-1}, \theta' - \theta), \tag{11.64}$$

as this follows from the very *group structure* of

$$\text{dom}\, D^1 = \mathcal{S}^1 \equiv \check{C}^1(\mathcal{U}, \mathcal{A}^{\cdot}) \oplus \check{C}^0(\mathcal{U}, \Omega) \tag{11.65}$$

(see also (11.27), or even (11.32)). So this proves already our last claim, concerning the map (11.12), and, in turn, *the proof* of Theorem 18.2 of Chapter VI, *is* thus *completely established,* as desired. ▮

Yet, an appropriate *physical interpretation* of the content of the same theorem, as above, will be given later on; see thus Chapter X, Section 5 in the sequel.

CHAPTER VIII

Curvature

"Matter tells space how to curve."

C. W. Milner - K. S. Thorne - J. A. Wheeler in "Gravitation" (Freeman, 1973). p. 5.

"...Mathematics aims to understand, to manipulate, to develop, and to apply those aspects of the universe which are formal."

S. Mac Lane in "Mathematics Form and Function" (Springer - Verlag, 1986). p. 456.

We consider in this Chapter the notion of *"curvature"*, which, under supplementary suitable assumptions for a given differential triad, can be associated, in a natural way, with any given $\mathcal{A}$-connection. Thus, in complete contrast with the situation one encounters with the case of $\mathcal{A}$-connections, which, as we know already (cf. Chapter VI in the preceding), do not always exist, we realize here that an $\mathcal{A}$-connection in the case there does exist is, instead, always accompanied with a curvature. This fact, as we shall see later on (cf. thus Chapter X in the sequel), can acquire an appropriate physical interpretation, as well. Yet, as was the case hitherto, one still witnesses here that several fundamental notions and results of the classical theory can also be recasted within the present abstract (axiomatic) framework, that has been advocated so far by this study.

So we start with establishing straight away the necessary basic terminology.

1. Preliminaries. Basic definitions

Before we proceed, we first comment a bit on a change of the notation applied hitherto, that will be employed hereafter:

So the notation for a *differential triad* used heretofore (cf. Chapt. VI; (3.1)), is now changed to

$$(\mathcal{A}, \partial, \Omega^1). \tag{1.1}$$

Again $(X, \mathcal{A})$ stands here for a given **C**-*algebraized space* on an arbitrary *topological space* X, as before, while we now set;

$$\Omega \equiv \Omega^1 \equiv \bigwedge^1 \Omega, \tag{1.2}$$

for the given $\mathcal{A}$-*module* Ω, range of the **C**-*derivation* ∂, as above. Thus, Ω^1 stands for the *1st exterior power of* Ω, the given $\mathcal{A}$-module on X, as was actually the case hitherto. Indeed, the reason for (1.2) is that we are going to consider right below *higher exterior powers of* $\Omega \equiv \Omega^1$ (see e.g. (1.6) in the sequel).

On the other hand, to distinguish the present situation from the *Whitney sum* notation, occasionally employed so far, we retain, henceforth, for the latter the (usual, for that matter) symbolism;

$$\Omega^{(n)} := \underbrace{\Omega \oplus \cdots \oplus \Omega}_{n\text{-times}}, \quad n \in \mathbf{N}. \tag{1.3}$$

Thus, to start with, consider a *differential triad*

$$(\mathcal{A}, \partial, \Omega^1) \tag{1.4}$$

(cf. (1.1)) on a given **C**-*algebraized space*

$$(X, \mathcal{A}). \tag{1.5}$$

Furthermore, consider the *2nd exterior power of* Ω^1;

$$\Omega^2 := \Omega^1 \wedge \Omega^1 \equiv \bigwedge^2 \Omega \equiv \bigwedge^2 \Omega^1 \tag{1.6}$$

(cf. (1.2), along with Chapt. IV; (7.1), (7.3) and (7.8). As already said the above is an instance justifying the previous change of notation, as in (1.1)).

On the other hand, apart from ∂, as in (1.4), we further assume that we are given a second **C**-*linear morphism*

$$d^1 \equiv d : \Omega^1 \longrightarrow \Omega^2 \tag{1.7}$$

between the **C**-*vector space sheaves* concerned, satisfying the two following conditions:

i) *The "differential operators"* ∂ *and* $d^1 \equiv d$, as above, *satisfy the relation;*

$$d \circ \partial = 0. \tag{1.8}$$

ii) *For any (local) sections* $\alpha \in \mathcal{A}(U)$ *and* $s \in \Omega^1(U)$, with U *open* in X *one has;*

$$d(\alpha \cdot s) = \alpha \cdot ds - s \wedge \partial\alpha = \alpha \cdot ds + \partial(\alpha) \wedge s. \tag{1.9}$$

Now, by extending the standard terminology, we call the previous map $d^1 \equiv d$, the *1st exterior derivation* (or even *1st exterior derivative operator*) of (1.4). Yet, by still mimicking the classical case, we also call occasionally the elements of the $\mathcal{A}$-*module* on X,

$$\Omega^n := \underbrace{\Omega^1 \wedge \cdots \wedge \Omega^1}_{n\text{-times}} \equiv \bigwedge^n \Omega \equiv \bigwedge^n \Omega^1, \quad n \in \mathbf{N}. \tag{1.10}$$

(cf. also (1.2)), "*n-forms*" on X. (Indeed, the terminology refers actually to *sections* of the latter $\mathcal{A}$-module).

Of course, (1.8) *is equivalent with* the relation;

$$\operatorname{im} \partial \subseteq \ker d. \tag{1.11}$$

Yet, as a first application of (1.8), (1.9), one also concludes, by the very definitions, that

$$d \circ \tilde{\partial} = 0, \tag{1.12}$$

as well, that is, *equivalently,*

$$\operatorname{im} \tilde{\partial} \subseteq \ker d. \tag{1.13}$$

Indeed, one has, *for any* $\alpha \in \mathcal{A}^{\cdot}(U) = \mathcal{A}(U)^{\cdot}$ and *open* $U \subseteq X$,

$$(1.14) \qquad \begin{aligned} d(\tilde{\partial}(\alpha)) &= d(\alpha^{-1}\partial\alpha) = \alpha^{-1}d(\partial(\alpha)) + \partial(\alpha^{-1})\wedge\partial(\alpha) \\ &= -\alpha^{-2}\partial(\alpha)\wedge\partial(\alpha) = -\alpha^{-2}(\partial\alpha\wedge\partial\alpha) = 0, \end{aligned}$$

which proves (1.12). (See also Chapt. VI; (1.12)). ∎

So we come now to the following basic notion for the sequel, as exhibited by the next.

Definition 1.1. Suppose we have a **C**-algebraized space $(X, \mathcal{A})$, endowed with a differential triad $(\mathcal{A}, \partial, \Omega^1)$ (cf. (1.1)), along with a **C**-linear morphism

$$(1.15) \qquad d : \Omega^1 \longrightarrow \Omega^2 \equiv \Omega^1 \wedge \Omega^1,$$

satisfying (1.8) and (1.9). Then, the quintuple

$$(1.16) \qquad (\mathcal{A}, \partial, \Omega^1, d, \Omega^2)$$

is called a *curvature datum* on X. Yet, for short, we also speak in that case of a *curvature space* X.

As follows from the above definition, we have a curvature datum, once we are given, of course, the *pair*

$$(1.17) \qquad (\partial, d)$$

of "differential operators", as before, *such that (1.8) and (1.9) be satisfied.*

Furthermore, based on (1.9), and for reasons that will be made apparent presently below, we may also call the previously given operator $d^1 \equiv d$, as in (1.15), the *1st prolongation of* ∂.

So our next objective is to define an analogous *"prolongation"* of a given $\mathcal{A}$*-connection* D. Indeed, one has the following.

Lemma 1.1. Suppose we are given a *curvature space* X (see Definition 1.1), and let $\mathcal{E}$ be an $\mathcal{A}$*-module* on X, endowed with an $\mathcal{A}$*-connection* D. Then, one defines the *1st prolongation of* D, as the map

$$(1.18) \qquad D^1 : \Omega^1(\mathcal{E}) \equiv \mathcal{E} \otimes_{\mathcal{A}} \Omega^1 \longrightarrow \mathcal{E} \otimes_{\mathcal{A}} \Omega^2 \equiv \Omega^2(\mathcal{E}),$$

such that

$$(1.19) \qquad D^1(s \otimes t) := s \otimes dt - t \wedge Ds = s \otimes dt + Ds \wedge t,$$

for any $s \in \mathcal{E}(U)$ *and* $t \in \Omega^1(U)$, with U *open* in X.

Moreover, D^1 is *the only* **C**-*linear morphism,* as in (1.18), *satisfying* (1.19).

Proof. At first, by looking at the corresponding *presheaves,* generating the $\mathcal{A}$-*modules* appeared in (1.18) (see Chapt. II; (5.5) and Chapt. IV; (7.1), (7.2)), one defines, *for every open* $U \subseteq X$, a **C**-*bilinear map*

$$\begin{aligned}(1.20)\quad &\rho_U : \mathcal{E}(U) \times \Omega^1(U) \longrightarrow \mathcal{E}(U) \otimes_{\mathbf{C}} \Omega^2(U) \hookrightarrow \mathcal{E}(U) \otimes_{\mathcal{A}(U)} \Omega^2(U) \\ &= (\mathcal{E}(U) \otimes_{\mathcal{A}(U)} \Omega^1(U)) \wedge \Omega^1(U) : (s,t) \longmapsto \rho_U(s,t) := s \otimes dt + Ds \wedge t,\end{aligned}$$

with $s \in \mathcal{E}(U)$ and $t \in \Omega^1(U)$. (Concerning the range of ρ_U, see, for instance, N. Bourbaki [3: Chap. 2; p.53, Proposition 2]). Fuerthermore, *the same map (1.20) is* $\mathcal{A}(U)$-*balanced.* That is, for every $\alpha \in \mathcal{A}(U)$ and s, t, as before, one has the relation;

$$(1.21)\qquad \rho_U(\alpha s, t) = \rho_U(s, \alpha t),$$

as one easily sees by (1.20). Indeed, one has (cf. also (1.9) and Chapt. VI; (3.6));

$$\begin{aligned}(1.22)\quad \rho_U(\alpha s, t) &= (\alpha s) \otimes dt + D(\alpha s) \wedge t = \alpha s \otimes dt + (\alpha\, Ds + s \otimes \partial\alpha) \wedge t \\ &= \alpha s \otimes dt + (s \otimes \partial\alpha) \wedge t + \alpha\, Ds \wedge t = s \otimes (\alpha\, dt + \partial\alpha \wedge t) + \alpha\, Ds \wedge t \\ &= s \otimes d(\alpha t) + Ds \wedge (\alpha t) = \rho_U(s, \alpha t),\end{aligned}$$

as just was to be proved. Accordingly,

(1.23) *the map* ρ_U, as given by (1.20), *can be extended,* by virtue of (1.21), *uniquely, to a* **C**-*linear map*

$$(1.23.1)\qquad \bar{\rho} : \mathcal{E}(U) \otimes_{\mathcal{A}(U)} \Omega^1(U) \longrightarrow \mathcal{E}(U) \otimes_{\mathcal{A}(U)} \Omega^2(U).$$

Now, *these various* **C**-*linear maps* ρ_U, when U runs over the open subsets of X, *satisfy the* corresponding *compatibility conditions* for a *presheaf morphism* (see Chapt. I; (6.5), (6.7)); in turn, this entails, still *uniquely,* a corresponding *sheaf morphism* between the sheaves generated by the presheaves involved (see e.g. (1.23.1), along with Chapt. I; (8.5) and Lemma 8.1). However, due to (1.23), the latter map is, in effect, a **C**-*linear morphism* of the **C**-*vector space sheaves* involved, which, *by definition,* is now the **C**-*linear morphism* D^1, as in (1.18), we were looking for; the same is, of course, *uniquely defined,* as well, as follows from (1.23) and the previous argument, and this terminates the proof. ∎

As a first application of the above, one obtains the following *two properties of the 1st prolongation* D^1 *of an* $\mathcal{A}$*-connection* D, which will be applied presently below. Thus, we have that:

For any (local) sections $\alpha \in \mathcal{A}(U)$, $s \in \mathcal{E}(U)$ *and* $t \in \Omega^1(U)$, with U *open* in X, *one obtains;*

$$(1.24)\qquad \begin{aligned} D^1(\alpha\cdot(s\otimes t)) &= \alpha\cdot D^1(s\otimes t) - (s\otimes t)\wedge\partial\alpha \\ &= \alpha\cdot D^1(s\otimes t) + \partial\alpha\wedge(s\otimes t), \end{aligned}$$

as well as,

$$(1.25)\qquad D^1(\alpha\cdot Ds) = \alpha\cdot D^1(Ds) - Ds\wedge\partial\alpha = \alpha\cdot D^1(Ds) + \partial\alpha\wedge Ds.$$

Indeed, one has (cf. also (1.21), (1.23), together with (1.19) and (1.9));

$$(1.26)\qquad \begin{aligned} D^1(\alpha\cdot(s\otimes t)) &= D^1(\alpha s\otimes t) = D^1(s\otimes\alpha t) = s\otimes d(\alpha t) - \alpha t\wedge Ds \\ &= s\otimes(\alpha\, dt - t\wedge\partial\alpha) - \alpha t\wedge Ds = s\otimes(\alpha\, dt) - (s\otimes t)\wedge\partial\alpha - \alpha t\wedge Ds \\ &= \alpha(s\otimes dt - t\wedge Ds) - (s\otimes t)\wedge\partial\alpha = \alpha\cdot D^1(s\otimes t) - (s\otimes t)\wedge\partial\alpha, \end{aligned}$$

That is, *the rel. (1.24), as desired.* Now, to prove (1.25), we first remark that *(1.24) is* already *one aspect of (1.25), in the case of decomposable tensors* $s\otimes t \in \mathcal{E}(U)\otimes_{\mathcal{A}(U)}\Omega^1(U)$. Hence, the general case of an element $D(s) \in \mathcal{E}(U)\otimes_{\mathcal{A}(U)}\Omega^1(U)$, with $s \in \mathcal{E}(U)$, that is, the rel. (1.25), *follows* now *from (1.24) and the* **C**-*linearity of* D^1 (Lemma 1.1). ∎

Now, for convenience, as well as, as a particular instance of a more general situation, which we deal with later on (see thus Section 7 in the sequel), we still write (1.19), the defining relation of the *1st prolongation* D^1 *of the* given $\mathcal{A}$*-connection* D, in the form;

$$(1.27)\qquad D^1 := 1_{\mathcal{E}}\otimes d^1 - 1_{\Omega^1}\wedge D \equiv 1\otimes d - 1\wedge D = 1\otimes d + D\wedge 1$$

where $1_{\mathcal{E}}$ and 1_{Ω^1} have the obvious meaning.

2. Curvature of an $\mathcal{A}$-connection

Based on the material of the preceding Section, we are now in the position to give the following fundamental definition, concerning the concept in the heading of this Section. So we have.

Definition 2.1. Suppose we are given a curvature space X (cf. Definition 1.1), as well as, an $\mathcal{A}$-module $\mathcal{E}$ on X, endowed with an $\mathcal{A}$-connection D. Then, the *curvature* of the given $\mathcal{A}$-connection D, denoted by

$$R(D) \equiv R, \tag{2.1}$$

is defined by the relation;

$$R(D) := D^1 \circ D, \tag{2.2}$$

where D^1 stands for the 1st prolongation of D (cf. Lemma 1.1). In other words, one has the following defining triangle for R (see (1.18));

$$\begin{array}{ccc} \mathcal{E} & \xrightarrow{D} & \Omega^1(\mathcal{E}) \\ & {\scriptstyle R \equiv D^1 \circ D} \searrow & \swarrow {\scriptstyle D^1} \\ & \Omega^2(\mathcal{E}). & \end{array} \tag{2.3}$$

Note 2.1.– In complete contrast with what happens in the case of *an $\mathcal{A}$-connection,* which, as we know (cf. Chapt. VI; (11.2)), *does not exist, in general,* we remark here that:

(2.4) *one can always define the curvature of a given $\mathcal{A}$-connection,*

provided, of course, *one has* the appropriate framework, to formulate this notion; hence, according to the preceding, *a curvature space* (cf. (1.16)).

Having now at our disposal the classical notion of curvature, we say that a given $\mathcal{A}$-connection D is *flat,* whenever one has;

$$R(D) = 0 \tag{2.5}$$

(see also the previous Note 2.1, concerning the relevant context for the formulation of (2.5)).

So, by looking at what we have called hitherto the *standard flat $\mathcal{A}$-connection* ∂ (cf. Chapt. VI; Subsection 4.(a), in conjunction with (1.1), (1.2) above), we see that, according to our hypothesis (1.8), one has;

$$R(\partial) := d^1 \circ \partial = 0. \tag{2.6}$$

That is, ∂ *is* really *a flat $\mathcal{A}$-connection,* in the sense of (2.5). Of course, we have used in (2.6) the fact that (the given exterior derivative operator) $d^1 \equiv d$ (cf. (1.7)) may be considered as the *1st prolongation of* ∂: Indeed, one has, *by definition* (cf. also Chapt. II; (5.15)),

$$d^1 : \Omega^1 = \Omega^1(\mathcal{A}) \longrightarrow \Omega^2(\mathcal{A}) = \Omega^2, \tag{2.7}$$

such that (1.9) *be valid;* however, the same rel. (1.9) corresponds to (1.19), when $\mathcal{E} \equiv \mathcal{A}$. That is, one can have here for (the $\mathcal{A}$-connection) ∂ an analogue of Lemma 1.1, pertaining, of course, for the case at hand to the *uniqueness of* (the given!) $d^1 \equiv d$ *only.* ∎

Remark 2.1.– Alternatively, concerning (2.6), we further note that one still obtains the relation;

$$D^1|_{\Omega^1} = d^1 \equiv d, \tag{2.8}$$

in the sense that one has then $\mathcal{E} \equiv \mathcal{A}$, as above. That is, we have by (1.19) (see also Chapt. VI; (1.7));

$$D^1(1 \otimes s) = 1 \otimes ds - s \wedge \partial(1) = 1 \otimes ds - 0 = ds, \tag{2.9}$$

for any $s \in \Omega^1(U)$ and *open* $U \subseteq X$, which thus again leads to (2.6), in view of (2.2). ∎

Now, our next basic information, concerning the curvature of a given $\mathcal{A}$-connection, is included in the following.

Lemma 2.1. *Suppose we have a curvature space X, an $\mathcal{A}$-module $\mathcal{E}$ on X, and an $\mathcal{A}$-connection D of $\mathcal{E}$. Then, the curvature of D, $R(D) \equiv R$* (cf. (2.1), (2.2)), *is an $\mathcal{A}$-morphism between the $\mathcal{A}$-modules $\mathcal{E}$ and $\Omega^2(\mathcal{E})$ on X; viz. one has* (see also Chapt. II; (6.9))

$$R \in \mathrm{Hom}_{\mathcal{A}}(\mathcal{E}, \Omega^2(\mathcal{E})) = \mathcal{H}om_{\mathcal{A}}(\mathcal{E}, \Omega^2(\mathcal{E}))(X). \tag{2.10}$$

In particular, in the case of a vector sheaf $\mathcal{E}$ on X, one has

$$R \in \Omega^2(\mathcal{E}nd\,\mathcal{E})(X). \tag{2.11}$$

Proof. To prove (2.10), we just apply the definition of R, according to (2.2), based also on (1.25), (1.19) and (1.8), along with Chapt. VI; (3.6). So, *for any* given $\alpha \in \mathcal{A}(U)$ *and* $s \in \mathcal{E}(U)$, with U *open* in X, one obtains;

$$\begin{aligned} R(\alpha \cdot s) &= (D^1 \circ D)(\alpha \cdot s) = D^1(D(\alpha s)) = D^1(\alpha \cdot Ds + s \otimes \partial\alpha) \\ &= D^1(\alpha \cdot Ds) + D^1(s \otimes \partial\alpha) = \alpha \cdot D^1(Ds) - Ds \wedge \partial\alpha + s \otimes d(\partial\alpha) - \partial\alpha \wedge Ds \\ &= \alpha \cdot D^1(Ds) = \alpha \cdot R(s), \end{aligned} \tag{2.12}$$

as desired. On the other hand, if $\mathcal{E}$ is a *vector sheaf* on X, then, by virtue of (2.10) (see also Chapt. IV; (6.8) and Chapt. II; (6.28)), one has;

$$\begin{aligned} R \in \mathrm{Hom}_{\mathcal{A}}(\mathcal{E}, \Omega^2(\mathcal{E})) &= \mathcal{H}om_{\mathcal{A}}(\mathcal{E}, \Omega^2(\mathcal{E}))(X) \equiv \mathcal{H}om_{\mathcal{A}}(\mathcal{E}, \mathcal{E} \otimes_{\mathcal{A}} \Omega^2)(X) \\ &= (\mathcal{H}om_{\mathcal{A}}(\mathcal{E}, \mathcal{E}) \otimes_{\mathcal{A}} \Omega^2)(X) \equiv ((\mathcal{E}nd\, \mathcal{E}) \otimes_{\mathcal{A}} \Omega^2)(X) \equiv \Omega^2(\mathcal{E}nd\, \mathcal{E})(X), \end{aligned} \tag{2.13}$$

as was to be proved, which also terminates the proof. ∎

Now, assume again that $\mathcal{E}$ is a *vector sheaf* on X, a *curvature spave,* as before, such that

$$\operatorname{rk} \mathcal{E} = n \in \mathbf{N}, \tag{2.14}$$

and let

$$\mathcal{U} = (U_\alpha)_{\alpha \in I} \tag{2.15}$$

be a *local frame of* $\mathcal{E}$. Yet, suppose that $\mathcal{E}$ is endowed with an $\mathcal{A}$-*connection* D. Then, in view of (2.13) (see also Chapt. III; (4.25), (4.28) and (8.13)), one has;

$$R \in \Omega^2(\mathcal{E}nd\, \mathcal{E})(X) = Z^0(\mathcal{U}, \Omega^2(\mathcal{E}nd\, \mathcal{E})). \tag{2.16}$$

On the other hand, by looking at a *local gauge* $U_\alpha \in \mathcal{U}$ *of* $\mathcal{E}$, as in (2.15), one obtains;

$$\Omega^2(\mathcal{E}nd\, \mathcal{E})(U_\alpha) = M_n(\Omega^2)(U_\alpha) = M_n(\Omega^2(U_\alpha)) \tag{2.17}$$

(see (2.14) above, in conjunction with Chapt. VI; (9.4), (9.3) and Chapt. V; (8.7), (8.6)). Thus, by virtue of (2.16) and (2.17), one now obtains;

$$\begin{aligned} R &\in \Omega^2(\mathcal{E}nd\, \mathcal{E})(X) = Z^0(\mathcal{U}, \Omega^2(\mathcal{E}nd\, \mathcal{E})) \\ &\subseteq C^0(\mathcal{U}, \Omega^2(\mathcal{E}nd\, \mathcal{E})) = \textstyle\prod_\alpha \Omega^2(\mathcal{E}nd\, \mathcal{E})(U_\alpha) = \prod_\alpha M_n(\Omega^2(U_\alpha)), \end{aligned} \tag{2.18}$$

such that one has

(2.19) $$R = \left(R^{(\alpha)}\right)_{\alpha \in I} \equiv \left(\left(\omega_{ij}^{(\alpha)}\right)\right),$$

where we have set

(2.20) $$R^{(\alpha)} \equiv \left(\omega_{ij}^{(\alpha)}\right) \in M_n(\Omega^2)(U_\alpha) = M_n(\Omega^2(U_\alpha)),$$

for any $\alpha \in I$, as in (2.15). So one concludes that:

(2.21) the *curvature* $R(D) \equiv R$ of a given $\mathcal{A}$-*connection* D of a *vector sheaf* $\mathcal{E}$ (on a *curvature space* X), *with* $\operatorname{rk} \mathcal{E} = n \in \mathbf{N}$, is a *0-cocycle of (local)* $n \times n$ *matrices with entries from (local sections of)* $\Omega^2 \equiv \Omega^1 \wedge \Omega^1$ (viz. *(local) "2-forms"* on X).

A detailed *local treatment of* the *curvature* $R(D)$, as above, along with applications thereon, will further be considered straight away in the subsequent sections.

3. Local form of the curvature. Cartan's structural equation

At first, suppose, as also was done hitherto, that we are given a *curvature space* X; that is (see Definition 1.1), we assume that we have the following quintuple

(3.1) $$(\mathcal{A}, \partial, \Omega^1, d^1 \equiv d, \Omega^2),$$

the *topological space* X being the common *base* of the $\mathcal{A}$-*modules* involved in (3.1).

Furthermore, consider an $\mathcal{A}$-*module* $\mathcal{E}$ on X, endowed with an $\mathcal{A}$-*connection* D, and let

(3.2) $$e^U \equiv \{U; (e_1, \ldots, e_n)\}$$

be a *local gauge* of $\mathcal{E}$ over an *open* set $U \subseteq X$ of *rank* $n \in \mathbf{N}$ (see, for instance, Chapt. VII; (1.5)). Thus, *for any (local) section* $s \in \mathcal{E}(U)$, one has a *unique expression* of the form (ibid. (1.6.1));

$$(3.3)\qquad s = \sum_{i=1}^{n} s_i e_i,$$

such that $s_i \in \mathcal{A}(U)$, *with* $1 \leqq i \leqq n$.

On the other hand, we already know (cf. Chapt. VII; (1.8.1), (1.11) and (1.13) that *the* given $\mathcal{A}$-*connection* D, as above, *is locally expressed through the* corresponding to (3.2) *local* $\mathcal{A}$-*connection matrix*

$$(3.4)\qquad \omega^U \equiv \omega = (\omega_{ij}) \in M_n(\Omega^1(U)) = M_n(\Omega^1)(U),$$

such that one has;

$$(3.5)\qquad D(e_j) = \sum_{i=1}^{n} e_i \otimes \omega_{ij}, \quad 1 \leqq j \leqq n.$$

Thus, by further applying the curvature of D to the given $\mathcal{A}(U)$-*basis of* $\mathcal{E}(U)$,

$$(3.6)\qquad (e_i)_{1 \leqq i \leqq n} \subseteq \mathcal{E}(U),$$

as in (3.2), one obtains (see also (3.5) and (1.19), or even (1.27));

$$(3.7)\qquad \begin{aligned} R(e_j) &= (D^1 \circ D)(e_j) = D^1(D(e_j)) = D^1(\textstyle\sum_i e_i \otimes \omega_{ij}) \\ &= \textstyle\sum_i D^1(e_i \otimes \omega_{ij}) = \textstyle\sum_i (e_i \otimes d\omega_{ij} - \omega_{ij} \wedge D(e_i)) \\ &= \textstyle\sum_i e_i \otimes d\omega_{ij} + \textstyle\sum_i D(e_i) \wedge \omega_{ij}. \end{aligned}$$

Furthermore, one obtains (cf. (3.5))

$$(3.8)\qquad \begin{aligned} \textstyle\sum_i D(e_i) \wedge \omega_{ij} &= \textstyle\sum_i (\textstyle\sum_k e_k \otimes \omega_{ki}) \wedge \omega_{ij} \\ &= \textstyle\sum_k e_k \otimes (\textstyle\sum_i \omega_{ki} \wedge \omega_{ij}). \end{aligned}$$

Thus, by (3.7) and (3.8), and by also changing, appropriately, indices in (3.8), *one gets;*

$$(3.9)\qquad \begin{aligned} R(e_j) &= \textstyle\sum_i e_i \otimes d\omega_{ij} + \textstyle\sum_i e_i \otimes (\textstyle\sum_k \omega_{ik} \wedge \omega_{kj}) \\ &= \textstyle\sum_{i=1}^{n} e_i \otimes (d\omega_{ij} + \textstyle\sum_k \omega_{ik} \wedge \omega_{kj}) \in \mathcal{E}(U) \otimes_{\mathcal{A}(U)} \Omega^2(U), \end{aligned}$$

such that $1 \leqq j \leqq n$. Accordingly, *we* now *set;*

$$(3.10)\qquad R(e_j)_i \equiv R_{ij} := d\omega_{ij} + \sum_k \omega_{ik} \wedge \omega_{kj} \in \Omega^2(U),$$

for any $1 \leqq i, j \leqq n$, such that one has

(3.11) $$R \equiv (R_{ij}) \in M_n(\Omega^2(U)) = M_n(\Omega^2)(U),$$

a fact, already noticed by (2.20) above in the particular case of a *local gauge* of a *vector sheaf* on X (cf. also (2.21)). Thus, by applying *matrix notation,* one obtains from (3.10),

(3.12) $$R = d\omega + \omega \wedge \omega,$$

as an *element* (local section) *of the* $\mathcal{A}(U)$*-module,* appeared *in* (3.11). The last relation yields in our case the so-called *("second") Cartan's structural equation.*

Now, by specializing, for convenience, to the case of a vector sheaf on X, we summarize the preceding into the form of the following.

Theorem 3.1. *(Cartan's structural equation). Suppose we are given a curvature space X and a vector sheaf $\mathcal{E}$ on X, of rank $n \in \mathbf{N}$, endowed with an $\mathcal{A}$-connection D. Moreover, let*

(3.13) $$\mathcal{U} = (U_\alpha)_{\alpha \in I}$$

be a local frame of $\mathcal{E}$, and

(3.14) $$\omega \equiv \left(\omega^{(\alpha)}\right) = \left(\left(\omega_{ij}^{(\alpha)}\right)\right) \in C^0(\mathcal{U}, M_n(\Omega^1))$$

the corresponding to (3.13) 0-cochain of local $\mathcal{A}$-connection matrices of D (see Chapt. VII; (2.39.1)). *Then, the curvature of D is given by the following 0-cocycle of $\mathcal{U}$; viz. by a global section of $\Omega^2(\mathcal{E}nd\,\mathcal{E})$, as follows* (cf. also (2.18)):

(3.15) $$\begin{aligned} R(D) \equiv R = \left(R^{(\alpha)}\right) = \left(\left(R_{ij}^{(\alpha)}\right)\right) &\in Z^0(\mathcal{U}, \Omega^2(\mathcal{E}nd\,\mathcal{E})) \\ &= \Omega^2(\mathcal{E}nd\,\mathcal{E})(X) \hookrightarrow \textstyle\prod_\alpha M_n(\Omega^2(U_\alpha)), \end{aligned}$$

such that one has;

(3.16) $$R^{(\alpha)} \equiv \left(R_{ij}^{(\alpha)}\right) = d\omega^{(\alpha)} + \omega^{(\alpha)} \wedge \omega^{(\alpha)} \in M_n(\Omega^2(U_\alpha)),$$

for every $\alpha \in I$ (cf. (3.13), as well as, e.g. (3.10)). ∎

Occasionally, it is still useful to write (3.12) in the following form: *Maurer-Cartan formula;*

(3.17) $$R = d\omega + \frac{1}{2}[\omega, \omega].$$

That is, one has, where ω is given by (3.4);

$$2(\omega\wedge\omega) = \omega\wedge\omega + \omega\wedge\omega = \omega\wedge\omega - \omega\wedge\omega \equiv [\omega,\omega], \tag{3.18}$$

which thus yields (3.17), by virtue of (3.12). ∎

On the other hand, assume again that we have the *general framework, referred to (3.12),* and let us further apply the curvature R to an arbitrary (local) section $s \in \mathcal{E}(U)$, as the latter is given by (3.3); thus, since R *is an* $\mathcal{A}$*-morphism* (cf. (2.10)), one has (see also (3.9))

$$\begin{aligned} R(s) &= R(\textstyle\sum_{i=1}^n s_i e_i) = \sum_i R(s_i\cdot e_i) = \sum_i s_i\cdot R(e_i) \\ &= \textstyle\sum_i s_i(\sum_k e_k\otimes(d\omega_{ki} + \sum_l \omega_{kl}\wedge\omega_{li})) \\ &= \textstyle\sum_k e_k\otimes(\sum_i s_i(d\omega_{ki} + \sum_l \omega_{kl}\wedge\omega_{li})) \in \mathcal{E}(U)\otimes_{\mathcal{A}(U)}\Omega^2(U). \end{aligned} \tag{3.19}$$

Thereore, one obtains (cf. (3.9));

$$R_{ij}(s) = \sum_{j=1}^n s_j\left(d\omega_{ij} + \sum_k \omega_{ik}\wedge\omega_{kj}\right) \in \Omega^2(U),\quad 1\leqq i\leqq n, \tag{3.20}$$

or even in *matrix notation* one has (cf. (3.12))

$$R(s) \equiv s\cdot R = s\cdot(d\omega + \omega\wedge\omega), \tag{3.21}$$

that is, a *row-matrix,* whose individual terms are given by (3.20). (The abuse of notation, applied for convenience in (3.19) and (3.21), is certainly clear).

Now, in the case of a *line sheaf* on X, one obtains, in particular, the following.

Corollary 3.1. *Suppose we have a curvature space X and a line sheaf $\mathcal{L}$ on X, endowed with an $\mathcal{A}$-connection D. Moreover, assume that*

$$\mathcal{U} = (U_\alpha)_{\alpha\in I} \tag{3.22}$$

is a local frame of $\mathcal{L}$ and let

$$\omega \equiv (\omega_\alpha) \in C^0(\mathcal{U},\Omega^1) = \prod_\alpha \Omega^1(U_\alpha) \tag{3.23}$$

be the 0-cochain of local $\mathcal{A}$-connection "1-forms" of D. Then, the curvature of D is given by the 0-cocycle;

$$\begin{aligned} R(D) \equiv R = (d\omega_\alpha) &\in Z^0(\mathcal{U}, d\Omega^1) \subsetneqq Z^0(\mathcal{U},\Omega^2) \\ &= \Omega^2(X) \subseteq \textstyle\prod_\alpha \Omega^2(U_\alpha). \end{aligned} \tag{3.24}$$

Proof. The assertion follows immediately from (3.15), (3.16), by just remarking that $\mathcal{E}nd\,\mathcal{L} = \mathcal{A}$ (up to an $\mathcal{A}$-*isomorphism.* See Chapt. II; (6.32), as well as, (5.15)). ∎

4. Change of local gauge

As the heading indicates, we examine in this Section the way the local matrices of the curvature (see e.g. (3.16)) alter, when one changes the local gauges that produce these matrices. However, before we embark on the relevant discussion, we first have to comment a bit more on the notion of an *"adjoint representation"* (cf. thus Chapt. VI; (1.37)), which will be of use in our subsequent calculations.

4.(a). Adjoint representation of a sheaf of groups in a given $\mathcal{A}$-module.– Suppose we have a **C**-*algebraized space* $(X, \mathcal{A})$, with X an arbitrary *topological space,* $\mathcal{G}$ a *sheaf of groups* (not necessarily abelian), as well as, an $\mathcal{A}$-*module* $\mathcal{E}$ on X. In particular, we consider presently below the case of the *general linear group sheaf,* of *order* $n \in \mathbf{N}$, on X, $\mathcal{GL}(n, \mathcal{A})$ (see Chapt. IV; (1.26), (1.29)).

Thus, by the *adjoint representation of* $\mathcal{GL}(n, \mathcal{A})$ *in* $\mathcal{E}$, one means the following group sheaf morphism (see, for instance, Chapt. V; (8.43))

$$\mathrm{Ad} : \mathcal{GL}(n, \mathcal{A}) \longrightarrow (\mathcal{E}nd\, M_n(\mathcal{E}))^{\cdot}, \tag{4.1}$$

such that one has;

$$[\mathrm{Ad}(\alpha)](s) := \alpha s \alpha^{-1}, \tag{4.2}$$

for any $\alpha \in \mathcal{GL}(n, \mathcal{A})(U) = \mathrm{GL}(n, \mathcal{A}(U))$ and $s \in M_n(\mathcal{E})(U) = M_n(\mathcal{E}(U))$, with U open in X. (In the second member of (4.2) one considers the usual *matrix multiplication* (of sections), given that $M_n(\mathcal{E})$ *is an* $M_n(\mathcal{A})$-*bimodule,* according to the hypothesis for $\mathcal{A}$; see Chapt. VI; (1.36)). Of course, it is easy to see that *the map (4.1),* as given by (4.2), *has the following properties,* which, in effect, have been alluded to already, by its definition; viz. *one has*

$$\mathrm{Ad}(\alpha \cdot \beta) = \mathrm{Ad}(\alpha) \cdot \mathrm{Ad}(\beta), \tag{4.3}$$

for any $\alpha, \beta \in \mathcal{GL}(n, \mathcal{A})(U)$, as above, so that *one still obtains*

$$\mathrm{Ad}(\alpha^{-1}) = \mathrm{Ad}(\alpha)^{-1}, \tag{4.4}$$

for any $\alpha \in \mathrm{GL}(n, \mathcal{A}(U))$.

In particular, suppose now that we are given a *curvature space* X (cf. Definition 1.1), so that let us further consider the $\mathcal{A}$*-modules* Ω^1 and Ω^2, together with the given *exterior derivative operator*

$$d^1 \equiv d : \Omega^1 \longrightarrow \Omega^2 \tag{4.5}$$

(see (1.7), along with (1.16)). Now, it is clear that *the operator (4.5) can be extended* ("coordinate-wise") *to a similar one between the corresponding matrix spaces for* Ω^1 *and* Ω^2, as above (for simplicity, we retain the same notation for the extended map), so that *one obtains the* following *commutative diagram;*

$$\begin{array}{ccc} M_n(\Omega^1) & \xrightarrow{\ Ad(\alpha)\ } & M_n(\Omega^1) \\ {\scriptstyle d}\downarrow & & \downarrow{\scriptstyle d} \\ M_n(\Omega^2) & \xrightarrow{\ Ad(\alpha)\ } & M_n(\Omega^2), \end{array} \tag{4.6}$$

for any $\alpha \in \mathcal{GL}(n, \mathcal{A})(U)$, and U *open* in X. *Equivalently,* one has

$$d \circ \mathrm{Ad}(\alpha) = \mathrm{Ad}(\alpha) \circ d, \tag{4.7}$$

for every $\alpha \in \mathcal{GL}(n, \mathcal{A})(U) = \mathrm{GL}(n, \mathcal{A}(U))$, as before: Indeed, *for every* $s \in M_n(\Omega^1)(U) = M_n(\Omega^1(U))$, one has (cf. also (4.2), (1.9) and Chapt. VI; (1.27.2));

$$\begin{aligned} (d \circ \mathrm{Ad}(\alpha))(s) &= d(\mathrm{Ad}(\alpha)(s)) = d(\alpha s \alpha^{-1}) \\ &= \alpha\, d(s\alpha^{-1}) - (s\alpha^{-1}) \wedge \partial\alpha = \alpha((ds)\alpha^{-1} - s \wedge \partial(\alpha^{-1})) + \alpha^{-1}\partial\alpha \wedge s \\ &= \alpha(ds)\alpha^{-1} + (\alpha\partial(\alpha^{-1}) + \alpha^{-1}\partial\alpha) \wedge s \\ &= [\mathrm{Ad}(\alpha)](ds) + (\tilde{\partial}(\alpha^{-1}) + \tilde{\partial}(\alpha)) \wedge s = (\mathrm{Ad}(\alpha) \circ d)(s), \end{aligned} \tag{4.8}$$

that was actually to be proved. ∎

We come now to our main objective of this discussion, as this is exposed by the ensuing Subsection.

4.(b). Change of curvature under a gauge transformation.– Suppose that we have a *curvature space* X and an *$\mathcal{A}$-module* $\mathcal{E}$ on X, equipped with an *$\mathcal{A}$-connection* D. Moreover, assume that we are given two *local gauges of* $\mathcal{E}$, say,

$$e^U \equiv \{U; (e_i, \ldots, e_n)\} \tag{4.9}$$

and

$$f^V \equiv \{V; (f_i, \ldots, f_n)\} \tag{4.10}$$

over the *open sets* U *and* V of X, *such that*

$$U \cap V \neq \emptyset, \tag{4.11}$$

and of the *same (finite) rank* $n \in \mathbf{N}$ see also Chapt. VII; (2.1)). Thus, suppose further that

$$g \equiv (g_{ij}) \in \mathrm{GL}(n, \mathcal{A}(U \cap V)) = \mathcal{GL}(n, \mathcal{A})(U \cap V) \tag{4.12}$$

is the *local gauge transformation,* which realizes the change from e^U to f^V, such that one has;

$$f_j = \sum_{i=1}^{n} g_{ij} e_i, \quad 1 \leqq j \leqq n \tag{4.13}$$

(see Chapt. VII; (2.3), (2.5), along with Remark 2.1).

Now, denoting by ω and ω' the *local $\mathcal{A}$-connection matrices of* D, which are associated with the given local gauges e^U and f^V, respectively (see e.g. (3.4)), one has the relation;

$$\omega' \equiv g^*(\omega) = \mathrm{Ad}(g^{-1})\omega + \tilde{\partial}(g) \tag{4.14}$$

(*transformation law of potentials.* Cf. Chapt. VII; (2.23)).

Thus, our next objective is now to find the corresponding relation to (4.14), pertaining to the *curvatures* R *and* R', which are defined by ω and ω', respectively (cf., for instance, (3.12)), when one takes the previous gauge transformation g into account. So, by *"differentiating"* (4.14), viz. by applying to it the *1st exterior derivation* $d^1 \equiv d$ (see (1.7) and (3.1)), one obtains (cf. also (1.12), (4.7) and (3.12), concerning the calculations that follow);

$$
\begin{aligned}
d\omega' &= d(\mathrm{Ad}(g^{-1})\omega + \tilde{\partial}(g)) = d(\mathrm{Ad}(g^{-1})\omega) + d(\tilde{\partial}(g)) \\
&= d(\mathrm{Ad}(g^{-1})\omega) = \mathrm{Ad}(g^{-1})(d\omega) = \mathrm{Ad}(g^{-1})(R - \omega \wedge \omega) \\
&= \mathrm{Ad}(g^{-1})R - \mathrm{Ad}(g^{-1})(\omega \wedge \omega).
\end{aligned} \tag{4.15}
$$

On the other hand, one still obtains (see (3.12) and (4.14));

$$
\begin{aligned}
d\omega' &= R' - \omega' \wedge \omega' = R' - (\mathrm{Ad}(g^{-1})\omega + \tilde{\partial}(g)) \wedge (\mathrm{Ad}(g^{-1})\omega + \tilde{\partial}(g)) \\
&= R' - \mathrm{Ad}(g^{-1})(\omega \wedge \omega) - \mathrm{Ad}(g^{-1})\omega \wedge \tilde{\partial}(g) - \tilde{\partial}(g) \wedge \mathrm{Ad}(g^{-1})\omega \\
&\quad + \tilde{\partial}(g) \wedge \tilde{\partial}(g) = R' - \mathrm{Ad}(g^{-1})(\omega \wedge \omega).
\end{aligned} \tag{4.16}
$$

(In this context, we have also applied above the relations (cf. (4.2) and (4.12)): $\mathrm{Ad}(g^{-1})\omega \wedge \mathrm{Ad}(g^{-1})\omega \equiv g^{-1}\omega g \wedge g^{-1}\omega g = g^{-1}(\omega g \wedge g^{-1}\omega)g \equiv g^{-1}(\omega \wedge g \wedge g^{-1} \wedge \omega)g = g^{-1}(\omega \wedge (gg^{-1}) \wedge \omega)g = g^{-1}(\omega \wedge \omega)g \equiv \mathrm{Ad}(g^{-1})(\omega \wedge \omega)$). Thus, by (4.15) and (4.16), one finally has;

$$
g^*(R) \equiv R' = \mathrm{Ad}(g^{-1})R \equiv g^{-1}Rg. \tag{4.17}
$$

So the last relation represents the *local behaviour of the curvature* R of a given $\mathcal{A}$-connection, *under a gauge transformation.*

Thus, an immediate consequence of (4.17) is now the fact that:

(4.18) *any gauge transformation preserves* (the eventual) *flatness* of a given $\mathcal{A}$-connection (cf. (2.5)).

Now, suppose, in particular, that $\mathcal{E}$ is a *vector sheaf,* of *rank* $n \in \mathbf{N}$, on a *curvature space* X, as before, such that $\mathcal{E}$ is also endowed with an *$\mathcal{A}$-connection* D. Moreover, let

$$
\mathcal{U} = (U_\alpha)_{\alpha \in I} \tag{4.19}
$$

be a *local frame* of $\mathcal{E}$, and

$$
(g_{\alpha\beta}) \in Z^1(\mathcal{U}, \mathcal{GL}(n, \mathcal{A})) \tag{4.20}
$$

a *coordinate 1-cocycle of* $\mathcal{E}$, relative to $\mathcal{U}$ (see Chapt. VII; Lemma 1.1, or even Theorem 2.1). Thus, for any *(local) gauge transformation* of $\mathcal{E}$,

$$
g_{\alpha\beta} \in \mathcal{GL}(n, \mathcal{A})(U_{\alpha\beta}) = \mathrm{GL}(n, \mathcal{A}(U_{\alpha\beta})), \tag{4.21}
$$

such that

$$
U_{\alpha\beta} := U_\alpha \cap U_\beta \neq \emptyset, \tag{4.22}
$$

with $\alpha, \beta \in I$, as in (4.19) (see also Chapt. V; (1.3) – (1.5)), one can consider, according to (4.17), the corresponding *local expression of changing the curvature* $R(D) \equiv R$ of the given $\mathcal{A}$-connection D; viz. one has

$$g^*_{\alpha\beta}(R) \equiv R' = \mathrm{Ad}(g^{-1}_{\alpha\beta})R \equiv g^{-1}_{\alpha\beta} R g_{\alpha\beta}. \tag{4.23}$$

In this context, we also note (as was actually already the case with (4.17), cf. (4.12), as well) that R above *stands, in effect, for*

$$R^{(\alpha\beta)} \equiv \left(R^{(\alpha\beta)}_{ij}\right) \in M_n(\Omega^2(U_{\alpha\beta})) \tag{4.24}$$

(similar observation for R'; cf. also (3.16), while we further notice that $M_n(\Omega^2)$ *is an* $M_n(\mathcal{A})$*-bimodule* on X, a fact actually *true, for every* $M_n(\Omega^p)$, *with* $p \in \mathbf{Z}_+$).

In particular, consider a *line sheaf* $\mathcal{L}$ on X, which is further endowed with an $\mathcal{A}$*-connection* D; viz. we are given a *Maxwell field*

$$(\mathcal{L}, D) \tag{4.25}$$

on X (cf. thus Chapt. VI; (18.11)). Therefore, if

$$(g_{\alpha\beta}) \in Z^1(\mathcal{U}, \mathcal{A}^{\cdot}) \tag{4.26}$$

is a *coordinate 1-cocycle of* $\mathcal{L}$, associated with a given *local frame* $\mathcal{U}$ *of* $\mathcal{L}$ (cf., for instance, (4.19), as well as, (4.20) for $n = 1$, along with Chapt. V; (2.59), (2.60)), one gets (cf. (4.23))

$$R' = g^{-1}_{\alpha\beta} R g_{\alpha\beta}, \tag{4.27}$$

such that (see (3.24))

$$R \in \Omega^2(X), \tag{4.28}$$

while one still has (cf. (4.22) and (4.26))

$$g_{\alpha\beta} \in \mathcal{A}^{\cdot}(U_{\alpha\beta}) = \mathcal{A}(U_{\alpha\beta})^{\cdot}; \tag{4.29}$$

of course, by analogy with (4.24) (viz., for $n = 1$), one actually has for R (and analogously for R') in (4.27):

$$R^{(\alpha\beta)} \equiv R|_{U_{\alpha\beta}}. \tag{4.30}$$

Therefore, in view of (4.27), (4.29) and of the *commutativity of* (the $\mathbf{C}$-algebra sheaf) $\mathcal{A}$, one finally obtains;

(4.31) $$R' = R.$$

That is,

(4.32) in the case of a *Maxwell field* $(\mathcal{L}, D)$ on a *curvature space* X, the *curvature of* D *does not change* (form) *under* (the action of) *a gauge transformation.* So, said otherwise, the *curvature* created by (or corresponding to) a Maxwell field *is space invariant.*

In this regard, see also Chapter X; Sections 4, 5 in the sequel for an appropriate *physical interpretation* of the last conclusion.

5. Flat $\mathcal{A}$-connections (contn'd)

The purpose of this Section is to relate the notion of a *flat* $\mathcal{A}$*-connection* D, as given above, that is, when one has

(5.1) $$R(D) \equiv R = 0$$

(see (2.5)), with the various notions of *"flatness"*, pertaining to a given $\mathcal{A}$*-module,* or, in particular, to a *vector sheaf,* which have already been considered in the preceding (cf. Chapt. V; Section 5, Chapt. VI; Subsection 11.(a) and Chapt. VII; Section 6). As we shall see, *there are important* particular *instances,* including, of course, that one of the standard differential geometry, where *all notions of flatness,* as alluded to before, *are,* in fact, *equivalent.*

So, by supplementing our previous information, as in the places just quoted above, we first state the following.

Theorem 5.1. *Suppose we are given a differential triad* $(\mathcal{A}, \partial, \Omega^1)$ *on a topological space* X, *along with a vector sheaf* $\mathcal{E}$ *on* X, *of rank* $n \in \mathbf{N}$. *Then, the following three assertions are equivalent* (see Chapt. VII; Theorem 6.1):

i) $\mathcal{E}$ is ∂*-flat.*

ii) $\mathcal{E}$ has an *integrable* $\mathcal{A}$*-connection,* and

iii) $\mathcal{E}$ has an $\mathcal{A}$-connection, supplying a *complete parallelism.*

In particular, if X, as above, is a *curvature space* (cf. Definition 1.1), then *any one of the* previous *three* equivalent assertions *implies* that;

iv) $\mathcal{E}$ has a *flat $\mathcal{A}$-connection* (cf. (5.1)).

Proof. Assume, for example, that the given vector sheaf $\mathcal{E}$ *has an $\mathcal{A}$-connection D, providing a complete parallelism.* Therefore (cf. Chapt. VII; Definition 6.1), there exists a *(parallel) local frame of $\mathcal{E}$*, say,

$$\mathcal{U} = (U_\alpha)_{\alpha \in I}, \tag{5.2}$$

such that the 0-cochain of local $\mathcal{A}$-connection matrices of D, which is associated with $\mathcal{U}$,

$$\omega \equis \left(\omega^{(\alpha)}\right) \in C^0(\mathcal{U}, M_n(\Omega^1)), \tag{5.3}$$

is trivial; viz. one has

$$\omega^{(\alpha)} \equiv \left(\omega_{ij}^{(\alpha)}\right) = 0 \in M_n(\Omega^1(U_\alpha)), \tag{5.4}$$

for every $\alpha \in I$ (cf. Chapt. VII; (6.7), (6.18) and (6.19)). Accordingly, by virtue of *Cartan's structural equation* (cf. (3.16)), one obtains

$$R^{(\alpha)} = 0 \in M_n(\Omega^2(U_\alpha)), \quad \textit{with } \alpha \in I, \tag{5.5}$$

so that *one gets* $R = 0$, as well (cf. also (2.19), (2.20) and (5.4) above); thus, the above proves actually that:

(5.6) *any $\mathcal{A}$-connection, providing a complete parallelism is flat.*

This, of course, yields now the assertion of the theorem, and the proof is complete. ∎

Scholium 5.1.– We have already seen (cf. Chapt. VII; (6.12)) that *every flat vector sheaf is ∂-flat* (see Chapt. V; Definition 5.1 and Chapt. VII; Definition 6.2). Consequently, as follows from the previous Theorem 5.1, one further obtains that:

(5.7) *every flat vector sheaf* (on a given curvature space) *admits a flat $\mathcal{A}$-connection.*

This, of course, justifies also the terminology applied so far. True, *in the classical case the two notions that appear in (5.7) are,* in effect, *equivalent.* However, *this is not necessarily the case here,* a fact that will become more apparent through the ensuing discussion. On the other hand, as we shall presently see, something better is actually valid in the particular case of a *line sheaf* over a suitable curvature space (see thus Theorem 5.2 in the sequel).

Now, the question of *whether flatness of* a given *$\mathcal{A}$-connection implies any one of the three equivalent notions,* which *appear in* the previous *Theorem 5.1,* is encountered by the next Subsection.

5.(a). Frobenius integrability condition.– We start with the following useful result, providing also half of the answer to the above question. So we have the next.

Lemma 5.1. *Suppose we are given a curvature space X and let $\mathcal{E}$ be an $\mathcal{A}$-module on X, endowed with an $\mathcal{A}$-connection D. Moreover, let*

$$e^U \equiv \{U; (e_i, \ldots, e_n)\} \tag{5.8}$$

be a local gauge of $\mathcal{E}$, of rank $n \in \mathbf{N}$, over an open set $U \subseteq X$ (see, for instance, (4.9) in the preceding, or even Chapt. VII; (2.1), *and*

$$g \in \mathrm{GL}(n, \mathcal{A}(U \cap V)) = \mathcal{GL}(n, \mathcal{A})(U \cap V) \tag{5.9}$$

a gauge transformation, sending (5.8) to another local gauge of $\mathcal{E}$, (of the same rank, as (5.8), over an open $V \subseteq X$, with $U \cap V \neq \emptyset$), *this being also a parallel one, with respect to the given $\mathcal{A}$-connection D* (see Chapt. VII; (6.3), along with (4.10), (4.12) and (4.13) in the preceding).

Then, D restricted to $U \cap V$ (cf. Chapt. VI; (6.22)) *is flat; that is, one has:*

$$R(D|_{U \cap V}) = R(D)|_{U \cap V} \equiv R|_{U \cap V} = 0 \tag{5.10}$$

(cf. also (2.11) and (3.11), as well as, (9.28) in the sequel).

Proof. Suppose that

$$f^V \equiv \{V; (f_j)_{1 \leqq j \leqq n}\} \tag{5.11}$$

is a *parallel local gauge of $\mathcal{E}$*, with respect to D (as in the statement of the theorem), arising from a *gauge transformation g*, like (5.9), acting on (5.8). Thus, by further considering the *local $\mathcal{A}$-connection matrices of D*, that are associated with (5.8) and (5.11), say,

$$\omega_U \equiv \omega \in M_n(\Omega^1(U)) \quad and \quad \omega_V \equiv \omega' \in M_n(\Omega^1(V)), \tag{5.12}$$

respectively, one has, according to the *transformation law of potentials* (cf. Chapt. VII; Lemma 2.1), the relation;

$$\omega' = \mathrm{Ad}(g^{-1})\omega + \tilde{\partial}(g) \tag{5.13}$$

(of course, all the *section-matrices* in (5.13) are defined *over* $U \cap V \neq \emptyset$). However, in view of the *hypothesis for (5.11)*, one has (cf. Chapt. VII; (6.3))

$$\omega' = 0, \tag{5.14}$$

so that one now obtains, by (5.13),

$$\mathrm{Ad}(g^{-1})\omega + \tilde{\partial}(g) = 0. \tag{5.15}$$

In other words, one has

$$g^{-1}\omega g + g^{-1}\partial(g) = 0, \tag{5.16}$$

or even

$$\omega g + \partial(g) = 0. \tag{5.17}$$

Thus, one further obtains;

$$\partial(g) = -\omega \cdot g. \tag{5.18}$$

Now, we next apply the operator $d^1 \equiv d$ (see e.g. (3.1)) to the above rel. (5.17), alias, we "differentiate" (5.17). Thus, we obtain (cf. also (1.8), (1.9), (5.18) and (3.12));

$$\begin{aligned} 0 &= d(\omega g + \partial g) = d(\omega g) + d(\partial g) = d(\omega g) \\ &= (d\omega)g - \omega \wedge \partial g = (d\omega)g - \omega \wedge (-\omega g) = (d\omega)g + (\omega \wedge \omega)g \\ &= (d\omega + \omega \wedge \omega) \cdot g = R \cdot g. \end{aligned} \tag{5.19}$$

That is, one finally has

$$R = 0, \tag{5.20}$$

as actually was to be proved. ∎

Scholium 5.2.- Concerning (5.18), we further note, for use presently below, that:

the above rel. (5.18) *is* still *equivalent with the relation*

$$\tilde{\partial}(g^{-1}) = \omega. \tag{5.21}$$

(See also Chapt. VI; (1.27.2), as well as, (1.42) therein, concerning the notation applied in (5.21)). Thus, *one* finally *concludes that:*

$$\omega \in \operatorname{im}\tilde{\partial}. \tag{5.22}$$

Now, the last relation characterizes, in fact, what we are going to call, right below, *trivial $\mathcal{A}$-connections.* See thus (5.47) and (5.49) in the sequel, along with (5.53) for *another justification of (5.10), based on (5.22).*

Thus, by summarizing the preceding, we are led to the following comments:

(5.23) Suppose we are given an *$\mathcal{A}$-module $\mathcal{E}$* on a *curvature space X*, endowed with an *$\mathcal{A}$-connection D*. Then, (5.10), viz. the *vanishing of the curvature,* is a *necessary condition for the existence of a* (local) *gauge transformation* (like (5.9)), *sending a given local gauge of $\mathcal{E}$ to a parallel one,* with respect to D.

On the other hand, the *sufficiency of (5.10),* to get the above situation, as described by (5.23) (or even by the *hypothesis of Lemma 5.1*), is actually, as it concerns the classical context, the *Frobenius integrability condition;* in other words,

(5.24) (5.10), viz. the *vanishing of the curvature, is a sufficient condition* too, *in order the equation* (cf. (5.17))

$$\omega g + \partial(g) = 0,$$

or, *equivalently* (see (5.21)), *the equation*

$$\tilde{\partial}(g^{-1}) = \omega,$$

to have a solution, with respect to g, as in (5.9), *when ω* (the *"potential" of D*) *is known.*

In this context, we are able to give a positive answer to the above question, by further imposing suitable *topological and cohomological conditions on* $\mathcal{A}$, as we shall see later on (cf. thus Chapter XI in the sequel. Yet, concerning the classical case, see, for instance, M. Crampin - F.A.E. Pirani [1: p. 146, and p. 282f], or even S. Kobayashi [1: p. 5]).

However, by restricting ourselves to *line sheaves,* we can have a complete answer to our previous question, through other means, that we already have at our disposal; yet, this by considering a suitable *restriction on the curvature space* X involved, as above. So one has the following.

Theorem 5.2. *Suppose we have a curvature space* X*, being also a sheaf exponential* (cf. Chapt. VII; (7.1)) *and paracompact (Hausdorff) space. Moreover, let* $\mathcal{L}$ *be a line sheaf on* X*. Then, the following assertions are equivalent:*

1) $\mathcal{L}$ is *∂-flat.*
2) $\mathcal{L}$ has an *integrable* $\mathcal{A}$*-connection.*
3) $\mathcal{L}$ admits an $\mathcal{A}$*-connection, yielding a complete parallelism,* and
4) $\mathcal{L}$ has a *flat* $\mathcal{A}$*-connection.*

Proof. We already know, by Theorem 5.1, that *the first three assertions are equivalent, while any one of them implies the fourth.* So we prove now, for instance, that *4) implies 1),* as well; viz. *if* $\mathcal{L}$ *has a flat* $\mathcal{A}$*-connection* $\widetilde{D}$, then, $\mathcal{L}$ *is ∂-flat* (cf. Chapt. VII; Definition 6.2):

Thus, assume that

$$\mathcal{U} = (U_\alpha)_{\alpha \in I} \tag{5.25}$$

is a *local frame of* $\mathcal{L}$, and let

$$(g_{\alpha\beta}) \in Z^1(\mathcal{U}, \mathcal{A}^{\cdot}) \tag{5.26}$$

be a *coordinate 1-cocycle of* $\mathcal{L}$, relative to $\mathcal{U}$. Furthermore, suppose that

$$\theta \equiv (\theta_\alpha) \in C^0(\mathcal{U}, \Omega^1) \tag{5.27}$$

is a *0-cochain of local* $\mathcal{A}$*-connection "1-forms"* of the given $\mathcal{A}$-connection $\widetilde{D}$ of $\mathcal{L}$, with respect to $\mathcal{U}$, while

$$\delta(D_\alpha) \in Z^1(\mathcal{U}, \Omega^1) \tag{5.28}$$

is the corresponding *Levi-Civita* $\mathcal{A}$*-connection 1-cocycle of* $\mathcal{L}$, associated with $\mathcal{U}$. So one has

$$\delta(D_\alpha) = \delta(\theta_\alpha) = \tilde{\partial}(g_{\alpha\beta}) \tag{5.29}$$

(see Chapt. VI; (11.8), in conjunction with Theorem 10.1. Yet, see also, for example, Chapt. VII; (2.59)).

On the other hand, according to our hypothesis for X (cf. Chapt. VII; (7.1), and Chapt. V; (6.1)), one has the following *short exact exponential sheaf sequence;*

$$0 \longrightarrow \mathbf{Z} \longrightarrow \mathcal{A} \stackrel{e}{\longrightarrow} \mathcal{A}^{\cdot} \longrightarrow 1. \tag{5.30}$$

Thus, since, by hypothesis, X is a *paracompact (Hausdorff)* space too, given the 1-cocycle (5.26), *there exists a 1-cochain*

$$(f_{\alpha\beta}) \in C^1(\mathcal{V}, \mathcal{A}) \tag{5.31}$$

such that one has

$$g_{\alpha\beta} = e(f_{\alpha\beta}) \tag{5.32}$$

(*"liftable cohomology";* cf. Chapt. III; Lemma 5.1). Moreover, one obtains (loc. cit.), in view also of (5.30);

$$\delta(f_{\alpha\beta}) \in Z^2(\mathcal{V}, \ker e = \mathbf{Z}). \tag{5.33}$$

Furthermore, according to the *exponential sheaf diagram,* that is associated with (5.30), by hypothesis for X (cf. Chapt. VII; (7.1)), one has the following *commutative triangle,* supplementing (5.30);

$$\begin{array}{ccc} \mathcal{A} & \stackrel{e}{\longrightarrow} & \mathcal{A}^{\cdot} \\ & {\scriptstyle\partial}\searrow \quad \swarrow{\scriptstyle\frac{1}{2\pi i}\tilde{\partial}} & \\ & \Omega^1. & \end{array} \tag{5.34}$$

Therefore, one has, in view of (5.32),

$$\tilde{\partial}(e(f_{\alpha\beta})) = \tilde{\partial}(g_{\alpha\beta}) = 2\pi i \cdot \partial(f_{\alpha\beta}). \tag{5.35}$$

Moreover, one still obtains;

$$\partial(f_{\alpha\beta}) \in Z^1(\mathcal{V}, \partial\mathcal{A}) \stackrel{\subseteq}{\rightarrow} Z^1(\mathcal{V}, \Omega^1). \tag{5.36}$$

Indeed, one has

(5.37) $$\delta(\partial(f_{\alpha\beta})) = \partial(\delta(f_{\alpha\beta})) = 0,$$

as immediately follows from (5.33) and Chapt. VI; (1.5), which thus justifies (5.36). Therefore, one gets a cohomology class

(5.38) $$[\partial(f_{\alpha\beta})] \in H^1(X, \Omega^1),$$

that will be of use right below.

Namely, consider now the following *short exact sequence* (of **C**-*vector space sheaves*),

(5.39) $$0 \longrightarrow \ker d \longrightarrow \Omega^1 \longrightarrow d\Omega^1 \longrightarrow 0,$$

along with the corresponding *long exact sequence in (Čhech) cohomology,*

(5.40) $$\cdots \longrightarrow \Gamma(X, d\Omega^1) \longrightarrow H^1(X, \ker d) \longrightarrow H^1(X, \Omega^1) \longrightarrow \cdots$$

(cf. Chapt. III; Theorem 5.3. See also Theorem 8.1 therein). Thus, the exactness of the last sequence and (1.8) yield now for (5.38) the relation;

(5.41) $$\begin{aligned}[\partial(f_{\alpha\beta})] &\in \ker(H^1(X, \ker d) \longrightarrow H^1(X, \Omega^1)) \\ &= \operatorname{im}(\Gamma(X, d\Omega^1) \longrightarrow H^1(X, \ker d)).\end{aligned}$$

On the other hand, based on the hypothesis for (5.27) and on (3.24), one obtains;

(5.42) $$R = (d\theta_\alpha) \in Z^0(\mathcal{U}, d\Omega^1) = \Gamma(X, d\Omega^1)$$

(cf. also Chapt. III; (4.55) and (8.13)). At the same time, one infers from (5.41) that (cf. Chapt. III; (5.21), for the notation applied)

(5.43) $$\delta^*(R) = [\partial(f_{\alpha\beta})] \in H^1(X, \ker d).$$

Furthermore, by applying again Lemma 5.1 of Chapter III to (5.42), in conjunction with (5.39), one concludes the existence of a *0-cochain*

(5.44) $$(\theta'_\alpha) \in C^0(\mathcal{V}, \Omega^1),$$

such that one has;

(5.45) $$R = (d\theta'_\alpha) \text{ and } \delta^*(R) = [(\delta\theta'_\alpha)] \in H^1(X, \ker d).$$

Therefore, based on (5.45), (5.41) (or even (5.43)), along with (5.29) and (5.35), we conclude that

(5.46) $$\delta(\theta'_\alpha) = \partial(f_{\alpha\beta}) = \frac{1}{2\pi i}\delta(\theta_\alpha) = \frac{1}{2\pi i}\tilde{\partial}(g_{\alpha\beta}).$$

Therefore, since *by* our *hypothesis*, $R = 0$ (and δ^*, as in (5.43) or (5.45), is a **C**-*linear map,* cf. Chapt. III; (5.20)), one has from (5.46)

(5.47) $$[\tilde{\partial}(g_{\alpha\beta})] = 0 \in H^1(X, \Omega^1).$$

However, the last relation implies already that

(5.48) $$\partial(g_{\alpha\beta}) = 0$$

(*Selesnick's Lemma;* see Chapt. VII; Lemma 7.2). That is, the given line sheaf $\mathcal{L}$ *is ∂-flat,* that actually was our claim, and this terminates the proof. ∎

In this context, a byproduct of the previous proof, which is of a particular significance for the sequel, as well (see thus Theorem 11.1 below), is now the following.

Lemma 5.2. *Suppose we have a sheaf exponential (Hausdorff) paracompact curvature space X. Moreover, let $\mathcal{L}$ be a line sheaf on X, endowed with an $\mathcal{A}$-connection D. Then, the curvature of D, $R(D) \equiv R$, supplies a 2-dimensional integral cohomology class of X. That is, denoting the latter class by $c(R)$, one has;*

(5.49) $$c(R) \in H^2(X, \mathbf{Z}).$$

Scholium 5.3.– By looking at the *canonical embedding*

(5.50) $$\mathbf{Z} \hookrightarrow \mathbf{C}$$

of the *constant sheaves* concerned, along with the corresponding map in cohomology,

(5.51) $$H^p(X, \mathbf{Z}) \longrightarrow H^p(X, \mathbf{C}), \quad p \in \mathbf{Z}_+,$$

one calls *(p-dimensional) integral cohomology class,* any element in the image of (5.51).

So this explains now the terminology applied before for (5.49). Yet, since by our hypothesis in the previous Lemma 5.2, X is a (Hausdorff) *paracompact* space, one may consider, of course, in the same relation *Čech cohomology* (cf. Chapt. III; (8.11), or even Theorem 8.1).

Proof of Lemma 5.2. Having, by hypothesis, the framework of the above Theorem 5.2, we first remark that (see (5.42))

$$R = (d\theta_\alpha) \in \Gamma(X, d\Omega^1) = Z^0(\mathcal{U}, d\Omega^1). \tag{5.52}$$

Furthermore, as a consequence of (5.52) and the exactness of (5.39), one also obtains (see (5.45))

$$R = (d\theta'_\alpha) = (d\theta_\alpha), \tag{5.53}$$

such that one has (cf. (5.46));

$$\delta(\theta'_\alpha) = \partial(f_{\alpha\beta}) = \frac{1}{2\pi i}\tilde{\partial}(g_{\alpha\beta}), \tag{5.54}$$

while, we still have (see (5.33)):

$$\delta(f_{\alpha\beta}) \equiv (\lambda_{\alpha\beta\gamma}) \in Z^2(\mathcal{V}, \mathbf{Z}). \tag{5.55}$$

Thus, based now on the last relation, we further obtain

$$c(R) = \left[\frac{1}{2\pi i}R\right] = \frac{1}{2\pi i}[R] := [(\lambda_{\alpha\beta\gamma})] \in H^2(\mathcal{V}, \mathbf{Z}), \tag{5.56}$$

which is just our claim. (See also the next Scholium 5.4, concerning the last definition in (5.56); yet, cf. Chapt. IX; (5.15) in the sequel). ∎

Scholium 5.4.- Concerning the last defining relation in (5.56), this is justified as follows:

First, one has (see (5.54))

$$\begin{aligned} \delta(\theta'_\alpha) = \partial(f_{\alpha\beta}) \in (\partial\mathcal{A})(U_{\alpha\beta}) &\subseteq (\ker d)(U_{\alpha\beta}) \\ &\subseteq \Omega^1(U_{\alpha\beta}) \end{aligned} \tag{5.57}$$

(see also e.g. (4.22) for the notation applied above). Hence, in particular, one obtains;

$$\partial(f_{\alpha\beta}) \in Z^1(\mathcal{U}, \partial\mathcal{A}) \overset{\subseteq}{\rightarrow} Z^1(\mathcal{U}, \Omega^1). \tag{5.58}$$

Yet, since (cf. (5.53))

$$R = (d\theta'_\alpha) = (d\theta_\alpha), \tag{5.59}$$

one now defines the *2-dimensional cohomology class of* R (a *closed "2-form"*, as we shall see right below), by the relation;

$$[R] := [2\pi i \cdot \delta(f_{\alpha\beta})] = 2\pi i \cdot [\delta(f_{\alpha\beta})], \tag{5.60}$$

such that, in view of (5.33), one obtains

$$[\delta(f_{\alpha\beta})] \equiv [(\lambda_{\alpha\beta\gamma})] \in H^2(X, \mathbf{Z}). \tag{5.61}$$

So this clarifies the situation in (5.56). (In this regard, see also Chapt. IX; Section 5. Yet, the appearance of the *"coefficient* $2\pi i$*"* in (5.60), or (5.56), is essentially due to the classical definition of the "exp" *function,* in our case, *section;* cf. also Chapter XI in the sequel).

Of course, connections of the above with what we have called in the preceding *Selesnick-Chern class of* $\mathcal{L}$, $c_{\mathcal{A}}(\mathcal{L})$, with $\mathcal{L}$ a given *line sheaf,* are obviously in force; so see, for instance, Chapt. VII; Theorem 7.1 and its Corollary 7.1.

Now, although not directly connected with the material of this particular Subsection, we do give, for convenience of reference, the following result, being also a further application of the aforementioned *Selesnick's lemma,* in conjunction with *Atiyah's criterion* about the existence of an $\mathcal{A}$-connection (cf. Chapt. VI; Theorem 11.1, or even Corollary 12.1). Thus, we have.

Proposition 5.1. *Suppose we are given a curvature space* X*, which is further endowed with a commutative triangle, like (5.34), where, in particular, the (exponential) morphism* e *is also surjective* (take, for instance, a *sheaf exponential curvature space* X). *Moreover, let* $\mathcal{L}$ *be a line sheaf on* X*, admitting an* $\mathcal{A}$*-connection. Then,* $\mathcal{L}$ *has a flat* $\mathcal{A}$*-connection too.*

Proof. Since, $\mathcal{L}$ admits, by hypothesis, an $\mathcal{A}$-connection, the rel. (5.47) is satisfied (*Atiyah's criterion;* loc. cit.). Hence (*Selesnick's lemma,* cf. Chapt. VII; Lemma 7.2), (5.48) is still in force; in other words, (see Chapt. VII; Definition 6.2), *the* given *line sheaf* $\mathcal{L}$ *is* ∂*-flat,* so that, according to the previous Theorem 5.1, it also admits a flat $\mathcal{A}$-connection. ∎

6. Trivial $\mathcal{A}$-connections

We consider below another useful application of the notion of *flatness;* for the classical counterpart see, for instance, B.A. Dubrovin - A.T. Fomenko - S.P. Novikov [1: p. 444, Theorem 41.2.2].

Thus, suppose that we are given a *curvature space* X and let $\mathcal{E}$ be a *vector sheaf* on X, such that

$$\mathrm{rk}\,\mathcal{E} = n \in \mathbf{N}. \tag{6.1}$$

(We could also take, instead, more generally, an $\mathcal{A}$-*module,* as, for instance, in Section 3; cf. e.g. (3.2)). Furthermore, consider a *local frame*

$$\mathcal{U} = (U_\alpha)_{\alpha \in I} \tag{6.2}$$

of $\mathcal{E}$, as well as, an $\mathcal{A}$-*connection* D of $\mathcal{E}$, with

$$\omega \equiv \left(\omega^{(\alpha)}\right) \equiv \left(\left(\omega_{ij}^{(\alpha)}\right)\right) \in C^0(\mathcal{U}, M_n(\Omega^1)) \tag{6.3}$$

the *0-cochain of local* $\mathcal{A}$-*connection matrices of* D, associated with $\mathcal{U}$ (cf. Chapt. VII; (1.13), or even (2.39)).

Now, we say that D is a *trivial* $\mathcal{A}$-*connection,* whenever, concerning (6.3), as above, one has, *first;*

$$\begin{aligned} \omega \equiv \left(\omega^{(\alpha)}\right) \in \tilde{\partial}(C^0(\mathcal{U}, \mathcal{GL}(n, \mathcal{A}))) &= C^0(\mathcal{U}, \tilde{\partial}(\mathcal{GL}(n, \mathcal{A}))) \\ &\subseteq C^0(\mathcal{U}, M_n(\Omega^1)). \end{aligned} \tag{6.4}$$

That is, there exists a *0-cochain,*

$$g \equiv \left(g^{(\alpha)}\right) \in C^0(\mathcal{U}, \mathcal{GL}(n, \mathcal{A})), \tag{6.5}$$

such that one further sets;

$$\omega := \tilde{\partial}(g^{-1}), \tag{6.6}$$

thus, an *equivalent expression of (6.4).*

Now, by considering further the $n \times n$ *(section-)matrix*

$$g^{(\alpha)} \equiv \left(g_{ij}^{(\alpha)}\right) \in \mathcal{GL}(n, \mathcal{A})(U_\alpha) = \mathrm{GL}(n, \mathcal{A}(U_\alpha)), \tag{6.7}$$

as in (6.5), one actually obtains an element of

$$(\mathcal{A}ut\mathcal{E})(U_\alpha) = \mathrm{Aut}(\mathcal{E}|_{U_\alpha}) \tag{6.8}$$

(see Chapt. VI; (17.24)); thus, if $U_{\alpha\beta} \equiv U_\alpha \cap U_\beta \neq \emptyset$, with $\alpha, \beta \in I$, as in (6.2), and one sets (cf. (6.7))

$$g_{\alpha\beta} := g^{(\alpha)}|_{U_{\alpha\beta}} \equiv \left(g_{ij}^{(\alpha)}|_{U_{\alpha\beta}}\right) \in \mathrm{GL}(n, \mathcal{A}(U_{\alpha\beta})), \tag{6.9}$$

then, *second,* we also assume that the *1-cochain of* $\mathcal{U}$, as given by (6.9), yields, in effect, a *gauge transformation of* $\mathcal{E}$ (loc. cit., comments following (17.30)), such that, in particular,

$$\tilde{g} \equiv (g_{\alpha\beta}) \in Z^1(\mathcal{U}, \mathcal{GL}(n, \mathcal{A})) \tag{6.10}$$

is a *coordinate 1-cocycle of* $\mathcal{E}$. (See, for example, the proof of Theorem 6.1 below, in particular, (6.38)).

Thus, by looking now at the corresponding to $\tilde{g}$, as in (6.10), *transformation law of potentials* (cf. Chapt. VII; (2.43), or even (2.51)), one has the following. (See also (6.6), (6.9) above, together with Chapt. VI; (1.33), (1.30) and (1.7)):

$$\begin{aligned} \tilde{g}^*\left(\omega^{(\alpha)}\right) &\equiv \omega^{(\beta)} = \mathrm{Ad}(g_{\alpha\beta}^{-1})\omega^{(\alpha)} + \tilde{\partial}(g_{\alpha\beta}) \\ &= \mathrm{Ad}(g_{\alpha\beta}^{-1})\tilde{\partial}(g_{\alpha\beta}^{-1}) + \tilde{\partial}(g_{\alpha\beta}) = \tilde{\partial}(g_{\alpha\beta}^{-1} \cdot g_{\alpha\beta}) = \tilde{\partial}(1) = 0. \end{aligned} \tag{6.11}$$

Accordingly, the corresponding local form of the *curvature* of the given $\mathcal{A}$-connection D (cf. (6.3)) has now as follows;

$$\begin{aligned} \tilde{g}^*\left(R^{(\alpha)}\right) &= R^{(\beta)} = d\omega^{(\beta)} + \omega^{(\beta)} \wedge \omega^{(\beta)} = 0 \\ &= g_{\alpha\beta}^{-1} R^{(\alpha)} g_{\alpha\beta} \end{aligned} \tag{6.12}$$

(see also (3.16) and (4.23)). Hence, one obtains

$$R^{(\alpha)} = 0, \quad \text{for every } \alpha \in I, \tag{6.13}$$

that is (cf. (2.19), (2.20), or even (3.15)), finally, one has;

$$R = 0. \tag{6.14}$$

In other words (cf. (2.5)), one thus concludes that:

(6.15) *every trivial $\mathcal{A}$-connection is flat.* (It concerns here an *$\mathcal{A}$-connection* of a *vector sheaf* $\mathcal{E}$ on a given *curvature space* X).

Scholium 6.1.– The previous conclusion (6.15) furnishes also another way of inferring that the *"canonical flat $\mathcal{A}$-connection"*

$$\partial : \mathcal{A} \longrightarrow \Omega^1 \tag{6.16}$$

(cf. Chapt. VI; (4.2)) *is indeed flat,* by showing that ∂ *is trivial,* in the sense of (6.6) (while (6.10) is trivially satisfied): Thus, by taking the *identity section of* $\mathcal{A}$ (alias, canonical *section-basis* of the *line sheaf* $\mathcal{A}$), one has

$$\tilde{\partial}(1) = \partial(1) = 0. \tag{6.17}$$

(See Chapt. VI; (1.7), as well as, e.g. Chapt. VII; (1.11)). So (6.17) proves our claim, hence, by virtue of (6.15), ∂ *is a flat* $\mathcal{A}$*-connection of* $\mathcal{A}$, in the sense that,

$$R(\partial) = 0. \tag{6.18}$$

(Of course, we already know (6.18), according to (2.6), while one comes to the *same conclusion,* by virtue of (6.17) and (3.12)).

On the other hand, the *converse of (6.15)* is still related with the aforementioned *Frobenius integrability condition;* see thus (5.24) in the foregoing, as well as, Chapter XI in the sequel. It seems, namely, that some *extra conditions have to be imposed on* $\mathcal{A}$*, for the converse of (6.15).* Indeed, this is demonstrated here in the particular case of a *line sheaf,* where an additional *cohomological condition* is put *on* $\mathcal{A}$ (cf. (6.19) below), the latter being otherwise just a $\mathbf{C}$*-algebra sheaf,* as was the case hitherto. Precisely, one thus obtains the following.

Theorem 6.1. *Suppose we have a curvature space* X (see (1.16)), *such that* (cf. (1.8), (1.11))

$$\operatorname{im}\partial = \ker d; \tag{6.19}$$

viz., equivalently, we assume the following short exact sequence

$$0 \longrightarrow \partial\mathcal{A} \longrightarrow \Omega^1 \xrightarrow{d^1 \equiv d} d\Omega^1 \longrightarrow 0. \tag{6.20}$$

Furthermore, suppose we are given the following commutative triangle;

(6.21)
$$\begin{array}{ccc} \mathcal{A} & \xrightarrow{e} & \mathcal{A}^{\cdot} \\ & \partial \searrow \quad \swarrow \frac{1}{2\pi i}\tilde{\partial} & \\ & \partial\mathcal{A}. & \end{array}$$

Finally, let $\mathcal{L}$ be a line sheaf on X. Then, a given $\mathcal{A}$-connection D of $\mathcal{L}$ is trivial if, and only if, D is flat.

Proof. Based on (6.15), it suffices only to prove the *"if"* part of the assertion; that is, assume that *the* given *$\mathcal{A}$-connection D* of *$\mathcal{L}$ is flat,* so that one obtains (cf. (3.24), or even (5.42))

(6.22) $$R = (d\omega_\alpha) = 0 \in \Gamma(X, d\Omega^1) = Z^0(\mathcal{U}, d\Omega^1).$$

Therefore, concerning the *local $\mathcal{A}$-connection "1-form" of D*, as in (6.22),

(6.23) $$\omega \equiv (\omega_\alpha) \in C^0(\mathcal{U}, \Omega^1),$$

where

(6.24) $$\mathcal{U} = (U_\alpha)_{\alpha \in I}$$

is a given *local frame* of $\mathcal{L}$, one has, by virtue of (6.22), that;

(6.25) $$d\omega \equiv d(\omega_\alpha) := (d\omega_\alpha) = 0,$$

so that, *equivalently,* one obtains

(6.26) $$\omega \equiv (\omega_\alpha) \in \ker d,$$

or even (cf. (6.19))

(6.27) $$\omega_\alpha \in \ker d = \operatorname{im} \partial, \quad \alpha \in I.$$

Now, since the *section functor* Γ is *left exact* (cf. Chapt. II; (2.59)), one concludes from the exactness of (6.20), an analogous *exactness of the sequence*

(6.28) $$0 \longrightarrow \Gamma(\partial\mathcal{A}) \xrightarrow{\partial} \Gamma(\Omega^1) \xrightarrow{d} \Gamma(d\Omega^1)$$

(loc. cit.; (2.57), (2.58)), where, of course, we still have set, for convenience,

(6.29) $$\Gamma(\partial) \equiv \partial \ \textit{and} \ \Gamma(d) \equiv d$$

(explaining thus the notation in (6.27), as well; see also Chapt. I; (13.3)). Hence, one obtains from (6.27)

$$\omega_\alpha \in (\ker d)(U_\alpha) = (\operatorname{im} \partial)(U_\alpha) \equiv (\partial \mathcal{A})(U_\alpha), \tag{6.30}$$

for every $\alpha \in I$. Accordingly, since *a sheaf morphism,* like ∂, *is fiber-preserving* (Chapt. I; Definition 1.2), one finds a *(local) section*

$$f_\alpha \in \mathcal{A}(V_\alpha), \quad \text{for some } \textit{open } V_\alpha \subseteq U_\alpha, \tag{6.31}$$

in such a manner that

$$\omega_\alpha = \partial(f_\alpha)|_{V_\alpha}, \quad \alpha \in I, \tag{6.32}$$

(see also Chapt. I; (2.9), (2.12)).

Therefore, by further applying (6.21), one first sets;

$$g_\alpha := e\left(\frac{1}{2\pi i} f_\alpha\right)^{-1} \in \mathcal{A}^{\boldsymbol{\cdot}}(V_\alpha) = \mathcal{A}(V_\alpha)^{\boldsymbol{\cdot}}, \quad \alpha \in I, \tag{6.33}$$

so that, by virtue of the hypothesis for (6.21), one has

$$\begin{aligned} \tilde{\partial}(g_\alpha^{-1}) &= \tilde{\partial}\left(e\left(\tfrac{1}{2\pi i} f_\alpha\right)\right) = 2\pi i \cdot \partial\left(\tfrac{1}{2\pi i} f_\alpha\right) \\ &= 2\pi i \cdot \tfrac{1}{2\pi i} \partial(f_\alpha) = \partial(f_\alpha) \in \partial(\mathcal{A}(V_\alpha)) \subseteq (\partial\mathcal{A})(V_\alpha) \end{aligned} \tag{6.34}$$

(see also Chapt. I; Proposition 2.1, concerning the last relation in (6.34)). Hence, in view of (6.32), one has

$$\omega_\alpha = \tilde{\partial}(g_\alpha^{-1})|_{V_\alpha}, \quad \alpha \in I. \tag{6.35}$$

In other words, one thus arrives at a *0-cochain*

$$g \equiv (g_\alpha) \in C^0(\mathcal{V}, \mathcal{A}^{\boldsymbol{\cdot}}) \tag{6.36}$$

(see (6.33), such that one has, by (6.35),

$$\omega = \tilde{\partial}(g^{-1}), \tag{6.37}$$

that is, the desired relation (6.6) for the *triviality of* D. (Yet, concerning (6.36), $\mathcal{V}$ stands there for the *local frame of* $\mathcal{L}$, defined by the V_α's, as in (6.31), in fact, a *refinement* of the given $\mathcal{U}$, as in (6.24); in this regard, see e.g. Chapt. II, proof of (4.9)). On the other hand, one further verifies (6.10), by just setting;

$$\tilde{g} \equiv (g_{\alpha\beta}) := \delta(g^{-1}) \equiv (g_\alpha g_\beta^{-1}) \in Z^1(\mathcal{V}, \mathcal{A}^{\boldsymbol{\cdot}}), \tag{6.38}$$

which thus, together with (6.37), prove that the given $\mathcal{A}$-connection D *is trivial,* as was our claim, and the proof is complete. ∎

As an immediate consequence of the preceding, one has now the following.

Corollary 6.1. *Suppose we have the framework of the above Theorem 5.2, while we also assume* (cf. (6.19)) *that;*

$$\ker d = \operatorname{im} \partial. \tag{6.39}$$

Furthermore, suppose that $\mathcal{L}$ is a line sheaf on X, equipped with an $\mathcal{A}$-connection D.

Then, triviality of D (cf. (6.6), (6.10)) *is a fifth further equivalent condition to the rest four ones for D, as in Theorem 5.2.* ∎

Note 6.1.– It is worth remarking here, in anticipation, that:

(6.40) *the context of the previous Corollary 6.1 is what one actually needs to prove "Weil's integrality theorem".*

(See thus Section 11 in the sequel). That is, one considers hereby a

(6.41) *paracompact (Hausdorff) sheaf exponential curvature space, such that (6.39) be satisfied.*

In this context, due to the aforementioned Weil's theorem (see Theorem 11.1 in the sequel), we are thus tempted to call a topological space X, satisfying (6.41), a *Weil space.* (In fact, this is in disagreement, somewhat, with our teminology previously applied in A. Mallios [6: p. 192, Definition 7.1], the present term being thus weakened, hence, improved).

7. Bianchi's identity

We consider below the counterpart in our case of the classical identity referred to in the title of this Section. Yet, by analogy with the standard theory, we examine in the next Chapter IX an important application of the same identity, by defining *characteristic classes* of vector sheaves.

So assume that we have a *curvature space* X and an $\mathcal{A}$-*module* $\mathcal{E}$ on X, endowed with an $\mathcal{A}$-*connection* D. Moreover, suppose that we are given a *local gauge*

$$e^U \equiv \{U; (e_1, \ldots, e_n)\} \tag{7.1}$$

of $\mathcal{E}$ over an *open* set $U \subseteq X$, of *rank* $n \in \mathbf{N}$ (cf. Chapt. VII; (1.5)). Now, we already know that the given $\mathcal{A}$-connection D and its *curvature* $R(D) \equiv R$ have *local expressions* on U, through the corresponding *local matrix-valued "forms"* (see Chapt. VII; (1.13), along with (3.11) above).

Thus, by considering the *curvature of* D, *restricted to* U, as above, as an *element of* $M_n(\Omega^2(U)) = M_n(\Omega^2)(U)$, we further look at the form, which is attained to by R, if one assumes, in addition, the action of a *second exterior derivation* (cf. also (1.10))

$$d^2 : \Omega^2 \longrightarrow \Omega^3. \tag{7.2}$$

Namely, suppose further that we are given a $\mathbf{C}$-*linear morphism*

$$d^2 \equiv d, \tag{7.3}$$

as in (7.2) above, between the $\mathbf{C}$-*vector space sheaves* concerned, in such a manner that one actually sets, by definition,

$$d^2(s \wedge t) := d^1 s \wedge t - s \wedge d^1 t \equiv ds \wedge t - s \wedge dt, \tag{7.4}$$

for any $s, t \in \Omega^1(U)$, while we still *assume that* (cf. also (8.9) below);

$$d^2 \circ d^1 \equiv d \circ d = 0. \tag{7.5}$$

Note 7.1.- Concerning the defining relation (7.4), as above, one easily proves that the map (7.2), as given by (7.4), is indeed a $\mathbf{C}$-*linear morphism of* Ω^1 *into* Ω^2. Yet, an analogous argument to that one used for the map (1.20) is here still in force. (See also the relevant comments about the map (1.7) in the preceding). Thus, *the extra condition* here, against a curvature datum (cf. Definition 1.1), *is,* in effect, *the rel. (7.5).*

Thus, for convenience of reference, we further set the following.

Definition 7.1. By a *Bianchi datum* on a topological space X, one means a curvature datum (see Definition 1.1), such that, in addition, we are given a **C**-linear morphism (7.2), defined by (7.4), in such a manner that (7.5) be satisfied.

We denote the above data by

$$\{(\mathcal{A}, \partial, \Omega^1, d^1 \equiv d, \Omega^2);\ d^2 \equiv d\}. \tag{7.6}$$

Moreover, for simplicity's sake, we also apply for the same data the following notation;

$$(\mathcal{A}, \partial, \Omega^1, d^1 \equiv d, \Omega^2, d^2 \equiv d) \equiv (\partial, d^1, d^2). \tag{7.7}$$

Finally, we say that X, as before, is a *Bianchi space*.

Thus, to be clear, a *Bianchi space* is a curvature space X (Definition 1.1), for which we further consider an exterior derivative operator $d^2 \equiv d$, as before, satisfying (7.5) (cf. also the previous Note 7.1).

Now, this same operator is "coordinate-wise" *extended to* a similar operator (we retain for simplicity the same symbol for the extended map too) between *the* following $\mathcal{A}$-*modules*

$$d^2 \equiv d : M_n(\Omega^2) \longrightarrow M_n(\Omega^3) \tag{7.8}$$

(cf. also Chapt. V; (8.7), for the notation applied). Thus, we can further apply the last *extended operator* to the curvature R (restricted to U, as in (7.1)), as the latter is given by *Cartan's structural equation* (cf. (3.12)). So one obtains (see (7.4), (7.5));

$$\begin{aligned} d^2 R \equiv dR &= d(d\omega + \omega\wedge\omega) = d(d\omega) + d(\omega\wedge\omega) \\ &= d(\omega\wedge\omega) = d\omega\wedge\omega - \omega\wedge d\omega \\ &= (R - \omega\wedge\omega)\wedge\omega - \omega\wedge(R - \omega\wedge\omega) \\ &= R\wedge\omega - \omega\wedge\omega\wedge\omega - \omega\wedge R + \omega\wedge\omega\wedge\omega, \end{aligned} \tag{7.9}$$

so that one finally has;

$$dR = R\wedge\omega - \omega\wedge R \equiv [R, \omega]. \tag{7.10}$$

We call the last relation the *("second") Bianchi's identity*.

Thus, still for convenience of reference, we collect now the preceding, formally, into the following.

Proposition 7.1. *Suppose we are given a Bianchi space X and an $\mathcal{A}$-module $\mathcal{E}$ on X, endowed with an $\mathcal{A}$-connection D. Moreover, let e^U be a given* *local gauge of* $\mathcal{E}$ (cf. (7.1)) *such that*

$$\omega \in M_n(\Omega^1(U)) \quad \text{and} \quad R \in M_n(\Omega^2(U)) \tag{7.11}$$

are the local (section) matrix-valued "forms" of D and R on U, respectively (see Chapt. VII; (1.13), as well as, (3.11) above). *Then, one obtains (second Bianchi's identity);*

$$dR = R \wedge \omega - \omega \wedge R \equiv [R, \omega]. \blacksquare \tag{7.12}$$

Now, suppose that we have a *Bianchi space* X, as above, and let

$$(\mathcal{E}, D) \tag{7.13}$$

be a given *pair, as in* the previous *Proposition 7.1.* Thus, one can further define another **C**-*linear morphism*

$$D^2 : \Omega^2(\mathcal{E}) \longrightarrow \Omega^3(\mathcal{E}), \tag{7.14}$$

according to the relation (cf. also (8.17) below, for $p = 2$);

$$\begin{aligned} D^2(s \otimes t) &:= s \otimes d^2 t + t \wedge Ds \equiv s \otimes dt + t \wedge Ds \\ &= s \otimes dt - Ds \wedge t, \end{aligned} \tag{7.15}$$

for any $s \in \mathcal{E}(U)$ and $t \in \Omega^2(U)$, with U *open* in X. (In this regard, see also the comments on (1.20) in the preceding).

We call the above map D^2, as defined by (7.14), (7.15), the *2nd prolongation of the* given $\mathcal{A}$-*connection* D; yet, by extending the classical terminology, we call D^2 the *2nd exterior covariant derivative operator* (or just *derivation*).

Thus, our next objective is to *evaluate* $D^2 R$ (cf. (7.14) and (2.3)) in the case of a given pair $(\mathcal{E}, D)$, as in (7.13), when, in particular, $\mathcal{E}$ is a *vector sheaf* on our *Bianchi space* X.

Thus, supposing again that e^U is a *local gauge of* $\mathcal{E}$ (cf., for instance, (7.1)), assume that

$$(e^U)^* \equiv \{U; (e_i^*)_{1 \leqq i \leqq n}\}, \tag{7.16}$$

with $n = \operatorname{rk} \mathcal{E} \in \mathbf{N}$, is the *dual local gauge of* e^U, viz. a corresponding to e^U *local gauge over the* same *open* set $U \subseteq X$ of the *dual vector sheaf* of $\mathcal{E}$,

$$\mathcal{E}^* \equiv \mathcal{H}om_{\mathcal{A}}(\mathcal{E}, \mathcal{A}) \tag{7.17}$$

(see Chapt. VII; (4.2) - (4.7)). On the other hand, due to the hypothesis for

$\mathcal{E}$, one has (cf. Chapt. II; (6.28) and Chapt. IV; (6.1))

$$\mathcal{E}nd\,\mathcal{E} \equiv \mathcal{H}om_{\mathcal{A}}(\mathcal{E},\mathcal{E}) = \mathcal{E} \otimes_{\mathcal{A}} \mathcal{E}^*. \tag{7.18}$$

Therefore, by considering the curvature R, restricted to U, one obtains in view also of the hypothesis for U (cf., for example, Chapt. V; (8.6));

$$\begin{aligned} R|_U \equiv R \in \Omega^2(\mathcal{E}nd\,\mathcal{E})(U) &\equiv ((\mathcal{E}nd\,\mathcal{E}) \otimes_{\mathcal{A}} \Omega^2) \\ = ((\mathcal{E} \otimes_{\mathcal{A}} \mathcal{E}^*) \otimes_{\mathcal{A}} \Omega^2)(U) &= (\mathcal{E} \otimes_{\mathcal{A}} \mathcal{E}^*)(U) \otimes_{\mathcal{A}(U)} \Omega^2(U) \\ &(\mathcal{E}(U) \otimes_{\mathcal{A}(U)} \mathcal{E}^*(U)) \otimes_{\mathcal{A}(U)} \Omega^2(U). \end{aligned} \tag{7.19}$$

On the other hand, by hypothesis for (7.1) and (7.16), the double sequence

$$(e_i \otimes e_j^*)_{1 \leqq i,j \leqq n} \tag{7.20}$$

yields a *local gauge of the vector sheaf* $\mathcal{E} \otimes_{\mathcal{A}} \mathcal{E}^*$ over the *open* $U \subseteq X$, of *rank* n^2 (see Chapt. IV; (5.19) and Chapt. II; (5.27)). Accordingly (see also e.g. W. Greub [1: p.7, Lemma 1.5.2]), one obtains, by (7.19),

$$R|_U \equiv R = \sum_{i,j} (e_i \otimes e_j^*) \otimes R_{ij}, \tag{7.21}$$

for some *uniquely defined* (sections)

$$R_{ij} \in \Omega^2(U), \quad 1 \leqq i,j \leqq n = \operatorname{rk} \mathcal{E}, \tag{7.22}$$

for given R, as in (7.19), along with (7.20), so that one actually gets;

$$(R_{ij}) \in M_n(\Omega^2(U)) = M_n(\Omega^2)(U). \tag{7.23}$$

Now, based on (7.14) and (7.15), we can further consider the $\mathbf{C}$-*linear morphism*

$$D^2 \equiv D : \Omega^2(\mathcal{E} \otimes_{\mathcal{A}} \mathcal{E}^*) \longrightarrow \Omega^3(\mathcal{E} \otimes_{\mathcal{A}} \mathcal{E}^*), \tag{7.24}$$

which, by analogy with (7.15), is defined by the relation;

$$D^2 \equiv D := 1_{\mathcal{E} \otimes_{\mathcal{A}} \mathcal{E}^*} \otimes d^2 + 1_{\Omega^2} \wedge D_{\mathcal{E} \otimes_{\mathcal{A}} \mathcal{E}^*}, \tag{7.25}$$

where, of course,

$$D_{\mathcal{E} \otimes_{\mathcal{A}} \mathcal{E}^*} : \mathcal{E} \otimes_{\mathcal{A}} \mathcal{E}^* \longrightarrow \Omega^1(\mathcal{E} \otimes_{\mathcal{A}} \mathcal{E}^*) \tag{7.26}$$

stands for the $\mathcal{A}$-*connection of* (the *vector sheaf*) $\mathcal{E} \otimes_{\mathcal{A}} \mathcal{E}^*$. The latter corresponds to the given $\mathcal{A}$-*connection* D of $\mathcal{E}$ and its *dual* $\mathcal{A}$-*connection*

$$D_{\mathcal{E}^*} \equiv D^* : \mathcal{E}^* \longrightarrow \Omega^1(\mathcal{E}^*), \tag{7.27}$$

as this is given by Chapt. VI; (5.31); thus, one has (loc. cit.; (5.38))

$$D_{\mathcal{E}\otimes_{\mathcal{A}}\mathcal{E}^*} := D \otimes 1_{\mathcal{E}^*} + 1_{\mathcal{E}} \otimes D^*. \tag{7.28}$$

Note 7.2.– Concerning (7.24), we further remark, in anticipation (see thus the next Section 9), that the operator at issue is, in effect, the *2nd prolongation of the $\mathcal{A}$-connection (7.26) of $\mathcal{E} \otimes_{\mathcal{A}} \mathcal{E}^*$*; in this regard, see also (8.17), for $p = 2$, in the sequel.

In this context, we still recall, for convenience, that if

$$\omega \equiv (\omega_{ij}) \in M_n(\Omega^1(U)) = M_n(\Omega^1)(U) \tag{7.29}$$

is the *local $\mathcal{A}$-connection matrix of D*, associated with (7.1) (see Chapt. VII; (1.13)), then the corresponding one for D^*, as in (7.27), is given by

$$\omega^* \equiv (\omega^*_{ij}) = -{}^t\omega \equiv -(\omega_{ji}) \in M_n(\Omega^1(U)) \tag{7.30}$$

(cf. (7.29), together with Chapt. VII; (4.20)), such that one has (see Chapt. VII; (4.19), as well as, e.g. Chapt. VI; (5.36));

$$D^*(e^*_j)(e_i) = \omega^*_{ij} = -\omega_{ji}, \quad 1 \leqq i, j \leqq n, \tag{7.31}$$

or even, *equivalently*,

$$D^*(e^*_j) = \sum_{i=1}^{n} e^*_i \otimes \omega^*_{ij} = -\sum_i e^*_i \otimes \omega_{ji}, \quad 1 \leqq j \leqq n. \tag{7.32}$$

So we are now in the position to formulate our main result in this Section. That is, we have.

Theorem 7.1. *Let X be a Bianchi space* (cf. Definition 7.1), *$\mathcal{E}$ a vector sheaf on X, with* $\operatorname{rk}\mathcal{E} = n \in \mathbf{N}$, *and U an open set in X, domain of definition of a local gauge of $\mathcal{E}$, while we also assume that $\mathcal{E}$ has an $\mathcal{A}$-connection D. Then, the two following assertions about the curvature of D, $R(D) \equiv R$, are equivalent:*

1) One has (Bianchi's identity; see (7.12));

$$dR = R \wedge \omega - \omega \wedge R \equiv [R, \omega]. \tag{7.33}$$

2) The following relation holds true ("differential" Bianchi's identity);

(7.34) $$DR = 0 \text{ (i.e., } D^2_{\mathcal{E}nd\,\mathcal{E}}(R) = 0)$$

(cf. also (7.18), (7.24) and Note 7.2).

Proof. First, by applying the operator (7.24) to (7.21), one gets;

(7.35) $$\begin{aligned} D^2(R|_U) \equiv DR &= D(\textstyle\sum_{i,j}(e_i \otimes e_j^*) \otimes R_{ij}) \\ &= \textstyle\sum_{i,j} D((e_i \otimes e_j^*) \otimes R_{ij}). \end{aligned}$$

On the other hand, based on (7.25), one has (cf. also (7.3) and (7.22));

(7.36) $$D((e_i \otimes e_j^*) \otimes R_{ij}) = (e_i \otimes e_j^*) \otimes dR_{ij} + R_{ij} \wedge D_{\mathcal{E}\otimes_{\mathcal{A}}\mathcal{E}^*}(e_i \otimes e_j^*).$$

However, by virtue of (7.28) and (7.32), in conjunction with Chapt. VII; (1.11), one further obtains;

(7.37) $$\begin{aligned} D_{\mathcal{E}\otimes_{\mathcal{A}}\mathcal{E}^*}(e_i \otimes e_j^*) &= D(e_i) \otimes e_j^* + e_i \otimes D^*(e_j^*) \\ &= (\textstyle\sum_k e_k \otimes \omega_{ik}) \otimes e_j^* + e_i \otimes \textstyle\sum_k e_k^* \otimes \omega_{jk}^* \\ &\approx \textstyle\sum_k (e_k \otimes e_j^*) \otimes \omega_{ik} - \textstyle\sum_k (e_i \otimes e_k^*) \otimes \omega_{kj}. \end{aligned}$$

(The notation "$\approx$" indicates here the appropriate change, that can be made *on the* relevant *(sheaf) spaces,* according to Chapt. II; (5.14), (5.16)). Therefore, based now on (7.37), one obtains;

(7.38) $$\begin{aligned} &\textstyle\sum_{i,j} R_{ij} \wedge D_{\mathcal{E}\otimes_{\mathcal{A}}\mathcal{E}^*}(e_i \otimes e_j^*) \\ &= \textstyle\sum_{i,j} R_{ij} \wedge (\textstyle\sum_k (e_k \otimes e_j^*) \otimes \omega_{ik}) - \textstyle\sum_{i,j} R_{ij} \wedge (\textstyle\sum_k (e_i \otimes e_k^*) \otimes \omega_{kj}) \\ &= \textstyle\sum_{i,j} (e_i \otimes e_j^*) \otimes (\textstyle\sum_k \omega_{ik} \wedge R_{kj} - \textstyle\sum_k R_{ik} \wedge \omega_{kj}). \end{aligned}$$

Consequently, one now obtains, in view also of (7.35) and (7.36),

(7.39) $$DR = \sum_{ij} (e_i \otimes e_j^*) \otimes \left(dR_{ij} + \sum_k \omega_{ik} \wedge R_{kj} - \sum_k R_{ik} \wedge \omega_{kj} \right),$$

so that one has

(7.40) $$(DR)_{ij} = dR_{ij} + \sum_k \omega_{ik} \wedge R_{kj} - \sum_k R_{ik} \wedge \omega_{kj}$$

for any $1 \leqq i, j \leqq n$. Thus, by employing *matrix notation* to (7.40), one finally obtains;

(7.41) $$DR = dR + \omega \wedge R - R \wedge \omega \equiv dR + [\omega, R],$$

which thus proves our assertion. ∎

8. Higher exterior derivations and prolongations

By analogy with the classical theory, we further consider below, within our abstract (viz. axiomatic) framework, the case of *exterior derivations of higher order,* along with the resulted thereby *prolongations.*

Thus, suppose that we are given a *differential triad*

$$(\mathcal{A}, \partial, \Omega) \tag{8.1}$$

on a *topological space* X. (We knowingly adopt here anew the notation of Chapt. VI; (3.1), concerning (8.1), which, however, will be justified straight away). So we call an (Ω) $\mathcal{A}$-*sequence* on X, the following *(cochain) complex of* $\mathbf{C}$-*vector space sheaves* (under the *constant* ($\mathbf{C}$-vector space) *sheaf* $\mathbf{C}$);

$$\begin{aligned}(\Omega) : 0 \longrightarrow \mathbf{C} \xrightarrow{\varepsilon} \Omega^0 \equiv \mathcal{A} &\xrightarrow{d^0 \equiv \partial} \Omega^1 \equiv \Omega \\ &\xrightarrow{d^1} \Omega^2 \xrightarrow{d^2} \Omega^3 \longrightarrow \cdots \longrightarrow \Omega^p \\ &\xrightarrow{d^p} \Omega^{p+1} \longrightarrow \cdots,\end{aligned} \tag{8.2}$$

where we have set (see also (1.10))

$$\Omega^p := \underbrace{\Omega \wedge \cdots \wedge \Omega}_{p-\text{times}} \equiv \wedge^p \Omega \equiv \wedge^p \Omega^1, \quad p \in \mathbf{N}, \tag{8.3}$$

while we also assumed that (cf. (8.2));

$$\Omega^0 \equiv \mathcal{A} \ \ and \ \ \Omega \equiv \Omega^1. \tag{8.4}$$

Thus, the individual terms in the given sequence (8.2) are, apart from the *constant sheaf* $\mathbf{C}$, in effect, $\mathcal{A}$-*modules* (by their very definition; cf. (8.1), (8.3) and Chapt. IV; (7.3), (7.8)). Hence, in particular, $\mathbf{C}$-*vector space sheaves,* where, as already noted, the *constant one* $\mathbf{C}$ is also included (cf. Chapt. II; (1.38)); yet, the connecting maps appeared in (8.2) are, by definition, $\mathbf{C}$-*linear morphisms.* Furthermore, we first remark that, by hypothesis for (8.1), one obtains (cf. Chapt. VI; Lemma 1.1)

$$\partial \circ \varepsilon = 0; \tag{8.5}$$

on the other hand, *we* also *assume,* by virtue of (8.2), *the existence of* the $\mathbf{C}$-linear morphism d^1, *such that*

$$d^1 \circ \partial = 0 \tag{8.6}$$

(cf. (1.7), (1.8)). Moreover, we define the operators d^p in (8.2), section-wise, by the relation;

$$d^{p+q}(s \wedge t) := d^p s \wedge t + (-1)^p s \wedge d^q t, \tag{8.7}$$

for any $p, q \in \mathbf{N}$, *with* $s \in \Omega^p(U)$, $t \in \Omega^q(U)$, and every *open* $U \subseteq X$.

Remark.– Since the *exterior powers* in (8.3) are viewed *with respect to* $\mathcal{A}$ (cf. Chapt. IV; (7.4), (7.5)), we actually use in (8.7) an *analogous argument* to that one, pertaining *to (1.20)* above.

Thus, *we can prove* now *that;*

$$d^{p+1} \circ d^p = 0, \quad \textit{for every} \ \ p \geqq 2, \tag{8.8}$$

while *we* finally *assume* that

$$d^2 \circ d^1 = 0, \tag{8.9}$$

as well. So the preceding rels. (8.6), (8.8) and (8.9) justify now our previous terminology for (8.2). In other words, we thus have

$$d^{p+1} \circ d^p = 0, \quad \textit{for every} \ \ p \in \mathbf{Z}_+. \tag{8.10}$$

Yet, for convenience, we just put (8.10) into the form;

$$d \circ d \equiv d^2 = 0. \tag{8.11}$$

Note.– In this context, we should also remark that: *We could have (1.9)* by suitably *extending (8.7) to the pair* of indices $(1,0) \in \mathbf{Z}_+$; that is, one then obtains

$$\begin{aligned} d^1(\alpha \cdot s) &\equiv d^{0+1}(\alpha \wedge s) = d^0\alpha \wedge s + \alpha \wedge d^1 s \\ &\equiv \partial\alpha \wedge s + \alpha \cdot d^1 s = \alpha \cdot ds - s \wedge \partial\alpha, \end{aligned} \tag{8.12}$$

for any $\alpha \in \mathcal{A}(U)$ *and* $s \in \Omega(U) \equiv \Omega^1(U)$, with U *open* in X, which is just our claim. ∎

Thus, the previous complex (Ω), as given by (8.2), yields, in particular, a *curvature datum*

$$(\mathcal{A}, \partial, \Omega^1, d, \Omega^2) \tag{8.13}$$

on X, by further setting $d^1 \equiv d$ (see (1.7)); viz. (Ω) furnishes already what we have called in the preceding a *curvature space* X (cf. Definition 1.1). So assume now that we are also given a pair

$$(\mathcal{E}, D), \tag{8.14}$$

consisting of an $\mathcal{A}$-*module* $\mathcal{E}$ on X and an $\mathcal{A}$-*connection* D *of* $\mathcal{E}$. Thus, consider now the following $\mathbf{C}$-*sequence* on X (cf. also (8.4));

$$\begin{gathered}\mathcal{E} \xrightarrow{D^0 \equiv D} \Omega^1(\mathcal{E}) \xrightarrow{D^1} \Omega^2(\mathcal{E}) \xrightarrow{D^2} \Omega^3(\mathcal{E}) \longrightarrow \cdots \\ \cdots \longrightarrow \Omega^p(\mathcal{E}) \xrightarrow{D^p} \Omega^{p+1}(\mathcal{E}) \longrightarrow \cdots,\end{gathered} \tag{8.15}$$

where, as usual, we have set

$$\Omega^p(\mathcal{E}) := \mathcal{E} \otimes_{\mathcal{A}} \Omega^p, \quad p \in \mathbf{Z}_+ \tag{8.16}$$

(see also (8.4), along with Chapt. II; (5.15)). Yet, one also defines

$$D^p := 1_{\mathcal{E}} \otimes d^p + (-1)^p 1_{\Omega^p} \wedge D^0 \tag{8.17}$$

(cf. (8.2)), while we have already set (see (8.15))

$$D^0 := D. \tag{8.18}$$

Thus, writing (8.17) in a section-wise manner, one has, *by definition* (see also (8.16));

$$D^p(s \otimes t) := s \otimes d^p t + (-1)^p t \wedge Ds, \tag{8.19}$$

for any $s \in \mathcal{E}(U)$ *and* $t \in \Omega^p(U)$, with U *open* in X.

Scholium 8.1.- In practice, the *open* set $U \subseteq X$, as used above, will be, in effect, (the domain of definition of) a local gauge of the $\mathcal{A}$-module (in fact, *vector sheaf*) $\mathcal{E}$ involved, so that one can then apply, concerning the definition of D^p, as in (8.19), an analogous argument to Chapt. VII; (1.10) (see thus e.g. (7.19) in the preceding).

On the other hand, by considering a *section of* Ω^p *over* U, as above, one actually looks at a *similar section of the* corresponding *presheaf generating* Ω^p (see Chapt. IV; (7.2), along with (7.47) therein). So it is, in effect, this point of view that one actually applies, when we have written, hitherto, by usually *abusing notation,* $\Omega^p(U)$. Yet, concerning the same definition (8.19), as above, an analogous argument to that one applied to the previous map (1.23), in conjunction with (1.19), is here still in force.

Now, we call the above *operator* D^p, $p \in \mathbf{N}$, (cf. (8.15)), as given by (8.17), or even (8.19), the *p-th prolongation of the* $\mathcal{A}$*-connection* D ($\equiv D^0$; cf. (8.18)).

Thus, (8.17) proves exactly that D^1 *is,* indeed, the *1st prolongation of the* given $\mathcal{A}$*-connection* D, as the latter has already been defined in the preceding, by (1.18) or (1.19). On the other hand, it is worth remarking here that:

(8.20) *the* **C***-sequence (8.15) does not, in general, yield a complex,* in the sense that *we do not always have;*

(8.20.1) $$D^p \circ D^{p-1} = 0, \quad p \in \mathbf{N}.$$

Indeed, we have already set in the preceding, by defining the *curvature of* D (cf. (2.2));

(8.21) $$R(D) \equiv R := D^1 \circ D \equiv D^1 \circ D^0$$

(see (8.18)). Hence, even the relation $D^1 \circ D = 0$ *is not valid, in general, unless* the given $\mathcal{A}$-connection D *is flat* (cf. (2.5)). In this context, *one proves,* indeed, *that;*

(8.22) $$(D^p \circ D^{p-1})(s \otimes t) = t \wedge R(s),$$

for any $p \in \mathbf{N}$, *and* $s \in \mathcal{E}(U)$, $t \in \Omega^{p-1}(U)$, with U *open* in X. Therefore, one concludes that;

(8.23) *the curvature* R *of a given* $\mathcal{A}$*-connection* D ($\equiv D^0$, see (8.18)) *measures the obstruction of being the above* **C***-sequence (8.15) a complex* (of **C**-vector space sheaves and **C**-linear morphisms, such that (8.20.1) be satisfied).

Now, *to prove (8.22),* one employs a *local gauge of* (the $\mathcal{A}$*-module*) $\mathcal{E}$, as e.g. in (3.2), in conjunction with the corresponding *local forms of* R *and* D. (See thus (3.19) above, along with Chapt. VI; (1.11) and (1.15). ■ (In this regard, we also observe that one can still recapture (8.21) from (8.22), by virtue of (8.4) and (2.11)).

8.(a). Trace of curvature.– Suppose we are given the previous *complex* (Ω), as in (8.2), on a *topological space* X (cf. also (8.6), (8.8) and (8.9)). Moreover, assume that $\mathcal{E}$ is an $\mathcal{A}$*-module* on X, which is further endowed with an $\mathcal{A}$*-connection* D (cf., for instance, (8.13) above for the relevant context). Finally, suppose that

$$e^U \equiv \{U; (e_i)_{1 \leqq i \leqq n}\} \tag{8.24}$$

is a *local gauge of* $\mathcal{E}$ (cf. (3.2)). Thus, our objective by the following discussion, as already indicated by the title of this Subsection, is to consider the *trace of the curvature* (operator) R of D, when restricted to the *open* $U \subseteq X$, as above.

Namely, given that one has

$$R|_U \equiv R = (R_{ij}) \in M_n(\Omega^2(U)) = M_n(\Omega^2)(U), \tag{8.25}$$

one further sets;

$$\operatorname{tr} R \equiv \operatorname{tr}(R|_U) := \operatorname{tr}(R_{ij}) = \sum_{i=1}^{n} R_{ii} \in \Omega^2(U). \tag{8.26}$$

On the other hand, let now

$$R^2 \equiv R \wedge R \in M_n(\Omega^4(U)), \tag{8.27}$$

where R is still given by (8.25). In particular, one obtains;

$$d^4(\operatorname{tr} R^2) \equiv d(\operatorname{tr} R^2) = \operatorname{tr}(dR^2) \tag{8.28}$$

(as easily follows from (8.26) and the **C**-*linearity of* $d \equiv d^4$). However, one further obtains, in view of (8.27) and (8.7) (for $p = q = 2$);

$$dR^2 \equiv d(R \wedge R) = dR \wedge R + R \wedge dR = 0. \tag{8.29}$$

Now, in this context, one further obtains (cf. (8.27), (8.26));

$$\begin{aligned} &\operatorname{tr}(R^2) \equiv \operatorname{tr}((R^2_{ij})) \equiv \operatorname{tr}((R \wedge R)_{ij}). \\ &= \operatorname{tr}((\textstyle\sum_{k=1}^{n} R_{ik} \wedge R_{kj})) = \sum_{i=1}^{n}(\sum_{k=1}^{n} R_{ik} \wedge R_{ki}) \equiv \sum_{i,k} R_{ik} \wedge R_{ki}, \end{aligned} \tag{8.30}$$

so that one has

$$
\begin{aligned}
d(\operatorname{tr} R^2) &\equiv d^4(\operatorname{tr} R^2) = d(\textstyle\sum_{i,k} R_{ik} \wedge R_{ki}) \\
&= \textstyle\sum_{i,k} d(R_{ik} \wedge R_{ki}) = \sum_{i,k}(dR_{ik} \wedge R_{ki} + R_{ik} \wedge dR_{ki}) = 0.
\end{aligned} \tag{8.31}
$$

Accordingly, as a result of (8.31), or even of (8.29), one concludes for the *curvature* R of the given $\mathcal{A}$-connection D, as above, that:

(8.32) *the trace of* $R^2 \equiv R \wedge R$ (cf. (8.27)) *is a "closed" 4-form; viz.* one has

(8.32.1)
$$d(\operatorname{tr} R^2) = 0.$$

The above is reminiscent, of course, of the analogous classical fact, pertaining to the so-called *"topological charge"*. Yet, as we shall see in the next Chapter IX, the same *"4-form"*

$$\operatorname{tr}(R^2) \equiv \operatorname{tr}(R \wedge R) \in \Omega^4(U) \tag{8.33}$$

(see (8.27), as well as, Scholium 8.1 in the preceding) is an example of what we shall call in the sequel a *characteristic class of* (the *vector sheaf*) $\mathcal{E}$; thus, as already seen (cf. (8.32.1)), it is a *"closed 4-form"* (on the *open* set $U \subseteq X$, so that one actually needs a *0-cocycle of such,* as (8.33) yields, for that matter, in the case of a *vector sheaf* $\mathcal{E}$, the local gauge (8.24), as above, being then just a member of a given local frame of $\mathcal{E}$; cf. Chapter IX). Furthermore, it is also *"gauge invariant"*, in a sense to be made precise (loc. cit.). In this regard, concerning the classical case see, for instance, B.A. Dubrovin - A.T. Fomenko - S.P. Novikov [1: p.452, Definition 42.2].

9. Curvatures of induced $\mathcal{A}$-connections

We examine below curvatures of $\mathcal{A}$-connections, which are induced by given ones (see Chapter VI), in relation to the curvatures of the initially given $\mathcal{A}$-connections. So we start with the

Curvature of a Whitney sum of $\mathcal{A}$-connections (cf. Chapt. VI; Subsection 5.(a)): For simplicity's sake, we just consider two *$\mathcal{A}$-modules* $\mathcal{E}$ and $\mathcal{F}$, endowed with *$\mathcal{A}$-connections* $D_{\mathcal{E}}$ and $D_{\mathcal{F}}$, respectively.

Thus, by taking the *Whitney sum* $\mathcal{E} \oplus \mathcal{F}$ of the previous $\mathcal{A}$-modules, this is further equipped with an *$\mathcal{A}$-connection* too, given by

$$D_{\mathcal{E}\oplus\mathcal{F}} = D_{\mathcal{E}} \oplus D_{\mathcal{F}} \tag{9.1}$$

(see Chapt. VI; (5.3)). Therefore, based now on (3.4) and (3.12), one easily sees, by virtue of (9.1), that the *curvature* of the latter $\mathcal{A}$-connection is given by;

$$R(D_{\mathcal{E}\oplus\mathcal{F}}) = R(D_{\mathcal{E}}) \oplus R(D_{\mathcal{F}}). \tag{9.2}$$

Yet, as alluded to already at the beginning of this Subsection, one further remarks that;

(9.3) the above argument, concerning (9.2), *is valid,* of course, *for any finite,* or even *infinite, many summands.* (Cf. also Chapt. VI; Scholium 5.1).

Now, by looking again at a *differential triad* $(\mathcal{A}, \partial, \Omega)$ on a *topological space* X, as also was the case before, then, as an immediate conclusion from (9.2) (cf. (9.3) too), together with Chapt. VI; (5.4), (5.5) and (2.6) in the preceding, one obtains that;

$$R(\partial^n) = R(\partial \oplus \cdots \oplus \partial) = R(\partial) \underbrace{\oplus \cdots \oplus}_{n-\text{times}} R(\partial) = 0, \tag{9.4}$$

for every $n \in \mathbf{N}$. In other words, one thus concludes that:

(9.5) ∂^n, $n \in \mathbf{N}$, *is a flat $\mathcal{A}$-connection of* $\mathcal{A}^n$ (see also Chapt. VI; (5.6)).

More generally, one actually infers from (9.2) (in fact, from its extension to *any finite number of summands,* see (9.3)) that;

(9.6) *the Whitney sum of flat $\mathcal{A}$-connections is* still *flat.*

On the other hand, we next examine the

Curvature of a tensor product of $\mathcal{A}$-connections (see Chapt. VI; Subsection 5.(b)): Let us consider again the same framework, as before; then, by analogy with (9.2), one can still apply *Cartan's structural equation* (cf. (3.12) in the preceding) to compute the *curvature of the $\mathcal{A}$-connection of*

$\mathcal{E} \otimes_{\mathcal{A}} \mathcal{F}$, *viz.* *of*

$$D_{\mathcal{E}\otimes_{\mathcal{A}}\mathcal{F}} = D_{\mathcal{E}} \otimes 1_{\mathcal{F}} + 1_{\mathcal{E}} \otimes D_{\mathcal{F}} \tag{9.7}$$

(see Chapt. VI; (5.10)). Thus, first, based on (9.7), one easily sees that the corresponding *local (matrix-)form of the $\mathcal{A}$-connection* (9.7) is given by the relation

$$\omega(D_{\mathcal{E}\otimes_{\mathcal{A}}\mathcal{F}}) \equiv \tilde{\omega} = \omega \otimes 1 + 1 \otimes \omega' \equiv \omega + \omega', \tag{9.8}$$

where, of course, we have set $1_{\mathcal{E}} \equiv 1$ and $1_{\mathcal{F}} \equiv 1$, along with $\omega \equiv \omega(D_{\mathcal{E}})$ and $\omega' \equiv (D_{\mathcal{F}})$ (see also e.g. Chapt. VII; (1.13)). Thus, in view of (3.12) and (9.8), one now obtains;

$$\begin{aligned} R(D_{\mathcal{E}\otimes_{\mathcal{A}}\mathcal{F}}) &= d\tilde{\omega} + \tilde{\omega} \wedge \tilde{\omega} \\ &= d(\omega \otimes 1 + 1 \otimes \omega') + (\omega \otimes 1 + 1 \otimes \omega') \wedge (\omega \otimes 1 + 1 \otimes \omega') \\ &= d\omega \otimes 1 + (\omega \wedge \omega) \otimes 1 + 1 \otimes d\omega' + 1 \otimes (\omega' \wedge \omega') + (\omega \otimes 1) \wedge (1 \otimes \omega') \\ &\qquad + (1 \otimes \omega') \wedge (\omega \otimes 1) \\ &= (d\omega + \omega \wedge \omega) \otimes 1 + 1 \otimes (d\omega' + \omega' \wedge \omega'). \end{aligned} \tag{9.9}$$

In this context, we should also notice that, for convenience, we have taken above that *both ω and ω' are elements of the same $\mathcal{A}(U)$-module, $M_n(\Omega(U)) = M_n(\Omega)(U)$* (cf. also Chapt. VII; (1.17)).

Accordingly, *one* finally *obtains the relation* (see the same (3.12));

$$R(D_{\mathcal{E}\otimes_{\mathcal{A}}\mathcal{F}}) = R(D_{\mathcal{E}}) \otimes 1 + 1 \otimes R(D_{\mathcal{F}}) \equiv R(D_{\mathcal{E}}) + R(D_{\mathcal{F}}). \tag{9.10}$$

Hence, as an immediate consequence of (9.10), one infers that;

(9.11) *the tensor product of flat $\mathcal{A}$-connections is a flat $\mathcal{A}$-connection,* as well.

Yet, by recapitulating (9.6) and (9.11), for convenience, one thus concludes that:

(9.12) the *Whitney sum and* the *tensor product preserve flatness of $\mathcal{A}$-connections.*

Now, a similar conclusion can be deduced, concerning the *dual* of a given (flat) *$\mathcal{A}$-connection:* Thus, by taking a pair $(\mathcal{E}, D)$, as before, along with a *local gauge e^U of $\mathcal{E}$* (cf. e.g. (8.24)), let us further look at the *dual $\mathcal{A}$-mo-*

dule of $\mathcal{E}$ and the corresponding *dual* $\mathcal{A}$*-connection of* D (see Chapt. VI; Subsection 5.(d)). Then, the *local* $\mathcal{A}$*-connection matrix (-"form") of the latter,* with respect to e^U, is given by the relation (cf. Chapt. VII; (4.20))

$$\omega(D^*) \equiv \omega^* = -{}^t\omega, \tag{9.13}$$

with $\omega \equiv \omega(D)$, while we have also set, as usual, $D_{\mathcal{E}^*} \equiv D^*$ (see thus Chapt. VI; (5.31), (5.36)). Therefore (cf. (3.12)), one has;

$$\begin{aligned} R(D^*) &= d\omega^* + \omega^* \wedge \omega^* = d(-{}^t\omega) + (-{}^t\omega) \wedge (-{}^t\omega) \\ &= -{}^t d\omega + {}^t\omega \wedge {}^t\omega, \end{aligned} \tag{9.14}$$

or even, by further setting, $R(D^*) \equiv R^*$, one finally obtains for the *curvature of* the *dual* $\mathcal{A}$*-connection,*

$$R^* = -{}^t d\omega + {}^t(\omega \wedge \omega). \tag{9.15}$$

In particular, by considering a *line sheaf* $\mathcal{L}$ on X, equipped with an $\mathcal{A}$-connection D, together with the *dual line sheaf* $\mathcal{L}^*$ and the *dual* $\mathcal{A}$*-connection* $D^* \equiv D_{\mathcal{L}^*}$, one obtains for the *curvature of* D^* (see (9.15));

$$R(D^*) \equiv R^* = d\omega^* = d(-\omega) = -d\omega. \tag{9.16}$$

In other words, one thus obtains;

$$R^* = -R. \tag{9.17}$$

On the other hand, as already alluded to at the beginning, it is easy to prove that:

(9.18) *the dual of a flat* $\mathcal{A}$*-connection is* still *flat.*

Indeed, assuming that $R(D) = d\omega + \omega \wedge \omega = 0$ (cf. (3.12), (2.5)), one has;

$$d\omega = -\omega \wedge \omega = \omega \wedge \omega \tag{9.19}$$

(see also, for instance, (3.18) in the foregoing for the meaning of the last equality in (9.19)). So, based now on (9.13) and (9.19), one obtains;

$$\begin{aligned} d\omega^* &= d(-{}^t\omega) = -d({}^t\omega) = -{}^t(d\omega) = -{}^t(\omega \wedge \omega) \\ &= {}^t\omega \wedge {}^t\omega = -{}^t\omega \wedge (-{}^t\omega) = \omega^* \wedge \omega^*, \end{aligned} \tag{9.20}$$

so that, by applying a similar argument, as in (9.19), one gets

(9.21) $$R^* \equiv R(D^*) = 0,$$

that was our claim. ∎

Finally, we further consider the situation which arises, when one looks at the *pull-back* of a given $\mathcal{A}$-*connection* (see Chapt. VI; Section 6), as it concerns the corresponding *curvature:*

Thus, suppose that we are given a *differential triad* $(\mathcal{A}, \partial, \Omega^1)$ on a *topological space* Y, as well as, a *continuous map*

(9.22) $$f : X \longrightarrow Y,$$

while, we still assume that we have a pair $(\mathcal{E}, D)$ *on* Y, as e.g. in (8.14). Therefore, by further considering the *pull-back of the* given *differential triad* on Y, one gets a *similar one on* X (cf. Chapt. VI; (6.3)). Consequently, the pair on X,

(9.23) $$f^*(\mathcal{E}, D) \equiv (f^*(\mathcal{E}), f^*(D))$$

has now an analogous meaning, as the given one on Y (loc. cit.; (6.13)). Now, based on the very definitions, one can prove, first, that:

(9.24) *the pull-back of a curvature space,* via a continuous map, like (9.22), *is a space of the same sort;* viz.

(9.24.1) $$(f^*(\mathcal{A}), f^*(\partial), f^*(\Omega^1), f^*(d), f^*(\Omega^2))$$

is a *curvature space on* X.

(Concerning the $f^*(\mathcal{A})$-*module* on X, $f^*(\Omega^2)$, cf. also Chapt. IV; (7.4), (7.6), as well as, Chapt. VI; Lemma 6.1). Thus, based further on the same definitions and on Cartan's structural equation (cf. (3.12)), one can prove now that;

(9.25) $$R(f^*(D)) = f^*(R(D)),$$

that is,

(9.26) the *curvature is preserved by the pull-back.*

In particular, one thus infers that:

(9.27) the *pull-back preserves flatness;* precisely, *the pull-back of a flat* $\mathcal{A}$-*connection is a flat* $f^*(\mathcal{A})$-*connection* (cf. (9.22), (9.24.1)).

On the other hand, as a particular instance of the above, one can consider the *restriction of a* given *curvature space* X (cf. (1.16)) to an *open* subset U of X (see Chapt. VI; (6.21)). So (9.25) yields then the following relation (*restriction of the curvature* R), which we already used, in anticipation, in the preceding (see thus (5.10));

$$R(D|_U) = R(D)|_U. \tag{9.28}$$

10. Torsion of an $\mathcal{A}$-connection. Local form

Assume, as usual, that we are given a *differential triad*

$$(\mathcal{A}, \partial, \Omega^1) \tag{10.1}$$

on a *topological space* X, while we still suppose that

$$\theta^U \equiv \{U; (\theta_1, \ldots, \theta_n)\} \tag{10.2}$$

is a *local gauge* of the given $\mathcal{A}$-*module* Ω^1 on X, of *rank* $n \in \mathbf{N}$, over the *open* $U \subseteq X$, such that

$$(\theta_i)_{1 \leqq i \leqq n} \subseteq \Omega^1(U) \tag{10.3}$$

is a *section-basis* of the *free* $\mathcal{A}(U)$-*module*

$$\Omega^1(U) = \mathcal{A}^n(U) = \mathcal{A}(U)^n \tag{10.4}$$

(see, for instance, Chapt. VII; (1.3), (1.4), (1.5)).

On the other hand, consider the *dual* $\mathcal{A}$-*module of* Ω^1; that is, we set

$$\mathcal{E} \equiv (\Omega^1)^* := \mathcal{H}om_{\mathcal{A}}(\Omega^1, \mathcal{A}). \tag{10.5}$$

Thus,

$$(\theta^U)^* \equiv e^U \equiv \{U; (e_1, \ldots, e_n)\} \tag{10.6}$$

is a *local gauge of* $\mathcal{E}$ *(dual of (10.2)) over* U, of the *same rank* n, as (9.2), such that

$$e_i := \theta_i^*, \quad 1 \leqq i \leqq n, \tag{10.7}$$

with

$$e_i(\theta_j) := \delta_{ij}, \quad 1 \leqq i, j \leqq n \tag{10,8}$$

(see Chapt. VII; (4.3), (4.6). In practice (cf. e.g. the classical theory), $\mathcal{E}$ is a given *vector sheaf* on X, whose *dual* (vector sheaf, then, too) *is* assumed to be Ω^1; cf., for instance, Chapt. VII; (5.3) and Scholium 5.1). Thus, we are led to the following.

Definition 10.1. Suppose we have a curvature space X, and let $\mathcal{E}$ be an $\mathcal{A}$-module on X, with $(\Omega^1)^* = \mathcal{E}$. Moreover, let D be an $\mathcal{A}$-connection of $\mathcal{E}$, while θ^U and $e^U \equiv (\theta^U)^*$ are local gauges of the $\mathcal{A}$-modules concerned, as in (10.2) and (10.6), respectively. Then, the *local torsion of D over U* is given by the relation;

$$\Theta|_U \equiv \Theta := d\theta + \omega \wedge \theta, \tag{10.9}$$

where $\omega \in M_n(\Omega^1(U))$ stands for the local $\mathcal{A}$-connection matrix of D, which is associated with the given local gauge e^U of $\mathcal{E}$ (cf. Chapt. VII; (1.13)).

> **Note.**– Concerning the notation in (10.9), the term θ stands for the *section-basis* of the given *local gauge* θ^U of Ω^1, as in (10.3); viz. we have set
>
> $$\theta \equiv (\theta_1, \ldots, \theta_n) \in \Omega^1(U)^n, \tag{10.10}$$
>
> considering it as a *"column vector"* of $\Omega^1(U)^n$.

The same relation (10.9) constitutes, in fact, the counterpart in our case of the so-called *Cartan's first structural equation* of the classical theory. (See, for instance, M. Crampin - F.A.E. Pirani [1: p. 278]).

On the other hand, suppose now that, in addition to the previously given curvature space X, we have, in particular, the framework of a *Bianchi space* X, in the sense of Definition 7.1 in the preceding (see also the comments following the same definition). Thus (loc. cit.), assume that we are given further the *second exterior derivation*

$$d^2 \equiv d : \Omega^2 \longrightarrow \Omega^3, \tag{10.11}$$

as this is given by (7.4), in such a manner that (7.5) *hold true.* So we are led now to the following classical result (see e.g. M. Crampin - F.A.E. Pirani [1: p. 279; *Bianchi identities,* or even p.281, Ex. 35]).

Theorem 10.1. *(Ricci's Lemma). Suppose we are given a Bianchi space* X (cf. Definition 7.1) *and an* $\mathcal{A}$*-module* $\mathcal{E}$ *on* X*, such that* $\mathcal{E} = (\Omega^1)^*$*, which is further endowed with an* $\mathcal{A}$*-connection* D*, while we also assume the existence of a local gauge of* Ω^1*, hence, of* $\mathcal{E}$ *too, as in (10.2) and (10.6) above, respectively. Then, concerning the (local) torsion of* D *(over* U*,* cf. (10.9)), *one has the relation;*

$$d\Theta = R\wedge\theta - \omega\wedge\Theta \tag{10.12}$$

(first Bianchi's identity), with ω and θ, as in (10.9) (see also (10.10)).

Proof. Based on (7.5), (7.4) and (3.12), and by applying $d^2 \equiv d$, as in (10.11), one obtains, in view of the same relation;

$$\begin{aligned} d\Theta &= d(d\theta + \omega\wedge\theta) = d(d\theta) + d(\omega\wedge\theta) = d(\omega\wedge\theta) \\ &= d\omega\wedge\theta - \omega\wedge d\theta = (R - \omega\wedge\omega)\wedge\theta - \omega\wedge(\Theta - \omega\wedge\theta) \\ &= R\wedge\theta - \omega\wedge\omega\wedge\theta - \omega\wedge\Theta + \omega\wedge\omega\wedge\theta = R\wedge\theta - \omega\wedge\Theta, \end{aligned} \tag{10.13}$$

as desired. ▮

11. Weil's integrality theorem

We finally come, by this Section, to the full proof of the theorem in the heading. Indeed, one half of it, namely, the *"only if"* part (cf., for instance, Theorem 11.1′ below) has already been established by Lemma 5.2 in the preceding. So we give right away the statement of the theorem in question.

Theorem 11.1. *Suppose that we are given a sheaf exponential (Hausdorff) paracompact curvature space* X*. Moreover, let* $\mathcal{L}$ *be a line sheaf on* X*, endowed with an* $\mathcal{A}$*-connection* D*. Then, the curvature of* D*,* $R(D) \equiv R$*, yields a 2-dimensional integral cohomology class of* R*,* $c(R)$*; viz. one has* (see (5.49), (5.51))

(11.1) $$c(R) \in H^2(X, \mathbf{Z}).$$

Conversely, assume that our space X, as before, satisfies the condition

(11.2) $$H^1(X, \mathcal{A}) = H^2(X, \mathcal{A}) = 0,$$

while we still suppose that;

(11.3) *every line sheaf on X, admits an $\mathcal{A}$-connection*

(take, for instance, $H^1(X, \Omega^1) = 0$, or even, concerning both (11.2) and (11.3), assume that $\mathcal{A}$ is a fine sheaf. See e.g. Chapt. III; (8.24)). *Then, given a 2-dimensional integral cohomology class of X, there exists a (uniquely defined) line sheaf on X, admitting an $\mathcal{A}$-connection, whose curvature yields the given (integral) cohomology class.*

Proof. As already noted, the first part of the statement is exactly Lemma 5.2 in the foregoing. So consider now a *2-dimensional integral cohomology class of X*, say,

(11.4) $$z \in H^2(X, \mathbf{Z})$$

(cf. also the above Scholium 5.3, regarding the meaning of (11.4)). Now, by hypothesis for X, one has the following *short exact exponential sheaf sequence*

(11.5) $$0 \longrightarrow \mathbf{Z} \longrightarrow \mathcal{A} \xrightarrow{e} \mathcal{A}^{\cdot} \longrightarrow 1$$

(see Chapt. VII; (7.1) and Chapt. V; (6.1)). Thus, by further considering the corresponding long exact sequence in cohomology one obtains (cf., for instance, Chapt. III; Theorem 5.3);

(11.6) $$\cdots \longrightarrow H^1(X, \mathcal{A}) \longrightarrow H^1(X, \mathcal{A}^{\cdot}) \xrightarrow{\delta} H^2(X, \mathbf{Z}) \longrightarrow H^2(X, \mathcal{A}) \longrightarrow \cdots.$$

Therefore, taking now (11.2) into account, one gets the exact sequence

(11.7) $$0 \longrightarrow H^1(X, \mathcal{A}^{\cdot}) \xrightarrow{\delta} H^2(X, \mathbf{Z}) \longrightarrow 0,$$

so that one finally obtains;

(11.8) $$H^1(X, \mathcal{A}^{\cdot}) = H^2(X, \mathbf{Z}),$$

within an *isomorphism of* (the abelian) *groups* concerned. (The *Bockstein operator* δ, as in (11.6), is called, in this context, the *Chern isomorphism,* as well; see also e.g. Chapt. V; (6.33), (6.35)). Thus, one now sets (cf. (11.4))

$$\mathcal{L} := \delta^{-1}(z) \in H^1(X, \mathcal{A}^{\cdot}), \tag{11.9}$$

this being, as we shall presently see, the *line sheaf* on X, that we are looking for (cf. Chapt. V; (2.59), (2.60)): Indeed, suppose that

$$(g_{\alpha\beta}) \in Z^1(\mathcal{U}, \mathcal{A}^{\cdot}) \tag{11.10}$$

is a representative *(coordinate) 1-cocycle of* $\mathcal{L}$ (see loc. sit.; (2.43), for $n = 1$), such that

$$\mathcal{U} = (U_\alpha)_{\alpha \in I} \tag{11.11}$$

is a *local frame* of $\mathcal{L}$. On the other hand, based on (11.3), one has for the respective *Atiyah class of* $\mathcal{L}$ the relation;

$$\mathfrak{a}(\mathcal{L}) \equiv \left[\tilde{\partial}(g_{\alpha\beta})\right] = 0 \in H^1(X, \Omega^1) \tag{11.12}$$

(see Chapt. VI; (10.28.1), for $n = 1$, along with Corollary 12.1). Therefore, one obtains

$$\tilde{\partial}(g_{\alpha\beta}) = \delta(\theta_\alpha), \tag{11.13}$$

for some *0-cochain* of *"1-forms"*

$$(\theta_\alpha) \in C^0(\mathcal{U}, \Omega^1). \tag{11.14}$$

Furthermore, in view of the short exact sequence (11.5) and the hypothesis for X, the 1-cocycle (11.10) provides a *1-cochain*

$$(f_{\alpha\beta}) \in C^1(\mathcal{V}, \mathcal{A}), \tag{11.15}$$

such that one has

$$g_{\alpha\beta} = e(f_{\alpha\beta}), \tag{11.16}$$

where $\mathcal{V}$ in (11.15) stands for an eventual refinement of $\mathcal{U}$ (cf. Chapt. III; Lemmas 5.1, 5.2). Therefore, based now on the commutative triangle (5.34), according to our hypothesis for X (see Chapt. VII; (7.1)), one obtains, by (11.16),

$$\tilde{\partial}(g_{\alpha\beta}) = \tilde{\partial}(e(f_{\alpha\beta})) = 2\pi i \cdot \partial(f_{\alpha\beta}) = \partial(2\pi i \cdot f_{\alpha\beta}), \tag{11.17}$$

so that, in view of (11.13), one has;

$$(11.18) \qquad \delta(\theta_\alpha) = \tilde{\partial}(g_{\alpha\beta}) = 2\pi i \cdot \partial(f_{\alpha\beta}).$$

(Of course, the first equality in the last relation expresses in our case the corresponding *transformation law of potentials;* cf. Chapt. VII; (2.59)).

Now, suppose that D is an *$\mathcal{A}$-connection* of $\mathcal{L}$ (cf. (11.3)), such that let

$$(11.19) \qquad D|_{U_\alpha} = D_\alpha - \theta_\alpha, \quad \alpha \in I,$$

be the *local form of* D, with respect to the local frame $\mathcal{U}$ of $\mathcal{L}$, as in (11.11) (see Chapt. VI; (8.13), (11.14), for $\widetilde{D} \equiv D$). On the other hand, the set of $\mathcal{A}$-connections of $\mathcal{L}$ being an *affine space,* modelled on $\Omega^1(X)$ (cf. Chapt. VI; Corollary 7.1), one may assume that the 0-cochain of "1-forms", $(\theta_\alpha) \in C^0(\mathcal{U}, \Omega^1)$, as appeared in (11.19), is that one defined by (11.13) and (11.14). Hence, the *curvature* of D, $R(D) \equiv R$, is given by the relation (cf. Corollary 3.1)

$$(11.20) \qquad R = (d\theta_\alpha).$$

So the last relation yields now a *closed 2-form on* X (loc. cit.), which can further be construed as that one supplying the given cohomology class $z \in H^2(X, \mathbf{Z})$, as in (11.4): Indeed, one has, by (11.18),

$$(11.21) \qquad \partial(f_{\alpha\beta}) = \frac{1}{2\pi i}\delta(\theta_\alpha) = \frac{1}{2\pi i}\tilde{\partial}(g_{\alpha\beta})$$

while one further obtains, by virtue of (11.5), (11.15) and (11.16), that

$$(11.22) \qquad \delta(f_{\alpha\beta}) \equiv (\lambda_{\alpha\beta\gamma}) \in Z^2(\mathcal{V}, \mathbf{Z} = \ker e),$$

which thus explains our claim about z and R, as the latter is given by (11.20) (see also Scholium 5.4 in the preceding), and this terminates the proof. ∎

Due to its significance (see e.g. Chapter X in the sequel), we restate the above result in the following more compact form. That is, we have.

Theorem 11.1′. *Let X be a Hausdorff paracompact sheaf exponential curvature space, such that (11.2) and (11.3) hold true. Then, the only 2-dimensional integral cohomology classes of X are* (those associated with) *field strengths of Maxwell fields* (viz. curvatures of $\mathcal{A}$-connections of line sheaves) *on X.* ∎

CHAPTER IX

Characteristic classes

"...an axiomatic approach ... emphasizes the mathematical structures involved..."

Sir Michael F. Atiyah in "The Geometry and Physics of Knots" (Cambridge Univ. Press, 1990). p. 12.

We consider below (Chern) characteristic classes of vector sheaves, the analogue in our case of the same classes of vector bundles (over a $\mathcal{C}^{\infty}$-manifold). In particular, we apply the counterpart here of the (differential-geometric) *Chern-Weil description* of these classes, *in terms of the curvature* (tensor). So, as happens in the standard case, the *Bianchi identity* (see the previous Chapter, Theorem 7.1) becomes here a key-result. On the other hand, one has to assume further, within this abstract treatment, the existence of an *appropriate analogue of de Rham's complex,* to ensure, by analogy with the standard case, a cohomology class, which is thus (canonically) associated with any given *"closed form"*. So we start with the necessary preliminary material.

1. Preliminaries

Suppose that we are given, as usual, a **C**-*algebraized space*

$$(1.1)\qquad (X, \mathcal{A})$$

on an arbitrary *topological space* X (cf. Chapt. II; Scholium 1.2). Furthermore, we assume that X is endowed with a *differential triad*

$$(1.2)\qquad (\mathcal{A}, \partial, \Omega^1)$$

(see Chapt. VI; (3.1) and Chapt. VIII; (1.1), (1.2)).

On the other hand, consider a pair

$$(1.3)\qquad (\mathcal{E}, D),$$

consisting of a *vector sheaf* $\mathcal{E}$ on X, of *rank* $n \in \mathbf{N}$, and a given $\mathcal{A}$-*connection* D *of* $\mathcal{E}$.

Now, if the *rank* of $\mathcal{E}$, as before, is *at least* 2, viz. in the case one has

$$(1.4)\qquad \operatorname{rk}_{\mathcal{A}}(\mathcal{E}) \equiv \operatorname{rk}\mathcal{E} = n \geqq 2,$$

then, by extending to the present abstract setting, the classical terminology, connected with elementary particle physics, a pair as in (1.3) above, will also be called a *Yang-Mills field* on X. In a similar manner, we have named in the preceding an analogous pair $(\mathcal{L}, D)$, where now $\mathcal{L}$ is a *line sheaf* on X, viz. when $\operatorname{rk}\mathcal{L} = 1$, a *Maxwell field* on X (cf. Chapt. VI; (18.11). See also Chapter X in the sequel).

Now, suppose that

$$(1.5)\qquad \mathcal{U} = (U_\alpha)_{\alpha \in I}$$

is a *local frame of* our *vector sheaf* $\mathcal{E}$. Then, as already known, *the* given $\mathcal{A}$-*connection* D *of* $\mathcal{E}$ *is determined* by a *0-cochain of local* $\mathcal{A}$-*connection matrices,* associated with $\mathcal{U}$,

$$(1.6)\qquad \omega \equiv \left(\omega^{(\alpha)}\right) = \left(\left(\omega_{ij}^{(\alpha)}\right)\right) \in C^0(\mathcal{U}, M_n(\Omega^1))$$

such that

$$(1.7)\qquad \omega^{(\alpha)} \equiv \left(\omega_{ij}^{(\alpha)}\right) \in M_n(\Omega^1(U_\alpha)) = M_n(\Omega^1)(U_\alpha),$$

for every $\alpha \in I$, as in (1.5) (cf. Chapt. VII; (2.39.1) and (2.21)). Yet, (1.6) has to satisfy the corresponding *transformation law of potentials;*

$$\omega^{(\beta)} = \mathrm{Ad}(g_{\alpha\beta}^{-1})\omega^{(\alpha)} + \tilde{\partial}(g_{\alpha\beta}) \tag{1.8}$$

(ibid.; Theorem 2.1. See also Theorem 3.2 therein). In this context,

$$(g_{\alpha\beta}) \in Z^1(\mathcal{U}, \mathcal{GL}(n, \mathcal{A})) \tag{1.9}$$

stands here for a *coordinate 1-cocycle of* $\mathcal{E}$, associated with $\mathcal{U}$, as in (1.5), and yielding the *gauge transformation* in (1.8).

On the other hand, let us assume now that X is, in particular, a *curvature space,* viz. (cf. Chapt. VIII; Definition 1.1) suppose further that we are given a *curvature datum*

$$(\mathcal{A}, \partial, \Omega^1, d^1, \Omega^2). \tag{1.10}$$

Thus, by considering a pair $(\mathcal{E}, D)$, as in (1.3), one has naturally associated with the given $\mathcal{A}$-connection D its *curvature* $R(D) \equiv R$, the latter being a *global section* of the *vector sheaf* $\Omega^2(\mathcal{E}nd\,\mathcal{E})(X)$. That is, we have

$$\begin{aligned} R(D) \equiv R \equiv \left(R^{(\alpha)}\right) &\in \Omega^2(\mathcal{E}nd\,\mathcal{E})(X) = Z^0(\mathcal{U}, \Omega^2(\mathcal{E}nd\,\mathcal{E})) \\ &= Z^0(\mathcal{U}, M_n(\Omega^2)), \end{aligned} \tag{1.11}$$

so that, in particular, one obtains;

$$R^{(\alpha)} \equiv \left(R_{ij}^{(\alpha)}\right) \in M_n(\Omega^2)(U_\alpha) = M_n(\Omega^2(U_\alpha)), \tag{1.12}$$

for every $\alpha \in I$ (see Chapt. VIII; (2.16) – (2.20), along with (2.21)). Yet, one has, by virtue of (1.12);

$$R_{ij}^{(\alpha)} \in \Omega^2(U_\alpha), \tag{1.13}$$

for every $\alpha \in I$, *and* $1 \leq i, j \leq n = \mathrm{rk}\,\mathcal{E}$, (In this regard, cf. also, for instance, Scholium 8.1 in the previous Chapter VIII). Furthermore, one has the analogue of (1.8), pertaining now to the corresponding *transformation law of curvature* (cf. also Chapt. VIII; (4.23)),

$$R^{(\beta)} = \mathrm{Ad}(g_{\alpha\beta}^{-1})R^{(\alpha)}, \tag{1.14}$$

which is thus associated with the given local frame $\mathcal{U}$ of $\mathcal{E}$ and the corresponding coordinate 1-cocycle $(g_{\alpha\beta})$ of it, as in (1.5), (1.9).

On the other hand, consider now the following $\mathcal{A}$-*module* on X

$$\Omega^{\bullet}_{\text{even}} \equiv \Omega^{\bullet}_{\text{ev}} := \bigoplus_{m=0}^{\infty} \wedge^{2m}\Omega^1 \equiv \bigoplus_{m \geqq 0} \Omega^{2m} \tag{1.15}$$

(see also e.g. Chapt. VIII; (8.3)); viz. one takes the *Whitney sum of the even exterior powers of the* given *$\mathcal{A}$-module* Ω^1 (cf. Chapt. IV; (7.9), (7.11.1)). Yet, we still set in (1.15);

$$\wedge^0\Omega^1 \equiv \Omega^0 := \mathcal{A} \tag{1.16}$$

(see, for example, Chapt. VIII; (8.4)). Thus, one now concludes that;

(1.17) *the* above *$\mathcal{A}$-module* on X, $\Omega^{\bullet}_{\text{ev}}$, as given by (1.15) and (1.16), *when* further *endowed with the* corresponding *exterior multiplication* (of *sections,* cf. e.g. Chapt. II; (1.67)), *provides a commutative unital* **C**-*algebra sheaf* on X.
Indeed, $\Omega^{\bullet}_{\text{ev}}$ is an *$\mathcal{A}$-algebra sheaf* on X. Cf. Chapt. II; Definition 1.4, 1.6 and Chapt. IV; (7.10).

In this context, it is still useful to remark that, in view of the relation,

$$\Omega^n_x \equiv (\wedge^n\Omega^1)_x = \wedge^n(\Omega^1_x), \quad n \in \mathbf{N}, \tag{1.18}$$

which is valid, up to an *$\mathcal{A}_x$-isomorphism, for every* $x \in X$, one can work with the corresponding *presheaf of sections, generating (1.15)* (cf. Chapt. IV; (7.11) and (7.14), (7.14′), along with Chapt. VIII; Scholium 8.1). Indeed, this is what we essentially employ in the sequel. Thus, as a direct consequence of the last remarks, we further note that, by virtue of (1.12) and (1.15) (see also e.g. Chapt. IV; (3.23) and Chapt. II; (5.18), (3.14)), one now obtains;

$$\begin{aligned} R^{(\alpha)} &\equiv \left(R^{(\alpha)}_{ij}\right) \in M_n(\Omega^2(U_\alpha)) = M_n(\Omega^2)(U_\alpha) \\ &= (\Omega^2)^{n^2}(U_\alpha) = \Omega^{2n^2}(U_\alpha) \overset{\subseteq}{\to} \left(\textstyle\bigoplus_{m=0}^{\infty} \Omega^{2m}\right)(U_\alpha) \\ &= \textstyle\bigoplus_{m \geqq 0} \Omega^{2m}(U_\alpha) \equiv \mathcal{A}(U_\alpha) \oplus \Omega^2(U_\alpha) \oplus \cdots \\ &= \Omega^{\bullet}_{\text{ev}}(U_\alpha) \overset{\subseteq}{\to} (\Omega^{\bullet}_{\text{ev}})(U_\alpha), \end{aligned} \tag{1.19}$$

for any $\alpha \in I$ (cf. (1.5)), and *any given* $n \in \mathbf{N}$. (Concerning the last inclusion in (1.19), see also Chapt. IV; (7.47), together with Chapt. I; (11.36′)). *Warning!* In this regard, it is still to be noticed that, in general, one has

$$(\Omega^{\bullet}_{\text{ev}})(U) \neq \Omega^{\bullet}_{\text{ev}}(U), \tag{1.20}$$

for a given *open* set U in X. (However, see also (1.18) above and the relevant comments after it, as well as, Chapt. VIII; Scholium 8.1).

Thus, we are now ready to embark on our main theme in this chapter, as indicated by its title. However, for convenience, we still recall in the next Section 2 certain basic facts, pertaining to *"invariant polynomials"*, that will be of essential use throughout the sequel.

2. Invariant polynomials

The polynomials in the title of this Section, that are going to be considered by the ensuing discussion are, in fact, *complex polynomials,* whose "indeterminates" will be elements of the *unital commutative* **C**-*algebra*

$$\Omega^{\cdot}_{\mathrm{ev}}(U_\alpha), \tag{2.1}$$

as above; see thus (1.5), (1.17), as well as, (1.19) and (1.20) of the previous Section. Notice that, in view of (1.19) (see also Chapt. II; (1.34)), *one* always *has;*

$$\mathbf{C} \underset{\varepsilon}{\overset{\subsetneq}{\rightarrow}} \mathcal{A}(U_\alpha) \overset{\subsetneq}{\rightarrow} \Omega^{\cdot}_{\mathrm{ev}}(U_\alpha), \tag{2.2}$$

for every $\alpha \in I$, as in (1.5).

So consider next the *polynomial algebra,*

$$\mathbf{C}[(\lambda_{ij})], \quad 1 \leqq i,j \leqq n \in \mathbf{N}. \tag{2.3}$$

That is, one thus takes a *commutative unital* **C**-*algebra,* consisting of all the *complex polynomials* in the n^2 many *indeterminates* (λ_{ij}), $1 \leqq i,j \leqq n$, so that let us consider an element

$$p \in \mathbf{C}[(\lambda_{ij})]. \tag{2.4}$$

Hence, one can further look at the respective *polynomial function,* relative to (2.1) (cf. (1.19), as well);

$$\tilde{p} : (\Omega^{\cdot}_{\mathrm{ev}}(U_\alpha))^{n^2} = M_n(\Omega^{\cdot}_{\mathrm{ev}}(U_\alpha)) \longrightarrow \Omega^{\cdot}_{\mathrm{ev}}(U_\alpha) \overset{\subsetneq}{\rightarrow} (\Omega^{\cdot}_{\mathrm{ev}})(U_\alpha), \tag{2.5}$$

for every $\alpha \in I$, as in (1.5), *and* $n \in \mathbf{N}$ (cf. also Chapt. V; (8.7) and Chapt.

VI; (9.3)). Yet, due to our previous *warning,* about (1.20), we also have, in general,

$$M_n(\Omega^{\cdot}_{ev})(U_\alpha) \neq M_n(\Omega^{\cdot}_{ev}(U_\alpha)). \tag{2.6}$$

But, see also (2.36) below; however, the above inconvenience *does not affect,* actually, our argument in the sequel.

On the other hand, the *correspondence*

$$p \longmapsto \tilde{p}, \tag{2.7}$$

as defined by (2.4) and (2.5), *is,* in effect, *one-to-one,* so that we further set, for convenience,

$$\tilde{p} \equiv p \tag{2.8}$$

(see, for instance, N. Bourbaki [4: p.28, Remarque 1)]).

In particular, suppose that we are given a *curvature space* X, along with a pair

$$(\mathcal{E}, D), \tag{2.9}$$

consisting of a *vector sheaf* $\mathcal{E}$ on X, *with* $\operatorname{rk}\mathcal{E} = n \in \mathbf{N}$, and an $\mathcal{A}$-*connection* D *of* $\mathcal{E}$. Furthermore, let

$$\mathcal{U} = (U_\alpha)_{\alpha \in I} \tag{2.10}$$

be a *local frame of* $\mathcal{E}$; then, associated with $\mathcal{U}$, there is a *0-cochain,* defined by the *curvature of* D (cf. (1.11) and (1.19)),

$$\begin{aligned} R \equiv \left(R^{(\alpha)}\right) \equiv \left(\left(R^{(\alpha)}_{ij}\right)\right) &\in C^0(\mathcal{U}, M_n(\Omega^2)) \\ &= C^0(\mathcal{U}, (\Omega^2)^{n^2}) \hookrightarrow C^0(\mathcal{U}, \Omega^{\cdot}_{ev}). \end{aligned} \tag{2.11}$$

Accordingly, *for every polynomial*

$$p \in \mathbf{C}[(\lambda_{ij})], \tag{2.12}$$

as in (2.4), one gets a *0-cochain*

$$\tilde{p}(R) \equiv p(R) \equiv \left(p\left(R^{(\alpha)}\right)\right) \in C^0(\mathcal{U}, \Omega^{\cdot}_{ev}) \tag{2.13}$$

(cf. (2.5), (2.8), (2.11) and (1.17)).

Now, in this context, we still recall that a given (complex) polynomial p, as in (2.12), is said to be *invariant,* whenever one has

(2.14) $$p(\alpha\boldsymbol{\lambda}\alpha^{-1}) = p(\mathrm{ad}(\alpha)\boldsymbol{\lambda}) = p(\boldsymbol{\lambda}),$$

for every $\boldsymbol{\lambda} \equiv (\lambda_{ij}) \in M_n(\mathbf{C})$ and $\alpha \in \mathrm{GL}(n, \mathbf{C})$. (Of course, the multiplication, appeared in the first member of (2.14) is matrix-multiplication, within $M_n(\mathbf{C})$).

Example 2.1.- The following is classical, being also of an essential use in the following: Thus, given a *complex* $n \times n$ *matrix*

(2.15) $$\boldsymbol{\lambda} \equiv (\lambda_{ij}) \in M_n(\mathbf{C}),$$

the corresponding *determinant* of $\boldsymbol{\lambda}$ can be construed as a (complex) *polynomial* with *indeterminates the entries of* $\boldsymbol{\lambda}$, $\lambda_{ij} \in \mathbf{C}$, $1 \leqq i, j \leqq n$; namely, one has (cf. (2.3))

(2.16) $$\det(\boldsymbol{\lambda}) \in \mathbf{C}[(\lambda_{ij})].$$

Thus, it is now a direct consequence of the same properties of det that;

(2.17) *the determinant function*

(2.17.1) $$\det : M_n(\mathbf{C}) \longrightarrow \mathbf{C}$$

is an invariant polynomial (cf. also (2.8), concerning the map (2.17.1)).

That is, (2.14) is fulfilled (cf. e.g. C.W. Curtis [1: p.149]). Indeed, the above situation will be amply applied in the sequel. (See thus, for instance, Section 5 below).

Now, our first relevant result pertains to the above 0-cochain (2.13) which *is, in fact, a 0-cocycle, whenever* our polynomial p *is invariant,* in the previous sense. Precisely, one thus obtains the following.

Lemma 2.1. *Suppose we are given a curvature space* X, *a pair* $(\mathcal{E}, D)$, *as in (2.9), and an invariant polynomial* $p \in \mathbf{C}[(\lambda_{ij})]$, *such that* $1 \leqq i, j \leqq n = \mathrm{rk}\,\mathcal{E}$. *Then, the curvature 0-cochain* $p(R)$ (cf. (2.13)), *which is associated with a given local frame* $\mathcal{U}$ *of* $\mathcal{E}$ (cf. (2.10)), *is actually a 0-cocycle, viz. one has*

(2.18) $$p(R) \equiv \left(p\left(R^{(\alpha)}\right)\right) \in Z^0(\mathcal{U}, \Omega^{\cdot}_{\mathrm{ev}}) = \Omega^{\cdot}_{\mathrm{ev}}(X).$$

In particular, if

$$\deg p = m, \tag{2.19}$$

then, one obtains;

$$p(R) \in \wedge^{2m}\Omega^1(X) \hookrightarrow (\wedge^{2m}\Omega^1)(X) \equiv \Omega^{2m}(X) \hookrightarrow \Omega^{\bullet}_{ev}(X). \tag{2.20}$$

Proof. Suppose that

$$p \in \mathbf{C}[(\lambda_{ij})] \tag{2.21}$$

is an *invariant polynomial.* Then, by considering a *coordinate 1-cocycle of* $\mathcal{E}$, relative to the given local frame $\mathcal{U}$ of $\mathcal{E}$, say,

$$(g_{\alpha\beta}) \in Z^1(\mathcal{U}, \mathcal{GL}(n, \mathcal{A})), \tag{2.22}$$

one obtains (cf. (1.14), (2.14));

$$p\left(R^{(\beta)}\right) = p\left(\mathrm{Ad}(g_{\alpha\beta}^{-1})R^{(\alpha)}\right) = p\left(R^{(\alpha)}\right), \tag{2.23}$$

for any $\alpha, \beta \in I$, with $U_{\alpha\beta} \equiv U_\alpha \cap U_\beta \neq \emptyset$ (cf. (2.10)). That is, *the 0-cochain* $\left(p\left(R^{(\alpha)}\right)\right)$ *is a 0-cocycle,* as was to be proved. On the other hand, (2.20) is certainly clear, by virtue of (2.19) and (1.19). ∎

To proceed further, we first recall some standard facts from the classical *theory of polynomials* pertaining, in particular, to *formal derivations:*

So let us take again a complex *polynomial*

$$p \in \mathbf{C}[(\lambda_{ij})], \quad 1 \leqq i, j \leqq n \tag{2.24}$$

(of n^2 *indeterminates*). Furthermore, consider the *matrix of the (formal) first partial derivatives of* p,

$$\left(\frac{\partial p}{\partial \lambda_{ij}}\right) \in M_n(\mathbf{C}[(\lambda_{ij})]) \tag{2.25}$$

(see e.g. N. Bourbaki [3: Chap. 4; p. 38, Définition 2]). Hence, by taking now the *derivative* (or even *differential*) *of* p, one has;

$$dp = \sum_{i,j} \frac{\partial p}{\partial \lambda_{ij}} d\lambda_{ij} \tag{2.26}$$

(loc. cit.; p.38, Définition 1, as well as, rels. (2), (3)). On the other hand, setting

$$(2.27) \qquad p'(\boldsymbol{\lambda}) := {}^{t}\!\left(\frac{\partial p}{\partial \lambda_{ij}}\right) \in M_n(\mathbf{C}[(\lambda_{ij})]),$$

viz. the *transpose* (matrix) *of (2.25)*, one obtains, in view of (2.26),

$$(2.28) \qquad dp = \sum_{i,j} \frac{\partial p}{\partial \lambda_{ij}} d\lambda_{ij} = \mathrm{tr}\,(p'(\boldsymbol{\lambda})\, d\boldsymbol{\lambda}),$$

where, of course, $\boldsymbol{\lambda} \equiv (\lambda_{ij}) \in M_n(\mathbf{C})$, while "tr" stands for the *trace* of the (product) matrix concerned. (In this regard, see also e.g. L. Smith [1: p.249, *"trace scalar product"*]; yet, cf. J.W. Milnor - J.D. Stasheff [1: p.297, proof of *"Fundamental Lemma"*]).

In particular, by taking an *invariant polynomial* (cf. (2.14)),

$$(2.29) \qquad p \equiv p(\boldsymbol{\lambda}) \equiv p((\lambda_{ij})) \in \mathbf{C}[(\lambda_{ij})], \quad 1 \leqq i, j \leqq n,$$

such that

$$(2.30) \qquad \boldsymbol{\lambda} \equiv (\lambda_{ij}) \in M_n(\mathbf{C}),$$

one concludes that (cf. also (2.27));

$$(2.31) \qquad p'(\boldsymbol{\lambda}) \cdot \boldsymbol{\lambda} = \boldsymbol{\lambda} \cdot p'(\boldsymbol{\lambda}).$$

See the last Ref. above for the proof of (2.31). ∎

So we are now in the position to give the proof of the following basic result for what follows. It is thus the analogue in our case of the aforementioned "Fundamental Lemma" (loc. cit.). So the point is that, *for every invariant polynomial* p, as in (2.29), *the global "form"* $p(R)$, as given by (2.20), *is, moreover,* a *"closed"* one (see (2.36) below). Precisely speaking, one thus obtains the next.

Theorem 2.1. *(Fundamental Lemma). Suppose we are given a Bianchi space* X (cf. Chapt. VIII; Definition 7.1) *and a vector sheaf* $\mathcal{E}$ *on* X, *with* $\mathrm{rk}\,\mathcal{E} = n \in \mathbf{N}$, *endowed also with an* $\mathcal{A}$*-connection* D. *Moreover, let*

$$(2.32) \qquad \mathcal{U} = (U_\alpha)_{\alpha \in I}$$

be a local frame of $\mathcal{E}$ *and*

$$(2.33) \qquad p \equiv p(\boldsymbol{\lambda}) \equiv p((\lambda_{ij})) \in \mathbf{C}[(\lambda_{ij})]$$

an invariant polynomial, such that

(2.34) $$\lambda \equiv (\lambda_{ij}) \in M_n(\mathbf{C}).$$

Then, the curvature 0-cocycle, associated with $\mathcal{U}$ *and* p (Lemma 2.1),

(2.35) $$p(R) = p\left(\left(R^{(\alpha)}\right)\right) \in \Omega^{\cdot}_{\mathrm{ev}}(X),$$

yields a "closed form"; viz. one has

(2.36) $$dp(R) = 0.$$

Proof. First, based on (1.12) and (1.19), one obtains;

(2.37) $$R^{(\alpha)} \equiv \left(R^{(\alpha)}_{ij}\right) \in M_n(\Omega^2(U_\alpha)) \hookrightarrow M_n(\Omega^{\cdot}_{\mathrm{ev}}(U_\alpha)) = M_n(\Omega^{\cdot}_{\mathrm{ev}})(U_\alpha),$$

for every $\alpha \in I$, as in (2.32), *where* also $1 \leqq i, j \leqq n = \mathrm{rk}\,\mathcal{E}$. Thus, by considering now the given (invariant) polynomial p, as in (2.33), one has (cf. (2.5), (2.8) and (2.18));

(2.38) $$p\left(R^{(\alpha)}\right) \equiv p\left(\left(R^{(\alpha)}_{ij}\right)\right) \equiv p(R) \in \Omega^{\cdot}_{\mathrm{ev}}(U_\alpha),$$

for every $\alpha \in I$, as before. (As already said, multiplication involved by (the form of) p is *exterior multiplication*. Yet, we should notice here the *abuse of notation* employed for $p(R)$ in (2.38) and (2.35) above, the obvious difference being, however, clear from the context). Thus, our previous rel. (2.31) yields now;

(2.39) $$p'(R)\wedge R = R\wedge p'(R).$$

In this context, we still recall here that one always understands in (2.39) and in the sequel too, $R \equiv \left(R^{(\alpha)}_{ij}\right)$, as in (2.37).

Hence, in view of (2.28), one further obtains,

(2.40) $$dp(R) = \mathrm{tr}\,(p'(R)\wedge dR).$$

On the other hand, *Bianchi's identity* (cf. Chapt. VIII; (7.12)) yields

(2.41) $$dR = [R, \omega] \equiv R\wedge\omega - \omega\wedge R,$$

so that, by virtue of (2.40) and (2.39), one now obtains;

$$\begin{aligned} dp(R) &= \mathrm{tr}\,(p'(R)\wedge dR) = \mathrm{tr}\,(p'(R)\wedge(R\wedge\omega - \omega\wedge R)) \\ &= \mathrm{tr}\,(p'(R)\wedge R\wedge\omega - p'(R)\wedge\omega\wedge R) \\ &= \mathrm{tr}\,((p'(R)\wedge R - R\wedge p'(R))\wedge\omega) = 0. \end{aligned} \tag{2.42}$$

Now, the last conclusion is valid, in fact, *for every* $\alpha \in I$, which thus, in view of (2.18), proves our assertion, viz. (2.36). ∎

It is still useful for the sequel to have the statement of the previous Theorem 2.1 less formally, combined too with the above Lemma 2.1 (cf. (2.20)). Thus, we have the next.

Corollary 2.1. *Suppose we have the framework of the above Theorem 2.1. Then, for every invariant polynomial*

$$p \in \mathbf{C}[(\lambda_{ij})] \tag{2.43}$$

(see (2.33)), *such that*

$$\deg p = m, \tag{2.44}$$

$p(R)$ (cf. (2.20)) *is a "closed 2m-form"; that is, one has the two following conditions fulfilled:*

$$p(R) \in \wedge^{2m}\Omega^1(X) \subsetneqq (\wedge^{2m}\Omega^1)(X) \equiv \Omega^{2m}(X) \subsetneqq \Omega^{\cdot}_{\mathrm{ev}}(X), \tag{2.45}$$

as well as,

$$dp(R) = 0. \tag{2.46}$$ ∎

Remark 2.1.– The *derivation* in (2.46) (the *same* holds true, of course, *for* (2.36)) is meant, in effect, for the operator

$$d^{2m} : \Omega^{2m} \longrightarrow \Omega^{2m+1}, \tag{2.47}$$

so that, as usual, *we have set;*

$$d^{2m} \equiv d. \tag{2.48}$$

Yet, we note that *the* same *operator does exist,* in view of the hypothesis for X (cf. Chapt. VIII; (8.7)).

3. Generalized de Rham spaces. Fundamental theorem

Our objective by the following discussion is to define, via (the *"closed form"*) $p(R)$, as above (see (2.45), (2.46)), an appropriate cohomology class of the base space X involved. In the classical case of (paracompact Hausdorff) $\mathcal{C}^\infty$*-manifolds,* this is accomplished through the corresponding *de Rham complex* of the manifold concerned (our base space X). Now, although, as we shall see in the sequel (cf. the succeeding Chapters X, XI), we do have particular examples, exhibiting analogous complexes, this is not valid, in general. Thus, keeping further the axiomatic point of view, that we have employed hitherto, we do give straight a way the following.

Definition 3.1. By a *generalized de Rham space of order* $m \in \mathbf{Z}_+$, or simply a *generalized de Rham* m*-space* (with m a positive integer), one means a paracompact (Hausdorff) space X, endowed with the following exact sequence of $\mathbf{C}$-vector space sheaves on X;

$$\begin{gathered} 0 \longrightarrow \ker \partial \longrightarrow \mathcal{A} \xrightarrow{d^0 \equiv \partial} \Omega^1 \xrightarrow{d^1 \equiv d} \Omega^2 \xrightarrow{d^2 \equiv d} \cdots \\ \cdots \longrightarrow \Omega^m \xrightarrow{d^m \equiv d} d\Omega^m \longrightarrow 0 \end{gathered} \tag{3.1}$$

(see also Scholium 3.1 below, concerning the notation applied in (3.1)).

In particular, if in the previous exact sequence one has

$$\ker \partial = \mathbf{C}, \tag{3.2}$$

then one speaks of a *de Rham* m*-space.*

On the other hand, in the case that (3.1) is valid, for every $m \in \mathbf{Z}_+$ (but, see also (3.5) below), then one speaks of a *generalized de Rham space,* or (if (3.2) is fulfilled) of a *de Rham space,* respectively.

Scholium 3.1.– By definition, a *generalized de Rham* m*-space* is a Hausdorff (paracompact) topological space X, endowed, first, with a *differential triad*

$$(3.3) \qquad (\mathcal{A}, \partial, \Omega^1),$$

in the usual sense, applied so far (cf., for instance, Chapt. VIII; (1.1)). In particular, X is a *curvature space*

$$(3.4) \qquad (\mathcal{A}, \partial, \Omega^1, d^1 \equiv d, \Omega^2)$$

(loc. cit.; (1.16)); hence, the respective **C**-*linear morphisms* ∂ and $d^1 \equiv d$ have the pertinent properties, as before (see the same Chapter VIII; Sections 1, 7, 8). Furthermore, the $\mathcal{A}$-*modules* Ω^i, $1 \leqq i \leqq m$, are defined as in Chapt. VIII; (8.3), together with the corresponding *higher derivative operators,* as the case may be, depending on the given $m \in \mathbf{Z}_+$, as in (3.1) (see also Chapt. VIII; (8.7)). Therefore, *if* $m \geqq 2$, our space X is a *Bianchi space* too (loc. cit.; (7.6)).

In this context, we should also remark that

$$(3.5) \qquad \textit{(3.1) is already exact at } \mathcal{A},$$

by the very definitions; *equivalently,* the following

$$(3.6) \qquad 0 \longrightarrow \ker\partial \longrightarrow \mathcal{A} \xrightarrow{\partial} \partial\mathcal{A} \longrightarrow 0$$

is a *short exact sequence* (of **C**-vector space sheaves on X).

Thus, *exactness of (3.1) refers,* in effect, to subsequent places of the sequence at issue, that is, *to* $m \geqq 1$.

Finally, the relation

$$(3.7) \qquad \ker\partial = \mathbf{C}$$

is a crucial restriction on $\mathcal{A}$ (nevertheless, important!, for its consequences), which might be called *Poincaré condition,* having to do in the standard case (again C^∞-*manifolds*) with the classical *Poincaré Lemma* (cf. Chapt. III; Lemma 8.2, together with (8.47)). Indeed, as we shall see later on (see the next Chapters X and XI), there do exist several particular *important examples* (apart, of course, from the aforementioned classical case), for which (3.7) *is satisfied.*

Thus, our first relevant result, which, in fact, was also the motive for the previous Definition 3.1, is now the following lemma, being also the counterpart in our case of the analogous standard situation. That is, we have.

Lemma 3.1. *Suppose we are given a generalized de Rham m-space X, as well as, a "closed m-form" ω on X* (with $m \in \mathbf{N}$); *that is, one has* (see (3.1))

$$\omega \in \Omega^m(X) \ \textit{and} \ d^m(\omega) \equiv d\omega = 0. \tag{3.8}$$

Then, one can associate with ω an m-dimensional (Čech) cohomology class of X with coefficients in (the **C**-*vector space sheaf* on X) $\ker \partial$, *denoted by $c(\omega)$; viz. one has*

$$c(\omega) \in H^m(X, \ker \partial) = \check{H}^m(X, \ker \partial). \tag{3.9}$$

Scholium 3.2.– For the proof of the above lemma we shall systematically employ Lemma 5.1 of Chapter III in the preceding, referring to the so-called *"liftable (Čech) cohomology"*, hence the appearance of the *paracompact* (Hausdorff) *space* X. Indeed, what we really apply is a strengthening of the aforementioned result, as given by Chapt. III; (5.52) (see thus loc. sit., Lemma 5.2, as well as, (5.86). In this context, one has to consider an eventual *refinement of the* initial *covering* $\mathcal{U}$ *of* X involved; for simplicity's sake, however, we do not mark it in the succeeding proof of Lemma 3.1, employing thus an obvious *abuse of notation* (see Chapt. III; (5.54)).

On the other hand, we further notice that, due to the *paracompactness* of X, *Čech and sheaf cohomology theories are* essentially *the same* (ibid.; Theorem 8.1 and (8.1)).

Thus, we are now ready to embark on the next.

Proof of Lemma 3.1. By hypothesis of the lemma, one assumes the *exactness of (3.1)*, so that let us, first, consider the following extract of the same sequence; viz. we take the next *short exact sequence,*

$$0 \longrightarrow \ker d^m \longrightarrow \Omega^m \xrightarrow{d^m \equiv d} d\Omega^m \longrightarrow 0, \tag{3.10}$$

along with the corresponding *long exact sequence in (Čech) cohomology* (cf.

also Chapt. III; Theorem 2.1, prop. 3), the comments before it, and Theorem 5.3),

$$0 \longrightarrow \Gamma(X, \ker d^m) \longrightarrow \Gamma(X, \Omega^m) \xrightarrow{\Gamma(d^m)} \Gamma(X, d\Omega^m) \longrightarrow \cdots . \tag{3.11}$$

Thus, in view of (3.6), one has

$$\omega \in \Omega^m(X) \equiv \Gamma(X, \Omega^m), \tag{3.12}$$

in such a manner, that one further has;

$$d^m(\omega) = \Gamma_X(d^m)(\omega) = \Gamma(d^m)(\omega) \equiv d\omega = 0. \tag{3.13}$$

Hence, by (3.13) and the exactness of (3.11) and (3.1), one now obtains;

$$\begin{aligned} \omega &\in \ker(\Gamma(X, \Omega^m) \longrightarrow \Gamma(X, d\Omega^m)) \\ &= \operatorname{im}(\Gamma(X, \ker d^m) \longrightarrow \Gamma(X, \Omega^m)) \\ &= \Gamma(X, \ker d^m) = \Gamma(X, \operatorname{im} d^{m-1}) = Z^0(\mathcal{U}, \operatorname{im} d^{m-1}) \\ &= Z^0(\mathcal{U}, d^{m-1}(\Omega^{m-1})) \equiv Z^0(\mathcal{U}, d\Omega^{m-1}), \end{aligned} \tag{3.14}$$

where $\mathcal{U}$ stands for an *open covering of* X. In conclusion, one thus obtains, according to (3.14), that;

$$\omega \in Z^0(\mathcal{U}, d\Omega^{m-1}) = \Gamma(X, d\Omega^{m-1}) \equiv (d\Omega^{m-1})(X) \subseteq \Omega^m(X), \tag{3.15}$$

a better information, of course, than the one supplied already, by the first of (3.8). (Concerning the above calculations in (3.14) and (3.15), cf. also Chapt. III; (4.55). In this context, we should still note that one could get (3.15) *directly from (3.8) and the exactness of (3.1),* according to the very definitions; so cf. also Chapt. I; (13.19) and Theorem 12.1, along with (2.9) and (2.12)).

Now, consider further the following *short exact sequence of* **C**-*vector space sheaves* on X;

$$0 \longrightarrow \ker d^{m-1} \longrightarrow \Omega^{m-1} \longrightarrow d^{m-1}\Omega^{m-1} \longrightarrow 0. \tag{3.16}$$

Accordingly, based now on (3.15) and Chapt. III; (5.86) (cf. also Scholium 3.2 above), one thus concludes, by employing *"liftable cohomology"*, that there exists a *0-cochain*

$$(\theta_\alpha) \in C^0(\mathcal{U}, \Omega^{m-1}), \tag{3.17}$$

such that one has, concerning the given ω, a *0-cocycle,* as in (3.15);

$$\omega = d^{m-1}((\theta_\alpha)) := (d^{m-1}(\theta_\alpha)) \equiv (d\theta_\alpha). \tag{3.18}$$

(Of course, one has in (3.17) an eventual *refinement* $\mathcal{V}$ of $\mathcal{U}$, the latter covering already appeared in (3.14), or even (3.15); however, for simplicity, as already mentioned above (see Scholium 3.2), we do not keep track of it here, as well as, in the sequel). On the other hand, (3.17) satisfies the relation (cf. Chapt. III; (5.86));

$$\delta(\theta_\alpha) \in Z^1(\mathcal{U}, \ker d^{m-1} = \operatorname{im} d^{m-2}). \tag{3.19}$$

Therefore, by the same token, one is finally led, inductively to an m-*cocycle*

$$z(\omega) \in Z^m(\mathcal{U}, \ker \partial) \tag{3.20}$$

(we recall that one has eventually here $\mathcal{U} = \mathcal{U}^{(m)}$, as explained above). Thus, the corresponding cohomology class to $z(\omega)$ will now be, *by definition,* the desired one, as in (3.9), supplied by the given closed m-form ω; that is, we set

$$c(\omega) := [z(\omega)] \in H^m(X, \ker \partial), \tag{3.21}$$

which thus terminates the proof. ∎

As an application of the preceding, we are now in the position to state the following *fundamental result.*

Theorem 3.1. *(Chern-Weil). Suppose we are given a generalized de Rham m-space X, such that*

$$m = 2k \in \mathbf{Z}_+. \tag{3.22}$$

Moreover, let $\mathcal{E}$ be a vector sheaf on X, with $\operatorname{rk} \mathcal{E} = n \in \mathbf{N}$, *endowed with an $\mathcal{A}$-connection D, and consider further an invariant polynomial*

$$p \in \mathbf{C}[(\lambda_{ij})], \quad \textit{with} \quad 1 \leqq i, j \leqq n, \tag{3.23}$$

and such that

$$\deg p = k. \tag{3.24}$$

Finally, let $R(D) \equiv R$ be the curvature of the $\mathcal{A}$-connection D of $\mathcal{E}$.

Then (see also (3.22) and (2.45) above),

(3.25) $$p(R) \in \wedge^m \Omega^1(X) \subsetneq (\wedge^m \Omega^1)(X) \equiv \Omega^m(X)$$

is a closed m-form on X (cf. (3.8)), *providing an m-dimensional cohomology class of X* (see (3.9)),

(3.26) $$c(p(R)) \equiv c(R) \in \check{H}^m(X, \ker \partial),$$

which, in addition, is independent of the given $\mathcal{A}$-connection D.

Proof. First, according to Corollary 2.1 (see (2.45), (2.46)), in conjunction with (3.22) and (3.24) (cf. also Scholium 3.1, concerning the relevant framework), one concludes that

(3.27) $$p(R) \in \wedge^m \Omega^1(X),$$

that is, *(3.25) is fulfilled,* while *the same "m-form" on X is closed,* as well. Hence (Lemma 3.1), $p(R)$, as above, supplies an *m-dimensional* (*Čech,* same Scholium 3.1) *cohomology class $c(R)$, as in (3.26)* (see (3.9)). So it only remains to prove that the latter cohomology class $c(R)$ *is,* in fact, *independent of the $\mathcal{A}$-connection D* involved:

Namely, consider the *unit interval*

(3.28) $$I \equiv [0,1] \subseteq \mathbf{R},$$

such that, *for every* $t \in I$, let us further consider the *continuous map*

(3.29) $$\phi_t : X \longrightarrow X \times I,$$

given by the relation;

(3.30) $$\phi_t(x) := (x,t).$$

Furthermore, consider the (canonical) *1st projection* of the range of ϕ_t,

(3.31) $$\psi \equiv \mathrm{pr}_1 : X \times I \longrightarrow X,$$

such that one has

(3.32) $$\psi(x,t) := x, \quad (x,t) \in X \times I.$$

Now, suppose that we are given two *$\mathcal{A}$-connections* of our vector sheaf $\mathcal{E}$ on X, say D_0 *and* D_1. Thus, by looking further at the *pull-back* of the given *differential triad* (3.3) on X, *via the* above (continuous) *map* ψ, let

(3.33) $$\mathcal{A}' \equiv \psi^*(\mathcal{A})$$

(cf. Chapt. VI; (6.3.1)). Furthermore, let

$$D'_i := \psi^*(D_i), \quad (i = 0, 1) \tag{3.34}$$

be the *induced* $\mathcal{A}'$*-connections of* (the *vector sheaf*) $\psi^*(\mathcal{E})$ on $X \times I$, provided by the previously two given ones of $\mathcal{E}$. (See thus loc. cit., Definition 6.1, along with Chapt. II; (4.14)). So let

$$\widetilde{D} := D'_0 + t(D'_1 - D'_0), \quad t \in [0,1] \equiv I, \tag{3.35}$$

be the *interpolation of* the above two $\mathcal{A}'$*-connections* of $\psi^*(\mathcal{E})$ (cf. (3.34)), yielding thus, for every $t \in [0,1] \subseteq \mathbf{R}$, *another* $\mathcal{A}'$*-connection* of the same vector sheaf $\psi^*(\mathcal{E})$ (cf. Chapt. VI; (7.28.1)).

On the other hand, by looking at the continuous maps ϕ_t and ψ, as above (see (3.29) and (3.31), respectively), one has, concerning the corresponding *pull-back* (*functors;* cf. Chapt. I; (14.45))

$$\phi_t^* \circ \psi^* = (\psi \circ \phi_t)^* = \mathrm{id}_X^*. \tag{3.36}$$

Therefore, based on (3.34), (3.35) and (3.36), one now obtains;

$$\phi_t^*(\widetilde{D}) = D_0 + t(D_1 - D_0) \equiv D_t, \quad t \in I, \tag{3.37}$$

that is, *another* $\mathcal{A}$*-connection of* $\mathcal{E}$, due to the hypothesis for D_0, D_1 and again to Chapt. VI; (7.28.1), as before. Thus, setting now

$$R(\widetilde{D}) \equiv \widetilde{R} \;\; and \;\; R(D_t) \equiv R_t, \tag{3.38}$$

for any $t \in I$, as in (3.37), one has the following relations (see (3.37), (3.38) and Chapt. VIII; (9.25). Yet, cf., for example, R. Godement [1: p.199, §4.16]);

$$\begin{aligned} \phi_t^*(c(p(\widetilde{R}))) &= c(p(\phi_t^*(\widetilde{R}))) = c(p(\phi_t^*(R(\widetilde{D})))) \\ &= c(p(R(\phi_t^*(\widetilde{D})))) = c(p(R(D_t))) = c(p(R_t)). \end{aligned} \tag{3.39}$$

In conclusion, *one* thus *obtains*

$$c(p(R_t)) = \phi_t^*(c(p(\widetilde{R}))), \tag{3.40}$$

for any $t \in [0,1]$.

However, *the continuous maps* ϕ_0 *and* ϕ_1 (cf. (3.28) – (3.30)) *are homotopic,* through the *identity map on* $X \times I$ (see, for instance, J. Dugundji [1: p.315, Definition 1.1]). Therefore, *they define the same functors in cohomo-*

logy (*"Homotopy axiom"*; see, for example, E.H. Spanier [1: p.240, 12], or even J.J. Rotman [1: p.379, Theorem 12.4]). Accordingly, one obtains, by virtue of (3.40) (for $t = 0, 1$, respectively);

$$c(p(R_0)) = \phi_0^*(c(p(\widetilde{R}))) = \phi_1^*(c(p(\widetilde{R}))) = c(p(R_1)). \tag{3.41}$$

In other words, one thus concludes that;

(3.42) $c(R) \equiv c(p(R)) \in H^m(X, \ker \partial)$ (cf. (3.26)), *is, in fact, independent of the $\mathcal{A}$-connection D*, considered,

once we are given $R \equiv R(D)$ (for some $\mathcal{A}$-connection D of $\mathcal{E}$), which actually was our claim, and this also completes the proof of the theorem. ∎

Scholium 3.3.– Concerning the first equality in (3.39), we have actually employed there the fact that for every *continuous map*

$$f : X \longrightarrow Y, \tag{3.43}$$

the corresponding *pull-back functor* (cf. Chapt. I; (14.44.1))

$$f^* : \mathcal{S}h_Y \longrightarrow \mathcal{S}h_X \tag{3.44}$$

commutes with (sheaf) *cohomology:* Namely, by considering, for instance, *cohomology as in (3.9)*, one obtains, via (3.44), a (canonical) *morphism in cohomology*

$$f^* : H^*(Y, \ker \partial) \longrightarrow H^*(X, \ker f^*(\partial)) \tag{3.45}$$

(for simplicity we retain the same notation with (3.44)), in such a manner, that *one has;*

$$f^*([z]) = [f^*(z)], \tag{3.46}$$

where $z \in Z^m(\mathcal{U}_Y, \ker \partial)$, *with* $m \in \mathbf{Z}_+$. (Again, see also R. Godement [1: p.199; §4.16]).

Scholium 3.4.– We comment below a bit more on the *cohomology* (**C**-*vector) spaces*, appeared e.g. *in (3.9):* For simplicity's sake, we also look at a *generalized de Rham space* X, so that, by hypothesis (cf. Definition 3.1), one has the following *exact sequence;*

$$\tilde{\Omega}^{\cdot}_{\mathrm{gdeR}} : 0 \longrightarrow \ker \partial \longrightarrow \Omega^0 \equiv \mathcal{A} \xrightarrow{d^0 \equiv \partial} \Omega^1 \xrightarrow{d^1 \equiv d} \Omega^2 \longrightarrow \cdots \longrightarrow \Omega^m \xrightarrow{d^m \equiv d} \Omega^{m+1} \longrightarrow \cdots. \tag{3.47}$$

Thus, by further employing here our previous notation in Chapter III (ibid.; (1.55′)), we still write (3.47) in the form;

$$0 \longrightarrow \ker \partial \longrightarrow \Omega^{\cdot}_{\mathrm{gdeR}}, \tag{3.48}$$

where we set

$$\Omega^{\cdot}_{\mathrm{gdeR}} : 0 \longrightarrow \mathcal{A} \equiv \Omega^0 \xrightarrow{\partial} \Omega^1 \xrightarrow{d} \cdots \longrightarrow \Omega^m \xrightarrow{d} \cdots, \tag{3.49}$$

this being thus the *(generalized) de Rham complex* of X (having the same *(**C**-linear) morphisms* with (3.47). In this regard, see also Chapt. III; (8.37′) for the analogous notation in the standard case of a $\mathcal{C}^{\infty}$*-manifold, viz. when* $\mathcal{A} \equiv \mathcal{C}^{\infty}_X$). Accordingly, (3.47), or even (3.48), is a *resolution of* $\ker \partial$, the latter object being, as already noted above, another **C**-*vector space sheaf* on X.

Now, given that, by hypothesis, our (base) space X is *paracompact (Hausdorff),* let us further assume that the coefficient **C**-*algebra sheaf* $\mathcal{A}$ is, in particular, *soft* (cf. Chapt. III; (8.25)). Thus, (3.49) *is* then Γ_X*-acyclic,* alias *the resolution (3.48) of* $\ker \partial$ *is* Γ_X-acyclic (cf. Chapt. III; (2.20), (8.26) and (8.56). Apart from the standard case, mentioned above, the previous hypothesis for $\mathcal{A}$ will be fulfilled in some particular concrete examples, we are going to consider later on (see Chapters X, XI in the sequel). Something more, $\mathcal{A}$ will be, in fact, a *fine* sheaf on a paracompact (Hausdorff) space X, as above). So, as a consequence of the *abstract de Rham theorem* (cf. Chapt. III; (3.24), (3.25)), one now infers that;

(3.50) in the case that $\mathcal{A}$ is a *soft* sheaf on (the *paracompact* (Hausdorff) space) X, *the sheaf cohomology of* X *with coefficients in* $\ker \partial$ *is* simply *the cohomology of* $\Gamma_X(\Omega^{\cdot}_{\mathrm{gdeR}})$, as above.

That is, *one has;*

$$H^*(X, \ker \partial) = \check{H}^*(X, \ker \partial) = h^*(\Gamma_X(\Omega^{\cdot}_{\mathrm{gdeR}})). \tag{3.51}$$

In particular, *one* thus *obtains*

$$(R^m\Gamma_X)(\ker\partial) \equiv H^m(X,\ker\partial) := h^m(\Gamma_X(\Omega^{\cdot}_{\mathrm{gdeR}})) := \ker\Gamma_X(d^m)/\mathrm{im}\,\Gamma_X(d^{m-1}), \tag{3.52}$$

for every *positive integer* m (loc. cit.; (3.5)).

4. Symmetric polynomials

For convenience of reference, we collect in this Section several standard facts, pertaining to the polynomials in the title, which will be of use right in the sequel. For simplicity, we also consider *complex polynomials* (the only ones, for that matter, that were and are going to be employed below, as well), although, concerning the material of this Section, any commutative ring with identity could be considered instead. (Cf., for example, S. Lang [1: p. 204]).

Thus, consider the *(complex) polynomial algebra*

$$\mathbf{C}[t_1,\dots,t_n], \tag{4.1}$$

with respect to the n *indeterminates* $t_1,\dots,t_n$ ($n\in\mathbf{N}$). Furthermore, setting

$$A \equiv \mathbf{C}[t_1,\dots,t_n] \tag{4.2}$$

(a *unital commutative* $\mathbf{C}$*-algebra*), consider now a *polynomial* p of one *indeterminate* x, with coefficients from A; viz. let

$$p\in A[x], \tag{4.3}$$

such that, in particular, we set

$$p(x) := \prod_{i=1}^n (x-t_i) \equiv (x-t_1)(x-t_2)\cdots(x-t_n) = x^n - s_1x^{n-1} + \cdots + (-1)^n s_n, \tag{4.4}$$

where, of course, we have

$$s_i \equiv s_i(t_1,\dots,t_n) \in \mathbf{C}[t_1,\dots,t_n],\quad 1\leqq i\leqq n. \tag{4.5}$$

The above polynomials (4.5), as they are given by (4.4), are called the *elementary symmetric polynomials,* with respect to (the indeterminates) $t_1, \ldots, t_n$.

Therefore, by virtue of (4.4), it is easy to see that one has the following *form of the* s_i's, as above;

(4.6)
$$\begin{aligned} s_1 &= t_1 + t_2 + \cdots + t_n \\ s_2 &= \textstyle\sum_{1 \leqq i<j \leqq n} t_i t_j \\ &\ldots\ldots\ldots \\ s_k &= \textstyle\sum_{1 \leqq i_1<i_2<\cdots<i_k \leqq n} t_{i_1} t_{i_2} \cdots t_{i_k} \\ &\ldots\ldots\ldots \\ s_n &= t_1 \cdot t_2 \cdots t_n. \end{aligned}$$

Now, a given polynomial $p \in \mathbf{C}[t_1, \ldots, t_n]$ is said to be *symmetric,* whenever it remains unchanged under any permutation of the independent variables (indeterminates); precisely speaking, in the case one has

(4.7)
$$p^{\pi}(t_1, \ldots, t_n) := p(t_{\pi(1)}, \ldots, t_{\pi(n)}) = p(t_1, \ldots, t_n),$$

or just,

(4.8)
$$p^{\pi} = p,$$

for every *permutation* $\pi \in \mathfrak{S}(n)$ (viz. the *symmetric group of order* n of all permutations of $[1, n] \subseteq \mathbf{N}$).

So it is an easy consequence of the preceding definition of the s_i's, $1 \leqq i \leqq n$, namely, as coefficients of the polynomial (4.4) (see thus (4.6)), that the "elementary symmetric functions", as above, are indeed *symmetric* in the previous sense.

Thus, the *fumdamental theorem* on symmetric polynomials has now as follows:

(4.9) *every symmetric polynomial in* $\mathbf{C}[t_1, \ldots, t_n]$ (cf. (4.8)) *is, in a unique way, a polynomial of the elementary symmetric polynomials* (cf. (4.6)).

That is, *for every symmetric* $p \in \mathbf{C}[t_1, \ldots, t_n]$, *there exists a unique* $f \in \mathbf{C}[y_1, \ldots, y_n]$, *such that one has;*

(4.9.1)
$$p(t_1, \ldots, t_n) = f(s_1(t_1, \ldots, t_n), \ldots, s_n(t_1, \ldots, t_n)).$$

For the proof of (4.9.1) see, for instance S. Lang [1: p.204, Theorem 9.1], or even A.I. Kostrikin [1: p.281, Theorem 1]. ∎ (Here one needs to consider, concerning the general case (cf. the comments at the introduction to this Section), an *integral domain,* where the polynomials involved have their coefficients).

Now, let us consider again the *polynomial algebra*

$$\mathbf{C}[(\lambda_{ij})], \quad 1 \leqq i, j \leqq n, \tag{4.10}$$

as in Section 2 in the preceding, as well as, the *invariant polynomial* (cf. (2.14))

$$\det(\mathbf{1} + t\boldsymbol{\lambda}), \tag{4.11}$$

such that $\boldsymbol{\lambda} \equiv (\lambda_{ij}) \in M_n(\mathbf{C})$, $\mathbf{1}$ stands for the respective $n \times n$ *identity matrix,* while $t \in \mathbf{C}$. (See thus, for instance, Example 2.1 above, in particular, cf. (2.17)). Thus, one has

$$\det(\mathbf{1} + t\boldsymbol{\lambda}) = \mathbf{1} + \sum_{i=1}^{n} t^i p_i(\boldsymbol{\lambda}), \tag{4.12}$$

in such a manner that;

(4.13) *each one of the* p_i*'s,* $1 \leqq i \leqq n$*, is a (homogeneous) invariant polynomial* (with respect to $\boldsymbol{\lambda} \equiv (\lambda_{ij})$, cf. (4.10)), *such that* $\deg p_i = i$, $1 \leqq i \leqq n$.

In this context, one actually proves that;

(4.14) *every invariant polynomial in* $\mathbf{C}[(\lambda_{ij})]$ *can be put (uniquely) in the form of a polynomial of the elementary symmetric polynomials in* $\mathbf{C}[t_1, \ldots, t_n]$ (see (4.6)).

For the proof of (4.14), consult e.g. R. Bott [1: p.77, Appendix A]. ∎

The previous result is based on the fact that *the restriction of an invariant polynomial of the algebra (4.10) to the diagonal elements of the same algebra yields,* in effect, *a symmetric polynomial of the polynomial algebra (4.1)* (loc. cit.); in turn, this is an immediate consequence of the fact that *every automorphism of* $M_n(\mathbf{C})$, *hence* a permutation of the diagonal entries of a matrix, as before, too, *is inner* (viz. it *comes from an inner automorphism,* defined *by an element of* $\mathrm{GL}(n, \mathbf{C})$. (Cf. also (2.14) and (4.7); yet, see e.g.

L. Chambadal - J.L. Ovaert [1: p.302, Remarque 2, and p.511, Ex. 158 A; I.4]).

Thus, by applying the above remarks to the invariant polynomial (4.11), one obtains (setting also $t = 1$);

$$
\begin{aligned}
&\det(\mathbf{1}+\boldsymbol{\lambda}) = \det(\operatorname{diag}(1+t_1,\ldots,1+t_n)) = \textstyle\prod_{i=1}^{n}(1+t_i)\\
&\equiv (1+t_1)(1+t_2)\cdots(1+t_n)\\
&= 1+(t_1+t_2+\cdots+t_n)+(t_1t_2+t_1t_3+\cdots t_{n-1}t_n)+\cdots\\
&\qquad\cdots+\textstyle\sum_{1\leqq i_1<i_2<\cdots<i_k\leqq n} t_{i_1}t_{i_2}\cdots t_{i_k}+\cdots+t_1\cdot t_2\cdots t_n\\
&= 1+s_1+s_2+\cdots+s_k+\cdots+s_n\\
&= 1+\operatorname{tr}\boldsymbol{\lambda}+\tfrac{1}{2}((\operatorname{tr}\boldsymbol{\lambda})^2-\operatorname{tr}\boldsymbol{\lambda}^2)+\cdots+\det\boldsymbol{\lambda}
\end{aligned}
\tag{4.15}
$$

(cf. also (4.6) above). Yet, concerning the last equality in (4.15), one employs here the so-called *Newton's relations,* connecting the elementary symmetric polynomials and the respective to (4.1) *power sums,* being also symmetric polynomials of the same algebra. (See, for instance, R. Kochendörffer [1: p.171 and p.176]).

Applications of the preceding will be made right below in the succeeding Section.

5. Generalized Chern classes

We have already at our disposal, as a result of the preceding discussion, the necessary material to define the particular (cohomology) classes, as in the title of this Section.

So, to fix the terminology applied, suppose that we are given a *curvature space* X (cf. Chapt. VIII; Definition 1.1), and let $\mathcal{E}$ be a *vector sheaf* on X, with

$$\operatorname{rk}\mathcal{E} = n \in \mathbf{N}, \tag{5.1}$$

while we further assume that $\mathcal{E}$ has an $\mathcal{A}$-*connection* D, whose *curvature* is $R(D) \equiv R$ (ibid.; (2.1)). Furthermore, suppose that

(5.2) $$\mathcal{U} = (U_\alpha)_{\alpha \in I}$$

is a *local frame of* $\mathcal{E}$, so that one has, concerning the curvature R, as before (see (2.11) in the foregoing);

(5.3) $$R \equiv \left(R^{(\alpha)}\right) \equiv \left(\left(R_{ij}^{(\alpha)}\right)\right) \in C^0(\mathcal{U}, M_n(\Omega^2)) \hookrightarrow C^0(\mathcal{U}, \Omega^{\cdot}_{\text{ev}}).$$

Thus, by further considering the *invariant polynomial* (4.11), as above, one now defines;

(5.4) $$c(\mathcal{E}) := \det\left(1 - \frac{1}{2\pi i} R\right).$$

We call (5.4) the *total Chern class* of the given *vector sheaf* $\mathcal{E}$. (The applied terminology will be fully justified right below).

In this context, we first remark that, in view of (4.11) and the previous Lemma 2.1, we actually have

(5.5) $$c(\mathcal{E}) \in \Omega^{\cdot}_{\text{ev}}(X)$$

(see (2.18)). Yet, given that

(5.6) $$\deg \det(\boldsymbol{\lambda}) = n, \quad \textit{for every } \boldsymbol{\lambda} \equiv (\lambda_{ij}) \in M_n(\mathbf{C}),$$

(cf. also (2.16)), one infers, in particular, that;

(5.7) $$c(\mathcal{E}) \in \Omega^{2n}(X) \hookrightarrow \Omega^{\cdot}_{\text{ev}}(X).$$

Accordingly (see also (1.15), (1.17)), the above polynomial (5.7) can be put in the form;

(5.8) $$c(\mathcal{E}) = 1 + c_1(\mathcal{E}) + c_2(\mathcal{E}) + \cdots,$$

such that one has

(5.9) $$c_i(\mathcal{E}) \in \wedge^{2i}\Omega^1(X) \hookrightarrow \Omega^{2i}(X), \quad i \in \mathbf{N}.$$

We call (5.9) the *i-th Chern class* of $\mathcal{E}$ (see (5.28) below for the justification of the terminology applied). Furthermore, we extend the latter definition, by still setting for the *0-th Chern class* of $\mathcal{E}$;

(5.10) $$c_0(\mathcal{E}) := 1 \in \mathcal{A}(X) \equiv \Omega^0(X),$$

viz. the *identity* (global) *section of* $\mathcal{A}$ (cf. also (1.16), as well as, Chapt. II; (1.22)).

In this regard, we should remark, of course, that *the sum in (5.8)* (finite, anyway, according to (5.7) and (1.15)) *is terminated at*

(5.11) $$c_n(\mathcal{E}) = \det\left(-\frac{1}{2\pi i}R\right),$$

where $n = \operatorname{rk}\mathcal{E}$ (see (5.1), (5.6), as well as, (5.22) below). That is, one has;

(5.12) $$c_i(\mathcal{E}) = 0, \quad \textit{for} \quad i > n = \operatorname{rk}\mathcal{E} = \deg\left(\det\left(-\frac{1}{2\pi i}R\right)\right).$$

Now, we can have at this point the following byproduct, when looking at a *line sheaf* $\mathcal{L}$ on X. Namely, in view of (5.12), one concludes that;

(5.13) in the particular case of a *line sheaf* $\mathcal{L}$ on X, *the only non-zero Chern class of* $\mathcal{L}$ (apart from the *0-th one* (see (5.10))) *is*

(5.13.1) $$c_1(\mathcal{L}) \in \Omega^2(X).$$

Therefore, based on (5.11), one further obtains;

(5.14) $$c_1(\mathcal{L}) = -\frac{1}{2\pi i}R \in \Omega^2(X)$$

(see also, for instance, Chapt. VIII; (3.24)). Hence, the corresponding *total Chern class* of the *line sheaf* $\mathcal{L}$ is given by

(5.15) $$c(\mathcal{L}) = 1 + c_1(\mathcal{L}) = 1 - \frac{1}{2\pi i}R$$

(see (5.4), or even (5.8), (5.13) and (5.14)).

In this context, we have still to remark that in the case the given $\mathcal{A}$-module Ω^1, as above, is, in particular, a *vector sheaf* on X too (this happens, of course, in the classical theory, cf. Chapt. VI; Section 2), then one gets a result similar to (5.12). Thus, if, for example, in that case one has,

(5.16) $$\operatorname{rk}\Omega^1 = m \in \mathbf{N},$$

then, in view of (5.9) (cf. also Chapt. IV; (7.20)), one gets again, concerning (5.8);

(5.17) $$c_i(\mathcal{E}) = 0, \quad \textit{for any} \quad 2i > m = \operatorname{rk}\Omega^1.$$

Apart from the classical case of *$\mathcal{C}^\infty$-manifolds,* as already noted above, we shall also see later on *further examples, for which (5.16) holds true* (cf. thus Chapters X, XI in the sequel).

Now, by direct application of our previous argument on (4.15) to the present context, one obtains;

$$g^{-1}\left(-\frac{1}{2\pi i}R\right)g \equiv \mathrm{Ad}(g^{-1})\left(-\frac{1}{2\pi i}R\right) = \mathrm{diag}(t_1,\dots,t_n) \equiv \boldsymbol{\lambda}, \tag{5.18}$$

where one has, of course (see also (5.2)),

$$g \equiv \left(g_{ij}^{(\alpha)}\right) \in \mathcal{GL}(n,\mathcal{A})(U_\alpha) = \mathrm{GL}(n,\mathcal{A}(U_\alpha)) \tag{5.19}$$

for some *coordinate 1-cocycle of $\mathcal{E}$,*

$$(g_{\alpha\beta}) \in Z^1(\mathcal{U},\mathcal{GL}(n,\mathcal{A})). \tag{5.20}$$

Accordingly, one has (cf. (5.4), (5.8) and (5.12))

$$c(\mathcal{E}) \equiv \det(1+\boldsymbol{\lambda}) = \det\left(1-\frac{1}{2\pi i}R\right) = 1 + c_1(\mathcal{E}) + \cdots + c_n(\mathcal{E}), \tag{5.21}$$

so that, by virtue of the same rel. (4.15), as above, in conjunction with (5.18), one now obtains;

$$\begin{aligned} c_0(\mathcal{E}) &= 1 \\ c_1(\mathcal{E}) &= \mathrm{tr}\,\boldsymbol{\lambda} = \mathrm{tr}\left(g^{-1}\left(-\tfrac{1}{2\pi i}R\right)g\right) = \mathrm{tr}\left(-\tfrac{1}{2\pi i}R\right) = -\tfrac{1}{2\pi i}\mathrm{tr}\,R \\ &\vdots \qquad \qquad \cdots\cdots\cdots \\ c_n(\mathcal{E}) &= \det\boldsymbol{\lambda} = \det\left(g^{-1}\left(-\tfrac{1}{2\pi i}R\right)g\right) = \det\left(-\tfrac{1}{2\pi i}R\right) = \left(-\tfrac{1}{2\pi i}\right)^n \det R. \end{aligned} \tag{5.22}$$

Scholium 5.1.– Concerning the above calculations, one can construed the *curvature R,* as a *local $n \times n$ matrix* (see e.g. (5.18)); viz. one has

$$\begin{aligned} R \equiv \left(R^{(\alpha)}\right) &\in M_n(\Omega^2(U_\alpha)) = M_n(\Omega^2)(U_\alpha) \\ &\hookrightarrow M_n(\Omega^{\cdot}_{\mathrm{ev}}(U_\alpha)) = M_n(\Omega^{\cdot}_{\mathrm{ev}})(U_\alpha), \end{aligned} \tag{5.23}$$

for every $\alpha \in I$ (cf. (5.2), as well as, e.g. (2.37)). Therefore (see also e.g. (2.17.1)), *one has*

(5.24) $$\det R \equiv \det\left(R^{(\alpha)}\right) \in \Omega^{\cdot}_{\mathrm{ev}}(U_\alpha), \quad \alpha \in I.$$

However, due to the fact that det is an *invariant polynomial* (cf. (2.17)), one actually concludes that (see Lemma 2.1);

(5.25) $$\det R \equiv \left(\det R^{(\alpha)}\right) \in Z^0(\mathcal{U}, \Omega^{\cdot}_{\mathrm{ev}}) = \Omega^{\cdot}_{\mathrm{ev}}(X).$$

In this context, we further recall the standard fact that, within the context of (2.3), (2.15),

the trace function

(5.26) $$\mathrm{tr} : M_n(\mathbf{C}) \longrightarrow \mathbf{C}$$

is an invariant polynomial.

See (2.14), as well as, e.g. S. Lang [1: p.443]. ∎

As a consequence of (5.26) one gets, of course, an analogous situation to (5.25) for (the *global section,* same Lemma 2.1) $\mathrm{tr}\, R$.

Thus, according to the preceding, we conclude now the following basic fact:

(5.27) *each one of the Chern classes* of the given *vector sheaf* $\mathcal{E}$, as they were given by (5.22), *yields a* (homogeneous) *invariant polynomial,* with respect to the *curvature matrix* R (cf. (5.23)), *whose degree is at most* $n = \mathrm{rk}\,\mathcal{E}$ (cf. (5.12), (4.13)).

Therefore, in the case our curvature space X, as given above, is in particular, a *(generalized) de Rham* $2n$*-space* (cf. Definition 3.1, along with (5.1)), then in view of the *Chern-Weil Theorem* (see Theorem 3.1), one further infers that:

(5.28) *each one of the above Chern classes* (see (5.22)) *supplies a* $2\deg p_i$*-dimensional,* that is (cf. (4.13)), *a* $2i$*-dimensional (Čech) cohomology class of* X, $1 \leqq i \leqq n$, *with coefficients in* $\ker\partial$ (see (3.26)). Namely, *one has;*

(5.28.1) $$c_i(\mathcal{E}) \equiv [c_i(\mathcal{E})] \in H^{2i}(X, \ker\partial),$$

with $1 \leqq i \leqq n = \mathrm{rk}\,\mathcal{E}$.

The previous rels. (5.28.1) provide also the aforementioned vindication of the nomenclature applied to the terms defined by (5.9).

In the classical theory the corresponding classes take their values from $\mathbf{R}$ ($\subsetneq$ $\mathbf{C}$) and, indeed, from $\mathbf{Z}$; that is, as we say, they are *integral,* in the sense that *each one lies in*

$$\operatorname{im}(\check{H}^{2i}(X,\mathbf{Z}) \longrightarrow H^{2i}(X,\mathbf{R})), \tag{5.29}$$

with $i = 1, 2, \ldots, n$, *respectively,* as the latter map is defined by the (canonical) *injection* $\mathbf{Z} \hookrightarrow \mathbf{R}$. In fact, this is a consequence of the so-called *splitting principle* (see e.g. R. Shanahan [1: p.191ff]).

In our case we do have a counterpart of the above for the particular instance of a *line sheaf,* through the *Weil integrality theorem* (see Chapt. VIII; Theorem 11.1, or even (5.49), (5.51) and (5.56)). On the other hand, concerning the previous principle, one could also keep track of the classical procedure in the general case of *vector sheaves,* as well, on the basis of the hitherto presented material. However, we do not carry it out here, leaving this matter, instead, to the interested reader.

6. Chern-Weil homomorphism

We consider below the analogous to our case *"functorial properties" of* the previously defined *(generalized) Chern classes* (see (5.28.1)) for a given *vector sheaf,* by analogy with the standard theory, pertaining, in effect, to the so-called *Weil homomorphism* (cf. (6.15) below) and its *"naturality".*

So to start with, suppose that we are given a *curvature space* X, together with a *vector sheaf* $\mathcal{E}$ on X, of *rank* $n \in \mathbf{N}$, which is further endowed with an $\mathcal{A}$-*connection* D, whose *curvature* is $R(D) \equiv R$.

On the other hand, consider the following *matrix algebra,* consisting of the *global sections* of the *(matrix)* $\mathbf{C}$-*algebra sheaf* involved (see also (1.15) and (2.1));

$$M_n(\Omega^{\cdot}_{\mathrm{ev}})(X) = Z^0(\mathcal{U}, M_n(\Omega^{\cdot}_{\mathrm{ev}})), \tag{6.1}$$

where, of course, as usual,

(6.2) $$\mathcal{U} = (U_\alpha)_{\alpha \in I}$$

stands for a given *local frame* of $\mathcal{E}$. Thus, we denote by

(6.3) $$\mathcal{I}^*(\mathcal{GL}(n, \mathcal{A}))$$

the *(graded)* **C**-*algebra of invariant polynomials,* in effect, *polynomial functions* on (6.1) (see, for example, (2.5), (2.8) and (2.14)), the *multiplication in (6.3) being,* by definition, the *exterior product.* Namely, one sets;

(6.4) $$\mathcal{I}^*(\mathcal{GL}(n, \mathcal{A})) = \bigoplus_{k=0}^{\infty} \mathcal{I}_k(\mathcal{GL}(n, \mathcal{A})),$$

where the summands in the second member of (6.4) correspond to the various *degrees of* the (invariant) *polynomials* concerned.

Now, let us consider further the *curvature matrix* of D, which is *associated with the* given *local frame* $\mathcal{U}$ of $\mathcal{E}$; viz. one has (see (1.11) and (1.12));

(6.5) $$\begin{aligned} R \equiv \left(R^{(\alpha)}\right) \equiv \left(\left(R^{(\alpha)}_{ij}\right)\right) &\in Z^0(\mathcal{U}, M_n(\Omega^2)) \\ &\subseteq \textstyle\prod_\alpha M_n(\Omega^2)(U_\alpha) = \prod_\alpha M_n(\Omega^2(U_\alpha)). \end{aligned}$$

In particular, one obtains;

(6.6) $$\left(R^{(\alpha)}_{ij}\right) \in M_n(\Omega^2)(U_\alpha) = M_n(\Omega^2(U_\alpha)), \quad \alpha \in I,$$

such that (cf. (1.13))

(6.7) $$R^{(\alpha)}_{ij} \in \Omega^2(U_\alpha), \quad \textit{with} \quad 1 \leqq i, j \leqq n = \operatorname{rk} \mathcal{E}, \quad \textit{and} \quad \alpha \in I.$$

Therefore, by looking now at an element

(6.8) $$p \in \mathcal{I}_k(\mathcal{GL}(n, \mathcal{A})) \hookrightarrow \mathcal{I}^*(\mathcal{GL}(n, \mathcal{A}))$$

(see (6.4)), that is, at an *invariant polynomial* of *degree* $k \in \mathbf{N}$, one obtains that (cf. also (2.20))

(6.9) $$\begin{aligned} p(R) \in \wedge^{2k}\Omega^1(X) \hookrightarrow (\wedge^{2k}\Omega^1)(X) \equiv \Omega^{2k}(X) &\hookrightarrow \Omega^{\cdot}_{\mathrm{ev}}(X) \\ \hookrightarrow M_n(\Omega^{\cdot}_{\mathrm{ev}}(X)) &= M_n(\Omega^{\cdot}_{\mathrm{ev}})(X). \end{aligned}$$

On the other hand, suppose now that our space X is, in particular, a *Bianchi space.* Thus (cf. Theorem 2.1), one then obtains;

(6.10) $$dp(R) = 0,$$

viz. $p(R)$, as in (6.9), is a *closed 2k-form on X*, with $k = \deg p \in \mathbf{N}$.

Furthermore, let us assume now that the same Bianchi space X, as before, is, in addition, a *(generalized) de Rham space* (no restriction on order; see Definition 3.1, together with Scholium 3.1. However, cf. (6.13) in the sequel, as well). Thus, in view of (6.10) and based further on Lemma 3.1 (cf. (3.25), (3.26)), one can associate with $p(R)$, as in (6.9), a *2k-dimensional (Čech) cohomology class of X, with coefficients in the* **C***-vector space sheaf* $\ker \partial$; viz. we have

$$\begin{aligned}(6.11)\qquad c(p(R)) \equiv c(R) &\equiv [p(R)] \in \check{H}^{2k}(X, \ker \partial) \\ &\hookrightarrow \bigoplus_{n=0}^{\infty} \check{H}^{2n}(X, \ker \partial) \equiv \check{H}^{**}(X, \ker \partial).\end{aligned}$$

The last item in (6.11) denotes the *(graded)* **C***-algebra of even dimensional (Čech) cohomology* **C***-vector spaces* of X, with *coefficients in* $\ker \partial$. In this concern, the *multiplication in the latter algebra* is still defined *by exterior multiplication* on the representatives (cocycles) of the cohomology classes involved (cf. (3.1) too).

Scholium 6.1.- Concerning the *order of the (generalized) de Rham space X*, as above (cf. Definition 3.1), it is to be noticed that *this cannot exceed* $\operatorname{rk} \Omega^1$, in the case the given $\mathcal{A}$*-module* Ω^1 *is,* in particular, a *vector sheaf* on X *too.* Thus, in tha latter case it is, in effect, enough to consider a *(generalized) de Rham m-space, with*

$$(6.12)\qquad m = \operatorname{rk} \Omega^1 \in \mathbf{N}.$$

Furthermore, based now on (6.9), one obtains;

$$(6.13)\qquad p(R) = 0, \quad \textit{if} \quad 2\deg p > m = \operatorname{rk} \Omega^1.$$

On the other hand, since

$$(6.14)\qquad \check{H}^{**}(X, \ker \partial) \hookrightarrow \check{H}^{*}(X, \ker \partial) := \bigoplus_{k \in \mathbf{Z}_+} \check{H}^{k}(X, \ker \partial)$$

(see also (6.11)), the multiplication in the second member of (6.14) being defined in an analogous manner to that one in (6.11), *one* finally *obtains the following map;*

(6.15) $$\chi_{\mathcal{E}} : \mathcal{I}^*(\mathcal{GL}(n, \mathcal{A})) \longrightarrow \check{H}^*(X, \ker \partial),$$

such that (see (6.11))

(6.16) $$\chi_{\mathcal{E}}(p) := c(p(R)) \equiv [p(R)] \in \check{H}^{\deg p}(X, \ker \partial),$$

the same map being also *well-defined,* by virtue of Theorem 3.1, *for any* given *pair* $(\mathcal{E}, D)$, as above.

The previous map (6.15), as given by (6.16), is named the *Chern-Weil homomorphism,* being thus associated with any vector sheaf having an $\mathcal{A}$-connection, when, of course, the appropriate context is provided, as well (cf. the following theorem).

Now, the above terminology for the map (6.15) is further justified according to the next.

Theorem 6.1. *Suppose we are given a (generalized) de Rham space X, and let $\mathcal{E}$ be a vector sheaf on X, endowed with an $\mathcal{A}$-connection D. Then, the corresponding Chern-Weil homomorphism $\chi_{\mathcal{E}}$* (cf. (6.15), (6.16)) *is a morphism of the respective (graded)* **C***-algebras* (see (6.4) and (6.14)).

On the other hand, for any given continuous map

(6.17) $$f : Y \longrightarrow X,$$

one obtains ("naturality of $\chi_{\mathcal{E}}$");

(6.18) $$\chi_{f^*(\mathcal{E})} = f^* \circ \chi_{\mathcal{E}},$$

where $f^*(\mathcal{E})$ denotes the vector sheaf on Y, pull-back, via f, of the given vector sheaf $\mathcal{E}$ on X.

Proof. The first assertion about $\chi_{\mathcal{E}}$ is an immediate consequence of the very definition of the multiplications in the **C**-algebras involved, as already explained at the pertinent places above (see the comments after (6.11) and (6.14), respectively, in the preceding).

Now, suppose that

(6.19) $$(g_{\alpha\beta}) \in Z^1(\mathcal{U}, \mathcal{GL}(n, \mathcal{A})), \quad \textit{with} \quad n = \operatorname{rk} \mathcal{E} \in \mathbf{N},$$

is a *coordinate 1-cocycle* of the *vector sheaf* $\mathcal{E}$, with respect to a *local frame of* $\mathcal{E}$,

(6.20) $$\mathcal{U} = (U_\alpha)_{\alpha \in I}.$$

Thus, it is an easy consequence of the very definitions that

$$f^*(\mathcal{U}) := (f^{-1}(U_\alpha))_{\alpha \in I} \tag{6.21}$$

is a *local frame of* $f^*(\mathcal{E})$, the latter sheaf being a *vector sheaf on* Y, *relative to* $f^*(\mathcal{A}) \equiv \mathcal{A}_Y$, such that

$$\operatorname{rk}_{f^*(\mathcal{A})} f^*(\mathcal{E}) = \operatorname{rk}_{\mathcal{A}} \mathcal{E} = n \tag{6.22}$$

(see Chapt. II; (4.14), (2.68) and (2.69), together with Chapt. I; (14.26)).

On the other hand, based on the *exactness of the pull-back functor* f^* (Chapt. II; (2.69)), one has transported to Y, through f, the corresponding framework on X, namely, that one of a *(generalized) de Rham space* and its paraphernalia, which, by hypothesis, is already given on X (see, for instance, Chapt. VIII; (9.24), in conjunction with Definition 3.1 in the preceding). Finally, *for every invariant polynomial* $p \in \mathcal{I}^*(\mathcal{GL}(n, \mathcal{A}))$ (see (6.3), (6.4)), *one has*

$$f^* \circ p = p \circ f^* \in \mathcal{I}^*(\mathcal{GL}(n, f^*(\mathcal{A}))), \tag{6.23}$$

so that one still obtains;

$$p(f^*(R)) = f^*(p(R)). \tag{6.24}$$

In this concern, we also recall that $f^*(R)$ is the *curvature of the* $f^*(\mathcal{A})$-*connection of* $f^*(\mathcal{E})$ (see Chapt. VIII; (9.25)). Therefore, *one has* (cf. (6.16));

$$f^*(\chi_{\mathcal{E}}(p)) = \chi_{f^*(\mathcal{E})}(p), \tag{6.25}$$

for every $p \in \mathcal{I}^*(\mathcal{GL}(n, \mathcal{A}))$, as desired, and this terminates the proof. ∎

PART THREE

EXAMPLES AND APPLICATIONS

CHAPTER X

Classical theory

"Progress in any aspect is a movement through changes in terminology."

W. Stevens in "Adagia" (included in "Opus Posthumous", ed. by Vintage Books, 1990). p.184.

We consider in this Chapter several particular instances, that have to do, essentially, with the classical theory of C^∞*-manifolds* (in particular, *differential geometry* of such), for which the framework, exhibited hitherto in the preceding Chapters, constitutes their abstract (axiomatic) description. Of course, this happens in the *finite-dimensional* case, while certain *infinite-dimensional examples* also are fitted in well, within the same context, as in the foregoing. A particularly interesting case is that one of the so-called *generalized manifolds,* or even *orbifolds,* which we are dealt with in Section 2. The case of *complex manifolds* is also discussed (Section 3). Finally, we also examine a potential application of the previous framework to the description of certain basic concepts of *theoretical physics,* including the case of *supermanifolds.* Further applications, as well as, examples of the preceding discussion, within the framework of *topological algebra theory,* will be given in the next Chapter XI.

1. Differential geometry of $\mathcal{C}^{\infty}$-manifolds. (Finite and infinite-dimensional cases)

We have already pointed out in the preceding several places, where the abstract framework employed so far specialized to the standard *differential geometry of finite-dimensional smooth manifolds:*

So, to avoid recurences, we simply indicate by the ensuing discussion concrete places in the foregoing, where one finds relevant references to the *standard case,* the latter being, for that matter, the *life-motif of our whole exposition.* In this context, we also note that just the headings too of the particular Sections in the second part of this treatise (viz. of *Part Two, Geometry,* in particular, Chapters VI – IX) indicate, of course, clearly the relation between the present material and the classical one; yet, the two aspects are linked up through the fundamental concept of a *differential triad* (Chapt. VI; (1.4), (3.1) and Chapt. VIII; (1.1)). In particular, here, for the case at issue, by the *standard differential triad*

$$(\mathcal{C}^{\infty}_{X}, d, \Omega^{1}_{X}), \tag{1.1}$$

with X a *(finite-dimensional)* $\mathcal{C}^{\infty}$*-manifold;* see thus Chapt. VI; Section 2.

On the other hand, we have already noted (same Chapt. VI; Subsection 4.(b)), how the present (axiomatic) point of view specializes to the standard *theory of connections* (viz. *covariant differentiation*) *of* $\mathcal{C}^{\infty}$*-(complex) vector bundles* (see also Note 1.1 below, concerning the more general case of *principal (fiber) bundles).* In this regard, we have also fully explained the fundamental fact that, in the particular case when our base $\mathcal{C}^{\infty}$-manifold X is *paracompact (Hausdorff),* then

(1.2) $\mathcal{C}^{\infty}$*-connections always exist*

(see Chapt. VI; (15.10), (16.18)), by contrast with what happens, in general, within our abstract framework; see thus, for instance, Chapt. III; Corollary

12.1, where one realizes that the existence of an *$\mathcal{A}$-connection* for a given *vector sheaf $\mathcal{E}$* is characterized by the *vanishing of* what we have called (generalized) *Atiyah class of $\mathcal{E}$.*

Now, as already indicated in the pertinent places of the preceding discussion, the important conclusion (1.2), for *(finite-dimensional paracompact) $\mathcal{C}^\infty$-manifolds,* is actually due to the basic fact that, for the case at issue, the *sheaf of coefficients,* or even the *structure sheaf,* viz. the $\mathbf{C}$-*algebra sheaf*

$$(1.3)\qquad {}^{\mathbf{C}}\mathcal{C}^\infty_X \equiv \mathcal{C}^\infty_X$$

(cf. Chapt. I; (4.39)) on a *smooth manifold* X, as before, is *fine* (Chapt. III; (8.55.1)). However, see also the next Chapter XI for a similar conclusion in certain particular instances in topological algebra theory, that also include the previous case.

On the other hand, *the local aspect of the classical theory has also been exposed, within the axiomatic framework* of the previous discussion (see thus Chapters VII and VIII). In this context, we should further remark that the *(partial) differential operator*

$$(1.4)\qquad \frac{\partial}{\partial x}$$

of the *standard theory* is to be replaced here by an expression of the form;

$$(1.5)\qquad \partial(s)(t)$$

(see, for example, Chapt. VII; (5.41) for the notation applied above, or even (5.63) therein). Analogously, one can consider *higher derivations,* by setting, for instance,

$$(1.6)\qquad \partial(s')(\partial(s)(t)),$$

as a substitute of the classical operator

$$(1.7)\qquad \frac{\partial^2}{\partial x\,\partial y}.$$

Here *one cannot expect, in general,* the usual *"symmetry"* of the second (partial) derivative, as in the classical theory, pertaining to the order of the (partial) differentiation considered. Of course, one still understands in (1.6) the relevant *context of Chapt. VII; Section 5.*

On the other hand, *metric ($\mathcal{A}$-)connections* have been considered in

Chapt. VII; Sections 8 and 9, extending thus, within the present abstract context, the classical subject of *Riemannian and hermitian connections* on a C^∞-manifold, as above. (Concerning the classical case, see e.g. R.O. Wells [1: p.76, Proposition 1.11]).

Now, another aspect that is also of interest, within the preceding framework, is the analogue in our case of the standard *de Rham complex;* thus, applying the notation of Chapt. IX; Definition 3.1, we consider the following *generalized de Rham complex*

$$
\begin{aligned}
&0 \longrightarrow \mathcal{A} \equiv \Omega^0 \xrightarrow{d^0 \equiv \partial} \Omega^1 \xrightarrow{d^1 \equiv d} \Omega^2 \longrightarrow \cdots \\
&\cdots \longrightarrow \Omega^n \xrightarrow{d^n \equiv d} \Omega^{n+1} \longrightarrow \cdots,
\end{aligned} \tag{1.8}
$$

being, *by hypothesis, a complex of* **C***-vector space sheaves* on X (see also Chapt. VIII; (8.2) – (8.4), as well as, Chapt. IX; Scholium 3.1). In particular, *we assume* that the following is an *exact sequence,* as before;

$$
\begin{aligned}
&0 \longrightarrow \ker \partial \longrightarrow \mathcal{A} \equiv \Omega^0 \xrightarrow{p} \Omega^1 \xrightarrow{d^1} \Omega^2 \longrightarrow \cdots \\
&\cdots \longrightarrow \Omega^n \xrightarrow{d^n} \Omega^{n+1} \longrightarrow \cdots.
\end{aligned} \tag{1.9}
$$

We have already seen in the previous Chapter IX the significance of (1.9), in order to get in our case the analogous procedure of *associating a cohomology class to a closed form* on X (ibid.; Lemma 3.1), and its impact, in turn, in proving the corresponding here *Chern-Weil Theorem,* the basic tool for extending to the present abstract framework the theory of *Chern (characteristic) classes* (loc. cit., Theorem 3.1. See also Scholium 3.4 therein, along with Section 5).

Furthermore, we can still consider the following *augmented generalized (strict) de Rham complex,* being also, by hypothesis, exact;

$$
0 \longrightarrow \mathbf{C} \xrightarrow{\varepsilon} \mathcal{A} \xrightarrow{\partial} \Omega^1 \xrightarrow{d} \cdots \xrightarrow{d} \Omega^n \xrightarrow{d} \cdots, \tag{1.10}
$$

the rest being, as in (1.8). Of course, when one looks at a C^∞*-manifold* X, the last complex (1.10) is the usual *(augmented) de Rham complex of* X of the classical theory, where one further employs **C***-valued functions* (cf. (1.3). See also Chapt. III; (8.37′) and (8.44)). In this context, we have already noted (loc. cit. Lemma 8.2) the significant rôle played here by *Poincaré's Lemma. In the general case,* however, as the one considered herewith, the aforementioned result, is *unattainable,* hence, the *exactness of (1.10),* as well. Nevertheless, as we shall see later on (cf. Chapter XI), we do have *exactness of (1.9) at the first place,* at least, in the following form;

$$0 \longrightarrow \mathbf{A} \longrightarrow \mathcal{A} \xrightarrow{\partial} \Omega^1 \longrightarrow \cdots, \tag{1.11}$$

were **A** stands for a suitable *topological algebra;* this is of a particular significance for *differential geometric applications* to the so-called **A**-*vector bundles,* pertaining, for instance, to their *flatness* (loc. cit.).

On the other hand, apart from the standard $\mathcal{C}^\infty$-manifolds, as above, we are going to consider right below *certain particular examples,* even *infinite dimensional,* where the *exactness of (1.10) holds true.* We thus exhibit in the sequel a number of such examples (in fact, *infinite-dimensional* ones), the rest being deferred for Section 3. But, before this, we still comment on another basic point of view, according to the classical theory, which, however, we do not examine by this treatise.

> **Note 1.1.**– We have treated heretofore, while this will also be the case in the sequel, *vector sheaves,* viz. *locally free* $\mathcal{A}$*-modules* (of *finite rank*), with respect to a given fixed **C**-*algebra sheaf* $\mathcal{A}$. As already explained (cf. Chapt. VI; (4.11)), these correspond to **C**-*vector bundles* of the classical theory.
>
> On the other hand, the counterpart in our case of the standard notion of a *principal fiber bundle* can also be examined within the abstract (axiomatic) framework, as above. Indeed, such a study has already been started by the recent work of E. Vassiliou [1] – [4].

1.(a). Infinite-dimensional examples.– We consider below examples of *infinite-dimensional* $\mathcal{C}^\infty$*-manifolds,* for which the *augmented de Rham complex* (1.10) yields an *exact sequence:*

The discussion that follows is based on the work of M.E. Verona [1]. Thus, consider a *projective limit topological space*

$$X = \varprojlim_{\alpha} X_\alpha \tag{1.12}$$

(see, for instance, N. Bourbaki [1: Chap. 1; p.28ff]), such that each one of the factor spaces X_α, $\alpha \in I$ (the latter being an upwards directed set) is a usual (finite dimensional) $\mathcal{C}^\infty$*-manifold, Hausdorff* and *2nd countable,* hence, *metrizable* and **a fortiori** *paracompact* (see e.g. J. Dugundji [1: p.195, Corollary 9.2 and p.186, Theorem 5.3]). Moreover, we assume that the connecting maps

$$\phi_{\alpha\beta} : X_\beta \longrightarrow X_\alpha, \quad \alpha \leqq \beta, \tag{1.13}$$

are *proper submersions;* thus, X is *locally compact* and *σ-compact,* hence, *paracompact* (see N. Bourbaki [1; p.77, Proposition 7] and J. Dugundji [1: p.241, Theorem 6.5]).

Now, one defines *"$\mathcal{C}^\infty$-functions"* on X, as follows: Taking an *open* $U \subseteq X$, a **C**-*valued map*

$$f : U \longrightarrow \mathbf{C} \tag{1.14}$$

is said to be a *$\mathcal{C}^\infty$-map on* U, whenever, for every $x \in U$, there exists an *open* $U_x \subseteq X$, with $x \in U_x$, and some $g \in \mathcal{C}^\infty(X_\alpha) \equiv {}^{\mathbf{C}}\mathcal{C}^\infty(X_\alpha)$ (that is, **C**-valued $\mathcal{C}^\infty$-functions on the smooth manifold X_α), where $\alpha \in I$, as above, such that;

$$f = g \circ \pi_\alpha|_{U_x}, \tag{1.15}$$

where $\pi_\alpha : X \longrightarrow X_\alpha$, $\alpha \in I$, stands for the corresponding canonical map. We denote the set of such maps on the *open* $U \subseteq X$ by

$$\mathcal{C}^\infty(U). \tag{1.16}$$

It is a *unital commutative* **C**-*algebra;* yet, one can consider, following standard reasoning, the **C**-*algebra sheaf* on X

$$\mathcal{A} \equiv \mathcal{C}^\infty_X, \tag{1.17}$$

viz. the *sheaf of germs of* **C**-*valued $\mathcal{C}^\infty$-functions on X.* (See also, for instance, Chapt. I; Subsection 4.(b)). It is proved that (1.17) *is* actually a *fine sheaf* on X (cf. the aforementioned work of M.E. Verona).

On the other hand, one further defines the *$\mathcal{A}$-module* on X of *"1-forms"*,

$$\Omega^1_X \equiv \Omega^1, \tag{1.18}$$

as follows; for every $x \in X$, one first considers the **C**-vector space of *point derivations at x of $\mathcal{C}^\infty(X)$* (cf. (1.16), for $U = X$). Then, the *dual of the* latter *space* is the **C**-*vector space of 1-forms of X at $x \in X$*, denoted by $\Omega^1_x(X)$. (It is further canonically associated with the *classical quotient* of the ideal I_x of $\mathcal{C}^\infty(X)$, consisting of those elements of $\mathcal{C}^\infty(X)$, which vanish at x; viz. one has $I_x/I_x^2 \cong \Omega_x^1(X)$, $x \in X$).

Now, for any open $U \subseteq X$, one defines the **C**-*vector space* (in fact, a $\mathcal{C}^\infty(U)$-*module,* cf. (1.16)) *of 1-forms on* U, as those maps

$$\omega : U \longrightarrow \bigcup_{x\in U} \Omega_x^1(X), \tag{1.19}$$

which are *locally* (with respect to U) *of the form;*

$$\omega = \sum_{i=1}^{n} g_i (df_i)_x |_{V \subseteq U, \text{ open}} \tag{1.20}$$

with $x \in V$, and $g_i, f_i \in \mathcal{C}^\infty(U)$, $1 \leqq i \leqq n$, such that $(df)_x := [f - f(x)] \in I_x/I_x^2 \equiv \Omega_x^1$, for any $f \in \mathcal{C}^\infty(U)$. (More precisely, one can actually define I_x, as those **C**-*valued* maps only *locally defined* and *smooth at* $x \in X$, see e.g. (1.16), *vanishing* also *at the same point;* thus, (1.20) acquires a more direct meaning). So we can further define Ω^1 (cf. (1.18)), as the *sheaf on* X, in effect, a $\mathcal{C}_X^\infty$-*module, generated by the 1-forms on the* various *open sets of* X, as before.

On the other hand, we still define (see also, for instance, Chapt. VIII; (8.3), (8.4))

$$\Omega^p := \wedge^p \Omega^1, \quad p \in \mathbf{N}, \tag{1.21}$$

while we also set (cf. (1.17));

$$\Omega^0 := \mathcal{A} \equiv \mathcal{C}_X^\infty. \tag{1.22}$$

Thus, one now concludes that

$$0 \longrightarrow \mathbf{C} \longrightarrow \Omega^0 \equiv \mathcal{A} \xrightarrow{d^0 \equiv d} \Omega^1 \xrightarrow{d^1} \Omega^2 \longrightarrow \cdots \tag{1.23}$$

is a *fine resolution of* **C**; that is, (1.23) is an *exact sequence of* **C**-*vector space sheaves,* such that *each one of the* $\mathcal{A}$-*modules* Ω^i, $i \in \mathbf{Z}_+$, *is a fine sheaf* on X (cf. (1.17), along with Chapt. III; (8.57)). Yet, the *"differential operators"* in (1.23), apart from $d^0 \equiv d$, already defined above (cf. (1.20)), can still be given on the basis of the same relation (1.20), following the classical pattern (see, for instance, Chapt. III; (8.33), (8.34)).

So, as a result of the preceding, in conjunction too with our discussion in the previous chapters, one concludes that;

(1.24) *the machinery developed so far, by the preceding exposition, is still available, within the particular case of the "generalized manifold" X, as in (1.12).*

But, there is, however, a *proviso*! Namely, the validity of the *exponential framework,* as the latter is given for instance, by Definition 6.1 of Chapter V. Indeed, as we shall see in the next Chapter XI, this is further assured by considering as our *structure sheaf* $\mathcal{A}$, an appropriate *topological algebra sheaf;* again, apart from the classical case (see (1.3)), this too specializes to certain particular examples (loc. cit.).

On the other hand, by considering a weakened version of a *"generalized manifold",* as above, to the extent, namely, that X *is not a locally compact* space, but, employing, however, *Banach manifolds* in place of the X_α's, as in (1.12), X can then be equipped with an appropriate *infinite dimensional manifold structure;* in particular, it can be modeled on a *Fréchet* (locally convex) *space.* Indeed, such a point of view is recently examined by the thesis of G. Galanis [1]. (In this context, still consult the recent work by M. Papatriantafillou [1]). Furthermore, concerning (1.12), see also, for instance, G.E. Bredon [1: p.69, §14].

Now, our *second* relevant *example,* where the framework of the preceding chapters can be realized, being still outside of the classical context, is supplied by the work of H.R. Fischer [1]: Thus, our $\mathcal{C}^\infty$*-manifold* X is here a given *locally convex space* E, on which one further considers $\mathcal{C}^\infty$*-functions* with values from another *complete locally convex space* F. (We thus have, so to speak, an *"affine" infinite dimensional* $\mathcal{C}^\infty$*-manifold;* the *"localization"* of this line of thoughts, according to the classical pattern, would be worthwhile. See, for instance, our next quotation right below). Thus, within the same context of ideas, one further defines a *de Rham complex*

$$\Omega^{\cdot}(E,F) \equiv (\Omega^p(E,F), d^p)_{p\in\mathbf{Z}_+} \tag{1.25}$$

of $\mathcal{C}^\infty$*-differentiable p-forms* ($p \in \mathbf{Z}_+$) *on* E, with values in F; in fact, *sheaves of germs* of such. Indeed, one has here again a *resolution of* $\mathbf{C}$, that is, an *exact sequence of* ($\mathbf{C}$*-vector space) sheaves*

$$0 \longrightarrow \mathbf{C} \longrightarrow \Omega^0 \longrightarrow \Omega^1 \longrightarrow \Omega^2 \longrightarrow \cdots . \tag{1.26}$$

Accordingly, one has here too a *conclusion similar to (1.24)* above.

Another *example* of the same nature, as before, is supplied within the context recently considered by J.-L. Brilynski [1]: Here by a suitably chosen E, as above, (in fact, an appropriate *projective limit of Hilbert spaces* is employed), one has that the corresponding *structure sheaf* $\mathcal{A}$ is *fine,* so that one finally obtains a *generalized de Rham theorem,* in the sense of Chapt. III; (3.24) (cf. also the above cited Ref.; p.35, Lemma 1.4.1, and p.39, Theorem 1.4.15. Thus, again

(1.27) a *conclusion analogous to that in (1.24) is here too in force.*

Yet, another example, within the same vein of ideas, as before, is the one appeared in the recent work of J. Bernstein – V. Lunts [1]; here one considers a *paracompact* (Hausdorff) *space*

$$X \cong \varinjlim X_n \tag{1.28}$$

homeomorphic with the *inductive limit space* of a suitable *increasing sequence of* standard $\mathcal{C}^\infty$-(paracompact) *manifolds.* Thus, by further taking the *de Rham complex* of each one of the individual manifolds X_n, $n \in \mathbf{N}$, as in (1.28), say $\Omega^{\cdot}_{X_n}$ (cf. Chapt. III; (8.44)), one defines the *de Rham complex of* X, by the relation,

$$\Omega^{\cdot}_X := \varprojlim \Omega^{\cdot}_{X_n}, \tag{1.29}$$

while the latter is proved to be a *(soft) resolution of the constant sheaf* $\mathbf{C}_X \equiv \mathbf{C}$ (ibid.; p.96, Proposition 12.2.3). Consequently, *a similar argument to (1.24) is still valid* in the case under consideration.

2. Holomorphy

We have already explained in the pertinent places, throughout our discussion in the preceding Chapters, the way the present point of view is related with the standard theory of *(finite-dimensional) complex manifolds;* thus, the whole treatment, concerning the *existence of $\mathcal{A}$-connections,* has been motivated, as already noted in the foregoing, by the relevant work of M.F. Atiyah

[1] in the case of *"holomorphic connections"*. (See, for example, Chapt. VI; §§10-13, where the so-called *Atiyah class* of a given *vector sheaf* plays a significant rôle; loc. cit. (10.28)). Precisely, we refer to Chapt. VI; Subsection 15.(b) for a concrete presentation of the relationship between our abstract framework, as exposed in the preceding, and that one of the classical theory. See thus, for instance, loc. cit. Definition 15.1, or even (15.36).

Furthermore, the connection of the above with the *exponential framework,* as the latter has been presented herein, was discussed in Chapt. V; Example 6.1, i) (cf., in particular, (6.4) therein). On the other hand, as already mentioned above, another realization of the same connection, as before, will be given in the next Chapter XI, through appropriate *topological algebra sheaves,* by still abstracting the corresponding *structure sheaf* $\mathcal{O}_X$ of a given *complex manifold* X. (See also, for example, Chapt. I; (4.40), Chapt. III; (8.65), (8.66) and Chapt. VI; (15.11)).

3. Generalized manifolds (contn'd)

We further consider in this Section certain particular instances of our axiomatic framework, as exposed in the preceding, which still correspond to suitable *generalizations of* the classical notion of *a (finite-dimensional) smooth* (viz. $\mathcal{C}^\infty$-)*manifold.* Indeed, as we shall see in the sequel, this extended point of view of the standard theory, which also fits well, within the preceding context, specializes to concrete cases and applications of the usual differential geometry of smooth manifolds (e.g. *Satake manifolds, orbifolds* and the like. See thus Subsection 3.(a) below).

The succeeding discussion corresponds to material based on earlier relevant work of J.W. Smith [1], as well as, on the more recent one by M.A. Mostow [1]; yet, the latter is also related with previous similar considerations in a series of papers by K.-T. Chen (see e.g. the said author [1]). So for details we also refer to the cited works; however, we deviate occasionally from the notation of the aforementioned authors, by employing instead a more contiguous notation to that one of our preceding exposition.

So we start with an arbitrary *topological space* X, along with a certain **C**-*algebra sheaf* $\mathcal{A}$ on X (cf. Chapt. II; Definition 1.4) such that one has

$$\mathcal{A} \subseteq \mathcal{C}_X \tag{3.1}$$

(see also Chapt. I; (4.10)); that is, $\mathcal{A}$ is a *subsheaf of the sheaf* of germs *of continuous* **C**-*valued maps on* X, which is further characterized by the following *"closure" condition:*

(3.2) *for any finite many global sections of* $\mathcal{A}$ *("smooth functions"),* $\alpha_1, \ldots, \alpha_n$ *and any* **C**-*valued* $\mathcal{C}^\infty$-*function* f *on* $\mathbf{R}^n$, *one has;*

$$f(\alpha_1, \ldots, \alpha_n) \in \mathcal{A}(X) \equiv \Gamma(X, \mathcal{A}). \tag{3.2.1}$$

On the other hand, one further defines the $\mathcal{A}$-*module*

$$\Omega^1_X \equiv \Omega^1 \tag{3.3}$$

on X, viz. the *sheaf of* (germs of) *"1-forms" on* X, as follows: First, one considers certain *local* X-*valued maps* on open subsets of some euclidean space, say $\mathbf{R}^n$ (*"plots"*), that are *"stable"*, with respect to *"$\mathcal{C}^\infty$-functions"* on their domain and range (cf. (3.2)); so these *plots* are then used as *"building blocks"* for the definition of the *"1-forms"* on X (or on any open subset of X. For details see M.A. Mostow [1: p.261], or even K.-T. Chen [2: p. 219]). Thus, the previous sheaf (3.3) is, by definition, that one, generated by the *presheaf of "1-forms"*, defined over the various open subsets of X. Furthermore, one can repeat here the classical argument, concerning the definition of *"p-forms"* on X, *for any* $p \in \mathbf{N}$. (In this regard, see also the relevant exposition by J.W. Smith [1: p. 119ff]).

On the other hand, one also defines an *exterior derivative operator*

$$d^p \equiv d : \Omega^p \longrightarrow \Omega^{p+1}, \quad p \in \mathbf{N}, \tag{3.4}$$

as well as, the corresponding one *for* $p = 0$, that is,

$$d^0 \equiv \partial : \Omega^0 \equiv \mathcal{A} \longrightarrow \Omega^1, \tag{3.5}$$

such that one finally gets a *de Rham complex* for X, yielding the *exact sequence;*

$$(3.6)\qquad \begin{aligned} 0 \longrightarrow \mathbf{C} \longrightarrow \mathcal{A} \equiv \Omega^0 \xrightarrow{\partial} \Omega^1 \xrightarrow{d} \Omega^2 \longrightarrow \cdots \\ \cdots \longrightarrow \Omega^p \xrightarrow{d} \Omega^{p+1} \longrightarrow \cdots . \end{aligned}$$

Precisely, one has (3.6) for a suitable *"differentiable space"*

$$(3.7)\qquad (X, \mathcal{A}),$$

as above; indeed, X is further supposed to be a (Hausdorff) *paracompact locally smoothly contractible* space and $\mathcal{A}$ a *fine sheaf* on X (see thus M.A. Mostow [1: p.268, Theorem 5.2]).

Now, within the same vein of ideas, on which our abstract context of the preceding Chapters is further specialized, one should also mention the earlier works by R. Sikorski [1], [2] and C.D. Marshall [1]; a comparison of these and other relevant studies (including the aforementioned ones by J.W. Smith and K.-T. Chen) has also been given by M.A. Mostow (loc. cit.). Finally, an early announcement by A. Verona [1] about *abstract prestratifications,* in conjunction with a *de Rham* type *theorem,* still belongs to the previous class of spaces. (An extension of the latter work was further given by M. Saralegi - G. Hector [1]; see also J.P. Brasselet - G. Hector - M. Saralegi [1], or even G. Hector - M. Saralegi [1]).

3.(a). Orbifolds (or Satake manifolds).– The term *V-manifold* is still applied for the type of (generalized) manifold we are dealt with below. This sort of manifolds is also contained in the class of spaces considered above (see, for example, J.W. Smith [1: p.134]), so that it further supplies another particular instance of the abstract (axiomatic) treatment, as this has been exhibited in the preceding.

Roughly speaking, a space of the above type is a *quotient* of a usual (finite dimensional $\mathcal{C}^\infty$-)*manifold* modulo a finite group of diffeomorphisms of X, that acts effectively on X.

Furthermore, *regular G-varieties,* in the sense of J.L. Koszul [1], fall also into the case of a generalized manifold, as above; again we have here the *quotient* space of a (finite dimensional) $\mathcal{C}^\infty$-*manifold* X, modulo a *compact Lie group* of transformations (viz. diffeomorphisms) of X. On the other hand, *orbit spaces* resulting from the action of *non-compact groups* still supply similar examples of generalized manifolds, as before. (In this context, see also J.W. Smith [1: p.131, comments following Theorem II]).

In this regard, it is further to be noticed that in all the preceding instances the base space X is *paracompact* (Hausdorff), while the *structure sheaf* $\mathcal{A}$ is *fine;* hence, all the respective $\mathcal{A}$-*modules* Ω^p, $p \in \mathbf{N}$, are also *fine* (cf. Chapt. III; (8.56)). As a result, one thus concludes that:

(3.8) *every vector sheaf* on such a space X *has always an* $\mathcal{A}$-*connection*. Hence, the same holds true, for every $\mathcal{C}^\infty$-*vector bundle* on X, while the corresponding theory of *Chern characteristic classes is* still *in force.*

See the preceding Chapter IX, as well as, for instance, T. Kawasaki [1: p.78].

4. Electromagnetism

It is a well known aspect that *quantization* arguments in elementary particle physics (e.g. electromagnetism) lead, via their mathematical formulation, to *cohomology* theory; it concerns here, of course, the so-called *geometric quantization* (see, for instance, J. Šniatycki [1], [2]).

So it is a standard procedure to describe *electromagnetism in terms of differential forms* on the *space-time manifold* (cf., for example, M. Göckeler - T. Schücker [1: p.43ff]). Indeed, a crucial item in the so-called *prequantization* of the *symplectic manifold,* representing the *phase space* of an *elementary (relativistic) particle,* for the case at issue, of an *electron,* is the classical *Weil's integrality theorem,* whose abstract formulation, within the present axiomatic framework, as exhibited in the preceding, has been given already by Chapt. VIII; Theorem 11.1 (see also ibid., Theorem 11.1′).

Thus, as we shall see right below (cf. the next Section 5), given that the *photon* (viz. the *"quantum"*, or even the *carrier* of the *electromagnetic field*) obeys *Bose-Einstein statistics,* equivalently, has *integral spin,* its *states* can be conceived as sections of a *line sheaf* (or in the standard terminology, equivalently, of a *line bundle*). Now, due to the *type of* X *and* $\mathcal{A}$, as above, the previous line sheaf always admits a *connection,* whose *curvature* supplies the required (for the prequantization) *2-dimensional integral cohomology class* (loc. cit.).

In this context, we further mention, in anticipation (see thus (5.13) below) that;

(4.1) *every* (free) *elementary particle can be conceived in a prequantized form.*

The preceding should also be compared with *Śniatycki's criterion* on the (pre)*quantization of (relativistic) charged* (elementary) *particles* (cf. e.g. J. Śniatycki [1]). In this respect, cf. also a recent account by M. Puta [1], or even the one by E. Prugovečki [2].

5. Elementary particles

We give in this Section a brief exposition of a potential application of our abstract framework to a description of the *elementary particles* and the accompanied *(pre)quantization,* in terms of *vector sheaves* and the associated *connections.* Indeed, the idea is based on an earlier study by S.A. Selesnick [3], to exhibit the so-called *second quantization* (alias *quantum field theory*), in terms of *projective finitely generated modules,* with respect to the algebra of *classical observables.* All that follows is more or less an adaptation, within the preceding abstract context, of the relevant account in the above cited work, along with the pertinent comments thereon.

Thus, one first considers *relativistic elementary particles,* for which the *space-time manifold* (or else *configuration space* of the physical system at issue) is a (Hausdorff) compact connected 4-dimensional *Lorentz ($\mathcal{C}^\infty$-)manifold* X, whose *lorentzian metric* represents the *gravitational field.* (We further assume that the various *differential geometric items,* which will be involved in the sequel, are *complexified;* cf., for instance, Chapt. VI; (4.16)). So the aforementioned *algebra of classical observables* of a physical system, as above, is defined to be

$$(5.1)\qquad \mathbf{A} := \mathcal{C}^\infty_{\mathbf{C}}(X) \equiv \mathcal{C}^\infty(X);$$

viz. the algebra (with point-wise defined operations) of $\mathbf{C}$-*valued* $\mathcal{C}^\infty$-*functions* on X. (As we shall see in the next Chapter XI, the previous algebra supplies an important example of a *topological algebra,* which is *not Banachable*!). Furthermore, the same algebra can be construed as the *algebra of global sections* of the *structure sheaf*

$$\mathcal{A} := {}^{\mathbf{C}}\mathcal{C}^{\infty}_{X} \tag{5.2}$$

(see also Chapt. I; (10.13), along with (11.11) and (11.12)). Yet, (5.2) provides an important instance of a *topological algebra sheaf,* in the sense of the next Chapter XI (cf. Theorem 1.1 therein).

Remark 5.1.– Indeed, what we actually consider is the so-called *phase space* (physics terminology) of the physical system, under discussion, viz. the *cotangent bundle* $\mathcal{T}^*(X)$ of the manifold X, as above. So it is, in effect, the algebra

$$\mathcal{C}^{\infty}(\mathcal{T}^*(X)), \tag{5.3}$$

which one considers as the *algebra of classical observables* of the system. (We can still assume *complexifications,* as before).

Now, according to an *axiomatic presentation of elementary particles,* due to S.A. Selesnick (loc. cit.), one concludes that:

(5.4) *bare* (viz. free) *particles have states that correspond to sections of suitable vector sheaves;* precisely, (bare) *bosons* are associated with *line sheaves,* while (bare) *fermions* with *vector sheaves,* viz. those of *rank greater that* 1.

As a matter of fact, by still following S.A. Selesnick [3: p.39], one realizes that:

(5.5) Ω^1 (viz. the sheaf of germs of sections of the complexified cotangent bundle of X, $\mathcal{T}^*(X)$) *represents* (by its *sections,* as before, *states* of) *fermions,* viz. *spin*-1/2 *particles.*

Now, by our hypothesis X is a *paracompact* (Hausdorff) space, while the *structure sheaf* $\mathcal{A}$ (see (5.2)) *is fine* (cf., for instance, (1.3)), so that *the same holds true for the* $\mathcal{A}$*-module* (in fact, *vector* sheaf) Ω^1 (cf. Chapt. III; (8.26) and Chapt. VI; (15.3), (15.4)).

Therefore, a *line sheaf* $\mathcal{L}$, as in (5.4), always *admits a* ($\mathcal{C}^{\infty} \equiv \mathcal{A}$-)*connection* D (cf. (1.2)); consequently, one infers that *the* (free) *boson* involved, as in (5.4), *may be construed, as yielding a pair*

(5.6) $$(\mathcal{L}, D)$$

on X, viz. a *Maxwell field.*

The above, also explains our previous terminology, concerning a pair, as in (5.6). On the other hand, as we shall presently see, the previous assertion also entails that;

(5.7) *the free* (elementary) *particle considered* (in our case, a *bare boson*) *has already, from the outset, a "prequantizing line bundle"* (equivalently, *line sheaf*), *associated with it.*

We have thus *circumvented,* through Selesnick's axiomatic (5.4), the need to resort to the standard *Hamiltonian formalism,* in order, namely, to find first the appropriate *quantum-mechanical* representative of the physical system at issue; that is, one thus avoids the *"correspondence principle"* (or even the *first quantization mechanism.* In this regard, see also e.g. D.J. Simms - N.M.J. Woodhouse [1]). Indeed, based on *Weil's integrality theorem* (see Chapt. VIII; Theorem 11.1′, or even Lemma 5.2 there), one concludes that *the curvature* of the $\mathcal{A}$-connection D, involved, $R(D) \equiv R$, *yields a* (2-dimensional) *integral cohomology class of* X, in the sense that one has (cf. Chapt. VIII; (5.56))

(5.8) $$\frac{1}{2\pi i}[R] \in H^2(X, \mathbf{Z}).$$

However, this is exactly the *prequantization condition* (or even *integrality condition*), viz. a criterion in order the *physical system,* under consideration, to be *susceptible of (pre)quantization.* (See the previously quoted authors, or even, for instance, N. Woodhouse [1: p.116ff]). In conclusion, we thus infer (see (4.1) in the preceding);

(5.9) *every* (free) *boson is prequantizable.*

Next, we are going to infer a *similar statement, for every (free) fermion,* as well: Thus, as already noted before (cf. (5.5)), such an elementary particle is represented (through its states) by (sections of) Ω^1. Hence, by further letting a boson act on Ω^1, one has a *new particle,* which *locally* is still Ω^1; indeed, one has (see Chapt. II; Lemma 5.1 and (5.15), along with (5.4) above)

$$\begin{aligned}\Omega^1(\mathcal{L})|_U \equiv \mathcal{L} \otimes_{\mathcal{A}} \Omega^1|_U = \mathcal{L}|_U \otimes_{\mathcal{A}|_U} \Omega^1|_U = \mathcal{A}|_U \otimes_{\mathcal{A}|_U} \Omega^1|_U \\ = (\mathcal{A} \otimes_{\mathcal{A}} \Omega^1)|_U = \Omega^1|_U.\end{aligned} \tag{5.10}$$

Here U stands, of course, for a *common local gauge* of Ω^1 and $\mathcal{L}$ (loc. cit.). In this context, the previous argument, which can also be construed as a kind of *"mediation of forces through the exchange of bosons"*, furnishes Ω^1 with the *prequantizable framework,* that we are looking for. Indeed, one has the following *commutative diagram;*

$$\begin{array}{ccc} \rho^*(\mathcal{L}) & \longrightarrow & \mathcal{L} \\ \big\downarrow & & \big\downarrow \\ \Omega^1 & \xrightarrow{\ \rho\ } & X\,, \end{array} \tag{5.11}$$

where ρ denotes the respective (canonical) *projection* (of the *sheaf* Ω^1 onto the *base space* X) and $\rho^*(\mathcal{L})$ the corresponding *pull-back of* $\mathcal{L}$, *via* ρ (cf. Chapt. I; (14.14)). Therefore, one obtains;

$$\frac{1}{2\pi i}[\rho^*(R)] = \rho^*\left(\frac{1}{2\pi i}[R]\right) \in H^2(X, \mathbf{Z}) \tag{5.12}$$

(see also Chapt. VIII; (9.25)). Accordingly, by the same token, as for (5.9), one arrives at the *analogous conclusion for spin*-1/2 *particles* too, so that, due to the standard *spin-classification* of elementary particles, one infers that;

(5.13) *every (free) elementary particle is prequantizable.* What amounts to the same thing, every such physical system is eo ipso (canonically) associated with a *prequantizing line sheaf.*

Indeed, the framework of the previous considerations is actually the classical differential geometry of $\mathcal{C}^\infty$-manifolds. Thus, their formulation here, in terms of the preceding terminology, is simply a further justification of our previous *sheaf-theoretic point of view* of treating the relevant notions. In this context, we also remark that the latter aspect seems to correspond to an *esoteric need* for more recent accounts *of quantum physics.* (See thus, for instance, Yu. Manin [1], or even S.A. Selesnick [5]). The classical *field of coefficients,* for example, that is, the complexes **C**, is now replaced by suitable **C**-*algebra sheaves,* such as, e.g. (5.2) above. Furthermore, as we shall see in the subsequent Chapter XI, the extra hypothesis of carrying the individual *section algebras* a *compatible topology* might be of interest, as well. On the other hand, this still makes explicit, for that matter, as the case may be, an analogous structure already existing in algebra sheaves of the form (5.2) (viz. *function algebra sheaves*).

Finally, within the same sheaf-theoretic context of a *differential triad,* seems also to fit well another application, that still refers to a mathematical formulation of *quantum field theory,* namely, that one derived from the so-called theory of *"supermanifolds".* We discuss it, in brief, right away by the next Section.

6. Supermanifolds

By the ensuing short discussion, we just want to point out that the framework of the theory, concerning the spaces in the heading is very akin to that one of an abstract *differential triad*

$$(\mathcal{A}, \partial, \Omega^1), \tag{6.1}$$

as this was the case in the preceding (see e.g. Chapt. VI; (1.4), (2.6) and (3.1)). Roughly speaking, every thing here *is* (appropriately) *graded* (in effect, $\mathbf{Z}_2$-*grading* is considered). In this cursory exposition we rather follow closely the recent account on the subject, as given by C. Bartocci, U. Bruzzo, D. Hernández-Ruipérez [1]:

Thus, the *base space* X is again a *topological space,* being in particular, a *graded space* over **R**. (However, one can always take *complexifications,* as we actually assume here; see also, for instance, loc. cit., p.164, Section 2). Precisely, the underlying space X, as above, is actually a *smooth* (viz. ($\mathcal{C}^\infty$-)*manifold,* in the usual sense, while (here comes the gradation), the structure sheaf $\mathcal{A}$ *is a sheaf of* (*graded,* as above) **C**-*algebras;* so, as this was the case hitherto in the preceding, one has at the basis what we have named in the foregoing (see, for instance, Chapt. II; (1.36)), a **C**-*algebraized space*

$$(X, \mathcal{A}). \tag{6.2}$$

However, as already mentioned, the deviation here from our previous context lies in that $\mathcal{A}$ is no more just *"commutative"*, so to speak, as was the case before, but now a *(unital) graded-commutative* **C**-*algebra sheaf* on X (in the sense, of course, that the corresponding local section algebras of $\mathcal{A}$ are of the said type). Specifically here, due to this hypothesis for $\mathcal{A}$, the above pair (6.2) is named a *graded space,* X being, otherwise, a smooth manifold, having thus too *its own structure sheaf* $\mathcal{C}^\infty_X \equiv \mathcal{S}$, which we shall also use right away. Indeed, one assumes (loc. cit., p.57) the following *short exact sequence of (**C**-algebra) sheaves;*

$$0 \longrightarrow \mathfrak{N} \longrightarrow \mathcal{A} \longrightarrow \mathcal{C}^\infty_X \longrightarrow 0, \tag{6.3}$$

where $\mathfrak{N}$ stands for the (2-sided) (graded) *ideal*) *sheaf of nilpotents of* $\mathcal{A}$. Moreover, we suppose that

$$\mathcal{E} \equiv \mathfrak{N}/\mathfrak{N}^2 \tag{6.4}$$

is a *vector sheaf,* with respect to $\mathcal{S}$ (viz. a *locally free* $\mathcal{S}$*-module of finite rank,* say n), in such a way that (loc. cit., p.58), for every point $x \in X$, there exists an open neighborhood U of x, such that

$$\mathcal{A}(U) = \Gamma(U, \wedge\mathcal{E}), \tag{6.5}$$

within an *isomorphism of graded-commutative* **C**-*algebras;* here $\wedge\mathcal{E}$ stands for the *exterior (**C**-)algebra sheaf of* $\mathcal{E}$, the previous vector sheaf, as in (6.4) (see also e.g. Chapt. IV; (7.9) in the preceding). The above constitute the issues in the definition of a *graded manifold* (loc. cit., p.57, Definition 1.2).

In this context, we still note, in passing, that one actually proves that (*Batchelor's theorem*)

$$\mathcal{A} = \wedge(\mathfrak{N}/\mathfrak{N}^2), \tag{6.6}$$

within an *isomorphism of graded-commutative* **C**-*algebra sheaves* (ibid., p.71, Corollary 1.9). The assertion is based on the fact that in the case of a *graded manifold* X, as above, *the structure sheaf* $\mathcal{A}$ *is fine* and hence acyclic (see e.g. Chapt. III; (8.26). Of course, we may still assume that the given smooth manifold X is, in particular, (Hausdorff) *paracompact*).

On the other hand, the situation changes entirely when one considers, more generally, *G-supermanifolds* (or even *graded-supermanifolds;* ibid., p.82ff). We shall not enter here into the technical details of the respective definition, our aim being, as already mentioned, just to point out the *relevance of the present discussion with our technique,* as exhibited in the preceding. Thus, roughly speaking, a crucial item here is in that the *constants* $\mathbf{C}$ (or $\mathbf{R}$) are now replaced by the *exterior* (alias *Grassmann*) $\mathbf{C}$-*algebra of* $\mathbf{R}^k$, $k \in \mathbf{N}$, *(complexified);*

$$C_k := (\wedge \mathbf{R}^k) \otimes_{\mathbf{R}} \mathbf{C}. \tag{6.7}$$

(As already said, we almost follow the notation and terminology of the previously quoted work). Thus, the corresponding pair $(X, \mathcal{A})$, as in (6.2), is now locally isomorphic to a pair based on C_k, having as structure sheaf an appropriate *graded-commutative* C_k*-algebra sheaf* on C_k (cf. ibid., p.85, Definition 4.1. See also, for instance, Chapt. II; (1.40) in the foregoing). Yet, we also assume that the base space X is a *(Hausdorff) paracompact* topological space. Moreover, $\mathcal{A}$ is suitably considered as a *subsheaf of the sheaf of* C_k*-valued continuous functions* on X. But, now $\mathcal{A}$ *is not, in general, fine* (cf. loc. cit., p.147, in particular, the comments at the end of p.148).

Accordingly, we do not have always what we have called in the preceding $\mathcal{A}$*-connections* (cf. Chapt. VI; Definition 3.1), a notion which has an analogous formulation for the case at hand (see the above cited work, p.160, Definition 1.1). Of course, one has first to define the $\mathcal{A}$*-module* Ω^1 of *(graded) differential 1-forms* on X. In fact, one sets

$$\Omega^1 := \mathcal{D}er^* \mathcal{A}, \tag{6.8}$$

viz. the *dual sheaf* of the sheaf of *(graded) derivations* of $\mathcal{A}$ (cf. also, for instance, Chapt. IV; (5.1), for the terminology applied). Yet, one considers the *differential operator*

$$d^0 \equiv \partial : \mathcal{A} \longrightarrow \Omega^1, \tag{6.9}$$

in effect, a *graded linear operator,* satisfying the corresponding *Leibniz condition.* Precisely, one sets;

$$\partial(s)(u) := (-1)^{|s| \cdot |u|} u(s), \tag{6.10}$$

for any *homogeneous (local) sections* $s \in \mathcal{A}(U)$ and $u \in \mathcal{D}er(\mathcal{A}(U))$, with U *open* in X. (Here $|\cdot|$ means the *degree* (of homogenuity) of the element (section) at issue, being, according to our assumption, always an *element of* $\mathbf{Z}_2 = \{0, 1\}$; see also loc. cit., p.127).

Furthermore, one defines the *differential operators*

$$d^n : \Omega^n \longrightarrow \Omega^{n+1}, \quad n \in \mathbf{N}, \tag{6.11}$$

by analogy with the classical case (cf., for instance, Chapt. III; (8.32), (8.33)). Of course, one always considers here the appropriate $\mathbf{Z}_2$*-grading,* as before (see the above cited book; p.115). Thus, one finally obtains a *(graded) de Rham complex,* which, in turn, yields a *resolution of the constant (*$\mathbf{C}$*-algebra) sheaf* C_k (cf. (6.7), along with Chapt. III; Definition 1.2); viz. one has the following *exact sequence of* C_k*-modules:*

$$0 \longrightarrow C_k \longrightarrow \mathcal{A} \xrightarrow{\partial} \Omega^1 \xrightarrow{d} \Omega^2 \longrightarrow \cdots \tag{6.12}$$

(cf. also the aforementioned work, p.145, Proposition 2.1).

On the other hand, we have already noted before that *the structure sheaf* $\mathcal{A}$, in the case *of a G-supermanifold* X, *is not, in general, acyclic.* Nevertheless, the situation can be remedied by taking the base space X to be a so-called *de Witt G-supemanifold* (loc. cit., p.122), a supermanifold very similar to a graded manifold, as considered at the beginning of this discussion. So in that case *the* corresponding *structure sheaf* $\mathcal{A}$ *is acyclic* (ibid., p.151, Proposition 3.2). In sum, one thus infers the following;

(6.13) the methodology of *the previous considerations,* in what concerns the *differential geometric point of view, are very akin to that one of the preceding* chapters, *modulo,* of course, the $\mathbf{Z}_2$-*grading* that is involved here. Indeed, the *graded-commutativity of the structure sheaf* $\mathcal{A}$, as above, must *replace* our assumption in the foregoing that the structure sheaf $\mathcal{A}$ of our algebraized space $(X, \mathcal{A})$, involved therein, had the corresponding (local) section algebras $\mathcal{A}(U)$, with U *open* in X, *commutative* (*"commutativity" of* the structure sheaf $\mathcal{A}$), in the usual sense, the same being, otherwise, unital $\mathbf{C}$-algebras, as happens here too. (A replacement, not particularly hard, for that matter; one can consider, for instance, in lieu of $\mathcal{A}$, a *non-necessarily commutative* $\mathcal{A}$).

Now, another feature, that is worth pointing out here, in connection with our considerations in the next Chapter XI, is that, for crucial reasons in the theory of supermanifolds (see, for example, loc. cit., as above, p.90, Proposition 4.5, together with p.93 and p.132, Axiom 6), one takes the structure sheaf $\mathcal{A}$, as before, to be, in effect, a *topological algebra sheaf,* in the sense of Chapter XI below (see Theorem 11.1 there); indeed, the individual (local) section algebras of $\mathcal{A}$ are, in fact *(Fréchet) locally m-convex (topological) algebras.* (See the above quoted book, p.62, Proposition 1.2, as well as, A. Mallios [4: p. 131; (4.17), along with the relevant comments therein]). So, as we shall see in the sequel, this, among other things, permits one to employ *the exponential function,* together with relevant desired properties of the latter: In particular, questions pertaining to *exponential sheaf sequences and applications* thereon, as was the case in the foregoing (see e.g. Chapters V, VII and VIII), can thus be dealt with in the framework of the previous discussion, as well. (See, for instance, the same citation, as above, on supermanifolds; p.165ff).

7. Some more applications

Further recent justifications of the point of view, that is advocated by the present treatise, come from the work of M. Heller and the rest researchers of the Polish group (see, for instance, M. Heller [1] – [5], or W. Sasin [1] – [3]), where there is applied an appropriate version of the notion of *"differential space"* à la Sikorski (see R. Sikorski [1], [2]). The above considerations fall also within the framework of our treatment. Yet, the same can be said about a recent study of S. Kobayashi [2]; in this context, see also our considerations in the next Chapter XI, where one finds, as already mentioned, a further application of the preceding in *topological algebra theory,* and vice-versa(!), as well as, the current work of M. Papatriantafillou [1], [2]. Potential applications of the present framework are also very likely to be found in the theory of PDOs, as this results, for instance, from the relevant work of S.I. Andersson [1] and A. Asada [1], [2]. Finally, *Lipschitz manifolds* and *"generalized functions"* (J.-F. Colombeau et al.) are still eventual areas of applications. (I am indebted to I. Colojoară and E.E. Rosinger, respectively, for drawing my attention, recently, to the latter subjects).

CHAPTER XI

Sheaves and presheaves with topological algebraic structures

"...[that] the utmost abstractions are the true weapons with which to control our thoughts of concrete fact."

A.N. Whitehead

"Reality is the motif."

W. Stevens in "Adagia" (included in "Opus Posthumous", ed. by Vintage Books, 1990). p.195

The purpose of the subsequent discussion is to show that, apart from the classical framework of differential ($\mathcal{C}^{\infty}$-)manifolds (generalized or not), as exposed in the preceding chapter, certain types of *topological algebras* too furnish further non-trivial examples, where one can apply the point of view of the present treatment, getting thus a considerable part of the standard machinery of the classical differential geometry. In this context, we still notice that, here too, the key-idea is to replace the standard sheaf of germs of ($\mathbf{C}$-valued) smooth functions, as appears in the classical theory (see e.g. Chapt. X; (1.3)), through the sheaf of germs of (continuous) sections of a suitable *topological algebra sheaf,* the aforementioned one of the classical theory constituting, in effect, simply a very special (important, nevertheless!) example of the latter type of sheaf; yet, one further seeks for a *de Rham complex* (generalized or not) analogous to the one already considered in the preceding (see, for instance, (1.8) or even (1.10) of the previous Chapter).

1. Preliminaries. Topological algebra sheaves

The notions of a *topological vector space presheaf,* respectively *sheaf,* are already standard, while that one of a *topological algebra presheaf,* or *sheaf,* is still standard, althouh less applied so far in the literature. Thus, we first recall the latter notions, as they will be exploited in the succeeding discussion. In this context, concerning the general theory of *topological algebras,* that will be used presently below, we shall steadily refer to A. Mallios [4], which, *for convenience, we quote henceforth* simply *by* [TA].

Now, as was the case hitherto, we always consider linear associative algebras over the complexes $\mathbf{C}$, while we briefly refer to such objects as $\mathbf{C}$-*algebras;* such an algebra will be further endowed with a *topology compatible with the* given *algebraic structure* in it. So one arrives at the notion of a *topological algebra.* (For the sake of generality, we also suppose that the *multiplication* in such an algebra is *separately continuous,* unless stated otherwise; ibid.). Thus, we start with the following basic notion.

1.(a). Topological algebra presheaves.– Now, following the categorical definition of a presheaf of sets (cf. Chapt. I; (5.10)), one has, by definition, that;

(1.1) by a *topological algebra presheaf* on a topological space X, one means a contravariant functor from the category of the open subsets of X to that one of topological algebras.

So in ordinary language (cf. Chapt. I; Definition 5.1) this means that;

a *topological algebra presheaf*

$$E \equiv \left(E(U), \rho_V^U\right)_{U \subseteq X,\,\text{open}} \tag{1.2.1}$$

on a topological space X, is a presheaf of sets on X, as above, such that;

(1.2) i) $E(U)$ is a *topological algebra,* for every open $U \subseteq X$, and
ii) each one of the (restriction) maps

(1.2.2) $$\rho_V^U : E(U) \longrightarrow E(V)$$

is a *continuous morphism of the topological algebras* concerned (U, V open sets in X, with $V \subseteq U$).

Examples of such mathematical objects will appear in abundance through the succeeding discussion. We just mention here that *all the concrete particular instances of (functional) presheaves of sets,* considered in the preceding, furnish such examples, when suitably topologizing the individual terms (algebras) of the presheaves involved (see, for instance, Chapt. I; (11.11) and Example 12.1).

1.(b). Topological algebra sheaves.– We come now to our main objective of this Section, namely, the sheaves in the heading of this Subsection. Thus, the following theorem supplies three equivalent definitions of the notion under consideration.

Theorem 1.1. *(Definition). Let $(\mathcal{E}, \pi, X)$ be a sheaf (of sets) over a topological space X. Then, the following assertions are equivalent:*

1) $\mathcal{E}$ is a *topological algebra sheaf;* viz. the set of local (continuous) sections of $\mathcal{E}$ over any *open* set $U \subseteq X$,

(1.3) $$\Gamma(U, \mathcal{E}) \equiv \mathcal{E}(U),$$

is a *topological algebra,* while the respective *restriction* (**C**-*algebra) morphisms* (cf. Chapt. I; (9.4), along with Chapt. II; (1.67) and (1.71)) are also *continuous.*

2) *$\mathcal{E}$ is (isomorphic to a sheaf) generated by a topological algebra presheaf.* (See (1.2) above, as well as, Chapt. I; (7.23)).

3) *Each local section set $\mathcal{E}(U)$ of $\mathcal{E}$,* associated with an *open* set $U \subseteq X$, is a **C**-*algebra* such that *the corresponding stalks $\mathcal{E}_x$, $x \in X$, of $\mathcal{E}$ are topological algebras;* in other words, $\mathcal{E}$ is a **C**-*algebra sheaf on X, whose stalks are topological algebras.*

Proof. By considering the (complete) *presheaf of* (continuous) *sections of* $\mathcal{E}$,

(1.4) $$\Gamma(\mathcal{E}) \equiv \left(\Gamma(U, \mathcal{E}), \sigma_V^U\right)$$

(see Chapt. I; (9.6), along with Lemma 11.1), one has the conditions of the previous definition (1.2) fulfilled, in the case we assume that 1) is valid. That is, (1.4) *is* then *a topological algebra presheaf,* while $\mathcal{E}$ is, of course, generated by it, in the sense that $\mathcal{E}$ *is* (isomorphic, as a sheaf, to that one generated by $\Gamma(\mathcal{E})$, or else to) *the sheafification of* $\Gamma(\mathcal{E})$. Namely, one has;

$$\mathcal{E} = \mathbf{S}(\Gamma(\mathcal{E})) \tag{1.5}$$

(up to a *sheaf isomorphism.* Cf. Chapt. I; Theorem 9.1). Thus, *1) implies 2).*

Now, suppose that 2) is valid; viz. we assume that *a* given *topological algebra presheaf,* as, for instance, (1.2.1) above, has *its sheafification,* say $\mathcal{E}'$, (sheaf) *isomorphic to* $\mathcal{E}$. Thus, consider next the sets;

$$\mathcal{E}'_x := \varinjlim_{U \in \mathcal{V}(x)} E(U), \quad x \in X, \tag{1.6}$$

stalks of the sheaf $\mathcal{E}'$, as before, where $\mathcal{V}(x)$ stands for an open basis of neighborhoods of $x \in X$ (cf. Chapt. I; (7.2), (7.6)). Therefore, $\mathcal{E}'_x$, $x \in X$, endowed with the corresponding *inductive limit* (vector space) *topology,* is made into a *topological algebra* (see [TA]; p.115, Lemma 2.2]; in this regard, we still remark that the type of the topology on $\mathcal{E}'_x$, as above, depends on that one of the particular (topological) algebras $E(U)$, $U \subseteq X$, open (loc. cit., p.115ff). Furthermore, we observe that $\mathcal{E}'_x$ *is a* $\mathbf{C}$*-algebra* (cf. Chapt. II; (1.55)), *so that,* by hypothesis (see Chapt. I; (1.21)), $\mathcal{E}_x$, *the stalk of the* given *sheaf* $\mathcal{E}$ *at* $x \in X$, *as well.* Thus, the same holds true, for each one of the local section sets of $\mathcal{E}$, $\Gamma(U, \mathcal{E})$, with U open in X (see also, for instance, Chapt. II; Remark 1.1). Accordingly, the given sheaf $\mathcal{E}$ itself *is a* $\mathbf{C}$*-algebra sheaf* on X (cf. also e.g. Chapt. II; (1.76), together with the comments following it). Thus, *2) implies 3),* as well.

Finally, we prove that *3) implies 1):* Indeed, if U is any *open set in* X, one obtains the following *bijection;*

$$\begin{aligned}\Gamma(U, \mathcal{E}) \cong \Big\{ (s_x)_{x \in U} \in \prod_{x \in U} \mathcal{E}_x : \forall\, x \in U \ \exists\, V \subseteq U, \text{ open, with} \\ x \in V, \text{ and } s \in \mathcal{E}(V), \text{ with } s(x) = s_x,\ x \in V \Big\} \subseteq \prod_{x \in U} \mathcal{E}_x.\end{aligned} \tag{1.7}$$

Now, the above is conceived on the basis of Chapt. I; Lemma 9.1, see, in particular (9.11) there. So the map

$$\tilde{s} \equiv (s_x)_{x \in U} : U \longrightarrow \mathcal{E} : x \longmapsto \tilde{s}(x) := s_x \in \mathcal{E}_x \subseteq \mathcal{E} \tag{1.8}$$

defines an element of $\mathcal{E}(U) \equiv \Gamma(U, \mathcal{E})$, and conversely (cf. also loc. cit.; (2.13)). Thus, the same rel. (1.7) proves now, in view of our hypothesis, that $\Gamma(U, \mathcal{E})$ *is a topological algebra* (see [TA: p.81; (1.6) and subsequent comments]). Moreover, for any *open sets* U, V in X, *with* $V \subseteq U$, *the* corresponding *algebra morphism*

$$\sigma_V^U : \mathcal{E}(U) \longrightarrow \mathcal{E}(V) \tag{1.9}$$

(cf. (1.4)) *is continuous,* as easily verified, by the same definition of the topology of $\mathcal{E}(U)$, as in (1.7). This proves already our assertion, and the proof of the theorem is complete. ∎

We specialize next to a particular topological algebra sheaf, that can be associated with any given topological algebra, being thus of a special importance for our subsequent discussion.

2. The Gel'fand sheaf of a topological algebra

Suppose that E is a given *topological algebra,* whose *spectrum* is $\mathfrak{M}(E)$; the latter is, by definition, the set of all *non-zero continuous morphisms of* E *into* $\mathbf{C}$, topologized by the *"weak* (or else *Gel'fand*) *topology",* defined on it by the following $\mathbf{C}$-valued maps;

$$\hat{x} : \mathfrak{M}(E) \longrightarrow \mathbf{C} : f \longmapsto \hat{x}(f) := f(x), \tag{2.1}$$

with $x \in E$ (cf. [TA: Chapt. V]). Thus, associated with E is the so-called *Gel'fand transform algebra of* E, viz. the algebra (of the *Gel'fand transforms of* the *elements of* E, as in (2.1))

$$E^{\wedge} \equiv \{\hat{x} : x \in E\} \subseteq \mathcal{C}_c(\mathfrak{M}(E)). \tag{2.2}$$

We topologize $E^{\wedge}$ with the relative topology from $\mathcal{C}_c(\mathfrak{M}(E))$; the latter denotes the *algebra of* $\mathbf{C}$*-valued continuous functions on* $\mathfrak{M}(E)$, endowed with the *compact-open topology* "c" in $\mathfrak{M}(E)$. Therefore, both sets appeared in (2.2) are (commutative) *locally m-convex* (topological) *algebras* (loc. cit.).

As already said, our purpose here is to associate with E a suitable $\mathbf{C}$-*algebra sheaf on* $\mathfrak{M}(E)$ (cf. Chapt. II; Definition 1.4), which at the end will appear to be, in effect, a *topological algebra sheaf* over $\mathfrak{M}(E)$, in the sense of the previous Section (cf. Theorem 1.1). Thus, we start with first defining a *(topological) algebra presheaf over* $\mathfrak{M}(E)$, as follows:

Given an *open* set $U \subseteq \mathfrak{M}(E)$, consider the following $\mathbf{C}$-*algebra;*

$$E^{\wedge}|_U := \{\hat{x}|_U : x \in E\}, \tag{2.3}$$

that is, the *restriction on* U *of the Gel'fand transform algebra of* E, as the latter is given by (2.1), (2.2). However, due to the definition of the topology of $\mathfrak{M}(E)$, each one of the maps $\hat{x}$, $x \in E$, is continuous (see e.g. (2.2)), so that the same holds true, of course, for their restrictions on the open set $U \subseteq \mathfrak{M}(E)$. Therefore, one actually gets;

$$E^{\wedge}|_U \subseteq \mathcal{C}(U), \tag{2.4}$$

for every open set $U \subseteq \mathfrak{M}(E)$, where $\mathcal{C}(U)$ stands, as usual hitherto, for the $\mathbf{C}$-*algebra of* $\mathbf{C}$-*valued continuous functions on* U. The latter algebra is endowed with the *compact-open topology in* U, we denote it by $\mathcal{C}_c(U)$, becoming thus a (commutative) *locally* m-*convex algebra;* therefore (see also [TA: Chapts. I, III]);

(2.5) *for every open* $U \subseteq \mathfrak{M}(E)$,

(2.5.1) $$E^{\wedge}|_U$$

(see (2.3)) *is a* (commutative) *locally* m-*convex algebra.*

Furthermore, *for any open sets* U, V *in* $\mathfrak{M}(E)$, *with* $V \subseteq U$, *the* corresponding *restriction morphism*

$$\rho_V^U : E^{\wedge}|_U \longrightarrow E^{\wedge}|_V \tag{2.6}$$

is continuous, as easily follows from (2.4). Consequently, based now on (1.2) one infers that;

(2.7) the family

(2.7.1) $$\mathbf{E} \equiv \left(E^{\wedge}|_U, \rho_V^U\right)$$

provides a *presheaf of topological algebras on* $\mathfrak{M}(E)$.

Indeed, what we actually need for the next notion is the fact that (2.7.1) yields just a presheaf of **C**-algebras on $\mathfrak{M}(E)$. Thus, we come now to the following.

Definition 2.1. Let E be a given topological algebra, whose spectrum is $\mathfrak{M}(E)$. Then, the *Gel'fand sheaf of* E is the (**C**-algebra) sheaf on $\mathfrak{M}(E)$, which is generated by the presheaf (2.7.1). We denote it by

$$(\mathcal{E}, \pi, \mathfrak{M}(E)), \tag{2.8}$$

or, for short, just by $\mathcal{E}$ (see Chapt. I; (7.7), as well as, (7.23)).

Now, by applying our previous terminology in Chapt. I; Section 3, we conclude that;

(2.9)

a *defining family of (continuous) sections for the Gel'fand sheaf* $\mathcal{E}$ *of* E, as above, is given by the maps

$$\tilde{x} : U \longrightarrow \mathcal{E}, \tag{2.9.1}$$

for any $x \in E$ *and open* $U \subseteq \mathfrak{M}(E)$, such that

$$\tilde{x}(f) := [\hat{x}]_f \in \mathcal{E}_f, \tag{2.9.2}$$

where we set

$$\mathcal{E}_f := \varinjlim_{U \in \mathcal{V}(f)} (E^\wedge|_U), \tag{2.9.3}$$

stalk of $\mathcal{E}$ *at* $f \in \mathfrak{M}(E)$, with $\mathcal{V}(f)$ a basis of open neighborhoods of f in $\mathfrak{M}(E)$. (See (2.7), together with Definition 2.1 above and Chapt. I; (7.2) and (9.16)).

Note.– Since $\hat{x}|_{\mathfrak{M}(E)} = \hat{x}$, for every $x \in E$ (see (2.1)), the *germ of* $\hat{x}|_U \in E^\wedge|_U$ *in* $\mathcal{E}_f$, with U open in $\mathfrak{M}(E)$, is given by

$$[\hat{x}|_U]_f = [\hat{x}]_f, \tag{2.10}$$

which thus justifies the previous notation in (2.9.2).

In other words, our claim in (2.9) means that *the family of the (local) maps*

$$\{\tilde{x}\}_{x \in E}, \tag{2.11}$$

as given by (2.9.1) and (2.9.2), *satisfies the* respective *conditions of the analogous family in* Chapt. I; (3.2), concerning the set (sheaf) here $\mathcal{E}$ and its fibers $\mathcal{E}_f$, $f \in \mathfrak{M}(E)$, as above, so that one has;

$$\mathcal{E} = \sum_{f \in \mathfrak{M}(E)} \mathcal{E}_f. \tag{2.12}$$

The assertion follows straightforwardly from the very definitions (see e.g. (2.9.3) above and loc. cit. (3.1), (3.4)). Thus, our claim about the *continuity of the* previous *maps* (2.9.1) follows now from Chapt. I; Theorem 3.1 (cf. also Lemma 3.1 therein).

In this context, one further infers that;

$$E\hat{}|_U \underset{\rho_U}{\hookrightarrow} \Gamma(U, \mathcal{E}) \equiv \mathcal{E}(U), \tag{2.13}$$

for any *open* $U \subseteq \mathfrak{M}(E)$, where *one sets* (cf. (2.9.1), (2.9.2))

$$\rho_U(\hat{x}|_U) := \tilde{x}, \tag{2.14}$$

for every $x \in E$. Precisely, one has that

(2.15) *the map* ρ_U, as given by (2.14), being also, in fact, a $\mathbf{C}$-*algebra morphism, is one-to-one.*

Indeed, suppose that

$$\rho_U(\hat{x}|_U) \equiv \tilde{x} = 0, \tag{2.16}$$

so that, by virtue of (2.9.2) and (2.10), one obtains

$$\tilde{x}(f) = [\hat{x}]_f = [\hat{x}|_U]_f = 0, \tag{2.17}$$

for every $f \in U \subseteq \mathfrak{M}(E)$. Therefore, *one has* (cf. (2.9.3))

$$\hat{x}|_V = 0, \tag{2.18}$$

for some open $V \subseteq U$, *with* $f \in V$; that is, one has $(\hat{x}|_V)(f) = \hat{x}(f) = 0$, concerning the particular, otherwise arbitrary (cf. (2.17)), element $f \in U$. Hence, one obtains $\hat{x}|_U = 0$, which thus proves our assertion in (2.15). ∎

Thus, as a consequence of (2.15) (in fact, is an equivalent assertion), one concludes that (see also Chapt. I; (11.14), p. 50);

(2.19) (2.7.1), the *Gel'fand presheaf* of E, *is a monopresheaf.*

(See also Chapt. I; Remark 11.1, i), together with (11.36′)). In particular, setting in (2.13) $U = \mathfrak{M}(E)$, *one obtains;*

$$E^{\wedge} \underset{\rho}{\hookrightarrow} \Gamma(\mathfrak{M}(E), \mathcal{E}) \equiv \mathcal{E}(\mathfrak{M}(E)). \tag{2.19}$$

In other words, one thus concludes hitherto that;

(2.20) *the Gel'fand transform algebra of any given topological algebra is represented* (within an *isomorphism of* **C**-*algebras*) *as a subalgebra of the global section algebra* (viz. of the **C**-algebra of the global (continuous) sections) *of the respective Gel'fand sheaf of the given algebra.*

Specifically, suppose that the corresponding *Gel'fand map* of the given topological algebra E,

$$\mathcal{G} : E \longrightarrow E^{\wedge} : x \longmapsto \mathcal{G}(x) := \hat{x} \tag{2.21}$$

(cf. (2.1)) is *one-to-one;* viz. we assume that we are given a so-called *semisimple* (topological) algebra (see [TA: p.266, Definition 3.2]). Thus, one has, by definition,

$$E \underset{\mathcal{G}}{=} E^{\wedge}, \tag{2.22}$$

within an *isomorphism of* **C**-*algebras,* established by $\mathcal{G}$. So (2.19) specializes now to the following relation;

$$E \underset{\mathcal{G}}{=} E^{\wedge} \underset{\rho}{\hookrightarrow} \Gamma(\mathfrak{M}(E), \mathcal{E}) \equiv \mathcal{E}(\mathfrak{M}(E)). \tag{2.23}$$

In other words, one thus infers by the last relation that;

(2.24) *any given semisimple topological algebra admits,* via its Gel'fand sheaf, *a faithful sectional representation.*

Of course, our final objective here is to find, as already said, suitable conditions for the topological algebras involved, so that the above *faithful representation,* as in (2.23), be also *full;* thus, $E^{\wedge}$ will then be identified with the global section algebra of $\mathcal{E}$, $\mathcal{E}(\mathfrak{M}(E))$. However, to this end, we first look at *another* useful *version of* $\mathcal{E}$, viz. of the Gel'fand sheaf of the given topological algebra, that we presently explain. So one considers

(2.25) *the Gel'fand sheaf* $\mathcal{E}$ of a given topological algebra E *as the sheaf of germs of* **C***-valued maps, locally belonging to* $E^{\wedge}$ (viz. to the *Gel'fand transform algebra* of E, see (2.2)).

Indeed, as we shall see right away (cf. (2.44) below), the above asserted identification is accomplished *modulo a sheaf isomorphism* between $\mathcal{E}$ and the sheaf descibed in (2.25); precisely, one means by the latter terminology the following: Take an *open* set $U \subseteq \mathfrak{M}(E)$, along with the set;

$$\tilde{E}_U = \{\alpha : U \longrightarrow \mathbf{C} : \forall f \in U \ \exists \text{ open } V \subseteq \mathfrak{M}(E), \text{ with } f \in V \subseteq U, \text{ together with } x \in E, \text{ such that } \alpha = \hat{x}|_V\}. \tag{2.26}$$

So $\tilde{E}_U$ is, by definition, the *set of* **C***-valued maps on* U*, which locally belong to* $E^{\wedge}$. It is easy to see that the latter set is, in effect, *a* **C***-algebra, consisting of (***C***-valued) continuous functions on* U. That is, one has;

$$\tilde{E}_U \subseteq \mathcal{C}(U) \tag{2.27}$$

(see also (2.4)). Thus, one actually obtains a *presheaf of* **C***-algebras on* $\mathfrak{M}(E)$,

$$\left(\tilde{E}_U, \tau_V^U\right), \tag{2.28}$$

by letting U varying over the open subsets of $\mathfrak{M}(E)$ and taking the corresponding restriction (**C**-algebra) morphisms. So we denote by

$$(\tilde{\mathcal{E}}, \pi', \mathfrak{M}(E)) \tag{2.29}$$

the **C***-algebra sheaf* on $\mathfrak{M}(E)$, generated by the (**C**-algebra) presheaf (2.28) (see also Chapt. II; (1.54)). Now, our aim is to show that one has;

$$\mathcal{E} = \tilde{\mathcal{E}}, \tag{2.30}$$

within a *sheaf isomorphism.*

Indeed, what we actually prove is that *the* above *presheaf (2.28) is isomorphic to* $\Gamma(\mathcal{E})$, the (complete) presheaf of (continuous) sections of $\mathcal{E}$ (see (1.4)). So one has the following. (But see also (4.4) in the sequel).

Lemma 2.1. *Let* E *be a given topological algebra with spectrum* $\mathfrak{M}(E)$ *and* $\mathcal{E}$ *the corresponding Gel'fand sheaf of* E *over* $\mathfrak{M}(E)$*. Then, for every open set* $U \subseteq \mathfrak{M}(E)$*, one obtains;*

$$\widetilde{E}_U = \Gamma(U, \mathcal{E}) \equiv \mathcal{E}(U), \tag{2.31}$$

within a **C**-*algebra isomorphism.*

Proof. For every open $U \subseteq \mathfrak{M}(E)$ and *for every map* $\alpha : U \longrightarrow \mathbf{C}$, *locally in* $E^{\wedge}$, so that (cf. (2.26)) $\alpha \in \widetilde{E}_U$, one obtains a *continuous section of* $\mathcal{E}$ *over* U, that is, *an element* $\tilde{\alpha} \in \mathcal{E}(U)$, as follows;

$$\tilde{\alpha} : U \longrightarrow \mathcal{E} : f \longmapsto \tilde{\alpha}(f) := [\hat{x}]_f \in \mathcal{E}_f \subseteq \mathcal{E} \tag{2.32}$$

(see also e.g. (2.12)), where the element $x \in E$ is determined, according to our hypothesis for α and the same condition (2.26). Moreover, the previous section $\tilde{\alpha}$ *is,* as asserted, *continuous,* since, due again to (2.26), one has (see also (2.9.2)),

$$\tilde{\alpha} = \tilde{x}|_V, \tag{2.33}$$

for some open $V \subseteq \mathfrak{M}(E)$, with $f \in V \subseteq U$; but, we have seen already that $\tilde{x}|_V \in \mathcal{E}(V)$ (see (2.9)). (At this point, we should also remark that the map $\tilde{\alpha}$ in (2.32) is well-defined, concerning the element $x \in E$, due to the same definition of $x \in E$, as in (2.26), together with (2.9.2) and (2.9.3). An analogous argument holds certainly true for the last relation (2.33)).

Accordingly, *for every open* $U \subseteq \mathfrak{M}(E)$, one obtains a map;

$$h_U : \widetilde{E}_U \longrightarrow \Gamma(U, \mathcal{E}) \equiv \mathcal{E}(U) : \alpha \longmapsto h_U(\alpha) := \tilde{\alpha} \tag{2.34}$$

(see (2.32)). Now, it is easy to see, according to the very definitions, that the previous map (2.34) *yields,* in fact, *a* **C**-*algebra morphism,* which also is *one-to-one:* Indeed, if $\tilde{\alpha} = 0$, for some $\alpha \in \widetilde{E}_U$, and $f \in U$, then (see (2.32)), $\tilde{\alpha}(f) = [\hat{x}]_f = 0$; therefore (cf. (2.33), along with (2.9.1), (2.9.2) and (2.26)), one concludes that $\alpha(f) = 0$, which thus proves our claim. Finally, we prove that *(2.34) is an onto map;* thus, if $s \in \mathcal{E}(U)$, consider the map

$$\overline{s} : U \longrightarrow \mathbf{C} : f \longmapsto \overline{s}(f) := \hat{x}(f) \equiv [\hat{x}]_f(f), \tag{2.35}$$

such that

$$s(f) = [\hat{x}]_f = [\hat{x}|_V]_f = \rho_V^f(\hat{x}|_V) = \tilde{x}(f), \tag{2.36}$$

(see (2.9.2), (2.9.3), (2.10) and (2.14), as well as, Chapt. I; (7.9), (7.10)). So one obtains (see also (2.14) and Chapt. I; (2.13))

$$s = \tilde{x}|_W, \tag{2.37}$$

with W an *open neighborhood of* f in V, hence, *in* $\mathfrak{M}(E)$ too. Therefore, in view of (2.35), (2.36) and (2.37), one obtains;

$$\overline{s} = \hat{x}|_W, \tag{2.38}$$

that is (cf. (2.26)), one finally has that

$$\overline{s} \in \widetilde{E}_U. \tag{2.39}$$

Thus, based further on (2.32), (2.33) and (2.37), one obtains (see also (2.34));

$$h_U(\overline{s}) = \widetilde{\overline{s}} = s, \tag{2.40}$$

which thus proves our assertion, and this completes the proof of the lemma. ∎

Now, as a result of (2.31) (see also Chapt. I; (6.10), Lemma 11.1 and (11.44)), one further obtains that:

(2.41) (2.28) yields a *complete presheaf of* **C**-*algebras* on $\mathfrak{M}(E)$, such that one actually has (loc. cit., Proposition 11.1)

$$\widetilde{E}_U = \Gamma(U, \widetilde{\mathcal{E}}) = \Gamma(U, \mathcal{E}) \equiv \mathcal{E}(U), \tag{2.41.1}$$

within an *isomorphism of* **C**-*algebras, for every open* set $U \subseteq \mathfrak{M}(E)$ (cf. also (2.29)).

In particular, *for* $U = \mathfrak{M}(E)$, one gets;

$$\widetilde{E}_{\mathfrak{M}(E)} = \Gamma(\mathfrak{M}(E), \widetilde{\mathcal{E}}) = \Gamma(\mathfrak{M}(E), \mathcal{E}) \equiv \mathcal{E}(\mathfrak{M}(E)), \tag{2.42}$$

up to an *isomorphism of* the **C**-*algebras* concerned. The first term in (2.42) is, by definition (cf. (2.26)), the *algebra of* (*continuous,* cf. (2.27)) **C**-*valued maps on* $\mathfrak{M}(E)$, *which locally belong to* $E^{\wedge}$.

On the other hand, the same relation (2.41.1) yields now, as already promised (cf. (2.30)), that;

$$\mathcal{E} = \widetilde{\mathcal{E}}, \tag{2.43}$$

within a *sheaf isomorphism* (see also Chapt. I; (13.7)). Furthermore, due to the fact that (2.7.1) is a presheaf of topological algebras, while the same holds true for (2.28) (see, for instance, (2.27)), one concludes, in view of Theorem 1.1, that:

(2.44) *the Gel'fand sheaf* $\mathcal{E}$ of a given topological algebra E, *as well as, the sheaf* $\widetilde{\mathcal{E}}$, *are,* in effect, *topological algebra sheaves on* $\mathfrak{M}(E)$, the spectrum of the initial algebra E.

For the moment, we shall not use the last conclusion, until Section 4 below. Instead, as already said, our purpose now is to provide suitable conditions for *the topological algebra E involved,* as above, so that the latter can be expressed *as the algebra of global sections of* $\mathcal{E}$ (viz. of the *Gel'fand sheaf* of E), hence, of $\widetilde{\mathcal{E}}$ too (cf. (2.43)), a fact, that will be of use presently below. So we undertake this task in the next Section.

3. Geometric topological algebras

The algebras in the title of this Section are those which will mostly be of concern to us throughout the sequel, while they also comprise the standard ones of the classical theory (viz. smooth and holomorphic functions). Moreover, they provide the type of topological algebras, alluded to at the end of the previous Section; it is, in effect, this latter property, which motivated their name, according to the classical terminology (see e.g. P. de Bartolomeis [1]). So we start with the following basic.

Definition 3.1. By a *geometric topological algebra,* one means a topological algebra E, which is ($\mathbf{C}$-algebra isomorphic with) the algebra of global continuous sections of a $\mathbf{C}$-algebra sheaf over (a topological space X, homeomorphic with) the spectrum $\mathfrak{M}(E)$ of the initial algebra.

Thus, our purpose by the ensuing discussion is to exhibit *sufficient conditions,* which enable a given topological algebra to be a geometric one, in the previous sense. In this context, the key-result is the so-called *Local Theorem,* a standard conclusion already from the classical Banach algebra theory (see, for instance, I.M. Gelfand - D. Raikov - G. Shilov [1: p.201, Theorem 1]). Indeed, one has the following basic result.

Lemma 3.1. *(Local Theorem). Let E be a strongly hereditarily complete regular topological* $\mathbf{C}$*-algebra, having an identity element and an equiconti-*

nuous spectrum $\mathfrak{M}(E)$. *Then, every* **C**-*valued map on* $\mathfrak{M}(E)$, *which locally belongs to* $E^{\wedge}$ (viz. to the Gel'fand transform algebra of E), *belongs globally to it, as well.*

For convenience, we first have to comment a bit on the terminology applied in the statement of the previous lemma:

Thus, a given topological algebra E is said to be *strongly hereditarily complete,* whenever, for every closed (2-sided) ideal I of E, the respective quotient algebra

$$E/I, \tag{3.1}$$

endowed with the quotient topology, is a *complete* topological algebra, while the same holds true, for every such quotient topological algebra, as in (3.1).

It is clear that every *Banach algebra* and, more generally, every *Fréchet,* or even, every *Pták (topological) algebra* is an algebra of the previous type. (See [TA] for the terminology applied herewith). Furthermore, it is an immediate consequence of the above definition, that *a topological algebra of the above type is itself complete.*

Now, the *proof of* the above *Lemma 3.1* can actually be patterned after that one of its classical (Banach algebra) analogue (loc. cit.). Indeed, a version of it for (Pták) *locally m-convex* (topological) *algebras* has been given already in [TA: p. 348, Theorem 5.1], while its present form appears in A. Mallios [7: p. 307, Lemma 2.1]. In this context, a crucial point is the fact that;

(3.2) *a topological algebra E* having the properties of that one, *as in the statement* of Lemma 3.1 *is,* in fact, *normal.* (Thus, by definition, $E^{\wedge}$, the Gel'fand transform algebra of E (cf. (2.2)), *"separates closed sets" in* $\mathfrak{M}(E)$, the spectrum of E; so, not only "closed sets and points" (*regular*). See also [TA] for the relevant terminology).

As already mentioned, the above property that acquires E is a decisive ingredient of the proof in the classical case (loc. cit., p.200, Corollary, and p.201, proof of the Lemma). For an idea of the proof of (3.2), cf. also [TA: p.345, Corollary 4.5], given there still for *Pták* (locally m-convex) *algebras,* while the same result is an extension, within the present context (viz. *no convexity* assumptions for the topological algebras concerned), of an earlier conclusion of M. Rosenfeld [1] for *Fréchet-Šilov* (locally m-convex) *algebras.* So, what one actually needs here, is the *type of completeness of E, as above* (cf. (3.1) and subsequent comments there). ∎

Therefore, as a consequence of the preceding, one now infers that:

(3.3) given a *topological algebra E, satisfying the conditions of Lemma 3.1,* one then obtains (see also (2.42));

(3.3.1) $$E^{\wedge} = \widetilde{E}_{\mathfrak{M}(E)},$$

where $\mathfrak{M}(E)$ is the *spectrum of E.*

Thus, by further specializing to *semisimple* (topological) *algebras* (see (2.22)), one gets the awaited so far result, according to the following.

Theorem 3.1. *Let E be a unital topological* **C***-algebra, which is strongly hereditarily complete regular and semisimple, having also an equicontinuous spectrum* $\mathfrak{M}(E)$. *Moreover, let* $\mathcal{E}$ *be the Gel'fand sheaf of E over* $\mathfrak{M}(E)$ (cf. (2.8)). *Then, one has*

(3.4) $$E = \Gamma(\mathfrak{M}(E), \mathcal{E}) \equiv \mathcal{E}(\mathfrak{M}(E)),$$

within a **C***-algebra isomorphism.*

Proof. Obvious, according to our hypothesis for E, in conjunction with (2.22), (3.3.1) and (2.42). ∎

To fix the terminology employed so far, we further set the following definition. (The relevant motivation will appear later on; cf. thus Theorem 4.1, or even (4.18) below). So we have.

Definition 3.2. By a *soft algebra,* we mean a unital topological **C**-algebra E with an equicontinuous spectrum $\mathfrak{M}(E)$, such that E is also strongly hereditarily complete regular and semisimple. (In other words, any algebra satisfying the conditions of Theorem 3.1).

Thus, in view of the previous Definitions 3.1, 3.2 and of Thorem 3.1, one concludes that;

(3.5) *every soft algebra is,* relative to its Gel'fand sheaf, *a geometric topological algebra.*

Note 3.1.– If E is a *soft* (topological) *algebra*, then (Definition 3.2), its *spectrum* $\mathfrak{M}(E)$ is *equicontinuous*, so that *the* corresponding *Gel'fand map of* E (see (2.21)) *is continuous* (cf. [TA: p. 184, Corollary 1.4]). Thus, one further obtains

$$\mathfrak{M}(E) = \mathfrak{M}(E^{\hat{}}), \tag{3.6}$$

within a *homeomorphism* of the topological spaces concerned (see R.I. Hadjigeorgiou [1: Chapt. II; Theorem 2.1], or even [2: Theorem 3.1]). Of course, here $E^{\hat{}}$ is a *normed algebra*, when endowed with the relative topology from $\mathcal{C}_u(\mathfrak{M}(E))$ ($\mathfrak{M}(E)$ being also *compact* (Hausdorff)); yet, $\mathfrak{M}(E^{\hat{}})$ *is equicontinuous*, $E^{\hat{}}$ being unital too, by hypothesis for E (see e.g. [TA: p. 146, Lemma 2.2]). Finally, $E^{\hat{}}$ *is* (always) *semisimple*, as well as, *regular*, according to the hypothesis for E and to (3.6). Nevertheless, *no information* is available, in general, concerning the *completeness* of $E^{\hat{}}$. However, in view of (3.6), (3.3.1), (2.42) and (2.43), *one has;*

$$E^{\hat{}} = \widetilde{E}_{\mathfrak{M}(E)} = \Gamma(\mathfrak{M}(E^{\hat{}}), \mathcal{E}). \tag{3.7}$$

Therefore,

(3.8) *for any given soft algebra* E, hence (cf. (3.5)), a geometric one, its Gel'fand transform algebra $E^{\hat{}}$ *is* also *a geometric topological* (in fact, *normed*) *algebra.*

In this context, we still remark that, due to the semisimplicity of $E^{\hat{}}$, $\mathcal{E}$ (viz. the Gel'fand sheaf of E) may also be viewed as the Gel'fand sheaf of $E^{\hat{}}$, by an obvious adaptation to the present case of the respective terminology for the presheaf (2.7.1) above. Anyway, our claim in (3.8) is justified, according to Definition 3.1 and to the **C**-*algebra isomorphism*

$$E^{\hat{}} = \Gamma(\mathfrak{M}(E), \mathcal{E}), \tag{3.9}$$

in view of (3.6).

On the other hand, by specializing to *locally* m-*convex algebras*, one proves that:

(3.10) every *unital Fréchet regular and semisimple* (for short, *Fréchet-Šilov*) *locally* m-*convex algebra is a geometric topological algebra,* relative to its Gel'fand sheaf.

The preceding is based on an earlier result of R.M. Brooks [1], pertaining to a corresponding version of *the Local Theorem* (see Lemma 3.1 above), *in terms of the topological algebras, as in (3.10)* (see loc. cit.; p.271, Theorem 2.6). In this context, we also remark that the *spectrum* $\mathfrak{M}(E)$ of the topological algebra involved here *is no longer assumed* to be *equicontinuous, hence* (Alaoglou - Bourbaki), *compact.* However, *it is* actually a *paracompact* (Hausdorff) space (see the above quoted work, p.266), a fact that will be of use in the sequel. ∎

Notice that some of the *important examples of geometric topological algebras* in the classical theory *are of the last type, as* namely *in (3.10);* for instance, the (topological) algebras $\mathcal{C}^\infty(X)$ (with X *not compact*) and $\mathcal{O}(X)$ (cf. e.g. Chapt. II; (10.13) and (10.14)). Concerning this kind of examples, we refer to A. Mallios [7: p. 315, Section 5]; we can find there too another example of a *Pták-Šilov locally m-convex Q-algebra,* hence, of a *geometric* one (cf. (3.5), along with [TA: p. 75, Proposition 7.1]), through the holomorphic functions on a *compact Stein set* of a *Stein manifold* X. (See thus A. Mallios [7: p. 318, Theorem 5.1], as well as, (11.40) in the sequel).

Now, looking at the above relation (3.4), as the defining condition that E be a *geometric* topological *algebra,* we can also express, roughly speaking, our previous information in (3.5) or (3.10), by saying that;

(3.11) *the Gel'fand sheaf of a geometric topological algebra* (of either form (3.5) or (3.10)) *can be reconstructed* (equivalently, *be determined*) *through the* **C**-*algebra of its global (continuous) sections* (the latter algebra being actually isomorphic to the given geometric topological algebra).

Yet, for convenience of reference, we also gather here our previous information (see comments following (3.10)) about the *spectra of geometric* topological *algebras,* as above. So, as already said, we have that:

(3.12) *the spectrum of every soft algebra* (cf. Definition 3.2) *is* a *compact* (Hausdorff) space, while *that one of a Fréchet-Šilov locally m-convex algebra* (cf. (3.10)) a *paracompact* (Hausdorff) space.

On the other hand, before we end this discussion, we still point out a further property of soft algebras, or that ones in (3.10), in connection with being the algebras in question geometric, in the sense of (3.4). Yet, the terminology involved will be of use right in the next Section 4. First, for convenience, we recall a relevant notion, which we shall need presently below.

Thus, suppose we are given a *topological algebra* E, whose *spectrum* is $\mathfrak{M}(E)$. Then, we say that an *ideal* I in E is a *kernel* of E, in the case one has;

$$I = k(h(I)) \equiv \bigcap_{f \in \mathfrak{M}(E),\, \ker f \supseteq I} \ker f. \tag{3.13}$$

It is clear that I is then a *closed 2-sided ideal of* E (being thus, by definition the *kernel of its hull;* see [TA: Chapter IX]). So we set now the following.

Definition 3.3. A given topological algebra E is called a *hereditarily geometric algebra,* whenever the given algebra E itself, as well as, every quotient algebra of it, modulo a kernel ideal of E is (in the quotient topology) a geometric topological algebra, with respect to the corresponding Gel'fand sheaf of the topological algebra, under consideration.

On the other hand, the significance of the preceding notion, as formulated by the previous definition, will become apparent after a while through the next Section 4; however, in the mean-time we do remark the following, that will also be of use below.

(3.14) *every soft algebra* (cf. Definition 3.2), as well as, *every Fréchet-Šilov locally m-convex algebra* (see (3.10)) *is a hereditarily geometric algebra.*

What one actually proves here is the fact that;

(3.15) *both types of topological algebras in (3.14) yield, modulo a kernel ideal, quotient algebras of the same type, as the initial ones; viz. the said algebras are "hereditarily" such.*

But, the latter are geometric algebras, according to (3.5) and (3.10). On the other hand, our claim in (3.15) is provided by standard results of the general topological algebra theory. (See [TA: p.339, Theorem 4.1, Corollary 4.1 and p.340, Corollaries 4.2 and 4.3]). ∎

We prove now, right in the next Section, the important property that the above algebras share, in having the Gel'fand sheaves *soft* and, in effect, *fine.*

4. Softness of the Gel'fand sheaf

Our objective by the following discussion has already been mentioned in the preceding, being, for that matter, clearly indicated by the heading of this Section.

Thus, to start with, suppose we are given a *topological algebra* E, whose *spectrum* is $\mathfrak{M}(E)$. Now, we say that a set $B \subseteq \mathfrak{M}(E)$ is a *hull* (of $\mathfrak{M}(E)$), in the case one has;

$$B = h(k(B)) \equiv \left\{ f \in \mathfrak{M}(E) : \ker f \supseteq k(B) \equiv \bigcup_{g \in B} \ker g \right\} \tag{4.1}$$

(see also [TA: Chapter IX]). Furthermore, by extending the preceding notation in (2.26), one sets, *for any subset,* say B, *of* $\mathfrak{M}(E)$;

$$\begin{aligned} \widetilde{E}_B := \Big\{ \alpha : B \longrightarrow \mathbf{C} : \forall f \in B \ \exists \ \textit{open} \ V \subseteq \mathfrak{M}(E), \ \textit{with} \ f \in V, \\ \textit{and} \ x \in E, \ \textit{such that} \ \alpha = \hat{x}|_{V \cap B} \Big\}. \end{aligned} \tag{4.2}$$

That is, again one considers *the set* (in fact, **C**-*algebra*) *of* **C**-*valued maps on* B, *that locally belong to* $E^\wedge$; as the analogous relation to (2.27), it is also clear that;

$$\widetilde{E}_B \subseteq \mathcal{C}(B). \tag{4.3}$$

On the other hand, if $\mathcal{E}$ is the *Gel'fand sheaf of* E over $\mathfrak{M}(E)$ (see Definition 2.1), then by applying a similar argument to that one for (2.31), *one* still *obtains that;*

$$\widetilde{E}_B = \Gamma(B, \mathcal{E}) \equiv \mathcal{E}(B), \tag{4.4}$$

for every $B \subseteq \mathfrak{M}(E)$, within an *isomorphism of* the **C**-*algebras* involved.

Thus, one gets through (4.4) a genuine *extension of* our previous *Lemma 2.1 to an arbitrary* $B \subseteq \mathfrak{M}(E)$, which we shall need presently below. Now, by further considering some B, as before, let

$$I = k(B) := \bigcap_{f \in B} \ker f \tag{4.5}$$

be its *kernel,* thus a *closed 2-sided ideal* in E (loc. cit.). So we can further take the corresponding *quotient topological algebra*

$$F = E/I, \tag{4.6}$$

whose *spectrum* is given by (see also (4.1), (4.5))

$$\mathfrak{M}(F) \equiv \mathfrak{M}(E/I) \underset{\rho}{=} h(I) \equiv h(k(B)) \subseteq \mathfrak{M}(E); \tag{4.7}$$

here ρ stands for a *homeomorphism* (cf. [TA: p.339, Theorem 4.1]). Precisely, if $\phi : E \longrightarrow E/I \equiv F$ denotes the (canonical) quotient map, the map ρ is given by

$$\rho \equiv {}^t\phi : \mathfrak{M}(E/I) \longrightarrow \mathfrak{M}(E), \tag{4.8}$$

its image in $\mathfrak{M}(E)$ being $h(I)$, viz. the *hull of* I.

In particular, if the set $B \subseteq \mathfrak{M}(E)$ involved hitherto is a *hull,* as in (4.1), then, in view of (4.7), one obtains

$$\mathfrak{M}(E/I) \equiv \mathfrak{M}(F) = B, \tag{4.9}$$

up to a *homeomorphism,* established by the map ρ, as above.

Specifically, by considering the (quotient) *topological algebra* F, as in (4.6), and by still assuming B *a hull,* one concludes that the $\mathbf{C}$-*algebra* $\widetilde{F}_{\mathfrak{M}(F)}$ (cf., for instance, (2.26) or even (2.42)) consists of those maps (cf. (4.9))

(4.10) *$\alpha : \mathfrak{M}(F) = B \longrightarrow \mathbf{C}$, such that, for every $g \in \mathfrak{M}(F)$, there exists an open neighborhood V of $\rho(g) \in B \subseteq \mathfrak{M}(E)$, along with an element $y \in F$, such that one has;*

$$\alpha = \hat{y}|_{\rho^{-1}(V)}. \tag{4.10.1}$$

Thus, as an application of the above, one is led to the following conclusion, having a particular interest **per se** (cf. Scholium 4.1 below), being also the crucial ingredient in the proof of our main result in the sequel. So we have.

Lemma 4.1. *Let E be a topological algebra with spectrum $\mathfrak{M}(E)$ and $B \subseteq \mathfrak{M}(E)$ a hull, with kernel $I = k(B)$* (cf. (4.1), (4.5)), *while $F \equiv E/I$ is the corresponding quotient topological algebra. Finally, let $\mathcal{E}$ and $\mathcal{F}$ be the Gel'fand sheaves of the topological algebras E and F, respectively. Then, one has;*

$$\Gamma(B,\mathcal{E}) = \Gamma(B,\mathcal{F}), \tag{4.11}$$

within a **C***-algebra isomorphism.*

Proof. Based on (4.4) and (2.42), it is enough to prove the following *isomorphism of* **C***-algebras;*

$$\widetilde{E}_B = \widetilde{F}_B = \widetilde{F}_{\mathfrak{M}(F)} = \Gamma(B,\mathcal{F}). \tag{4.12}$$

Here one looks at the given hull $B \subseteq \mathfrak{M}(E)$, as the *spectrum of* F, via the *homeomorphism* ρ (see (4.8), (4.9)).

Thus, our assertion in (4.12) is now a consequence of (4.2) and (4.10), in conjunction with the relation

$$\alpha = \widehat{\phi(x)} = \hat{x} \circ \rho|_{\rho^{-1}(V)}, \tag{4.13}$$

which holds true, *for any map* α, *as in (4.10), with* $y = \phi(x) \in E/I \equiv F$, and this terminates the proof of the lemma. ∎

Scholium 4.1.– The algebra $\widetilde{E}_B$, with $B \subseteq \mathfrak{M}(E)$, may also be considered, as the *localization of* E, *relative to* B. (We refer to A. Mallios [9] for a general account thereon). Thus, (4.12) or even the **C***-algebra isomorphism*

$$\widetilde{E}_B = (\widetilde{E/I})_B, \tag{4.14}$$

with $I = k(B)$ (see (4.5)), can now be expressed, as follows:

(4.15) by looking at the *localization* of a given topological algebra E, with respect to a *hull* of its spectrum, *the* (Gel'fand transforms of the) *elements of* E *that vanish on this hull* (hence, those in its kernel) *do not count.*

We come now to the promised assertion, concerning the *softness of a* given *Gel'fand sheaf.* Indeed, one has the following result (see also Definition 3.3 above).

Theorem 4.1. *The Gel'fand sheaf of a regular hereditarily geometric* (topological) *algebra is soft.*

Proof. Let E be a given topological algebra with spectrum $\mathfrak{M}(E)$ and Gel'fand sheaf $\mathcal{E}$, satisfying the conditions of the theorem (cf. Definition

3.3), while let B ba a *closed* subset of $\mathfrak{M}(E)$. So one has to prove that *the* corresponding *restriction map* (in fact, $\mathbf{C}$-*algebra morphism*)

$$\rho_B : \Gamma(\mathfrak{M}(E), \mathcal{E}) \longrightarrow \Gamma(B, \mathcal{E}) \tag{4.16}$$

is surjective (see also e.g. Chapt. III; (8.25)): Indeed, since, by hypothesis, E is regular, *the closed set* $B \subseteq \mathfrak{M}(E)$ is still closed, with respect to the *hull-kernel topology* in $\mathfrak{M}(E)$, therefore, *a hull in* $\mathfrak{M}(E)$ (cf. [TA: p.332, Theorem 2.1, along with p.330, Lemma 1.1 and p.331, Definition 1.2]). Consequently, if $I = k(B)$ denotes the *kernel of* B (see (4.5)), one has (cf. (4.1)) $B = h(k(B)) \equiv h(I)$. Therefore, based now on our hypothesis for E (see Definition 3.3), and on (4.11), one obtains;

$$\begin{aligned} \Gamma(\mathfrak{M}(E), \mathcal{E}) = E \xrightarrow{\phi} E/I \equiv F = \Gamma(\mathfrak{M}(F), \mathcal{F}) \\ = \Gamma(B, \mathcal{F}) = \Gamma(B, \mathcal{E}), \end{aligned} \tag{4.17}$$

that is, exactly our assertion for (4.16), and this completes the proof. ∎

On the other hand, we have already remarked in the previous Section (see (3.14)) that soft, as well as, Fréchet-Šilov (locally m-convex) algebras are herediterily geometric ones; moreover, both type of the latter algebras are, by their definition, *regular* (cf. Definition 3.2 and (3.10)). Thus, the last remark, together with the previous Theorem 4.1, provide now the promised assertion, alluded to by the heading of this Section. Namely, one concludes that;

(4.18) *the Gel'fand sheaf of any soft algebra, or* that one *of any Fréchet-Šilov* (locally m-convex) *algebra, is soft.*

The above also justifies the terminology adopted thus, in anticipation, in the preceding by Definition 3.2. In this context, something more than (4.18) is actually true; namely, due to the topological properties of the spectra of the topological algebras appeared in (4.18) (cf. (3.12)), as well as, to the topological-algebraic (viz. *"geometric"*) structure of the same algebras (see (3.2) above, and S.A. Selesnick [1: p.319, Proposition 2.4, (iii)]), one concludes that;

(4.19) *the Gel'fand sheaf of either type of* the *topological algebras* appeared *in (4.18) is fine.* (See e.g. Chapt. III; Definition 8.1). Hence (loc. cit., (8.26)), *the same sheaf is soft* too.

The significance of the previous remark is, of course, evident, concerning *cohomological questions* (cf. Chapt. III; Section 8), connected with the Gel'fand sheaf, as we shall see in the sequel.

5. De Rham-Kähler complex of a topological algebra

Our purpose by the following discussion is to associate a (de Rham-) *Kähler complex of differentials* with a given topological algebra, the latter being, as we shall presently see, of a fairly general type. Precisely, we thus assume that;

(5.1) the *topological algebras* involved in the sequel are *locally convex* ones with (jointly) *continuous multiplication* [TA], being also *unital* and *commutative* **C**-*algebras*. Furthermore, if E is such an algebra, we always assume, as usual, that $\mathfrak{M}(E)$ stands for its *spectrum* (loc. cit.).

Of course, we are based here upon the standard theory of *Kähler differentials*, which is actually valid for any unital commutative ring (see, for instance, N. Bourbaki [2: Chap. 3; p.134], or even E. Kunz [1]). However, we consider instead, as already mentioned (cf. (5.1)), *unital commutative algebras over the complexes*, for short, **C**-*algebras*, which are also *suitably topologized*. The interference here of a toplology, compatible with the algebraic structure involved, is due, as we shall see right below, to our desire to cope with certain *analytical aspects*, as, for instance, with the existence of the *exponential map*; thus, the latter is actually supplied by a particular type of topological algebras, general enough, however, to cover the relevant applications (see Section 7 in the sequel). On the other hand, as we have already seen in our abstract treatment in the preceding (see e.g. Chapter VIII), the map in question has important *geometrical applications*; so we have employed it, for instance, in our treatment of *line sheaves, in connection with Chern classes*, the counterpart here of the classical *Chern isomorphism* (for *line bundles*, see Chapt.

V; Section 6), or even in the proof of *Weil's integrality theorem* (cf. Chapt. VIII; Theorem 11.1. See also (7.19) below).

Thus, to start with, suppose that we are given a *topological algebra E, as in (5.1).* Furthermore, we consider the *locally convex* (tensor product) *algebra*

$$E \underset{\pi}{\otimes} E, \tag{5.2}$$

endowed with the *projective tensorial topology* π [TA: Chapter X] and the *multiplication morphism*

$$\mu : E \underset{\pi}{\otimes} E : x \otimes y \longmapsto \mu(x \otimes y) := xy, \tag{5.3}$$

being, in fact, a *continuous algebra morphism* of the topological algebras concerned. Thus, the *kernel of* μ is a *closed 2-sided ideal* of $E \underset{\pi}{\otimes} E$, so that by further setting

$$I \equiv \ker \mu \subseteq E \underset{\pi}{\otimes} E, \tag{5.4}$$

one has the following *continuous exact sequence*

$$0 \longrightarrow \ker \mu \longrightarrow E \underset{\pi}{\otimes} E \overset{\mu}{\longrightarrow} E \longrightarrow 0. \tag{5.5}$$

As a result, one obtains;

$$E \underset{\pi}{\otimes} E = E \oplus \ker \mu, \tag{5.6}$$

a *topological direct sum decomposition:* The assertion on (5.6) is based on the *exactness of (5.5)* and the *continuity of the projection* $i \circ \mu$ of $E \underset{\pi}{\otimes} E$, corresponding to (5.5), where one also employs the *continuous* (canonical) *injection* (in fact, a *section* of μ);

$$i : E \longrightarrow E \underset{\pi}{\otimes} E : x \longmapsto i(x) := x \otimes 1. \tag{5.7}$$

(In this context, see also e.g. J. Horváth [1: p.121, comments following Definition 2.1]. Yet, we still remark here that one can take in (5.6) the *complete algebra* $E \widehat{\underset{\pi}{\otimes}} E$ [TA], having again a similar decomposition, by assuming now E to be *complete;* see e.g. A. Mallios [8: p. 153; (4.5)]).

On the other hand, consider now the vector subspace of $E \otimes E$, generated by $I \cdot I$ (see (5.4)), denoted by $I^2 \equiv [I \cdot I]$, taking, in effect, an *ideal* of $E \otimes E$, such that $I^2 \subseteq I$; hence, by further looking at the quotient space I/I^2, endowed with the respective *quotient topology,* one obtains a *complete locally convex E-module,* by setting

$$\Omega^1_E := \widehat{I/I^2}. \tag{5.8}$$

(For technical reasons, having to do with *"analytical questions"* (see, for example, the comments following (5.1) above), we consider in (5.8) the corresponding *completed* space). Now, following classical terminology (see, for instance, R. Hartshorne [1: p.172], or even E. Kunz [1: p.39]),

(5.9) we call Ω^1_E, as given by (5.8), the *Kähler module of the 1st-order differentials of* E, or even simply the (*1st-order*) *Kähler module of* E.

Furthermore, setting

$$\wedge^p\Omega^1_E := \widehat{\wedge^p(I/I^2)} \equiv \Omega^p_E, \tag{5.10}$$

for any $p \in \mathbf{Z}_+$, where we also set

$$\wedge^0\Omega^1_E \equiv \Omega^0_E := E \quad and \quad \wedge^1\Omega^1_E := \Omega^1_E, \tag{5.11}$$

one finally obtains a *sequence of complete locally convex* E*-modules* (we assume, of course, due to the first of (5.11), that E is *complete* too). Thus, we now define (see (5.10));

$$\Omega^*_E \equiv \wedge\Omega^1_E := \widehat{\bigoplus_{p=0}^{\infty} \wedge^p(I/I^2)} \equiv \bigoplus_{p\geqq 0} \wedge^p\Omega^1_E \equiv \bigoplus_{p\geqq 0} \Omega^p_E, \tag{5.12}$$

viz. the *exterior* (or even *Grassmann*) *algebra of the* (complete convex) E-*module* Ω^1_E, as the latter is given by (5.8).

On the other hand, one can further consider on the above algebra Ω^*_E the *(locally convex) direct sum topology;* in other words, the finest locally convex (vector space) topology on it, making the canonical inclusion ($\mathbf{C}$-linear) maps (cf. (5.12))

$$\Omega^p_E \longrightarrow \Omega^*_E, \quad p \in \mathbf{Z}_+, \tag{5.13}$$

continuous. Thus, by virtue of (5.10), (5.11) and of our hypothesis for E, one concludes that;

(5.14) Ω^*_E *is a complete locally convex algebra with continuous multiplication and an identity element* (viz. that one of E, cf. (5.11)).

In this concern, see also, for instance, G. Köthe [1: p.212, (3)], along with A. Colojoară [1], [2]. ∎

Furthermore, by still following the classical algebraic counterpart (see e.g. N. Bourbaki [2: Chap. 3; p. 132], or even E. Kunz [1]), one defines a *continuous derivation of E into* (the E-module) Ω_E^1, as follows;

$$(5.15)\quad d : E \longrightarrow \Omega_E^1 : x \longmapsto d(x) := (1 \otimes x - x \otimes 1) + I^2 \equiv [(1 \otimes x - x \otimes 1)];$$

viz. d, as above, is a $\mathbf{C}$-*linear map,* which still satisfies the corresponding *Leibniz condition* (i.e., one has $d(x, y) = x \cdot dy + y \cdot dx$, for any x, y in E). Yet, setting

$$(5.16)\qquad \tilde{d} : E \longrightarrow E \otimes E : x \longmapsto \tilde{d}(x) := 1 \otimes x - x \otimes 1 \in \ker \mu \equiv I,$$

one infers that;

$$(5.17)\qquad I = \overline{[\operatorname{im} \tilde{d}]} \subseteq E \mathbin{\widehat{\otimes}_\pi} E.$$

In other words, one thus obtains that:

(5.18) $\tilde{d}(E) \subseteq I$ is a *set of topological generators of I,* the latter being considered as an *E-module;* thus, equivalently, every element of I can be approximated by elements of the form,

$$(5.18.1)\qquad \sum_i x_i \tilde{d}(y_i),$$

such that $\sum_i x_i \otimes y_i \in E \otimes E$.

In this regard, see also e.g. N. Bourbaki [2: Chap. 3; p.132, Lemme 1]. ∎

We call the $\mathbf{C}$-linear map (5.15), as above, the *Kähler derivation* (precisely, the *1st-order Kähler derivation*) *of* E.

Thus, our next objective is to find now a *(continuous) exterior derivation of Ω_E^* into itself,* satisfying the appropriate conditions, according to the classical pattern, provided by the *de Rham algebra of a smooth manifold:*

Indeed, given a *topological algebra E, as in (5.1),* one defines a *continuous $\mathbf{C}$-linear map*

$$(5.19)\qquad \delta : \Omega_E^1 \longrightarrow \Omega_E^2,$$

as follows; thus, one first sets (cf. (5.15), together with (5.10)),

$$(5.20) \qquad \delta\left(\sum_i x_i \otimes y_i\right) := \sum_i dx_i \wedge dy_i,$$

for any $\sum_i x_i \otimes y_i \in E \otimes E$, so that *one* actually *obtains* $I^2 \subseteq \ker\delta$. Therefore, due to the *continuity of the* **C**-*linear map, as given by (5.20),* one finally gets the desired map, as in (5.19), (we still denote the latter map by δ); viz. one has (cf. (5.10)),

$$(5.21) \qquad \delta : \widehat{I/I^2} \equiv \Omega^1_E \longrightarrow \Omega^2_E \equiv \wedge^2 \Omega^1_E \equiv \widehat{\wedge^2(I/I^2)}.$$

(In this regard, see also A. Mallios [8: p. 155, Lemma 4.1]). ∎ So the relevant situation is now exhibited by the following result, patterned after the standard algebraic case (see also loc. cit.; p.156, Theorem 4.1).

Theorem 5.1. *Let E be a unital commutative locally convex algebra with continuous multiplication. Then, there exists a continuous exterior derivation of degree one* (cf. (5.12))

$$(5.22) \qquad D : \Omega^*_E \longrightarrow \Omega^*_E,$$

viz. a continuous **C**-*linear map, as above, such that the following conditions hold true;*

$$(5.23) \qquad D(\Omega^p_E) \subseteq \Omega^{p+1}_E, \quad p \in \mathbf{Z}_+,$$

and, for any α, β in Ω^*_E, with $\alpha \in \Omega^p_E$, one has,

$$(5.24) \qquad D(\alpha \wedge \beta) = D(\alpha) \wedge \beta + (-1)^p \alpha \wedge D(\beta).$$

Moreover, the following relations are still valid (cf. (5.11), (5.12), (5.15));

$$(5.25) \qquad D|_E = d,$$

which thus implies the uniqueness of D (cf. also (5.19), (5.20)), *as well as,*

$$(5.26) \qquad D^2 = 0. \ \blacksquare$$

Thus, given a *topological algebra E*, as in the above Theorem 5.1, we are now justified, by virtue of the same theorem, and by following standard terminology, to call the pair (see (5.12), (5.22))

$$(5.27) \qquad (\Omega^*_E, D) \equiv (\wedge \Omega^1_E, D)$$

the *de Rham-Kähler complex of E.*

We are going now to look at the preceding, right in the next Section, within the appropriate sheaf-theoretic framework, that will be also of use in the sequel.

6. Sheafification of the de Rham-Kähler complex. Logarithmic derivation

Suppose we are given a *topological algebra sheaf* $\mathcal{E}$ on a *topological space* X (cf. Theorem 1.1) in such a manner that:

(6.1) for any *open* set $U \subseteq X$, the corresponding *(local) section algebra of* $\mathcal{E}$ *over* U,

(6.1.1) $$\Gamma(U, \mathcal{E}) \equiv \mathcal{E}(U)$$

is a *topological algebra, as in (5.1)* above.

Thus, by further employing the notation of (5.8) in the preceding, one can consider the following *presheaf on* X,

(6.2) $$U \longmapsto \Omega^1_{\Gamma(U,\mathcal{E})} := \widehat{I_U/I_U^2},$$

where U is varying over the open sets in X, while one also takes the obvious restriction morphisms. In other words, with any open $U \subseteq X$, one associates the corresponding *Kähler module of* $\Gamma(U, \mathcal{E})$. So we call the previous presheaf on X the *Kähler presheaf of* $\mathcal{E}$, while the respective sheaf on X, generated by the previous presheaf, is named the *Kähler sheaf* (or even the *sheaf of 1st-order Kähler differentials*) *of* $\mathcal{E}$ and is denoted by

(6.3) $$\Omega^1_{\mathcal{E}}.$$

Consequently, given that *each one of the sets (6.2),* as above, when U varies over the open sets in X, *is a* $\Gamma(U, \mathcal{E})$*-module* (the corresponding *restriction morphisms* being also such *of the pertinent type*), one obtains that (see also Chapt. II; (1.54));

(6.4) *the sheaf* $\Omega^1_{\mathcal{E}}$, as defined above, *is,* in fact, *an* $\mathcal{E}$*-module over* X.

(See Chapt. II; Definition 1.5, along with (1.66), or even Proposition 1.1 therein, taking also our hypothesis for $\mathcal{E}$ into account).

Now, one further defines the *Kähler derivation of* $\mathcal{E}$, as the following *morphism of sheaves,*

$$\partial \equiv (\partial_U) : \mathcal{E} \longrightarrow \Omega^1_{\mathcal{E}}, \tag{6.5}$$

such that one sets;

$$\partial_U(s) := (1 \otimes s - s \otimes 1) + I_U^2 \equiv [\tilde{d}_U(s)], \tag{6.6}$$

for any (local) *section* $s \in \mathcal{E}(U) \equiv \Gamma(U, \mathcal{E})$, *and* an arbitrary *open* $U \subseteq X$. (Here $\tilde{d}_U$ is defined analogously with (5.16); yet, concerning the same rel. (6.6), *one actually has* $\partial_U = d_U$ (cf. (5.15)). In this context, cf. also Chapt. I; Definition 1.2, along with (6.3), (8.5) and Lemma 8.1). Thus, by the very definitions, one infers that:

(6.7) ∂ *is a* $\mathbf{C}$*-linear morphism* of the $\mathbf{C}$*-vector space sheaves* involved in (6.5) (see Chapt. II; (1.38)), such that the following *Leibniz condition* holds true, as well;

$$\partial(s \cdot t) = (\partial s) \cdot t + s \cdot \partial t, \tag{6.7.1}$$

for any (continuous local) *sections* s, t *in* $\mathcal{E}(U)$, *and any open* $U \subseteq X$ (see also e.g. Chapt. VI; (1.2)).

The above thus justifies too the previously applied terminology, concerning the map (6.5) (cf. Chapt. VI; Definition 1.1). So, in other words, the triple

$$(\mathcal{E}, \partial, \Omega^1_{\mathcal{E}}), \tag{6.8}$$

as given above, supplies here, in effect, a particular example of a *differential triad,* in the sense of Chapt. VI; (1.4) (see also (3.1) there). We call it the *Kähler differential triad* of the given *topological algebra sheaf* $\mathcal{E}$, as above (cf. (6.1)). In fact, we are going right away to specialize the previous triple, as in (6.8), to the case of a *topological algebra* E, by employing the corresponding *Gel'fand sheaf* of E (cf. (2.8)).

Thus, by virtue of (2.41.1) and (2.27) (see also (2.44)), one concludes, in particular, that:

given a *topological algebra* E, *unital* (not necessarily commutative), whose *spectrum* is

(6.9.1) $$\mathfrak{M}(E) \equiv X,$$

the corresponding *Gel'fand sheaf* $\mathcal{E}$ of E (see (2.8)) yields a *topological algebra sheaf on* X, such that one has;

(6.9) *for any open* $U \subseteq X$, the respective *(local) section algebra of* $\mathcal{E}$ *over* U,

(6.9.2) $$\Gamma(U, \mathcal{E}) \equiv \mathcal{E}(U)$$

is a unital commutative locally m-convex (topological **C**-*)algebra.*

In other words (cf. Theorem 1.1), *the Gel'fand sheaf* $\mathcal{E}$ *is a unital commutative locally m-convex topological* **C**-*algebra sheaf on* X, the *spectrum* of the given topological algebra E, as above (cf. (6.9.1)).

Consequently, our previous *cond. (5.1) is fulfilled* for the topological algebras involved. (Of course, in view of the relation $\widetilde{E}_U \subseteq \mathcal{C}_c(U)$ (see (2.72)), one considers on $\widetilde{E}_U$ the *relative topology from* $\mathcal{C}_c(U)$; cf., for instance, (2.4) and the ensuing comments therein, or even (2.5)). *Warning*! At this place, we should also remark that the given topological algebra E *is not necessarily a geometric* (topological) *algebra* (Definition 3.1), so that *we do not have, in general, (3.4) fulfilled,* in either sense (viz. either algebraically or even topologically).

On the other hand, within the same vein of ideas, and based on the respective definitions in (5.19), (5.20), along with Theorem 5.1, one further defines *exterior differential operators*

(6.10) $$d^p \equiv d : \Omega^p_{\mathcal{E}} \longrightarrow \Omega^{p+1}_{\mathcal{E}}, \quad p \geqq 0,$$

where we have set (see also (5.10), together with Chapt. IV; (7.1), (7.3))

(6.11) $$\Omega^p_{\mathcal{E}} := \wedge^p \Omega^1_{\mathcal{E}}$$

(cf. (6.3) above), while we still set (see also (5.11));

(6.12) $$\wedge^0 \Omega^1_{\mathcal{E}} := \mathcal{E} \ \text{ and } \ \wedge^1 \Omega^1_{\mathcal{E}} := \Omega^1_{\mathcal{E}}.$$

Thus, one has, for instance, that (see (6.5) and (6.10) for $p = 1$);

$$d^1 \circ \partial \equiv d \circ \partial = 0 \tag{6.13}$$

(note that $d(1) = 0$; cf. (5.15). See also (5.26) in the preceding; so we have found again the relation that guarantees the *flatness of the $\mathcal{E}$-connection ∂ of $\mathcal{E}$*. See Chapt. VIII; (1.8), (2.6)). Of course, one can further formulate here the corresponding *sheaf-theoretic version of* the same *Theorem 5.1,* as above. Thus, one obtains the (sheaf-theoretic) *de Rham-Kähler complex of the* given *topological algebra sheaf* $\mathcal{E}$, as before; viz.

$$(\Omega^*_{\mathcal{E}}, d), \tag{6.14}$$

with the obvious meaning of the notation involved here; see also (5.27), with $D \equiv d$, along with (5.12), (6.10), (6.11) and (6.12).

6.(a). Logarithmic derivation.– The notion in the heading of this Subsection has been already examined in the preceding, within the appropriate abstract axiomatic context (cf. Chapt. VI; Definition 1.2). So here we assume that we are given a *topological algebra sheaf* $\mathcal{E}$, as above (see (6.1)), and let

$$\mathcal{E}^{\cdot} \tag{6.15}$$

be the corresponding *sheaf of units of* $\mathcal{E}$, an *abelian group sheaf* on X; thus, for every *open* $U \subseteq X$, we denote by $\mathcal{E}(U)^{\cdot}$ the *group of units* of the given algebra $\mathcal{E}(U) \equiv \Gamma(U, \mathcal{E})$ (cf. (6.1.1) and (5.1)), while one still obtains,

$$\mathcal{E}^{\cdot}(U) \equiv \Gamma(U, \mathcal{E}^{\cdot}) = \Gamma(U, \mathcal{E})^{\cdot} \equiv \mathcal{E}(U)^{\cdot}, \tag{6.16}$$

within a *group isomorphism* (see Chapt. IV; Lemma 1.1, in particular, (1.15)).

So we further define the following *sheaf morphism* (cf. (6.3))

$$\tilde{\partial} : \mathcal{E}^{\cdot} \longrightarrow \Omega^1_{\mathcal{E}}, \tag{6.17}$$

such that one sets;

$$\tilde{\partial}(s) := s^{-1} \cdot \partial(s), \tag{6.18}$$

for any $s \in \mathcal{E}^{\cdot}(U) = \mathcal{E}(U)^{\cdot}$, and U *open* in X.

Now, according to the general theory (see Chapt. VI; (1.27)), one thus concludes that the map (6.17) is a *morphism of sheaves of (abelian) groups;* viz. one has

(6.19) $$\tilde{\partial}(s \cdot t) = \tilde{\partial}(s) + \tilde{\partial}(t),$$

for any s, t in $\mathcal{E}^{\cdot}(U)$ and U open in X. Moreover, one gets, straightforwardly from (6.18), that;

(6.20) $$\ker \tilde{\partial} = (\ker \partial) \cap \mathcal{E}^{\cdot}.$$

In particular, we are going to apply the above in the sequel (see thus the next Section 7 below).

Finally, as was the case already in Chapter VI, due to (6.18), we call the map (6.17), for the case under consideration here, the *(kählerian) logarithmic derivation of* $\mathcal{E}$, while the element $\tilde{\partial}(s) \in \Omega^1_{\mathcal{E}}(U)$, as in (6.18), the *(kählerian) logarithmic derivative of* (the local section) $s \in \mathcal{E}^{\cdot}(U) = \mathcal{E}(U)^{\cdot}$ (see (6.16)).

7. The sheaf exponential sequence

We consider in this Section conditions for a given topological algebra, so that its Gel'fand sheaf (cf. (2.8)) yield an exact exponential sheaf sequence, in the sense already examined within our abstract context in the preceding (see Chapt. V; Definition 6.1).

Indeed, one first obtains the following preliminary result, valid under fairly general conditions for the topological algebras involved. Thus, for convenience, we recall, however, at this place that:

(7.1) given a *unital locally m-convex* (topological) **C**-*algebra* E, which is also σ-*complete* (i.e., every *Cauchy sequence* in E converges), then the *exponential function acts* (or *operates*) *on* E, *viz. one has*

(7.1) $$\exp x := \sum_{n=0}^{\infty} \frac{1}{n!} x^n \in E,$$

for every element $x \in E$.

(In this context, one just considers the corresponding numerical series $\sum_n \frac{1}{n!} p(x)^n$, for every (continuous) *submultiplicative seminorm* p on E, taken from a family of such, defining the topology of E; cf. [TA: p.18, Proposition 3.2]. See also A. Mallios [3: p. 492; (5.1)], or even [6: p. 196; (8.5)]). ∎

Thus, we come now right away to the promised result. That is, one has the following.

Lemma 7.1. *Let E be a unital σ-complete locally m-convex algebra with Gel'fand map continuous* (cf. (2.21), (2.2)). *Moreover, let $\mathcal{E}$ be the corresponding Gel'fand sheaf of E over $\mathfrak{M}(E)$, the spectrum of the given topological algebra E* (see Definition 2.1), *and $\mathcal{E}^{\cdot}$ the respective sheaf of units of $\mathcal{E}$* (cf. (6.15)). *Then, there exists the exponential sheaf morphism*

$$e : \mathcal{E} \longrightarrow \mathcal{E}^{\cdot}, \tag{7.2}$$

such that the following sequence (of sheaves of groups, cf. Chapt. V; (6.1), (6.2)) *is exact;*

$$0 \longrightarrow \mathbf{Z} \longrightarrow \mathcal{E} \xrightarrow{e} \mathcal{E}^{\cdot}. \tag{7.3}$$

Proof. Consider an *open* set $U \subseteq \mathfrak{M}(E)$, along with the following *(local) exponential function* (cf. also (2.3))

$$\begin{aligned} e_U : E^{\hat{}}|_U \longrightarrow (E^{\hat{}}|_U)^{\cdot} : \hat{x}|_U \longmapsto e_U(\hat{x}|_U) \\ := \exp 2\pi i(\hat{x}|_U) = \widehat{\exp 2\pi i x}|_U. \end{aligned} \tag{7.4}$$

Indeed, according to our hypothesis for E, *one has*

$$\widehat{\exp 2\pi i x} \in E^{\hat{}}, \tag{7.5}$$

for any element $x \in E$ (cf. (7.1.1)). But, in view of the same definition (7.1.1) and the *continuity of the Gel'fand map* (in fact, a *continuous* **C**-*algebra morphism*), one further obtains (see also (2.21));

$$\begin{aligned} \widehat{\exp 2\pi i x} &\equiv \mathcal{G}(\exp 2\pi i x) = \mathcal{G}\left(\lim_n \textstyle\sum_{k=0}^{n} \frac{1}{k!}(2\pi i x)^k\right) \\ &= \lim_n \mathcal{G}\left(\textstyle\sum_{k=0}^{n} \frac{1}{k!}(2\pi i x)^k\right) = \lim_n \textstyle\sum_{k=0}^{n} \frac{1}{k!}\, \widehat{2\pi i x}^{\,k} \\ &= \exp \widehat{2\pi i x} = \exp 2\pi i \hat{x} \in E^{\hat{}}. \end{aligned} \tag{7.6}$$

(The last relation proves, in effect, that *the Gel'fand transform algebra $E^{\hat{}}$ of*

E is still *"closed" with respect to the exponential function,* or even *the latter function acts* also *on* $E^{\wedge}$). Therefore, due, in turn, to the *continuity of the restriction* (**C**-*algebra) morphism*

$$\rho_U : E^{\wedge} \longrightarrow E^{\wedge}|_U, \tag{7.7}$$

for any open $U \subseteq \mathfrak{M}(E)$ (cf. also (2.2), (2.6)), one finally obtains;

$$\widehat{\exp 2\pi i x}|_U \equiv \exp(\widehat{2\pi i x}|_U) = \exp 2\pi i(\hat{x}|_U) \in E^{\wedge}|_U, \tag{7.8}$$

which thus completely justifies now (7.4).

On the other hand, by considering the *Gel'fand presheaf of* E *over* $\mathfrak{M}(E)$, as given in the preceding by (2.7.1), in conjunction with the above exponential functions, as in (7.4), one gets the following *commutative diagram;*

$$\begin{array}{ccc} E^{\wedge}|_U & \xrightarrow{\ e_U\ } & (E^{\wedge}|_U)^{\bullet} \\ {\scriptstyle \rho_V^U}\downarrow & & \downarrow{\scriptstyle \rho_V^U} \\ E^{\wedge}|_V & \xrightarrow{\ e_V\ } & (E^{\wedge}|_V)^{\bullet}, \end{array} \tag{7.9}$$

for any open U, V *in* $\mathfrak{M}(E)$, *with* $V \subseteq U$. Therefore, one obtains the following *morphism of* the *presheaves* concerned;

$$e_U : E^{\wedge}|_U \longrightarrow (E^{\wedge}|_U)^{\bullet} \stackrel{\subseteq}{\rightarrow} \Gamma(U, \mathcal{E})^{\bullet} = \Gamma(U, \mathcal{E}^{\bullet}), \tag{7.10}$$

for any open $U \subseteq \mathfrak{M}(E)$ (see (2.15) and (6.16)). Thus, one finally gets the desired sheaf morphism (7.2) (cf. Chapt. I; (8.4), (8.5)). Furthermore, it is clear, in view of (7.4), that;

$$\ker(e_U) \cong \mathbf{Z} \text{ with } U \text{ open in } \mathfrak{M}(E). \tag{7.11}$$

(We assume, of course, here that *the* given *topology of* E *is Hausdorff*). Hence, *one has*

$$\ker(e) \cong \mathbf{Z}, \tag{7.12}$$

as desired, **Z** being considered here as a *constant sheaf* (of abelian groups) *over* $\mathfrak{M}(E)$ (cf. Chapt. I; (4.1), Scholium 4.1 and (10.2), (10.8)). Thus, (7.12) stands, in fact, for an *isomorphism of* the *(abelian group) sheaves* concerned, and this terminates the proof. ∎

Now, to have the previous sequence (7.3) in the form of a *short exact sequence* (a fact, that is now our objective), one has to supplement appropriately our hypothesis for the topological algebra E involved in the above Lemma 7.1. Thus, the pertinent condition here seems to be the assumption that E is a *soft algebra,* being further a *self-adjoint* one.

Thus, for convenience, we recall at this point (see also, for instance, [TA: p.482, (1.4)]) that a given *topological algebra* E, which is further endowed with a *(continuous) involution* "$*$" (we then refer to a *topological* **-algebra*) is said to be *self-adjoint,* if the corresponding Gel'fand map of E (see (2.21)) is *involution* (or, as we also say, *-)*preserving;* viz. one has

$$\mathcal{G}(x^*) \equiv \widehat{x^*} = \overline{\hat{x}} \equiv \overline{\mathcal{G}(x)}, \tag{7.13}$$

for any $x \in E$.

Now, coming back to our abovementioned objective, viz. to get a *short exact (exponential sheaf) sequence* through (7.5), we note that a crucial step here is Lemma 7.2, which follows; its Banach algebra analogue has been employed already, within the appropriate framework, by S.A. Selesnick [2: p.58, Proposition 2.5], while the same is based further on a classical argument, still within the Banach algebra theory, as applied to abstract harmonic analysis. (See thus, Y. Katznelson [1: p.236, Lemma]). So we now have the next.

Lemma 7.2. *Suppose we have a soft self-adjoint locally m-convex *-algebra E* (cf. Definition 3.2 and (7.13)). *Moreover, let $f \in \mathfrak{M}(E)$* (the *spectrum* of E) *and U an open neighborhood of f in $\mathfrak{M}(E)$. Then, there exists an element $x \in E$ and an open neighborhood V of f in $\mathfrak{M}(E)$, such that one has;*

$$\begin{aligned} &\text{i) } 0 \leqq \hat{x} \leqq 1|_{\mathfrak{M}(E)}, \\ &\text{ii) } \operatorname{Supp} \hat{x} \subseteq U, \text{ and} \\ &\text{iii) } \hat{x} = 1_V. \end{aligned} \tag{7.14}$$

Proof. As a consequence of our hypothesis (see Definition 3.2, (3.2) and [TA: p.334, Definition 2.2]), E is a *normal* (locally m-convex) *algebra.* Now, this latter property of E is what one actually needs from the classical argument in Banach algebra theory (loc. cit. before the statement of the lemma). So the desired element $x \in E$ is given by

$$x := \sin^2 \frac{\pi}{2} y y^* \in E, \tag{7.15}$$

for some suitably chosen $y \in E$, according to our hypothesis, for f and U, by applying the *normality* of E (ibid.). In this context, we also note that *(7.15) defines,* of course, *an element of* E, as this follows from the convergence in E of the *power series* concerned. (At this place, we further note that, as follows from the hypothesis, E *is complete;* see thus our comments on (3.1) in the preceding. However, *σ-completeness* of E would also suffice for the existence of $x \in E$, as given by (7.15); see, for example, the analogous situation in (7.1.1) above).

On the other hand, $\mathfrak{M}(E)$ being, by hypothesis (Definition 3.2), *equicontinuous,* one concludes (see [TA: p.184, Theorem 1.2, or even Corollary 1.4]) that *the Gel'fand map of E is continuous.* Therefore, in view of (7.13) (see also e.g. (7.6)), one finally obtains that;

$$\hat{x} = \sin^2 \frac{\pi}{2} \hat{y}\bar{\hat{y}} \in E^{\hat{}}. \tag{7.16}$$

Now, it is easy to see (cf. also the classical argument for Banach algebras, as quoted at the beginning) that *the* above *element* $\hat{x} \in E^{\hat{}}$, *as given by (7.16), satisfies* all *the* three *conds. in (7.14),* and this terminates the proof of the lemma. ∎

Now, consider the *Gel'fand sheaf* $\mathcal{E}$ of a *topological algebra* E, *as in* the previous *Lemma 7.2,* and let $f \in \mathfrak{M}(E)$. So, by looking at the corresponding *fiber map*

$$e_f : \mathcal{E}_f \longrightarrow \mathcal{E}^{\cdot}_f, \tag{7.17}$$

as this is defined by (7.4) (see also (2.9.3)), one proves that e_f *is surjective:* Indeed, based on Lemma 7.2, one can repeat here the respective argument of S.A. Selesnick [2: p.58, Proposition 2.1], given there for the appropriate class of Banach algebras (viz. for *"self-adjoint (Banach-)Šilov algebras"*, ibid.). Therefore (see Chapt. I; (1.21)), one infers that *the (exponential) sheaf morphism*

$$e : \mathcal{E} \longrightarrow \mathcal{E}^{\cdot} \tag{7.18}$$

(Lemma 7.1) *is surjective,* a fact that we were looking for, of course. In conclusion, we can thus state the desired result, as follows, while its proof is already provided by Lemma 7.1 and by what has been said before. So we have now the next.

Theorem 7.1. *Let E be a soft self-adjoint locally m-convex *-algebra, $\mathcal{E}$ its Gel'fand sheaf, and $\mathcal{E}^{\cdot}$ the sheaf of units of $\mathcal{E}$. Then, one has the following short exact sequence of sheaves* (of abelian groups; *exact exponential sheaf sequence of E*)

$$(7.19) \qquad 0 \longrightarrow \mathbf{Z} \longrightarrow \mathcal{E} \xrightarrow{e} \mathcal{E}^{\cdot} \longrightarrow 1.$$

Scholium 7.1.– As already said, the class of topological algebras, appeared in the above Theorem 7.1, specializes to that one of *self-adjoint (Banach-)Šilov algebras,* already employed by S.A. Selesnick (loc. cit.). Yet, we have used in A. Mallios [6] *unital commutative Pták-Šilov self-adjoint (locally m-convex *-)algebras with equicontinuous spectra,* or else *Selesnick algebras,* as they have been named there (due to the abovementioned work of S.A. Selesnick in the case of Banach algebras), to which class the algebras involved in Theorem 7.1 have also a special bearing.

In this context, according to what has been already said in (3.12), we also remark that all the above classes of (topological) algebras have their *spectra compact (Hausdorff) spaces,* as follows from their very definitions. (We still recall here too that *equicontinuous spectra* (of topological algebras [TA]) *are always (relatively) compact Hausdorff spaces* (Alaoglou-Bourbaki)). On the other hand, we consider right away, by the next Theorem 7.2, another class of topological algebras, to which a similar argument to that one employed for (7.19), can still be applied and whose *spectra* are no more compact; however, the same are indeed *paracompact (Hausdorff) spaces* (see the same remark (3.12) in the preceding).

The last conclusions above, about the spectra of the (topological) algebras involved, are also of a particular importance, in conjunction with our previous remarks in (4.18) and (4.19), in what concerns certain *cohomological applications,* we are going to consider in the sequel. So we come first to employ the other class of topological algebras, than those in Theorem 7.1, that we alluded to above. That is, we have.

Theorem 7.2. *Let E be a unital commutative Fréchet regular and semisimple self-adjoint locally m-convex *-algebra;* thus, alias, a *self-adjoint Fréchet-Šilov (locally m-convex) algebra. Then,* by considering the Gel'fand sheaf $\mathcal{E}$ of E, *one obtains an exact exponential sheaf sequence of the form (7.19).*

Proof. By hypothesis E is a Fréchet (locally m-convex topological) algebra, hence, it is a *barrelled* (and **a fortiori** a *spectrally barrelled*) one, having thus the respective *Gel'fand map continuous* (see [TA: p.264, Scheme 2.1 and p.183, Corollary 1.1]). Therefore, Lemma 7.1 can be applied (the given topological algebra E being also *complete,* as by definition Fréchet). On the other hand, *any Fréchet-Šilov* (locally m-convex) *algebra is normal* (see the comments following (3.2) in the preceding); accordingly, one can further apply the argument in the proof of Lemma 7.2, hence, also that one concerning the *surjectivity of (7.18),* as above, a fact that now completes the proof. ∎

We come now, and we shall continue in the next Section 8, to a number of *applications of the* above *exact exponential sheaf sequence,* as given by (7.19):

Thus, we first have a variant, within the previous context, of the classical *Arens-Royden theorem,* initially formulated for (unital commutative) Banach algebras (see, for instance, E.L. Stout [1: p.99, Theorem 10.1]). The same result has been extended by R.M. Brooks [2] to *unital commutative Fréchet (locally m-convex) algebras* with *spectra k-spaces* (loc. cit., p.332, Corollary 3.5). To fix our terminology, we first set the following.

Definition 7.1. Let E be a unital commutative σ-complete locally m-convex algebra, whose spectrum is $\mathfrak{M}(E)$. Then, E is called an *Arens-Royden algebra,* whenever one has;

$$H^1(\mathfrak{M}(E), \mathbf{Z}) = E^{\cdot}/e(E), \tag{7.20}$$

within an *isomorphism of* the *abelian groups* concerned.

Here we still denote by

$$e : E \longrightarrow E^{\cdot} \tag{7.21}$$

the *exponential function* on E (with $E^{\cdot}$ the *group of units of* E), as given by (7.1.1); in particular, we set

(7.22) $$e(x) := \exp 2\pi i x, \quad x \in E,$$

(see also e.g. (7.4). So we actually apply here an obvious *abuse of notation,* by comparison with our previous one in (7.2)). Thus, our clain now is that;

(7.23) *both types of* (the locally m-convex) *algebras, appeared in* the above *Theorems 7.1 and 7.2, are Arens-Royden algebras* (Definition 7.1).

An analogous remark was made by S.A. Selesnick (loc. cit., as above, p.60), for the particular class of Banach algebras employed by the said author. (See, for instance, our previous Scholium 7.1 about that type of algebras).

Now, as already remarked in Scholium 7.1, the *spectra of the* (topological) *algebras, involved in Theorems 7.1 and 7.2,* are *compact,* respectively, *paracompact (Hausdorff) spaces,* so that (see, for instance, Chapt. III; Theorem 8.1) one may consider in the first member of (7.20) *Čech cohomology.* Yet, by virtue of the said theorems, both types of algebras yield a *short exact* (sheaf) *sequence of the form* (7.19), so that by looking at the corresponding *long exact sequence in (Čech) cohomology* (cf. Chapt. III; Theorem 5.3), one obtains;

(7.24) $$\begin{aligned} 0 \longrightarrow \Gamma(X, \mathbf{Z}) \longrightarrow \Gamma(X, \mathcal{E}) &\stackrel{e}{\longrightarrow} \Gamma(X, \mathcal{E}^{\cdot}) \\ &\stackrel{\delta}{\longrightarrow} H^1(X, \mathbf{Z}) \longrightarrow H^1(X, \mathcal{E}) \longrightarrow \cdots . \end{aligned}$$

For convenience, we have set here

(7.25) $$\mathfrak{M}(E) \equiv X,$$

concerning the *spectrum* of a topological algebra E of either one of the types in question. However, referring to the same sort of topological algebras, as before, we have already remarked (cf. (4.18)) that *their* respective *Gel'fand sheaves are soft, hence* (see Chapt. III; (8.26)), *acyclic,* so that one has;

(7.26) $$H^1(X, \mathcal{E}) = 0.$$

Therefore, in view of (7.24) and (7.26), one now obtains the following *exact sequence;*

(7.27) $$\cdots \longrightarrow \Gamma(X, \mathcal{E}) \stackrel{e}{\longrightarrow} \Gamma(X, \mathcal{E}^{\cdot}) \longrightarrow H^1(X, \mathbf{Z}) \longrightarrow 0,$$

hence, the relation

$$H^1(X, \mathbf{Z}) = \Gamma(X, \mathcal{E}^{\cdot})/e(\Gamma(X, \mathcal{E})), \tag{7.28}$$

up to an *isomorphism of* the (abelian) *groups* concerned.

On the other hand, the same types of (topological) algebras, as above, are geometric ones (see (3.14), in conjunction with Definition 3.3), so that one has (cf. Definition 3.1), along with Theorem 3.1);

$$E = \Gamma(X, \mathcal{E}) \equiv \mathcal{E}(X), \tag{7.29}$$

within a C-*algebra isomorphism.* Furthermore, by virtue of (6.16), by taking $U = \mathfrak{M}(E) \equiv X$, and of (7.29), one obtains

$$\Gamma(X, \mathcal{E}^{\cdot}) = \Gamma(X, \mathcal{E})^{\cdot} \equiv \mathcal{E}(X)^{\cdot} = E^{\cdot} \tag{7.30}$$

(isomorphisms of the (abelian) groups involved). Thus, in view of (7.28), (7.30) and (7.29), one finally obtains;

$$H^1(X, \mathbf{Z}) = E^{\cdot}/e(E), \tag{7.31}$$

as desired (cf. (7.20)). ∎

Now, as already mentioned, we continue presently below by the next Section our discussion on some further *applications of the exactness of (7.19).*

8. Line sheaves over the spectrum of a geometric topological algebra

We consider below certain further *implications of the short exact exponential sheaf sequence in (7.19).* Indeed, what we are actually going to do is to derive two particular *morphisms of short exact sequences* (one of which will be, of course, that one in (7.19)), which further will be of use right away in the ensuing discussion.

Thus, suppose that we are given the *framework of either one of* the above two *Theorems 7.1 or 7.2,* while we also adopt the *notation of (7.25).* So assume now, as usual (see Chapt. I; Subsection 4.(b)), that

$$\mathcal{C}_X \tag{8.1}$$

stands for the *sheaf* (of germs) *of* **C**-*valued continuous functions* on $X \equiv \mathfrak{M}(E)$, where the latter space is, as already said (cf. (7.25)), the *spectrum* of the given *topological algebra E, which satisfies the hypotheses of Theorem 7.1 or those of Theorem 7.2.* Thus, one first obtains the following *commutative diagram,* or even a *morphism of short exact sequences* (cf. also e.g. Chapt. III; (2.17), concerning the applied terminology);

$$\begin{array}{ccccccccc} 0 & \longrightarrow & \mathbf{Z} & \longrightarrow & \mathcal{E} & \xrightarrow{e} & \mathcal{E}^{\cdot} & \longrightarrow & 1 \\ & & \| & & \downarrow & & \downarrow & & \\ 0 & \longrightarrow & \mathbf{Z} & \longrightarrow & \mathcal{C}_X & \xrightarrow{e} & \mathcal{C}^{\cdot}_X & \longrightarrow & 1. \end{array} \tag{8.2}$$

Indeed, the assertion follows straightforwardly from the very definitions of the sheaves $\mathcal{E}$ (Definition 2.1) and $\mathcal{C}_X$ (cf. Chapt. I; (4.25.1). Yet, see loc. cit., (10.9) and (11.12.1), as well as, (8.4) and (8.5), along with (2.4) and (2.7.1) in the preceding).

On the other hand, we already know (cf. (4.19)) that the Gel'fand sheaf $\mathcal{E}$ of a topological algebra E, as above, is *fine* over the *paracompact* (or even compact) *Hausdorff space* $X \equiv \mathfrak{M}(E)$ (see the above Scholium 7.1). Furthermore, based on *Urysohn's Lemma* (see e.g. J. Dugundji [1: p.146, Theorem 4.1]) and on its use in obtaining (continuous) *partitions of unity* (loc. cit., p.173; (2), or even F. Hirzebruch [1: p.30, Theorem 2.8.4]), one also concludes that;

(8.3) *the sheaf* $\mathcal{C}_X$ *of germs of continuous* **C**-*valued functions on a paracompact (Hausdorff) space* X *is fine.*

See, for instance, F. Hirzebruch [1: p.35, Theorem 2.11.1]. ∎ However, as we know (cf. Chapt. III; (8.24)), *fine sheaves over paracompact spaces are* (Γ_X-) *acyclic* (viz. they yield *trivial (Čech) cohomology*). Thus, this fact applied to the sheaves $\mathcal{E}$ and $\mathcal{C}_X$, as above, in conjunction with the *long exact cohomology sequences,* which are associated with each one of the two short exact (sheaf) sequences in (8.2), respectively, (see, for instance, (7.24) above, concerning the sequence in question for $\mathcal{E}$), yield now the following *commutative diagram* (of exact sequences of abelian groups);

$$
(8.4)\qquad
\begin{array}{ccccccc}
 & & & & & & 0 \\
 & & & & & \nearrow & \\
0 & \longrightarrow & H^1(X,\mathcal{E}^{\cdot}) & \xrightarrow{\ \delta\ } & H^2(X,\mathbf{Z}) & \longrightarrow & 0 \\
 & & \big\downarrow & \nearrow_{\delta} & & & \\
 & & H^1(X,\mathcal{C}^{\cdot}_X) & & & & \\
 & \nearrow & & & & & \\
0 & & & & & &
\end{array}
$$

In other words, one has the next *isomorphisms of* the *abelian groups* concerned;

$$
(8.5)\qquad H^1(X,\mathcal{C}^{\cdot}_X) = H^2(X,\mathbf{Z}) = H^1(X,\mathcal{E}^{\cdot}).
$$

Before we proceed to the *geometric* interpretation of (8.5), we sum up the preceding discussion by stating that;

(8.6) given a *topological algebra* E *of* the *type appeared in* either one of the *Theorems 7.1 or 7.2,* the respective *Gel'fand sheaf* $\mathcal{E}$ *of* E (cf. (2.8)) *over the spectrum* $\mathfrak{M}(E) \equiv X$ *of the* given *algebra* yields the corresponding *sheaf of units* $\mathcal{E}^{\cdot}$ (see (6.15)) *such that* (8.5) (or, *equivalently,* (8.4)) *holds true.*

We come now straight away, as already said, to *explain (8.5) in geometric terms* (see thus Theorem 8.1 below):

Thus, first of all, the (abelian) group on the left-hand side of (8.5) classifies (cohomologically) *continuous* $\mathbf{C}$*-line bundles over* $X \equiv \mathfrak{M}(E)$*; equivalently,* since a (continuous) $\mathbf{C}$*-line bundle is (uniquely) determined by the* respective *sheaf (of germs) of continuous sections* (see also, for instance, Chapt. IV; 2.11), with $\mathbf{A} = \mathbf{C}$, and Chapt. VI; (4.11), taking $\mathcal{C}_X$ in place of $\mathcal{C}^{\infty}_X$), the latter being a $\mathcal{C}_X$*-line sheaf,* in our terminology (loc. cit. and Chapt. II; Definition 4.3), *the* same *group*

$$
(8.7)\qquad H^1(X,\mathcal{C}^{\cdot}_X)
$$

classifies $\mathcal{C}_X$*-line sheaves on* X (cf. also Chapt. V; Theorem 2.1 and (2.64), (2.65)). Of course, this is a classical result; see e.g. F. Hirzebruch [1: p.41,

Theorem 3.2.1], where one sets $G \equiv \mathbf{C}^{\cdot} := \mathbf{C} - \{0\}$. Namely, one has

$$H^1(X, \mathbf{C}^{\cdot}) = H^1(X, \mathcal{C}_X^{\cdot}), \tag{8.8}$$

within an *isomorphism of* the *abelian groups* concerned; see also (8.11) below, for $n = 1$ and $\mathbf{A} = \mathbf{C}$, along with (8.13), as well as, Chapt. V; (4.15).

Now, the above situation has been amply extended in A. Mallios [1] to *continuous* $\mathbf{A}$*-vector bundles over* X; viz. to locally trivial fiber bundles, in the standard sense, over an *arbitrary topological space* X, whose fibers are *finitely-generated projective* $\mathbf{A}$*-modules,* with $\mathbf{A}$ a *unital* (commutative) *locally* m*-convex* (topological) *algebra*. So suppose that

$$\mathcal{A} := \mathcal{C}_X^{\mathbf{A}}, \tag{8.9}$$

that is, consider the *sheaf of germs of continuous* $\mathbf{A}$*-valued maps on* X (see Chapter IV, Section 2), indeed, a $\mathbf{C}$*-algebra sheaf on* X (cf. Chapt. II; Definition 1.4), so that

$$\mathcal{GL}(n, \mathcal{A}) \tag{8.10}$$

is the associated with (8.9) *sheaf of germs of continuous* $\mathrm{GL}(n, \mathbf{A})$*-valued maps* on X, thus a *sheaf of groups* on X (*non-abelian,* unless $n = 1$, with $\mathbf{A}$ *commutative*); cf. Chapt. IV; (2.11). (In this context, see also A. Mallios [1: p. 460ff] for a systematic account on $\mathbf{A}$*-vector bundles,* as well as, A. Mallios [5: p. 419, Section 5], concerning the possibility of applying *more general topological algebras* $\mathbf{A}$, than the ones considered above). Thus, by looking now at

$$H^1(X, \mathrm{GL}(n, \mathbf{A})), \tag{8.11}$$

viz. at the *set of isomorphism classes of (continuous)* $\mathbf{A}$*-vector bundles over* X*, of rank* $n \in \mathbf{N}$, one obtains;

$$H^1(X, \mathrm{GL}(n, \mathbf{A})) = H^1(X, \mathcal{GL}(n, \mathcal{A})), \tag{8.12}$$

within a *(set-theoretic) bijection,* where $\mathcal{A}$ is given by (8.9) (see loc. cit. p.406, Theorem 1.1, as well as, concerning the second member of (8.12), Chapt. V; (1.11), (2.5) and Theorem 2.1). (*Warning*! In (8.11) $\mathrm{GL}(n, \mathbf{A})$ is considered as a *topological group,* in the relative topology from $M_n(\mathbf{A})$ (see Chapt. IV; Scholium 2.1), thus *not* necessarily *in the discrete topology;* however, see Chapt. V; Section 5, in particular, Scholium 5.1, concerning *flat* $\mathbf{C}$*-vector bundles,* or even A. Mallios [5; §2, p. 410ff] for *flat* $\mathbf{A}$*-vector bundles,* in general, $\mathrm{GL}(n, \mathbf{A})$ carrying now the discrete topologogy.

Thus, setting now in (8.12) $n = 1$, one obtains the following *isomorphism of* (abelian, when $\mathbf{A}$ commutative) *groups;*

$$H^1(X, \mathbf{A}^{\cdot}) = H^1(X, \mathcal{A}^{\cdot}), \tag{8.13}$$

such that one has, in view of (8.9) (cf. also the above *warning*),

$$\mathcal{A}^{\cdot} \equiv \left(\mathcal{C}_X^{\mathbf{A}}\right)^{\cdot} = \mathcal{C}_X^{\mathbf{A}^{\cdot}} \tag{8.14}$$

(in fact, a *group sheaf isomorphism*). Indeed, (8.14) follows from the fact that;

$$\begin{aligned} \mathcal{A}^{\cdot}(U) = \mathcal{A}(U)^{\cdot} \equiv \left(\mathcal{C}_X^{\mathbf{A}}\right)(U)^{\cdot} &= \mathcal{C}(U, \mathbf{A})^{\cdot} \\ = \mathcal{C}(U, \mathbf{A}^{\cdot}) &= \left(\mathcal{C}_X^{\mathbf{A}^{\cdot}}\right)(U), \end{aligned} \tag{8.15}$$

for every open set $U \subseteq X$ (see also Chapt. IV; Lemma 1.1 and Scholium 2.1, along with Chapt. I; (11.10) and (9.17)). ∎ However, we already know (Chapt. V; (2.60)) that *the second member of (8.13) classifies* $\mathcal{A}$*-line sheaves* on X (see also loc. cit., Theorem 4.1).

Now, after the above short deviation, pertaining to the *more general situation, supplied by (8.9)* (instead, namely, just of (8.1)), let us *look again at (8.5)*. Thus, since $X \equiv \mathfrak{M}(E)$, according to our hypothesis for E, is *paracompact Hausdorff* (cf. Scholium 7.1 above), one still obtains;

$$H^1(X, \mathcal{E}^{\cdot}) = H^1(X, \mathbf{E}^{\cdot}), \tag{8.16}$$

within an *isomorphism of* the *abelian groups* involved. Here $\mathbf{E}^{\cdot}$ stands for the *unit presheaf* (or even *presheaf of units*) of the corresponding *Gel'fand* (**C**-*algebra) presheaf* **E**, that is associated with the given topological algebra E, as above (see (2.7.1)); so one sets

$$U \longmapsto \mathbf{E}^{\cdot}(U) := (E^{\wedge}|_U)^{\cdot}, \tag{8.17}$$

for every open $U \subseteq X$, which thus, along with the obvious *restriction morphisms,* defines the *presheaf* $\mathbf{E}^{\cdot}$ at issue. Concerning (8.16), see also Chapt. III; (6.35), in particular, Theorem 6.2, in conjunction with (8.11) there. Therefore, based now on (8.5) and (8.16), one infers that;

(8.18) *any (continuous)* **C***-line bundle on* X, viz. *every element* $\xi \in H^1(X, \mathcal{C}^{\boldsymbol{\cdot}}_X)$ *corresponds (bijectively) to a 1-cocycle from* $\mathbf{E}^{\boldsymbol{\cdot}}$. That is, *one has;*

(8.18.1) $$\xi \longleftrightarrow (g_{\alpha\beta}) \in Z^1(\mathcal{U}, \mathbf{E}^{\boldsymbol{\cdot}}).$$

In the last relation above (see also e.g. Chapt. III; (6.7)), one considers a *local frame of* ξ *over* X, say, $\mathcal{U} = (U_\alpha)_{\alpha\in I}$ (cf. Chapt. II; Definition 4.2, in conjunction with (8.8) above), such that, *for any* U_α, U_β *in* $\mathcal{U}$, *with* $U_{\alpha\beta} \equiv U_\alpha \cap U_\beta \neq \emptyset$, one has

(8.19) $$g_{\alpha\beta} \in \mathbf{E}^{\boldsymbol{\cdot}}(U_\alpha \cap U_\beta) = (E^{\wedge}|_{U_{\alpha\beta}})^{\boldsymbol{\cdot}} \subseteq \mathcal{C}(U_{\alpha\beta})^{\boldsymbol{\cdot}} = \mathcal{C}^{\boldsymbol{\cdot}}_X(U_{\alpha\beta})$$

(see also (8.17) and (2.4), as well as, e.g. (6.16), concerning the last relation in (8.19)).

In other words, *any constituent* $g_{\alpha\beta}$ *of the 1-cocycle* $(g_{\alpha\beta})$, as in (8.18.1), *is an* (invertible) *element of the Gel'fand transform algebra* $E^{\wedge}$ of E, *restricted to* $U_{\alpha\beta}$, hence, according to our hypothesis for E (we assumed E *semisimple*), *of the algebra* E *itself.* Furthermore, based on the last issues of the same relation (8.19), as above, one infers that, *every such element of* E may be construed as an *invertible* (hence, *nowhere vanishing*) *continuous* **C***-valued map on* $U_{\alpha\beta} \subseteq X$, something that we knew, for that matter, due to the hypothesis for ξ and to (8.8); thus, our new information comes actually from the first items in (8.19).

The preceding supply now the proof of the next result. The corresponding statement, within the Banach algebra theory, has been given by S.A. Selesnick [2: p.60, Corollary 2.3]. So one has the following.

Theorem 8.1. *Suppose that* E *is either a self-adjoint soft, or a self-adjoint Fréchet-Šilov, locally m-convex algebra* (see e.g. Theorems 7.1, 7.2 above), *whose spectrum is* $\mathfrak{M}(E)$. *Then, every continuous* **C***-line bundle over a topological space* X*, homotopic to* $\mathfrak{M}(E)$*, can be expressed locally by elements of* E, viz. *the line bundle in question can be represented by a 1-cocycle, whose elements belong locally to* E. ∎

Warning! Concerning the previously applied terminology, the phrasing *"belong locally to* E*"* means, in fact, here a relation, like (8.19), taken also into account that $E^{\wedge} = E$ (within a **C***-algebra isomorphism,* provided by the Gel'fand map of E (cf. (2.21)), due to the *semisimplicity* of E). This terminology should thus not be mixed up with the analogous one, already employed in (2.35), (2.26), or even in (2.42).

On the other hand, we can give to *the previous Theorem 8.1* a more general form, valid *for any (continuous)* **C**-*vector bundle,* not necessarily of rank 1, as was the case before, by restricting ourselves, however, to *soft* (locally m-convex) *algebras,* which thus, as follows from their definition (cf. Scholium 7.1), have their *spectra compact (Hausdorff)* spaces. Our argument has been motivated by some similar considerations in the realm rather of *(Real) Analytic* and/or of *Algebraic Geometry* (cf., for instance, K. Lønsted [1], or even R. Bkouche [1]; the same result has been given already in A. Mallios [1: p.488, Theorem 6.1], within the context of *Pták* (locally m-convex) *algebras,* where we still refer for relevant details. Thus, we now have the following.

Theorem 8.2. *Let E be a soft locally m-convex* (topological) *algebra* (see Definition 3.2), *having spectrum $\mathfrak{M}(E) \equiv X$ and Gel'fand sheaf $\mathcal{E}$. Then, for every positive integer $n \in \mathbf{N}$, the following sets are (canonically) equivalent:*

(i) The set of isomorphism classes of vector sheaves, with respect to $\mathcal{E}$, of rank $n \in \mathbf{N}$, over X, denoted by $\mathbf{\Phi}^n_{\mathcal{E}}(X)$ (cf. Chapt. II; (4.15)).

(ii) The set of isomorphism classes of n-generated projective E-modules, denoted by $\mathcal{P}_n(E)$.

(iii) The set of isomorphism classes of n-generated projective $\mathcal{C}(X)$-modules, denoted by $\mathcal{P}_n(\mathcal{C}(X))$).

(iv) The set of isomorphism classes of continuous (complex) n-plane bundles over X.

In other words, one has the following (canonical) set-theoretic bijections:

$$\begin{aligned} H^1(X, \mathrm{GL}(n, E)) &= H^1(X, \mathcal{GL}(n, \mathcal{E})) = \mathbf{\Phi}^n_{\mathcal{E}}(X) = \mathcal{P}_n(E) = \mathcal{P}_n(\mathcal{C}(X)) \\ &= H^1(X, \mathcal{GL}(n, \mathcal{C}_X)) = H^1(X, \mathrm{GL}(n, \mathbf{C})). \end{aligned} \tag{8.20}$$

Proof. First, it is standard that one has

$$H^1(X, \mathcal{GL}(n, \mathcal{C}_X)) = \mathcal{P}_n(\mathcal{C}(X)), \tag{8.21}$$

for any $n \in \mathbf{N}$, and *compact (Hausdorff) space X,* within a (set-theoretic) *bijection* (*Serre-Swan Theorem;* see e.g. M. Karoubi [1: p.32, Theorem 6.18]). Thus, (8.21) identifies, in fact, the last three sets in the statement, the fifth one being identified with the 1st cohomology set, as in (8.21) (see, for instance, Chapt. V; Theorem 2.1, along with Chapt. VI; (4.11), applied to $\mathcal{C}_X$, in place of $\mathcal{C}_X^\infty$; cf. also (8.12) above, for $\mathbf{A} = \mathbf{C}$).

On the other hand, according to our hypothesis for E (see Definition 3.2), the algebra in question is *unital commutative complete with an equicontinuous spectrum* $\mathfrak{M}(E) \equiv X$; therefore (S. Warner [1: p.7, Theorem 6]), a *Q-algebra,* as well (viz. *the group of units of E is* an *open* subset of the latter algebra; see also e.g. [TA: p.187, Lemma 1.3 and p.184, Theorem 1.2]). Now, the following result has been obtained in A. Mallios [1: p.485, (5.6)] (an extension to the sort of topological algebras considered, as before, of an earlier result for *commutative unital Banach algebras,* due to O. Forster [1: p. 10, Satz 6]);

(8.22) *continuous (complex) n-plane bundles over a compact (Hausdorff) space X are "algebraic"* (cf. (8.23) below), *relative to any commutative unital complete locally m-convex Q-algebra E* (viz. to a *Waelbroeck algebra* [TA]), *whose spectrum $\mathfrak{M}(E)$ is homotopic to X.*

Thus, one obtains the following *bijections;*

$$\mathcal{P}_n(E) = \mathcal{P}_n(\mathcal{C}(\mathfrak{M}(E)))) = \mathcal{P}_n(\mathcal{C}(X)), \tag{8.23}$$

for every $n \in \mathbf{N}$. (See also A. Mallios [2: p.297, (1.1), and p.298, Theorem 1.1, as well as, p.305, Appendix]). So our assertion follows now from (8.21) and (8.23), in conjunction with the fact that, for any topological algebra E, as in the statement, one further obtains;

$$\mathcal{P}_n(E) = H^1(X, \mathcal{GL}(n, \mathcal{E})), \tag{8.24}$$

within a *bijection* (see A. Mallios [1: p.488, Theorem 6.1, in particular, (6.2)]. In this context, we still remark that it is actually enough to consider here just *soft algebras,* as above, in place of *Pták-Šilov* ones as was the case therein). Finally, we also notice that the second equality (in fact, set-theoretic bijection) in (8.20) is already a consequence of Chapt. V; Theorem 2.1 (being valid, in effect, for any **C**-*algebraized space* whatsoever), and this also terminates the proof. (See also (8.25) below, concerning the 1st member of (8.20)). ∎

Now, taking in the previous Theorem 8.2, $n = 1$, *we have* the preceding *Theorem 8.1, more generally, for any soft locally m-convex algebra.* (We need *no involution* here, in particular, self-adjointness of the algebra involved; however, the *compact spectrum* of the topological algebra considered is a crucial issue here. So the previous argument does not apply to the type of topological algebras employed by Theorem 7.2 in the foregoing (see also Theorem 8.1 above). Thus, applying now a similar argument to that one in (8.16), one obtains (see Chapt. III; Theorem 6.2);

$$\begin{aligned}\mathcal{P}_n(E) = \mathcal{P}_n(\mathcal{C}(X)) = H^1(X, \mathcal{GL}(n, \mathcal{C}(X)) &= H^1(X, \mathcal{GL}(n, \mathcal{E})) \\ &= H^1(X, \mathrm{GL}(n, \mathbf{E})).\end{aligned} \tag{8.25}$$

Here $\mathrm{GL}(n, \mathbf{E})$ stands for the *presheaf of groups* on X, defined by (cf. (2.3));

$$U \longmapsto \mathrm{GL}(n, \mathbf{E})(U) := \mathrm{GL}(n, E^{\wedge}|_U), \tag{8.26}$$

for any *open* $U \subseteq X \equiv \mathfrak{M}(E)$ (cf. also Theorem 8.2), along with the obvious *restriction morphisms.* Thus, taking first (2.19) into account (cf. also e.g. Chapt. I; (11.14)), it is easy to see that;

(8.27) *the above presheaf* $\mathrm{GL}(n, \mathbf{E})$, as given by (8.26), *is a mono-presheaf* (of groups) on X.

Furthermore, it is not hard to see, based on the very definitions, that;

$$\mathbf{S}(\mathrm{GL}(n, \mathbf{E})) = \mathbf{S}(\Gamma(\mathcal{GL}(n, \mathcal{E}))) = \mathcal{GL}(n, \mathcal{E}) \tag{8.28}$$

(see also Chapt. I; Theorem 9.1). In other words, one concludes that;

(8.29) *the sheafification of* $\mathrm{GL}(n, \mathbf{E})$ *is,* in effect, $\mathcal{GL}(n, \mathcal{E})$.

See thus Chapt. I; (7.23), (7.6) and (7.10), along with Chapt. IV; (1.28). Indeed, one proves that;

$$\mathrm{GL}(n, \mathbf{E})_f = \mathcal{GL}(n, \mathcal{E})_f, \tag{8.30}$$

for every $f \in \mathfrak{M}(E) \equiv X$ (loc. cit., along with (8.27) and (2.9.3) above). Yet, see Chapt. I; Theorem 3.1 and (2.14) in the foregoing, which thus justifies (8.28). ∎ So one finally concludes that;

(8.31) given a *soft locally m-convex algebra* E (Definition 3.2), *every continuous (complex) n-plane bundle* ξ *on* (a topological space *homotopic to*) $\mathfrak{M}(E) \equiv X$, the *spectrum of* E, can be *(uniquely) represented by a 1-cocycle, whose elements belong (locally) to*

(8.31.1) $$\mathrm{GL}(n, \mathbf{E})$$

(cf. (8.26)). That is, one has (cf. e.g. (8.18.1) and (8.19), concerning the notation applied);

(8.31.2) $$\xi \longleftrightarrow (g_{\alpha\beta}) \in Z^1(\mathcal{U}, \mathrm{GL}(n, \mathbf{E})).$$

In particular, by looking at (8.31.2), one has;

(8.32) $$\begin{aligned} g_{\alpha\beta} \in \mathrm{GL}(n, \mathbf{E})(U_{\alpha\beta}) &= \mathrm{GL}(n, E\hat{}|_{U_{\alpha\beta}}) \cong \mathrm{GL}(n, E|_{U_{\alpha\beta}}) \\ &\subseteq \mathrm{GL}(n, \mathcal{C}(U_{\alpha\beta})) = \mathcal{GL}(n, \mathcal{C}_X)(U_{\alpha\beta}), \end{aligned}$$

for any U_α, U_β *in* $\mathcal{U}$ (a given *local frame of* ξ), *with* $U_\alpha \cap U_\beta \equiv U_{\alpha\beta} \neq \emptyset$. Thus, one infers that;

(8.33) *the 1-cocycle* $(g_{\alpha\beta})$, as in (8.32.1), *representing* ξ, *consists (locally) of invertible* $n \times n$ *matrices, with entries from the* given (topological) *algebra* E ($\cong E\hat{}$, cf. (8.31)).

The previous conclusion is, in effect, an *equivalent way of expressing* the fact *that;*

(8.34) *continuous (complex) n-plane bundles over a compact (Hausdorff) space X are "algebraic"* (see (8.22), (8.23)).

Scholium 8.1.– In the case of *commutative unital Banach algebras* O. Forster (loc. cit., p.10) refers to the analogous result (cf. (8.22) above) as the *"Banach algebra analogue of Grauert's theorem"* (the latter classical result pertaining to a similar situation in the case of *holomorphic and continuous* **C**-*vector bundles on a Stein manifold;* cf. H. Grauert [1: p. , Satz]). Thus, following O. Forster (loc. cit.), one has, according to (8.22) (see also (8.23)),

(8.35) *a Waelbroeck (locally m-convex) algebra analogue of Grauert's theorem.*

Indeed, (8.35) might also be construed as the analogue, within the present context, of the classical *"Oka's principle"* (see, for instance, [TA: p.300]). Yet, see also J. Wagner [1: p.166, Proposition], as well as, M.E. Novodvorskii [1: p.490, Theorem 2] for another, respectively, independent, point of view, concerning the above mentioned Forster's result.

On the other hand, as a byproduct of the previous discussion and, in particular, of (8.23), for $n = 1$, *one obtains;*

$$\mathcal{P}_1(E) = \mathcal{P}_1(\mathcal{C}(X)), \tag{8.36}$$

with $X \equiv \mathfrak{M}(E)$ the *spectrum* of the given *topological algebra* E*, as in* (8.22). Thus, employing the notion of the *Picard group of* E, (8.36) is written, by definition, equivalently, in the form;

$$\operatorname{Pic}(E) = \operatorname{Pic}(\mathcal{C}(X)) \tag{8.37}$$

(see also e.g. Chapt. V; Definition 4.1, along with (4.15), concerning the previously applied terminology. Yet, cf., for instance, N. Bourbaki [4: Chapt. II, p. 115, Proposition 7, as well as, Remark (2)], pertaining to the concept of the *Picard group of a* commutative unital ring, in general). Therefore, taking now (8.5) and (8.21) (for $n = 1$) into account, one obtains

$$\operatorname{Pic}(E) = \operatorname{Pic}(\mathcal{C}(X)) = H^1(X, \mathcal{C}^{\bullet}_X) = H^2(X, \mathbf{Z}), \tag{8.38}$$

within *isomorphisms of* the *abelian groups* involved. Accordingly, one now arrives at the following result, an extension to our case of an earlier conclusion of O. Forster (loc. cit., p.15, Satz 8), for *unital commutative Banach algebras.* Thus, we have the next.

Theorem 8.3. *Let* E *be a Waelbroeck algebra* (viz. a *commutative unital complete locally m-convex Q-algebra*), *having spectrum* $\mathfrak{M}(E)$. *Then, one has;*

$$\operatorname{Pic}(E) = H^2(X, \mathbf{Z}), \tag{8.39}$$

up to an isomorphism of abelian groups, where X *stands for any paracompact (Hausdorff) space, homotopic to* $\mathfrak{M}(E)$. ∎

8.(a). Morphisms of short exact exponential sheaf sequences

(contn'd).– We continue below, as already promised, with our second morphism of short exact sequences, that can be derived from (7.19), having the latter sequence as one of its constituents (see e.g. (8.) in the sequel): Thus, suppose again, as in Theorem 7.1 above, that E is a *soft self-adjoint locally* m*-convex* **-algebra,* whose *Gel'fand sheaf* is $\mathcal{E}$, and let $\mathcal{E}^{\cdot}$ be the corresponding *sheaf of units of* $\mathcal{E}$ (cf. (2.8) and (6.15) in the foregoing). Moreover, let $\mathfrak{M}(E) \equiv X$ be the *equicontinuous spectrum* of E (cf. Definition 3.2), thus a *compact (Hausdorff) space* (see Scholium 7.1).

Furthermore, consider now the *exponential sheaf morphism*

$$(8.40) \qquad e : \mathcal{E} \longrightarrow \mathcal{E}^{\cdot}$$

(see Lemma 7.1), where thus e *is the sheafification of the presheaf maps*

$$(8.41) \qquad e_U : E^{\hat{}}|_U \longrightarrow (E^{\hat{}}|_U)^{\cdot}$$

with U *open* in X (see (7.10)), as they are defined by (7.4) (cf. also (7.8); in this context, see Chapt. I; Section 8, in particular, (8.4) and (8.10)). Thus, *for any (local) section* $s \in \Gamma(U, \mathcal{E}) \equiv \mathcal{E}(U)$, and U *open* in X, one sets;

$$(8.42) \qquad e(s) := \exp 2\pi i s \in \mathcal{E}^{\cdot}(U) = \mathcal{E}(U)^{\cdot}$$

(see (6.16)), *such that one has*

$$(8.43) \qquad e(s)(f) := \exp 2\pi i s(f) \in \mathcal{E}^{\cdot}_f = \mathbf{E}^{\cdot}_f,$$

for every $f \in U \subseteq X \equiv \mathfrak{M}(E)$; of course, here $\mathbf{E}^{\cdot}_f$ stands for the *stalk at* $f \in U \subseteq \mathfrak{M}(E)$ of the *unit presheaf* over $\mathfrak{M}(E)$ *of* $\mathbf{E}$, the *Gel'fand presheaf of the* given (topological) *algebra* E (see (2.7.1) and (8.17), along with Chapt. I; (9.16), concerning the last relation in (8.43)). Thus, by further considering the *"formal derivative"* of (8.43), one obtains (cf. also (7.1.1));

$$(8.44) \qquad (\exp 2\pi i s(f))' = 2\pi i \cdot \exp 2\pi i s(f) \in \mathcal{E}^{\cdot}_f, \quad f \in U.$$

Therefore, one concludes that;

$$(8.45) \qquad (\exp 2\pi i s)' = 2\pi i \cdot \exp 2\pi i s \in \mathcal{E}^{\cdot}(U),$$

for every $s \in \mathcal{E}(U)$, as in (8.42), the latter relation (8.45) *yielding* thus *the "formal derivative" of (8.43).* Accordingly (see also A. Mallios [8: p. 157, Lemma 5.1]), by applying now to (8.42) the corresponding *Kähler derivation of* $\mathcal{E}$ (cf. (6.5)), one obtains

(8.46) $$\partial(\exp 2\pi i s) = (\exp 2\pi i s)' \cdot \partial s = 2\pi i \cdot (\exp 2\pi i s) \cdot \partial s,$$

for any $s \in \mathcal{E}(U)$, as before. Hence, by considering, in turn, the *(kählerian) logarithmic derivation of* $\mathcal{E}$ (see (6.18)), one concludes, by virtue of (8.46) and (8.42), that;

(8.47) $$\tilde{\partial} \circ e = 2\pi i \cdot \partial.$$

Now, the previous relation supplies, in effect, the desired *commutative diagram,* or even *morphism of short exact (sheaf) sequences,* as follows;

(8.48) $$\begin{array}{ccccccccc} 0 & \longrightarrow & \mathbf{Z} & \longrightarrow & \mathcal{E} & \xrightarrow{\;e\;} & \mathcal{E}^{\cdot} & \longrightarrow & 1 \\ & & \Big\downarrow{\scriptstyle \varepsilon} & & \Big\| & & \Big\downarrow{\scriptstyle \tilde{\partial}} & & \\ 0 & \longrightarrow & \ker \partial & \longrightarrow & \mathcal{E} & \xrightarrow{\;2\pi i \cdot \partial\;} & \partial\mathcal{E} & \longrightarrow & 0 \end{array}$$

(see also (6.7), (6.8), as well as, Chapt. VI; Lemma 1.1 and Chapt. V; (6.13)). In conclusion, one thus infers that;

(8.49) *(8.48) holds true* for any *soft self-adjoint locally m-convex *-algebra,* or even for every *self-adjoint Fréchet-Šilov* (locally *m*-convex *-)*algebra.*

Thus, the types of topological algebras appeared in (8.49) provide, through their Gel'fand sheaves and the corresponding Kähler differential triads (see (6.8)), concrete *examples* of *exponential algebraized spaces* (see Chapt. V: Definition 6.1). It is also an immediate consequence of the very definitions that the same too are, in effect, *examples* of *strictly exponential algebraized spaces* (loc. cit., Definition 6.4), in the sense that one has the following *morphism of short exact (sheaf) sequences;*

(8.50) $$\begin{array}{ccccccccc} 0 & \longrightarrow & \mathbf{Z} & \longrightarrow & \mathbf{C} & \xrightarrow{\;e \equiv \exp 2\pi i\;} & \mathbf{C}^{\cdot} & \longrightarrow & 1 \\ & & \Big\| & & \Big\downarrow{\scriptstyle \varepsilon} & & \Big\downarrow & & \\ 0 & \longrightarrow & \mathbf{Z} & \longrightarrow & \mathcal{E} & \xrightarrow{\;e\;} & \mathcal{E}^{\cdot} & \longrightarrow & 1. \end{array}$$

Therefore, all in all, taking (8.48) and (8.50) together, one finally obtains the following *commutative diagram;*

(8.51)
$$\begin{array}{ccccccccc}
0 & \longrightarrow & \mathbf{Z} & \longrightarrow & \mathbf{C} & \xrightarrow{e\equiv\exp 2\pi i} & \mathbf{C}^{\cdot} & \longrightarrow & 1 \\
 & & \Vert & & \downarrow & & \downarrow{\scriptstyle \varepsilon} & & \\
0 & \longrightarrow & \mathbf{Z} & \longrightarrow & \mathcal{E} & \xrightarrow{e} & \mathcal{E}^{\cdot} & \longrightarrow & 1 \\
 & & \downarrow{\scriptstyle \varepsilon} & & \Vert & & \downarrow{\scriptstyle \tilde{\partial}} & & \\
0 & \longrightarrow & \ker\partial & \longrightarrow & \mathcal{E} & \xrightarrow{2\pi i\cdot\partial} & \partial\mathcal{E} & \longrightarrow & 0.
\end{array}$$

Furthermore, supplementing our previous information in (6.20), and based on (8.48), in conjunction with Chapt. VII; Lemma 7.1, one still obtains

(8.52)
$$\tilde{\partial}(\mathcal{E}^{\cdot}) = \partial(\mathcal{E}) = 2\pi i \cdot \partial(\mathcal{E}).$$

On the other hand, we already know (see (4.18), or even (4.19), in the preceding) that *the Gel'fand sheaf* $\mathcal{E}$ of a topological algebra E, as in (8.49) (in fact, something less is needed, loc. cit.) *is fine on the paracompact* (if not *compact,* when E is soft) *Hausdorff space* $X \equiv \mathfrak{M}(E)$, so that (cf. Chapt. III; (8.24)) $\mathcal{E}$ *is acyclic;* therefore, one has

(8.53)
$$H^1(X, \mathcal{E}) = H^2(X, \mathcal{E}) = 0.$$

So we have here again, via the topological algebras considered in (8.49), another *concrete example* of a *strict sheaf exponential space* (see Chapt. VII; Definition 7.2). Thus, by looking further at the appropriate context of that chapter, one can conclude, for instance, that

(8.54) given a *topological algebra* E, *as in* (8.49), having *spectrum* $\mathfrak{M}(E) \equiv X$ and *Gel'fand sheaf* $\mathcal{E}$, *a line sheaf* $\mathcal{L}$ *on* X *is* ∂-*flat if, and only if, one has*

(8.54.1)
$$c_{\mathcal{E}}(\mathcal{L}) = 0$$

(*Selesnick - Chern class of* $\mathcal{L}$; loc. cit.).

See Chapt. VII; Corollary 7.1, along with Definition 7.1. In this context, we also recall (ibid., (7.5), as well as, (7.20) and (7.23)) that the above class of $\mathcal{L}$ is given by the relation;

$$c_{\mathcal{E}}(\mathcal{L}) = \frac{1}{2\pi i}\tilde{\partial}(g_{\alpha\beta}), \tag{8.55}$$

where $(g_{\alpha\beta}) \in Z^1(\mathcal{U}, \mathcal{E}^{\cdot})$ is a *coordinate 1-cocycle of* $\mathcal{L}$ (see Chapt. V; (1.8), for $n = 1$. We still recall that we are here applying the *Kähler differential triad* (cf. (6.8) above), that is (canonically) associated with the given topological algebra E, as before). Furthermore, according to our previous Theorem 8.1, one may consider *the above cocycle* entirely as *being expressed (locally) through elements of the* (topological) *algebra* E at issue (cf. (8.19), as well as, e.g. (8.32), for $n = 1$).

Concerning the same conclusion (8.54), as above, we also refer to Chapt. VII; Theorem 6.1, or even *Chapt. VIII;* Theorem 5.1 and, in particular, *Theorem 5.2* for further relevant *geometric consequences of* (8.54) (in fact, for *equivalent interpretations* of the same assertion). In this respect, one also remarks that;

(8.56) by considering the *Kähler differential triad,* defined by a *topological algebra* E*, as in* (8.49) (see (6.8)), one realizes that *a Bianchi datum* (cf. Chapt. VIII: Definition 7.1) *is* actually *obtained.*

See thus (6.13), by employing, at the same time, a corresponding *sheafification of Theorem 5.1* above, in the spirit of Section 6 of this Chapter. ∎

Thus, the aforementioned *Theorem 5.2 of Chapter VIII* acquires a meaning, within the present *concrete framework,* which is derived, according to the preceding, from the particular *topological algebra* E at issue, *as in* (8.49). On the other hand, an *extension,* under appropriate presuppositions, *of the situation in* (8.54) above *to vector sheaves,* in general (viz. *not* necessarily *of rank* 1, as was the case before), is presently examined in the next Section 9.

9. Frobenius integrability condition (contn'd)

As already promised, we consider in the sequel the possibility of *extending,* under appropriate presuppositions, *Theorem 5.2 of Chapter VIII to* general *vector sheaves,* a fact that in the classical case (of complex $\mathcal{C}^{\infty}$-vector bundles, for instance) is assured by employing the integrability condition of the heading of this Section (cf., for example, S. Kobayashi [1: p. 5]). In the particular case of *line sheaves,* as was the case so far (see e.g. the proof of the aforesaid theorem, in particular, (5.30) there, or even (8.54) above), the corresponding result was attained, due mainly to the *existence of the short exact exponential (sheaf) sequence,* as e.g. in (7.19), the latter permitting thus one to circumvent the said condition. However, in the general case of vector sheaves this seems to be possible only (so far!) under some *extra assumptions* on the part of the differential triad involved (see (9.3) below), this also including the consideration of the respective *structure sheaf,* as a suitable *topological algebra sheaf* on our *base space* X (see (9.2) in the sequel).

Thus, suppose that we have, as in the preceding chapters, a *differential triad* (cf. Chapt. VI; (3.1), or even Chapt. VIII; (1.1))

$$(\mathcal{A}, \partial, \Omega^1) \tag{9.1}$$

on a *topological space* X, such that;

(9.2) $\mathcal{A}$ is a *topological algebra sheaf* (cf. Theorem 1.1), in such a manner that *the local section algebras of* $\mathcal{A}$, $\mathcal{A}(U) \equiv \Gamma(U, \mathcal{A})$, for every *open* $U \subseteq X$, *are unital* commutative σ*-complete locally* m*-convex* (topological) $\mathbf{C}$*-algebras.*

On the other hand, suppose also that X is a *curvature space* (cf. Chapt. VIII: Definition 1.1), while we still assume that

$$0 \longrightarrow \partial\mathcal{A} \longrightarrow \Omega^1 \xrightarrow{d^1 \equiv d} d\Omega^1 \longrightarrow 0 \tag{9.3}$$

is an *exact sequence of* ($\mathbf{C}$-vector space) *sheaves.* Now, one obtains, *for every* $n \in \mathbf{N}$, a *matrix analogue of (9.3)* in the form of the following *exact sequence*

$$0 \longrightarrow \partial M_n(\mathcal{A}) \equiv M_n(\partial\mathcal{A}) \longrightarrow M_n(\Omega^1) \longrightarrow dM_n(\Omega^1) \equiv M_n(d\Omega^1) \longrightarrow 0. \tag{9.4}$$

Thus, assuming that (9.3) is exact, we shall prove that the same holds true for (9.4), for every $n \in \mathbf{N}$; so one actually infers that:

(9.5) *exactness of (9.3) is equivalent to that one of (9.4), for every* $n \in \mathbf{N}$.

Indeed, one can prove the assertion *fiber-wise,* our claim being then a consequence of the hypothesis for (9.3) and a standard argument, pertaining to the direct sum decomposition of the individual spaces *(stalks),* corresponding to (9.4), for any given $n \in \mathbf{N}$. (See also Chapt. I; Lemmas 12.2 and 12.3, as well as, e.g. N. Bourbaki [2: Chap. 2; p. 13, Proposition 7]).

On the other hand, due to our hypothesis for $\mathcal{A}$ (see (9.2), in conjunction with (7.1) in the preceding), one concludes that;

(9.6) *the exponential function,* in fact, *sheaf morphism* e *acts on* $\mathcal{A}$ (viz., in effect, *on the local (continuous) sections* of the latter sheaf).

That is, one sets;

$$e(s) := \exp 2\pi i s, \tag{9.7}$$

for any (local) section $s \in \mathcal{A}(U)$ and U *open* in X. Indeed, one can obtain (9.6) *for every matrix algebra sheaf* $M_n(\mathcal{A})$, $n \in \mathbf{N}$, *with* $\mathcal{A}$ *as in* (9.2), by first remarking that;

(9.8) $M_n(\mathcal{A})$ is a *topological algebra sheaf on* X *of the same sort as* $\mathcal{A}$ (cf. (9.2)), apart, of course, of the commutativity, unless $n = 1$.

The assertion follows from the relation;

$$M_n(\mathcal{A})(U) = M_n(\mathcal{A}(U)), \tag{9.9}$$

valid, *for any open* $U \subseteq X$, *with* $n \in \mathbf{N}$ (cf. Chapt. IV; (1.9)), in conjunction with the fact that *the second member of (9.9) is* still *a topological algebra of the same type as* $\mathcal{A}(U)$, when on the same algebra one considers the corresponding *cartesian product topology* (cf. also A. Mallios [1: p.458], as well as, the same author [3: comments at the beginning of p.465]) ∎ Therefore, one obtains the following *exact (sheaf) sequence;*

$$0 \longrightarrow \mathbf{Z} \longrightarrow M_n(\mathcal{A}) \xrightarrow{e} M_n(\mathcal{A})^{\cdot} \equiv \mathcal{GL}(n, \mathcal{A}) \tag{9.10}$$

(see also Chapt. IV: (1.26)). Furthermore, based on (8.47), one still concludes that;

$$\tilde{\partial} \circ e = 2\pi i \cdot \partial, \tag{9.11}$$

with the obvious extension (*"section-wise"*) to $M_n(\mathcal{A})$ and $M_n(\mathcal{A})^{\cdot}$ of the sheaf morphisms involved in (9.11). (In this concern, see also A. Mallios [8: p. 157, Lemma 5.1]).

So after the preceding preliminary discussion, we are now in the position to formulate the desired *general form of Theorem 5.2 of Chapter VIII.* That is, one has the following.

Theorem 9.1. *Suppose we are given a curvature space X with corresponding curvature datum*

$$(\mathcal{A}, \partial, \Omega^1, d, \Omega^2) \tag{9.12}$$

(cf. Chapt. VIII; Definition 1.1), *whose structure sheaf $\mathcal{A}$ satisfies (9.2), while we also assume the following short exact sequence of* **C***-vector space sheaves on X;*

$$0 \longrightarrow \partial\mathcal{A} \longrightarrow \Omega^1 \longrightarrow d\Omega^1 \longrightarrow 0. \tag{9.13}$$

Moreover, let $\mathcal{E}$ be a vector sheaf on X, with $\operatorname{rk}\mathcal{E} = n \in \mathbf{N}$. *Then, the following are equivalent:*

i) $\mathcal{E}$ is ∂-flat.

ii) $\mathcal{E}$ admits a flat $\mathcal{A}$-connection.

Proof. We have already seen in Chapt. VIII; Theorem 5.1 that *prop. i)* of the statement *implies that $\mathcal{E}$ has a flat $\mathcal{A}$-connection,* under the much more general condition that X is just a *curvature space* (thus, without any supplementary assumption for the structure sheaf $\mathcal{A}$ and the exactness of the sequence in (9.13)). So we prove now that *ii) implies i),* as well.

Namely, suppose that

$$\omega \equiv (\omega_U) \in C^0(\mathcal{U}, M_n(\Omega^1)), \tag{9.14}$$

is an *$\mathcal{A}$-connection 0-cochain of 1-forms* on X, defining the given *$\mathcal{A}$-connecti-*

on, say D, of $\mathcal{E}$ (cf. Chapt. VII; (2.39)). Thus, our objective now is to find a *gauge transformation*

$$h \in C^0(\mathcal{U}, \mathcal{GL}(n, \mathcal{A})) \tag{9.15}$$

in such a manner that one has (see also Chapt. VI; (1.28), (1.29));

$$\tilde{\partial}(h^{-1}) = \omega, \tag{9.16}$$

where ω is given by (9.14). (Again, what our supplementary assumption assures here is that (9.16) *has a solution, given that,* by hypothesis, $R = 0$ (cf. Chapt. VIII; (2.5)), the latter relation being a *necessary condition, in order (9.16) to have a solution;* see thus Chapt. VIII; (5.23), (5.24)). Now, by applying d to (9.16) (or even, as we also say, by *"differentiating"* (9.16)), one obtains (cf. Chapt. VIII; (1.12))

$$d\omega = 0, \tag{9.17}$$

so that, in view of the exactness of (9.13), *one has*

$$\omega \in \ker d = \operatorname{im} \partial. \tag{9.18}$$

Therefore, *one obtains (locally)*

$$\omega = \partial(g), \tag{9.19}$$

for some $g \in M_n(\mathcal{A})(V) = M_n(\mathcal{A}(V))$, *with* V *open* in X (see also, for example, (9.9)). In this context, we still note that (9.19) *can be obtained by applying (9.18) fiber-wise.* Thus, one finally gets an *open covering of* X, for the individual members of which one has an analogous relation to (9.19), for any given ω, as in (9.14); of course, one can choose too a common *open covering of* X, *corresponding to both* the relations (9.14) *and* (9.19).

Accordingly, in view now of (9.11) and (9.19), one obtains

$$\tilde{\partial}\left(e\left(\frac{1}{2\pi i}g\right)\right) = 2\pi i \cdot \partial\left(\frac{1}{2\pi i}g\right) = \partial(g) = \omega; \tag{9.20}$$

hence, by further setting

$$h := e\left(\frac{1}{2\pi i}g\right)^{-1}\Big|_V \in \mathcal{GL}(n, \mathcal{A})(V) = \mathrm{GL}(n, \mathcal{A}(V)) \tag{9.21}$$

(see (9.15), along with our previous comments on (9.19), as well as, Chapt. IV; (1.28)), one finally obtains

(9.22) $$\tilde{\partial}(h^{-1}) = \omega,$$

which was actually looked for (cf. (9.16)). On the other hand, the last relation is equivalent, in effect, with the following one;

(9.23) $$\mathrm{Ad}(h^{-1})\omega + \tilde{\partial}(h) = 0$$

(cf. also, for instance, Chapt. VII; (2.17)). Therefore, h (see (9.15)) *can* now *be construed as that gauge transformation, which sends* ω, as the latter is given by (9.14), *to the 0-cochain*

(9.24) $$\omega' \equiv (\omega'_V) \in C^0(\mathcal{U}, M_n(\Omega^1)),$$

such that one has;

(9.25) $$\omega'_V = 0, \quad \textit{for every} \quad V \in \mathcal{U}$$

(see also the comments following (9.19) above). Our previous argument is still justified by the corresponding *transformation law of potentials,* as the latter is given here by (9.23) (cf. also Chapt. VII; (2.24)). Accordingly, ω' as defined by (9.24) and (9.25), provides now a *complete parallelism for the* given *vector sheaf* $\mathcal{E}$ (cf. Chapt. VII; Definition 6.1 and (6.37)), which *equivalently,* means that $\mathcal{E}$ *is* ∂*-flat* (see Chapt. VIII; Theorem 5.1), that actually was our claim, and this also completes the proof of the theorem. ∎

Now, by the same token we have also proved, in effect, that;

(9.26) a *vector sheaf* $\mathcal{E}$ on a *topological space* X, *satisfying the conditions of Theorem 9.1, admits a flat* $\mathcal{A}$*-connection if, and only if,* $\mathcal{E}$ *has a trivial* $\mathcal{A}$*-connection.*

See thus Chapt. VIII; (6.4), or even (6.6), along with (9.22) above. (In this respect, cf. also (9.27) right in the sequel). ∎ Therefore, under the *supplementary topological assumptions for* $\mathcal{A}$ (cf. (9.2)), the preceding provide also the already promised *extension of Theorem 6.1 in Chapter VIII to vector sheaves,* in general. Thus, in sum, one concludes that;

(9.27) under the *hypothesis of Theorem 9.1,* one obtains, all in all, *five equivalent expressions of flatness for* a given $\mathcal{A}$*-connection of a vector sheaf.*

See the same Theorem 5.1 of Chapter VIII, along with (6.4) there; yet, see also Corollary 6.1 of that chapter for the analogous situation in the particular case of *line sheaves* (no extra hypothesis for $\mathcal{A}$ is needed here). ∎

Note.- By looking more closely at the proof of the previous Theorem 9.1, one realizes that;

(9.28) the conclusion of *Theorem 9.1 holds true, under the assumption of (9.11), in place of (9.2).*

(See also, for instance, Chapt. VIII; (6.21)). Indeed, it was in effect (9.11), which we derived from the above discussion, through our extra hypothesis on $\mathcal{A}$, as in (9.2). (In this respect, see thus, in particular, (9.20) above).

Scholium 9.1.- Our previous remark in (9.28) can still be applied to the case of a *topological algebra sheaf,* as defined by the *Gel'fand sheaf* of a given *topological algebra* (cf. (2.44)), the latter being, in particular, of the type appeared, for instance, in (8.49), or even, more generally, in Lemma 7.1 above: Thus, according to that lemma, one has then the *exponential sheaf morphism, as in* (7.2) (cf. also (7.3)), so that *we* can *further obtain* (8.47). (Notice here that, concerning the last relation, we do not actually need the exponential morphism e, as before, to be *surjective,* therefore, the full hypothesis of Theorem 7.1, that is, *self-adjointness* of the topological algebras involved).

The above framework remedy thus the disadvantage that we might have, in the general case by taking a *Gel'fand sheaf,* as before, whose *local section algebras* are, namely, *not necessarily σ-complete* (as was, instead, the case with (9.2); see also (2.4) and (2.7.1)). Of course, we could start, in place of the Gel'fand presheaf, as in (2.7), with a *topological algebra presheaf* of the form;

$$U \longmapsto \overline{E^{\wedge}|_U} \subseteq \mathcal{C}_c(U), \tag{9.29}$$

along with the obvious *restriction* (algebra) *morphisms.* Here U varies over the *open* subsets of the *spectrum* $\mathfrak{M}(E) \equiv X$ of a given *topological algebra* E, while we also assume for X that it is, in particular, a *hereditarily k-space* (viz. we suppose that every open subset of X is a *k-space;* take, for instance, a locally compact, or a 1st countable space. See also e.g. J.L. Kelley [1: p.231, Theorems 12, 13], as well as, concerning (9.29), A. Mallios [8: p. 146, (1.12)]).

As a result of our discussion in the previous Scholium 9.1, one thus infers that;

(9.30) *(9.28) remains valid for a Gel'fand sheaf* $\mathcal{E}$, as before, (that is, corresponding to a *topological algebra* E, *as in Lemma 7.1, or* more specifically, *in Theorem 7.1*), *provided that* one has the relation

(9.30.1) $$\ker d = \operatorname{im} \partial,$$

that is, *equivalently, that (9.13) be exact.* (Of course, one applies here the *Kähler differential triad* (cf. (6.8)), which can be associated with E, as above).

Now, it is still our purpose to give in the subsequent Section 11 a further *example of topological algebras, where (9.30.1) holds true,* by suitably changing the (differential) complex, under consideration.

10. Flat A-vector bundles vis-à-vis to ∂-flatness of the associated vector sheaves

We have seen already in the preceding (cf. (8.12)) the relation between **A**-*vector bundles* and *vector sheaves* on a given *topological space* X; namely, an object of the first type is thus (canonically) associated, through its local (continuous) sections, to a vector sheaf on X. Indeed, *the two notions coincide, under the aforesaid correspondence, for every unital* (commutative) *locally* m*-convex* (topological **C**-)*algebra,* at least. (See also Chapt. IV; Scholium 2.1, along with A. Mallios [5: p. 419, Section 5, especially the comments at the beginning of p. 420, and rel. (5.1) there]).

In particular, *flat* **A***-vector bundles are classified through the 1st cohomology set*

(10.1) $$H^1(X, \mathrm{GL}(n, \mathbf{A})),$$

where one considers

(10.2) $$\mathrm{GL}(n, \mathbf{A}) \equiv \mathrm{GL}(n, \mathbf{A})_X \overset{\subsetneq}{\rightarrow} \mathcal{GL}(n, \mathcal{A})$$

as a *constant sheaf* (of groups) *on* X with fiber the *discrete group* $\mathrm{GL}(n, \mathbf{A}) := M_n(\mathbf{A})^{\cdot}$, viz. the *group of units* of the full *matrix-algebra* $M_n(\mathbf{A})$, while the $\mathbf{C}$-*algebra sheaf* $\mathcal{A}$ is given by (8.9). (Yet, as usual, $n \in \mathbf{N}$ stands here for the *rank* of the vector bundles, respectively sheaves considered). In this concern, see also, for instance, Chapt. V; (5.1), (5.3), regarding the above relation (10.2). Furthermore, by complete analogy with the particular case of (continuous) *flat* $\mathbf{C}$-*vector bundles* on X (take thus $\mathbf{A} = \mathbf{C}$; loc. cit.), one obtains that:

> *every flat* $\mathbf{A}$-*vector bundle on* X *can be determined by a 1-cocycle,* say $(g_{\alpha\beta})$, *consisting of locally constant* $\mathrm{GL}(n, \mathbf{A})$-*valued maps; viz. one has*
>
> (10.3) (10.3.1)) $$(g_{\alpha\beta}) \in Z^1(\mathcal{U}, \mathrm{GL}(n, \mathbf{A})),$$
>
> where $\mathcal{U} = (U_\alpha)_{\alpha \in I}$ stands for a *"trivializing" open covering of* X (the analogue here of a *local frame* for the corresponding vector sheaf).

(See loc. cit. (5.19), (5.20)). In this respect, we also note that $\mathrm{GL}(n, \mathbf{A})$ in (10.3.1) still stands for the corresponding *constant sheaf* on X, as in (10.2). On the other hand, it is a direct consequence of the very definitions that:

> (10.4) by considering the above group $\mathrm{GL}(n, \mathbf{A})$ as an ordinary *topological group* with the canonical topology it inherits from $M_n(\mathbf{A}) \cong \mathbf{A}^{n^2}$ (*cartesian product topology,* cf. Chapt. IV; (2.7), along with Scholium 2.1), the *sheaf of germs of locally constant* $\mathrm{GL}(n, \mathbf{A})$-*valued maps on* X (hence, continuous) *is the same, as the constant sheaf* (10.2), as above.

In this respect, see also Chapt. I; Theorem 11.1. (Of course, an analogous conclusion to the previous one in (10.4) is valid too, quite generally, *if in place of the topological group* $\mathrm{GL}(n, \mathbf{A})$, as in (10.4), *one considers any topological space* Y, concerning the *constant sheaf,* determined by the corresponding *discrete space* Y). ∎

Now, suppose that the *topological space* X considered is, in particular, *path-connected,* while $\mathbf{A}$ is still a *unital locally* m-*convex* (topological) *algebra.* Then, by analogy with the classical case (where one has $\mathbf{A} = \mathbf{C}$, see e.g. R.C.

Gunning [2: p.123ff]), one infers that;

(10.5) *for every flat* **A**-*vector bundle*

(10.5.1) $$\xi \in H^1(X, \mathrm{GL}(n, \mathbf{A}))$$

(cf. (10.1)), *there is associated an element*

(10.5.2) $$\tilde{\xi} \in \mathrm{Hom}(\pi_1(X), \mathrm{GL}(n, \mathbf{A})).$$

We call $\tilde{\xi}$, as above, the *characteristic representation of* ξ (see also (10.6) below).

Thus, "Hom" in (10.5.2) above stands for the *set of* the *group morphisms* concerned, where $\pi_1(X)$ denotes, of course, the *fundamental* (or even *Poincaré*) *group of* X. So the previous element $\tilde{\xi}$, as in (10.5.2), yields by its definition, a *representation of* $\pi_1(X)$ *in the* (free locally convex) **A**-*module* $\mathbf{A}^n$, in such a manner that $\tilde{\xi}$ *characterizes,* in fact, ξ, *within an inner automorphism of* $\mathrm{GL}(n, \mathbf{A})$. Speaking formally, one thus obtains the following *bijection;*

(10.6) $$H^1(X, \mathrm{GL}(n, \mathbf{A})) = \mathrm{Hom}(\pi_1(X), \mathrm{GL}(n, \mathbf{A}))^{\mathrm{GL}(n, \mathbf{A})},$$

the second member of (10.6) denoting the *set of isomorphism classes of representations of* $\pi_1(X)$ *in* $\mathbf{A}^n$, *modulo inner automorphisms of* $\mathrm{GL}(n, \mathbf{A})$. (In this concern, see also A. Mallios [5: p. 412f] for further relevant details, as well as, consequences of (10.6). Yet, cf. e.g. R.C. Gunning [1: p. 186, Lemma 27], where the corresponding result to (10.6), for *principal* G-*bundles,* is proved *for any group* G, whatsoever).

Accordingly, by further applying the above rel. (10.6) to the particular case of **A**-*line bundles* (viz., equivalently (cf. (8.12)), to *line sheaves,* with respect to the **C**-algebra sheaf $\mathcal{A}$, as the latter is defined by (8.9)) on (the *path-connected* space) X and for a *commutative* **A**, *as before,* one obtains the following (set-theoretic) *bijection*

(10.7) $$H^1(X, \mathbf{A}^{\boldsymbol{\cdot}}) = \mathrm{Hom}(\pi_1(X), \mathbf{A}^{\boldsymbol{\cdot}});$$

the last relation *characterizes* thus *flat* **A**-*line bundles on* X, with **A** and X, as above.

10.(a). ∂-flatness of A-vector bundles.– We have seen already, according to (10.1) and (8.12), that

(10.8) $$H^1(X, \mathrm{GL}(n, \mathbf{A})) \subsetneq H^1(X, \mathcal{GL}(n, \mathcal{A})),$$

the second member of the last relation classifying **A**-vector bundles, where $\mathcal{A}$ is given by (8.9) (cf. (8.12)), while the first one identifies flat **A**-vector bundles on X. (Of course, (10.8) indicates just the resulting *canonical map between the cohomology sets* concerned, *in view of the* corresponding *map in* (10.2); see thus, for instance, Chapt. III; (11.31)).

On the other hand, suppose now that we are given a *differential triad* (cf. Chapt. VIII; (1.1))

(10.9) $$(\mathcal{A}, \partial, \Omega^1)$$

on a *topological space* X, where $\mathcal{A}$ is still given by (8.9). (See, for instance, a variant of the situation at issue, as supplied by the ensuing Example 10.1). Furthermore, *we still assume* here (a crucial hypothesis, for that matter, pertaining to our main conclusion in (10.11) below) *that one has;*

(10.10) $$\mathbf{A} \subseteq \ker \partial$$

(a natural assumption, otherwise, for the case at hand, corresponding to Chapt. VI; Lemma 1.1. See also e.g. (10.13) below).

Thus, based already on (10.3), one concludes now that;

(10.11) *every flat* **A**-*vector bundle on* X *has the corresponding vector sheaf* (of its local sections, cf. (8.12) or even (10.8)) *on* X ∂*-flat.* Of course, *the converse holds true,* as well, *in the particular case* that one has (*generalized Poincar/'e condition,* see e.g. Chapt. IX; (3.7))

(10.11.1) $$\mathbf{A} = \ker \partial$$

(see the subsequent example).

Example 10.1.– Consider a (Hausdorff *connected*) $\mathcal{C}^\infty$*-manifold* X and the **C**-*algebra of* **A**-*valued* $\mathcal{C}^\infty$*-funtions* on X, where **A** is a *unital commutative* σ*-complete locally* m*-convex* **C**-*algebra.* Thus, the corresponding *structure sheaf* here is

(10.12) $$\mathcal{A} \equiv {}^{\mathbf{A}}\mathcal{C}^\infty_X = \mathcal{C}^\infty_X \varepsilon \mathbf{A},$$

where "ε" denotes the respective "ε-(topological) *tensor product*" (see A.

Mallios [1: p. 489] and [4: p. 392ff]). So one can prove now that

(10.13) $$\ker \partial = \mathbf{A};$$

indeed, one proves that *an* $\mathbf{A}$*-valued smooth funtion* on X *is locally constant if, and only if,* (the respective germ of) *it belongs to* $\ker \partial$. (See, for instance, L. Nachbin [2: p. 124, Proposition 19.1 and its proof]).

Therefore (cf. (10.11)), a given (smooth) $\mathbf{A}$*-vector bundle* on X *is flat if, and only if, it is* ∂*-flat;* hence (Chapt. VIII; Theorem 5.1) *if, and only if, it has a flat* $\mathcal{A}$*-connection.* In this context, we further remark that by assuming, in particular, the $\mathcal{C}^{\infty}$-manifold X to be *paracompact* (Hausdorff), one concludes that;

(10.14) *every* (smooth) $\mathbf{A}$*-vector bundle* on X *admits an* $\mathcal{A}$*-connection, where* $\mathcal{A}$ *is given by (10.12).*

Indeed, $\mathcal{A}$ *is a fine sheaf* on X, this being already the case for $\mathcal{C}^{\infty}_X$, so that *the same holds true, for every vector sheaf on* X (see Chapt. III; (8.56)). Therefore, (10.14) is already a consequence of Chapt. VI; Theorem 16.1. ∎

Finally, we still note that *every continuous flat* $\mathbf{A}$*-vector bundle on* X, as above, *can be construed as a smooth one,* as well, by virtue of (10.3). (See also A. Mallios [5: p. 412, Remark 2.1], for relevant details thereon).

11. Watts-Selesnick complex

We consider below another complex, different from the de Rham-Kähler one (cf. (5.27) and/or (6.14) in the preceding), that can be associated with a *topological algebra* of an appropriate type (see e.g. (11.34) or even (11.37) in the sequel). In its algebraic formulation the notion, we are going to be dealt with, is the standard *Watts complex,* introduced by C.E. Watts [1] and pertaining to the cohomology of algebras (essentially à la Amitsur) with an application to the ring (algebra) of $\mathbf{R}$-valued continuous functions on a compact space X. (In this respect, see also, for instance, A.T. Kitchen [1] and A.A. Korobov [1]). Now, as already said, the study of Watts is quite *algebraic,* involving, in particular, tensor products of the algebra concerned. On the other hand, independently of that work, an analogous complex has also been considered by S.A. Selesnick [1], who has employed, instead, (topological) *completed tensor products of Banach*(-Šilov) *algebras,* while was still looking at the *cohomology of the spectrum* of such algebras, viz. of the *compact space* X in the particular case of Watts' ring $\mathcal{C}(X)$ (cf. also [TA: p.233, Theorem 1.2]). We call in the sequel the complex thus obtained the *Watts-Selesnick complex* of the topological algebra involved (see (11.21) below).

More generally, we shall finally consider in the sequel (see thus (11.34) below) *soft* (topological) *algebras* (cf. Definition 3.2), although we do start straight away by employing *as general topological algebras, as the case* at issue *permits.* Yet, for convenience, we recall, in brief, the construction of the aforementioned complex, within the present more general context, referring, on the other hand, to the above cited work of Selesnick for further relevant details, that can also be adapted straightforwardly to the present framework.

Thus, suppose that E is a *unital locally convex* (topological) *algebra* (with *continuous multiplication* [TA]), so that one further defines the following sequence of *locally convex E-modules* and *continuous* $\mathbf{C}$*-linear morphisms;*

$$
(11.1) \qquad
\begin{aligned}
&E^0 := E\\
&E^1 := E^0 \underset{\pi}{\otimes} E = E \underset{\pi}{\otimes} E\\
&\cdots\cdots\cdots\cdots\cdots\cdots\\
&E^p := E^{p-1} \underset{\pi}{\otimes} E = E \underbrace{\otimes \cdots \otimes}_{p+1-\text{times}} E\\
&\cdots\cdots\cdots\cdots\cdots\cdots,
\end{aligned}
$$

with $p \in \mathbf{N}$, where π denotes the corresponding *projective tensor product topology* (cf. [TA: p.376, Lemma 3.1]). So each one of the previous spaces E^p, $p \in \mathbf{Z}_+$, is made, in the obvious way, into an *E-module,* indeed, into a *locally convex E-module* (see, for instance, loc. cit.; p.365, (2.5)).

On the other hand, one also defines the following $\mathbf{C}$*-linear operators* (on decomposable tensors, extending then by $\mathbf{C}$-linearity);

$$
(11.2) \qquad d^p : E^p \longrightarrow E^{p+1}, \quad p \in \mathbf{Z}_+,
$$

such that one sets

$$
(11.3) \qquad d^p := \sum_{i=0}^{p+1} (-1)^i d_i^p,
$$

where one further defines

$$d_i^p : E^p \longrightarrow E^{p+1}, \quad i = 0, \ldots, p+1, \quad \textit{and} \quad p \in \mathbf{Z}_+, \tag{11.4}$$

according to the following relations (cf. also (11.1));

$$d_i^p(x_0 \otimes \cdots \otimes x_p) := x_0 \otimes \cdots \otimes x_{i-1} \otimes 1 \otimes x_i \otimes \cdots \otimes x_p, \tag{11.5}$$

for any $0 < i \leqq p$, *with* $p \in \mathbf{N}$. (Of course, $1 \equiv 1_E$ stands here, and in the sequel too, for the *identity element* of the given algebra E). Thus, we still set;

$$d_0^p(x_0 \otimes \cdots \otimes x_p) := 1 \otimes x_0 \otimes \cdots \otimes x_p, \tag{11.6}$$

as well as,

$$d_{p+1}^p(x_0 \otimes \cdots \otimes x_p) := x_0 \otimes \cdots \otimes x_p \otimes 1, \tag{11.7}$$

for any $p \in \mathbf{N}$ (extending thus (11.5) *for* $i = 0$ *and* $p+1$, respectively). Yet, one finally sets;

$$d_0^0(x) := 1 \otimes x, \quad \textit{as well as}, \quad d_1^0(x) := x \otimes 1, \tag{11.8}$$

for every $x \in E$, extending thus the preceding *for* $p = 0$ (see also (11.2) and (11.3)).

Accordingly, we further note at this place, in anticipation, that by virtue of (11.3), one has, in particular;

$$d^0 := d_0^0 - d_1^0, \tag{11.9}$$

so that, in view of (11.8), one obtains

$$d^0(x) = 1 \otimes x - x \otimes 1, \tag{11.10}$$

for every $x \in E$.

Now, it is an easy consequence of the very definitions, along that one of the tensor product topology π (loc. cit.), that *all the previously defined* **C**-*linear maps are continuous;* hence, they can be extended by continuity to the respective *completed spaces.* Thus, one obtains the following extended *continuous* **C**-*linear maps:*

$$\widetilde{d^p} : \widetilde{E^p} \longrightarrow \widetilde{E^{p+1}}, \quad p \in \mathbf{Z}_+. \tag{11.11}$$

On the other hand, suppose that

$$\mathfrak{M}(E) \equiv X \tag{11.12}$$

is the *spectrum* of the given topological algebra E (see beginning of Section 2 in the preceding). Thus, for every *continuous character* $f \in \mathfrak{M}(E)$, one can consider the map;

$$f^p := f \underbrace{\otimes \cdots \otimes}_{p+1-\text{times}} f : \widetilde{E^P} \longrightarrow \mathbf{C}, \quad p \in \mathbf{Z}_+, \tag{11.13}$$

given by the relation

$$(f \otimes \cdots \otimes f)(x_0 \otimes \cdots \otimes x_p) := f(x_0) \cdots f(x_p) \in \mathbf{C}, \tag{11.14}$$

with $x_i \in E$, $0 \leqq i \leqq p$ (for decomposable tensors), and then extending by linearity and continuity to the whole of $\widetilde{E^p}$, proving, in effect, that f^p is also a *continuous character of the complete locally convex* $\mathbf{C}$-*algebra* $\widetilde{E^p}$. That is, one has (see (11.1))

$$\begin{aligned} f^p \in \mathfrak{M}(\widetilde{E^p}) = \mathfrak{M}(E^p) &= \mathfrak{M}(E) \underbrace{\times \cdots \times}_{p+1-\text{times}} \mathfrak{M}(E) \\ &\equiv \mathfrak{M}(E) \otimes \cdots \otimes \mathfrak{M}(E). \end{aligned} \tag{11.15}$$

In this context, the first equality above is referred to a *set-theoretic bijection,* which is also *continuous,* while the second one is a *homeomorphism,* with respect to the natural *(Gel'fand) topologies* of the spaces involved (see [TA: p.145, Lemma 2.1 and p.407, Lemma 1.1]; yet, see e.g. (2.1) in the preceding, concerning the aforesaid topologies). In particular, the same first equality in (11.15) above becomes, in fact, a *homeomorphism,* if $\mathfrak{M}(E)$ is *(locally) equicontinuous* (loc. cit., p.411, Theorem 1.2). Indeed, this will be, for instance, the case in the sequel when employing *soft algebras* (cf. Definition 3.2, as well as, (11.34) below). On the other hand, one has, of course (see (11.15)), that

$$\ker(f^p) \subseteq \widetilde{E^p} \tag{11.16}$$

is a *closed maximal* (2-sided) *ideal* in $\widetilde{E^p}$, so that, by still following S.A. Selesnick (loc. cit., p.317), we further denote by

$$\lambda_f^p : \widetilde{E^p} \longrightarrow (\widetilde{E^p})_{\ker(f^p)} \equiv (\widetilde{E^p})_{f^p} \tag{11.17}$$

the corresponding (canonical) *localizing map of* (the topological algebra) $\widetilde{E^p}$ at $f^p \in \mathfrak{M}(\widetilde{E^p})$, given by the relation

(11.18) $$\lambda_f^p(z) := z/\otimes^{p+1} 1 \equiv [z] \in (\widetilde{E^p})_{\ker(f^p)},$$

for every $z \in \widetilde{E^p}$. (In this concern, see also, for instance, A. Mallios [9] for a detailed account on *"localizations of topological algebras"*). Thus, we further set;

(11.19) $$N^p := \bigcap_{f \in \mathfrak{M}(E)} \ker(f^p) \subseteq \widetilde{E^p}, \quad p \in \mathbf{Z}_+,$$

getting a *closed* (2-sided) *ideal of* $\widetilde{E^p}$, such that one has (cf. (11.11))

(11.20) $$\widetilde{d^p}(N^p) \subseteq N^{p+1},$$

for every $p \in \mathbf{Z}_+$. (See also S.A. Selesnick [1: p.318, Proposition 2.4]). Hence, one can now consider the following (quotient) complex

(11.21) $$W_S(E) \equiv (\widetilde{E^p}/N^p, \widetilde{d^p})_{p \in \mathbf{Z}_+},$$

which we call the *Watts-Selesnick complex of* E, that can thus be associated, according to the preceding, with every E, a *unital locally convex* **C**-*algebra* with *continuous multiplication.*

On the other hand, following further Selesnick (loc. cit.) and based on a suitable *sheafification of the* previous *complex* (11.21), one can compute the (Čech) *cohomology of the spectrum* of an appropriate topological algebra, via the aforementioned sheafified complex, hence, entirely, *in terms of the* given *algebra itself.* Yet, the same sheafified complex, as before, yields also *another example of an* $\mathcal{E}$-*complex* (with $\mathcal{E}$ a **C**-*algebra sheaf*), providing an *exact sequence of de Rham type* (see (11.44) below). This becomes also of importance, in view of the potential (differential-) geometric applications we were considering in the preceding chapters. We discuss these topics straight away in the ensuing Subsection.

11.(a). Sheafification of the Watts-Selesnick complex.– Suppose we are given an *algebra* E, satisfying the following conditions:

(i) E is a *locally convex* (topological) **C**-*algebra* with *continuous multiplication* and an *identity element.*

(ii) The *spectrum* of E, $\mathfrak{M}(E) \equiv X$, is *locally equicontinuous* (viz. every $f \in \mathfrak{M}(E)$ has an *equicontinuous* neighborhood).

Therefore, by considering the *(topological tensor product) algebras* E^p, $p \in \mathbf{N}$, as in (11.1), one concludes that;

(11.22) *the completed algebras* $\widetilde{E^p}$, $p \in \mathbf{Z}_+$, are *(complete) locally convex algebras, which also satisfy the previous conds. (i) and (ii).*

See [TA: p.410, Lemma 1.2 and p.146, Lemma 2.2]. ▮ As a result, one thus obtains, concerning the *spectrum of* $\widetilde{E^p}$ (cf. also (11.12))

$$\mathfrak{M}(\widetilde{E^p}) = X \underbrace{\times \cdots \times}_{p+1-\text{times}} X \equiv X^{p+1}, \tag{11.23}$$

within a *homeomorphism* (of the *Gel'fand spaces* concerned; ibid., p.411, Theorem 1.2).

Thus, for each one of the preceding (complete) topological algebras $\widetilde{E^p}$, $p \in \mathbf{Z}_+$, one can consider the following *presheaf on* X; namely, for every *open* $U \subseteq X$, one defines (see also e.g. (2.26))

$$\begin{aligned}\mathbf{E}^p(U) := \left\{\alpha : U^{p+1} \longrightarrow \mathbf{C} : \alpha = \hat{z}|_{U^{p+1}}, \text{ for some } z \in \widetilde{E^p}\right\} \\ \subseteq \mathcal{C}(U^{p+1}),\end{aligned} \tag{11.24}$$

where, of course, we set (cf. (11.23))

$$U^{p+1} := U \underbrace{\times \cdots \times}_{p+1-\text{times}} U \subseteq X^{p+1} \equiv \mathfrak{M}(\widetilde{E^p}), \tag{11.25}$$

for every open set $U \subseteq X$, as above. So the above correspondence, as given by (11.24), together with the obvious *restriction morphisms,* supplies already a *presheaf of (unital commutative)* **C**-*algebras on* $X \equiv \mathfrak{M}(E)$. (The same can still be construed as a *presheaf of topological algebras on* X (in effect, *locally m-convex* ones), in view of the last issue in (11.24), by applying a similar argument here, as that one in (2.7) in the preceding; see also e.g. (2.27), (2.28). Hence (cf. Theorem 1.1), *an analogous remark is* actually *valid for the sheaf* (11.26), we are going to define presently below). Thus, we denote by

$$\mathcal{E}^p, \quad p \in \mathbf{Z}_+, \tag{11.26}$$

the *sheaf of* **C**-*algebras on* X, generated by the previous presheaf on X, as defined by (11.24). In particular, *for* $p = 0$, one obtains

$$\mathcal{E}^0 = \mathcal{E}, \tag{11.27}$$

where the second member of the last relation stands for the *Gel'fand sheaf of the* given *topological algebra* E *over* $\mathfrak{M}(E) \equiv X$ (cf. (2.3), (2.7.1), Definition 2.1 and Lemma 2.1).

Now, our next objective is to define the appropriate *"differentials"* (viz. $\mathbf{C}$-*linear morphisms*) between the previous sheaves, as in (11.26) above, the same sheaves being, in fact, according to their definition (see e.g. (11.24)), $\mathcal{E}$-*modules on* $X \equiv \mathfrak{M}(E)$, with $\mathcal{E}$ the *Gel'fand sheaf* of E (cf. (11.27)): Thus, we start by first defining the differentials, we are seeking for, between the elements of the *presheaf of* $\mathbf{C}$-*algebras* on X, as the latter is given by (11.24). That is, *for every open set* $U \subseteq X$, we set

$$(11.28) \qquad \delta^p : \mathbf{E}^p(U) \longrightarrow \mathbf{E}^{p+1}(U), \quad p \in \mathbf{Z}_+,$$

such that one defines (cf. (11.24), (11.25));

$$(11.29) \quad \delta^p(\alpha|_{U^{p+1}})(f_0, \ldots, f_{p+1}) := \sum_{i=0}^{p+1}(-1)^i(\alpha|_{U^{p+1}})(f_0, \ldots, f_{i-1}, f_{i+1}, \ldots, f_{p+1}).$$

Thus, one obtains, based on the very definitions (cf. also (11.11));

$$(11.30) \qquad \delta^p(\hat{z}|_{U^{p+1}}) = \widehat{d^p(z)}|_{U^{p+2}}, \quad p \in \mathbf{Z}_+,$$

for every $z \in \widetilde{E^p}$ (see also loc. cit., p.319, Proposition 2.5, (i)), while we still define (cf. (2.3) and (2.7.1))

$$(11.31) \qquad \varepsilon : \mathbf{C} \longrightarrow \mathbf{E}^0(U) = \mathbf{E}(U) = E^{\hat{}}|_U : \lambda \longmapsto \varepsilon(\lambda) := \lambda \cdot 1_{E^{\hat{}}}|_U.$$

(Here $\mathbf{C}$ stands for the respective *constant presheaf* on X, defined by the (discrete) *set* $\mathbf{C}$; cf. Chapt. I; (10.2)). Thus, it is clear, according to the same definitions, that (11.28) *defines a* $\mathbf{C}$-*linear map,* indeed, *a* $\mathbf{C}$-*linear morphism of the (*$\mathbf{C}$-*vector space) presheaves concerned* (see (11.29), in conjunction with Chapt. I; (6.3), (6.5)). Therefore (Chapt. I; (8.4), (8.5) and Lemma 8.1), one thus obtains $\mathbf{C}$-*linear morphisms of the* respective *sheafifications* of the same presheaves, viz. the maps

$$(11.32) \qquad \delta^p : \mathcal{E}^p \longrightarrow \mathcal{E}^{p+1}, \quad p \in \mathbf{Z}_+.$$

(We applied above an obvious *abuse of notation,* retaining the same symbol for the sheafifications of the maps in (11.28)). Of course, an analogous conclusion holds true for the map (11.31), referring now to the *constant sheaf* $\mathbf{C} \equiv \mathbf{C}_X$ *on* X (see also Chapt. I; (4.1), (4.4′) and (11.21)). Thus, one also obtains the following $\mathbf{C}$-*linear morphism* (cf. (11.27), along with the above remark on the notation applied);

$$(11.33) \qquad \varepsilon : \mathbf{C} \longrightarrow \mathcal{E}^0 = \mathcal{E}.$$

Now, to get further a suitable *exact sequence of the sheaves,* in fact, *$\mathcal{E}$-modules, defined by* (11.26), (11.27) (in particular, they are here considered as **C**-*vector space sheaves* on X), along with the previous **C**-*linear morphisms, as in* (11.32), (11.33), we shall restrict ourselves to an appropriate class of topological algebras. So *hereafter we assume that;*

(11.34) E is a *soft* topological *algebra* (cf. Definition 3.2), whose *spectrum* is $\mathfrak{M}(E) \equiv X$ (a *compact Hausdorff space* (cf. (3.12)).

As an application of (11.34), one obtains, for instance, the following useful relation (see also (11.36) below), pertaining to the *"local behaviour"* of E;

$$(11.35) \qquad \mathcal{E}_f = E_{\ker f},$$

for any $f \in \mathfrak{M}(E)$, the second member of (11.35) denoting the *localization of E at $f \in \mathfrak{M}(E)$.* (See A. Mallios [9], for the case considered by (11.34), or even S.A. Selesnick [1: p.314, Proposition 1.1 (ii)] for Banach algebras). On the other hand, we have already noticed that

(11.36) *the Gel'fand sheaf of E, $\mathcal{E} = \mathcal{E}^0$* (cf. (11.27)), *is a fine sheaf on X.*

See (4.18) and (4.19). ∎ Indeed, *the previous conclusion is* actually *valid for all the sheaves $\mathcal{E}^p$, $p \in \mathbf{N}$,* as in (11.26), as we shall presently see; however, *under* certain *additional assumptions for E.* Thus, *we* next *suppose that:*

(11.37) E is a *soft locally convex* (topological) *algebra* (cf. (11.34)), *closed under* the formation of *completed projective* (topological) *tensor products.*

Scholium 11.1.– We have already considered from the beginning of this Subsection *locally convex algebras, satisfying the conds. (i) and (ii),* so that the respective (topological) tensor product algebras, as in (11.1), along with their completions, be of the same type (a fact, that we applied above; see thus (11.22), as well as, for instance, (11.23)). On the other hand, for technical reasons, having to do, for example, with the application of Lemma 3.1 (*Local Theorem*), we also want the same (topological) tensor product algebras, as before (viz. the *completions* of the algebras appeared in (11.1), *for* $p > 0$), to *retain the softness* of the initial algebra: Thus, we refer to A. Mallios [4: Chapt. XIII] for details on this matter, within the context of (locally convex) topological algebras, in general; for example, *semisimplicity* is thus preserved, under *additional assumptions for the algebras* involved. Such an extra hypothesis is, for instance, the so-called *approximation property* (loc. cit., p.446), that has to be shared by the locally convex algebra, under consideration. Now, a familiar class of such (topological) algebras is, of course, the *nuclear* (locally convex) *algebras* (ibid., p.302f). Accordingly, a class of algebras satisfying (11.37) is that one of

(11.38) *nuclear Fréchet-Šilov* (locally convex) *Q-algebras.*

See also (3.10) in the preceding, concerning the terminology applied. An *example* of the above type of (topological) algebras is certainly supplied by the (*non-normed!*) algebra $\mathcal{C}^\infty(X)$, where X is a (connected Hausdorff) *2nd countable compact* $\mathcal{C}^\infty$*-manifold.* In this respect, we still note, in passing, that:

(11.39) a (finite-dimensional connected Hausdorff) *2nd countable* $\mathcal{C}^\infty$*-manifold* X *is compact if, and only if, the algebra* $\mathcal{C}^\infty(X)$ (endowed with the *Schwartz topology*) *is a Q-algebra.* Moreover, *the same topology is the only one making* $\mathcal{C}^\infty(X)$ *an FQ-algebra.*

(See also loc. cit.; p. 227, Theorem 2.1, as well as, p. 187, Lemma 1.3, p. 183, Corollary 1.1 and p. 143, Proposition 1.1. Yet, concerning the last part of the assertion, cf. M. Fragoulopoulou [1: p. 967, Proposition 3.3, or even p. 966, Corollary 2.8, (iii)]). ∎ On the other hand, there do exist *Banach(-Šilov) algebras having the approximation property,* so that the previous cond. (11.37) is still in force, for the particular case at hand. The latter algebras have been employed, for instance, by S.A. Selesnick [1]. Moreover, *Fourier algebras* of the type considered by the next Section 12 are also Banach algebras of the previous sort. However, (topological) *algebras satisfying (11.38) cannot be Banach, unless they are finite-dimensional.* (See, for instance, A. Grothendieck [1: Chap. II; p.38, Corollary 2]).

Finally, we also note, supplementing thus our previous discussion in Section 3 in the preceding (see the comments following (3.10)), that;

(11.40) the algebra

$$\mathcal{O}(K) := \varinjlim_{U \supseteq K} \mathcal{O}(U) \qquad (11.40.1)$$

of *"locally holomorphic"* $\mathbf{C}$*-valued functions* on a *Stein compact* set K of a (2nd countable) *Stein manifold* X (when endowed with the respective inductive limit *van Hove* topology) is a

(11.40.2) *nuclear Pták-Šilov (locally m-convex) Q-algebra.*

In this context, see also A. Mallios [7: p. 318, Theorem 5.1], as well as, A. Grothendieck [1: Chap. II; p.56, Corollaire and p.48, Corollaire 1]. Yet, it is worth noting here that *the van Hove topology on* $\mathcal{O}(K)$, as above, *is the only one making it an* $\mathcal{L}FQ$*-algebra.* See M. Fragoulopoulou [1: p. 966, Proposition 3.1, or even Corollary 2.8, (i)]. ▮ Thus, one obtains, through (11.40.1), another *example* of a topological algebra *satisfying,* by virtue of (11.40.2), the *cond.* (11.37). On the other hand, *the class of* (topological) *algebras of the type (11.40.2) is,* of course, *bigger than the one defined by* (11.38). (Namely, *every Fréchet* (locally convex) *space is Pták (Krein-Šmulian Theorem);* see, for instance, H.H. Schaefer [1: p.162]).

Thus, assuming now that the given topological algebra E *satisfies* (11.37), one first obtains an *extension of (11.35), for every* $p > 0$; that is, one has (cf. also (11.26) and (11.17))

$$\mathcal{E}^p_f = (\widetilde{E^p})_{f^p}, \quad p \in \mathbf{N}, \qquad (11.41)$$

for every $f \in \mathfrak{M}(E) \equiv X$, within a $\mathbf{C}$*-algebra isomorphism.* (See also S.A. Selesnick [1: p.320,Proposition 2.6]). On the other hand, as already said, one further concludes the following, supplementing thus (11.36);

(11.42) *each one of the sheaves* $\mathcal{E}^p$, $p \in \mathbf{N}$ (cf. (11.26)), is a *fine sheaf on* X.

See also loc. cit., p.319, Proposition 2.5. ▮ Furthermore (ibid.), one now proves that, *for every open* $U \subseteq X$, *the following sequence is exact* (see also (11.29), (11.31));

$$0 \longrightarrow \mathbf{C} \xrightarrow{\varepsilon} \mathbf{E}(U) \xrightarrow{\delta^0} \mathbf{E}^1(U) \xrightarrow{\delta^1} \mathbf{E}^2(U) \longrightarrow \cdots . \tag{11.43}$$

(In this context, we still observe that the relevant argument in loc. cit., p.319, proof of Proposition 2.5, (iii), can be based on [TA: p.447, Lemma 5.1], avoiding thus the use of *Fréchet algebras;* see also loc. cit., p.449, Scholium 5.1, as well as, the comments at the end of the above Scholium 11.1). Thus, one finally infers that *the next sequence of sheaves on X is exact* too;

$$\mathcal{E}^* : 0 \longrightarrow \mathbf{C} \xrightarrow{\varepsilon} \mathcal{E} \xrightarrow{\delta^0} \mathcal{E}^1 \xrightarrow{\delta^1} \mathcal{E}^2 \longrightarrow \cdots . \tag{11.44}$$

In other words, based thus on (11.36) and (11.42), one now concludes that;

(11.45) (11.44) *provides a fine* (torsionless) *resolution of (the constant sheaf)* $\mathbf{C}$.

Therefore, in view of the *compactness of* X (cf. (11.37)), one obtains a Γ_X-*acyclic resolution of* $\mathbf{C}$ (cf. Chapt. III; (8.24)). Accordingly, based on the *abstract de Rham Theorem* (ibid.; (3.24) or even (8.25)), one now comes to the following basic conclusion, extension to our case of the analogous result of S.A. Selesnick [1: p.317, Theorem 2.1], for *Banach-Šilov algebras* having the property that *completed projective tensor products* of such algebras *are* still *of the same type* (viz. the corresponding condition here to (11.37) above; thus, one may consider, for instance (see also the previous Scholium 11.1), *Banach-Šilov algebras with the* (Banach) *approximation property.* Cf. the next Section 12). Thus, one has the following.

Theorem 11.1. *Let E be a soft (locally convex) algebra, closed under the formation of completed projective* (topological) *tensor products* (see e.g. (11.38), (11.39), (11.40)), *with spectrum $\mathfrak{M}(E) \equiv X$. Moreover, let $\mathcal{E}^*$ be the fine resolution of $\mathbf{C}$ on X, as defined by (11.44), hence a Γ_X-acyclic resolution of $\mathbf{C}$. Then, one has;*

$$\check{H}^*(X, \mathbf{C}) = H^*(X, \mathbf{C}) = H^*(\Gamma_X(\mathcal{E}^*)) \equiv W_S^*(E), \tag{11.46}$$

within isomorphisms of the graded $\mathbf{C}$-algebras concerned.

Indeed, by considering the *constant ($\mathbf{C}$-vector space) sheaf* $\mathbf{C}$ and its *fine resolution* (11.44) (cf. (11.45)), our claim about (11.46) is now just a consequence, as already said, of Chapt. III; (3.24.2). Thus, according to that relation, one has;

(11.47) $$(R^n\Gamma_X)(\mathbf{C}) = H^n(X, \mathbf{C}) \cong h^n(\Gamma_X(\mathcal{E}^*)) = \check{H}^n(X, \mathbf{C}),$$

for every $n \in \mathbf{Z}_+$, within *isomorphisms of* the **C**-*vector spaces* concerned, the last relation being a consequence of Theorem 8.1 of that Chapter, which thus proves (11.46). ∎

We call the last item in (11.46) the *Watts-Selesnick cohomology of* E, the latter algebra being, as in (11.37). Specifically, one has

(11.48) $$W_S^*(E) \equiv H^*(\Gamma_X(\mathcal{E}^*)) := \{h^n(\Gamma_X(\mathcal{E}^*))\}_{n\in\mathbf{Z}_+}$$

(see also Chapt. III; (3.38.2), or even (1.13)).

On the other hand, concerning the *complex* $\Gamma_X(\mathcal{E}^*)$, one has (cf. Chapt. III; (3.8), along with (11.27) above)

(11.49) $$\begin{gathered} 0 \longrightarrow \Gamma(X, \mathcal{E}) \xrightarrow{\delta^0} \Gamma(X, \mathcal{E}^1) \xrightarrow{\delta^1} \Gamma(X, \mathcal{E}^2) \longrightarrow \cdots \\ \cdots \longrightarrow \Gamma(X, \mathcal{E}^p) \xrightarrow{\delta^p} \Gamma(X, \mathcal{E}^{p+1}) \longrightarrow \cdots . \end{gathered}$$

Therefore, one obtains (same Chapter; (3.5). See also (11.47), (11.48) above);

(11.50) $$(R^n\Gamma_X)(\mathbf{C}) = h^n(\Gamma_X(\mathcal{E}^*)) := \ker \Gamma_X(\delta^n)/\operatorname{im} \Gamma_X(\delta^{n-1}),$$

for every $n \in \mathbf{Z}_+$. However, one really concludes that;

(11.51) *the cohomology of the complex* $\Gamma_X(\mathcal{E}^*)$ in (11.49), as this is given by (11.50), *is actually that one of the Watts-Selesnick complex,* as the latter is given by (11.21).

The above also justifies the terminology applied in (11.48). *Indeed, one has*

(11.52) $$\Gamma(X, \mathcal{E}^p) = \widetilde{E^p}/N^p, \quad p \in \mathbf{Z}_+,$$

within an *isomorphism of* **C**-*algebras* (cf. also (11.19)); furthermore, one concludes here that

(11.53) $$N^p = \ker \tau^p,$$

where

(11.54) $$\tau^p : \mathbf{E}^p(X) \cong \widetilde{E^p} \longrightarrow \Gamma(X, \mathcal{E}^p)$$

is an *onto* **C**-*algebra morphism,* given by the relation

(11.55) $$\tau^p(z) := [z]_{f^p} \in (\widetilde{E^p})_{f^p} = \mathcal{E}^p_f,$$

for every $z \in \widetilde{E^p}(\cong \mathbf{E}^p(X))$ *and* $p \in \mathbf{Z}_+$. (See (11.15), (11.24), (11.25), as well as, (11.35), (11.41) and (11.17); yet, cf. also S.A. Selesnick [1: p.321, proof of Theorem 2.1 and p.320, Proposition 2.7]). ∎ In particular, one obtains

$$\Gamma(X, \mathcal{E}^0) = \Gamma(X, \mathcal{E}) = \tilde{E} = E, \tag{11.56}$$

based on (11.27) and the fact that the given *topological algebra E, as in* (11.37), is, by definition, *complete,* as well as, a *geometric* one, being, by our hypothesis, a *soft algebra* (see (3.14) and (3.4)).

Thus, recapitulating the preceding and taking also (11.51) into account, one can restate the previous Theorem 11.1 into the form;

(11.57) *the Watts-Selesnick cohomology of a topological algebra, satisfying* (11.37) (see (11.48)), *equals* (viz. is *isomorphic to*) *the Čech cohomology of its spectrum with complex coefficients.*

Finally, in this context and for other purposes, as well (see, for instance, A. Mallios [9]) we further remark that, in view of (11.35) (cf. also S.A. Selesnick [1: p.316, Corollary 1.6]), one concludes that;

(11.58) *the Gel'fand sheaf of a soft locally m-convex algebra* (see, for example, (11.40.1)) *is strictly henselian* (in the sense that its *stalks* are *strictly henselian local* **C**-*algebras*).

Now, another point that is worth pointing out at this place, in connection with our considerations in the preceding chapters, is the following fact. Namely,

(11.59) *the zero-th dimensional operator*

$$\delta^0 \equiv \partial : \mathcal{E} \longrightarrow \mathcal{E}^1 \tag{11.59.1}$$

(see (11.32)) *supplies,* in effect, *a* **C**-*derivation* (viz. a *flat* $\mathcal{E}$-*connection*) *of the* **C**-*algebra sheaf* $\mathcal{E}$ *into the* $\mathcal{E}$-*module* $\mathcal{E}^1$.

Indeed, by virtue of (11.10), one obtains

$$\partial(s \cdot t) = \partial(t) \cdot s + t \cdot \partial(s), \tag{11.60}$$

for any (local) sections s, t *in* $\mathcal{E}(U)$, where $U \subseteq X$ is *open;* thus, one actually applies (11.10) to the corresponding individual element of the *presheaf*

(11.61) $$\left\{\mathbf{E}^0(U) = E^{\hat{}}|_U; \sigma_V^U\right\}$$

(see (11.24) and (2.3)), verifying thus (11.60) for the case at hand, which in turn is obtained by *sheafification* (see also (11.35)). An analogous argument can be applied, of course, to prove that ∂ is a **C**-*linear morphism of* the **C**-*vector space sheaves* involved (see also Chapt. VI; Definition 1.1). ▮

Therefore, one concludes that;

(11.62) the triple

(11.62.1) $$(\mathcal{E}, \partial, \mathcal{E}^1)$$

yields a *differential triad,* in the sense of Chapt. VI; (1.4), *for which the* corresponding *sequence (11.44) is exact.*

We call (11.62.1) the *Watts-Selesnick differential triad,* which thus can be associated with any given *topological algebra, satisfying (11.37).*

Of course, the above statement makes it possible to apply, within the framework of (11.62.1), several of our considerations in the preceding chapters of a *differential-geometric nature* (see thus, for instance, Chapt. VI; (16.17)). In particular, by considering *soft topological algebras, as in* (11.37), which are, moreover, *self-adjoint locally m-convex *-algebras,* the relevant considerations of the foregoing, concerning *exponential sheaf spaces* and the like, are here too applicable (see also (7.13), or even Theorems 7.1 and 7.2 in the foregoing).

12. Further applications

We consider in this Section two more particular instances, where the technique developed in the foregoing could be applied. So we begin with the following example, pertaining to topological groups, with an eventual special *differential-geometric application to* the case of (compact) *Lie groups.* The same example supplies also, as already mentioned in the preceding, an important particular instance of a *Banach-Šilov algebra, having the approximation property;* as we have seen in the previous Section 11, the latter class of topological algebras, together with its pertinent generalizations, constituted the appropriate framework for that discussion. Yet, as already said, this class of (Banach) algebras, along with the example we discuss below, was also the case that has been considered earlier by S.A. Selesnick [1: p.321, §3].

12.(a). Fourier algebras.– For convenience, we first recall, in brief, the relevant basic terminology: So, to start with, suppose that we are given a *group* G (not necessarily abelian); thus, a $\mathbf{C}$-valued function ϕ on G is said to be *positive-definite* (or even a *function of positive type*), whenever one has;

$$\sum_{i,j} \lambda_i \bar{\lambda}_j \phi(x_i x_j^{-1}) \geqq 0, \tag{12.1}$$

for any (finite-many) elements $x_1, \ldots, x_n$ in G and $\lambda_1, \ldots, \lambda_n$ in $\mathbf{C}$. (Yet, we may restrict ourselves to *distinct elements* of G, as before).

In the sequel we denote by $P_+(G)$ the *set of* $\mathbf{C}$*-valued positive-definite functions* on a given group G, while we also set

$$P(G) = [P_+(G)] \tag{12.2}$$

for the $\mathbf{C}$-*vector space spanned by the set* $P_+(G) \subseteq \mathbf{C}^G$. (The latter item here stands, of course, for the $\mathbf{C}$-*vector space of* $\mathbf{C}$-*valued maps on* G, with *point-wise defined operations*).

On the other hand, it is not hard to see that;

(12.3) in the case of a *topological group* G, $\mathcal{P}(G)$, viz. the $\mathbf{C}$-*linear span of* the set of *all continuous positive-definite functions on* G, $\mathcal{P}_+(G)$, with point-wise defined operations (cf. (12.2)), is a *unital commutative self-adjoint* (relative to complex conjugation) *subalgebra of* $\mathcal{C}(G)$ (algebra of $\mathbf{C}$-valued continuous functions on G).

See, for instance, E. Hewitt-K.A. Ross [1: p.260, Theorem 32.10] ▮

On the other hand, a fact that also interests us here, is the interpretation of *the algebra* $\mathcal{P}(G)$, as in (12.3) above, *as* the space of *absolutely convergent Fourier series on* the given topological group G. Now, as already said at the beginning of this Subsection, G is an arbitrary *group* viz. *not necessarily abelian;* on the other hand, *we assume henceforth that;*

(12.4) G is a given *compact* (Hausdorff topological) *group*.

Thus, the familiar from the abelian case character group (i.e., *1-dimensional representations*) of G has to be replaced here by *finite-dimensional,* in general, *representations of* G; indeed, we know that *continuous irreducible (unitary) representations of a compact group are finite-dimensional* (see, for instance, M.A. Naĭmark [1: p. 442, Theorem 4]). So we denote by

$$\hat{G} \equiv K \tag{12.5}$$

the *set of equivalence classes* of such representations of G. (The name *dual object,* or just *dual,* of G is also of use for K). On the other hand, there exists a *one-to-one and onto correspondence* (modulo, of course, the appropriate identifications) *between continuous positive-definite functions on a locally compact group G and continuous unitary representations of G* (ibid.; p.400, Theorem 1), connecting thus the two notions in (12.3) and (12.5), as above, concerning, in particular, the given compact group G.

Thus, if $\alpha \in K$, as in (12.5), we denote by π_α any *representative of the class* α at issue, hence, a *continuous irreducible (unitary) representation* of the *compact group* G in, say, an n_α*-dimensional Hilbert space* H_α, *with* $n_\alpha \in \mathbf{N}$. Yet, we further denote by dx the (unique) *Haar measure* on G, which we choose thus *normalized,* viz. we assume that

$$\int_G dx = 1. \tag{12.6}$$

Now, for every $f \in L^1(G)$ (viz. the *group algebra* of G), one further defines the *Fourier series* of f

$$\sum_{\alpha \in K} n_\alpha \mathrm{tr}\,(T_\alpha \pi_\alpha) \tag{12.7}$$

(here "tr" stands for the *trace* of the corresponding operator (matrix) in $L(H_\alpha) \equiv \mathcal{L}(H_\alpha)$, cf. also (12.9) below), such that the *Fourier coefficient* is the following (continuous) endomorphism of H_α

$$T_\alpha = \int_G \pi_\alpha(x^{-1}) f(x)\, dx \in \mathcal{L}(H_\alpha), \tag{12.8}$$

where one sets;

$$T_\alpha(z) = \int_G \pi_\alpha(x^{-1})(z) f(x)\, dx \in H_\alpha, \tag{12.9}$$

for every $z \in H_\alpha$. (Of course, one has here a typical case of an *"operator-valued" integration;* see e.g. loc. cit., p.149, §19). Thus, we say that the *Fourier series* in (12.7) is *absolutely convergent,* whenever one has;

(12.10) $$\sum_{\alpha \in K} n_\alpha \mathrm{tr}\, |T_\alpha| < +\infty.$$

Here we have set (cf. also (12.8))

(12.11) $$|T_\alpha| := \sqrt{T_\alpha T_\alpha^*}$$

in the (finite-dimensional) C^**-algebra* $\mathcal{L}(H_\alpha)$ of the *(finite-dimensional) Hilbert space* H_α (viz. the *representation space of* π_α, as above). So in that case *one obtains;*

(12.12) $$f(x) = \sum_{\alpha \in K} n_\alpha \mathrm{tr}\,(T_\alpha \pi_\alpha(x)),$$

for every $x \in G$. (in this respect, we further note that here, as well as, in (12.10), *the set of those* $\alpha \in K$, *with* $|T_\alpha| > 0$, *is denumerable;* see E. Hewitt – K.A. Ross [1: p.331, Theorem 34.5, or even p.81, Corollary 28.41]). Thus, the fundamental result that interests us here, is the following fact:

(12.13) the set of *those* $f \in L^1(G)$ which possess *absolutely convergent Fourier series* on G (cf. (12.10)), constitutes a *unital commutative* **C**-*algebra,* being actually (in the *norm defined by* (12.10), see also (12.12)) a *Banach algebra;* the same algebra is also *self-adjoint regular* and *semisimple.* We denote it by

(12.13.1) $$A(G),$$

while its *spectrum* is given by

(12.13.2) $$\mathfrak{M}(A(G)) = G,$$

within a *homeomorphism.*

We call $A(G)$ the *Fourier algebra of the compact group* G, under consideration.

Thus, in other words, one concludes that;

(12.14) *the Fourier algebra* $A(G)$ *of a compact* (Hausdorff) *group* G *is a* (complex) unital commutative *self-adjoint Banach-Šilov algebra, whose spectrum is* (homeomorphic to) G.

Furthermore, connecting the preceding with (12.3), one also proves that

$$A(G) = \mathcal{P}(G), \tag{12.15}$$

for any given compact group G, so that the *operations in A(G)* can still be defined *point-wise*. (Details on all the foregoing and relevant proofs can be found in E. Hewitt – K.A. Ross [1: p.336, Theorem 34.13], concerning (12.15), while for (12.13) see ibid., p.502, Theorem 39.31, p.340, Theorem 34.20, along with the relevant comments at the end of its proof, as well as, p.260, Theorem 32.10). ∎ On the other hand, one finally concludes that;

(12.16) *the Fourier algebra* $A(G)$, as in (12.13.1), *satisfies the* (Banach) *approximation property.*

See (12.13), as well as, A. Grothendieck [1: Chap. I; p.187, Corollaire 2], in conjunction with the cited work of Hewitt and Ross; p.260, Theorem 32.10, p.341, Remark 34.22 and p.346, Theorem 34.26. ∎

Scholium 12.1. In the case that G is, in particular, a *compact abelian group,* then the above argument is reduced to the standard *Gel'fand description* of the corresponding situation: Namely, one can then consider the (complex) *unital commutative Banach algebra*

$$l^1(\widehat{G}), \tag{12.17}$$

(viz. the respective *group algebra* of $\widehat{G}$), where now $\widehat{G}$ stands for the *(discrete) character group* of G. Thus, by further considering the respective *Gel'fand map of* $l^1(\widehat{G})$ (cf. (2.21) in the preceding, along with (12.19) below) one now defines the *Fourier algebra* of G, $A(G)$ (we retain the same notation for the latter algebra, as before), according to the following relation (see also, for instance, (2.1) and (2.2));

$$A(G) := l^1(\widehat{G})^{\widehat{}} \subseteq \mathcal{C}(G). \tag{12.18}$$

Concerning the last item of the previous relation, one has, in effect, regarding the *spectrum of* $l^1(\widehat{G})$, the following *homeomorphisms;*

$$\mathfrak{M}(l^1(\widehat{G})) = \widehat{\widehat{G}} = G, \tag{12.19}$$

where the second one is due to *Pontryagin's duality.* (In this respect, see also, for instance, [TA: p. 237, Theorem 4.1], as well as, M.A. Naĭmark [1: p. 415ff, in particular, p. 425, Theorem 5]).

On the other hand, it is a standard result that *the group algebra of any locally compact* (Hausdorff) *abelian group,* like (12.17) above, *is regular and semisimple.* (See, for example, the above cited book of Naĭmark; p. 427, Theorem 7 and p.380, VII; in this respect, see also loc. cit., p. 376, concerning the *unital* group algebra (12.17), as before). Therefore, one infers that (cf. also (12.19));

(12.20) $l^1(\widehat{G})$, as in (12.17), is a *unital commutative Banach-Šilov algebra,* whose *spectrum* is (homeomorphic to the given compact group) G. Furthermore, *the same algebra satisfies the* (Banach) *approximation property.*

Concerning our last assertion in (12.20), see also e.g. A. Grothendieck [1: Chap. I; p. 185, Proposition 41]. ∎

Now, the *Fourier algebra of* G, $A(G)$, as defined by (12.18), *is* further *topologized by the relation*

(12.21) $$l^1(\widehat{G}) \underset{\mathcal{G}}{\cong} l^1(\widehat{G})^{\frown},$$

where $\mathcal{G}$ stands for the respective *Fourier(-Gel'fand) transform of* (the *semisimple* Banach algebra) $l^1(\widehat{G})$ (cf. (2.21), (2.22) amd (12.20) above). Thus, in view of (12.21), $A(G)$ becomes a *Banach algebra isomorphic* (as such) *to* $l^1(\widehat{G})$. Now,

(12.22) it is actually *under the previous topologization,* through (12.21) (in conjunction with (12.18)), that *one understands the Fourier algebra* $A(G)$ *of a* given *compact abelian group* G, *as a Banach algebra.*

(In this respect, see also e.g. H. Reiter [1: p.104, (4)]). Therefore, one infers that;

(12.23) *the Fourier algebra* $A(G)$, as before, *being* (by its very definition, as above) *Banach algebra-isomorphic to* $l^1(\widehat{G})$, *shares* the same properties with the latter algebra, hence, (12.20) *too.*

Of course, our previous discussion, throughout the present Scholium, has only the meaning of a more direct access to $A(G)$, than in (12.13.1), due now, according to our hypothesis, to the *commutativity* of the given compact group G.

Accordingly, based now on the foregoing, we conclude that;

(12.24) given a *compact group* G (abelian or not), its *Fourier algebra* $A(G)$ (cf. (12.18), (12.22), or (12.13.1)) is a *(complex) unital commutative self-adjoint Banach-Šilov algebra, having the* (Banach) *approximation* property, and whose *spectrum* is (homeomorphic to) G.

Thus, as an application of the preceding considerations, one realizes that, by looking at the *Fourier algebra* $A(G)$ of a *compact group* G, one obtains a very convenient indeed *"geometric"* algebra (even in the technical sense of the term, as in Section 3 above), in such a manner that the *differential geometric applications,* as exhibited in the previous chapters of this treatise, *are* here too *available.* Thus, suppose, in particular, that;

(12.25) G is a given *compact* (Hausdorff) *Lie group* (abelian or not).

Consequently, as an *application of the previous discussion,*

(12.26) one can formulate several *differential geometric notions and* obtain related *results, in terms of the,* eventually more convenient, *Fourier algebra of* G, $A(G)$, *than of the algebra* $C^{\infty}(G)$ (of $\mathbf{C}$*-valued* C^{∞}*-functions on* G), that traditionally is associated with G in the case, under consideration.

What amounts to the same thing, one thus concludes that;

(12.27) one can do *differential geometry on a* given *compact Lie group* (at least, to the extent we consider it in the preceding Chapters), *by means of absolutely convergent Fourier series,* rather, *than of* C^{∞}*-functions* (as usual) on the group.

12.(b). Formal power series algebras.- We terminate our discussion, by exhibiting one more *example* of a *topological algebra,* which indeed provides a *resolution of* $\mathbf{C}$, through an appropriate *"de Rham complex"*, that is associated with such an algebra, by constructing the pertinent to the algebra at issue *modules of "differentials"*.

Now, as the heading of this subsection indicates, the example we are going to consider below, contains as a particular instance the *algebra of,* say, *complex polynomials in n* (independent) *variables,*

(12.28) $$E \equiv \mathbf{C}[t_1, \ldots, t_n],$$

that also exhibits an analogous situation, as that one mentioned above: Namely, by taking, for example, the corresponding to *E de Rham-Kähler complex,* as in (5.27), one obtains a *resolution of* **C**, in the sense that *the following sequence of* **C**-*vector spaces is exact;*

(12.29) $$0 \longrightarrow \mathbf{C} \longrightarrow E \longrightarrow \Omega_E^1 \longrightarrow \cdots \longrightarrow \Omega_E^n \longrightarrow 0.$$

In this context, one considers on E, as given by (12.28), the *discrete topology,* so that E is just a **C**-*algebra.* (We do not make actually here any use of the aforementioned topology; see also S. Lang [1: p.612, Theorem 12.3]). ∎

On the other hand, one can consider on the same algebra E, as in (12.28), a canonical topology, making it into a *locally m-convex* **C**-*algebra,* via the relation;

(12.30) $$E \equiv \mathbf{C}[t_1, \ldots, t_n] \overset{\subsetneq}{\rightarrow} \mathcal{C}_c(\mathbf{C}^n),$$

such that, concerning the *spectrum* of E, one has

(12.31) $$\mathfrak{M}(E) \equiv \mathfrak{M}(\mathbf{C}[t_1, \ldots, t_n]) = \mathbf{C}^n,$$

within a *homeomorphism.* In this context, *we* still *remark, in passing, that;*

(12.32) *every maximal ideal of* (the unital commutative locally m-convex **C**-algebra) $E \equiv \mathbf{C}[t_1, \ldots, t_n]$, as the latter algebra is topologized by (12.30), *is closed,* given by a point of $\mathbf{C}^n$ (*Nullstellensatz*).

On the other hand, E *cannot be a Q-algebra.*

The last assertion is true, since, otherwise, E would have a *compact* (viz. closed and equicontinuous) *spectrum* [TA: p.187, Lemma 1.3]), a contradiction, of course, to (12.31). ∎

Now, the same (polynomial) *topological algebra E*, as in (12.30), is also *semisimple,* a direct consequence of the very definition, in conjunction with (12.30), (12.31) (the corresponding *Gel'fand map* of E being also *continuous*). Yet, E is *metrizable* (as a topological algebra, $\mathbf{C}^n$ being a *hemicompact* space) and *self-adjoint* (with respect to complex-conjugation); therefore, its *completion,* being, in effect, the whole algebra $\mathcal{C}_c(\mathbf{C}^n)$ (Stone-Weierstrass, cf. e.g. L. Nachbin [1: p.48, Corollary 2]) is a *self-adjoint Fréchet-Šilov (locally m-convex) algebra,* whose *spectrum* is again *(homeomorphic to)* $\mathbf{C}^n$ (cf. [TA: p.223, Theorem 1.2]). Therefore, one can further apply here our previous considerations in Section 7 (see, for instance, Theorem 7.2), concerning, for example, the *existence of a sheaf exponential sequence* (cf. (7.19)).

Thus, after the above prelude, we come back again to a description, in brief, of the situation one has in the more general case of a *formal power series* **C**-*algebra* (in *finite-many variables*). That is, of an algebra, say, of the form;

$$C_n \equiv \mathbf{C}[[t_1, \ldots, t_n]]. \tag{12.33}$$

Now, concerning the purely algebraic part of our exposition, see e.g. T.W. Hungerford [1: p.154ff], while for the rest we mainly follow the relevant account in the work of H. Grauert – R. Remmert [1]. So (12.33) is, by its very definition, first, a *unital commutative* **C**-*algebra,* being also a *local* **C**-*algebra.* Thus, the *unique maximal ideal of* C_n consists of those *formal power series without* (viz. having zero) *constant term,* the rest elements of C_n yielding its *group of units* (cf. e.g. Hungerford, loc. cit.; p.155, Corollary 5.10).

Indeed, we restrict ourselves below to a certain subalgebra of C_n, consisting of the so-called *convergent formal power series:* That is, following, as we have said already, the terminology of H. Grauert – R. Remmert [1: p.15ff], we set

$$E \equiv E_n := \mathbf{C}\langle t_1, \ldots, t_n\rangle = \bigcup_{\lambda \in \mathbf{R}_+^n} B_\lambda\langle t_1, \ldots, t_n\rangle, \tag{12.34}$$

such that one further defines (cf. (12.33))

$$B_\lambda\langle t_1, \ldots, t_n\rangle := \{f \in C_n : \|f\|_\lambda < \infty\}, \tag{12.35}$$

for any $\lambda \equiv (\lambda_1, \ldots, \lambda_n) \in \mathbf{R}_+^n$, where we also set

$$\|f\|_\lambda := \sum_{\nu \in \mathbf{N}^n} |\alpha_\nu| \lambda^\nu, \tag{12.36}$$

with $\nu \equiv (\nu_1, \ldots, \nu_n) \in \mathbf{N}^n$ and $\lambda^\nu := \lambda_1^{\nu_1} \cdots \lambda_n^{\nu_n}$, for every *formal power series,*

$$f = \sum_{\nu \in \mathbf{N}^n} \alpha_{\nu_1 \cdots \nu_n} t_1^{\nu_1} \cdots t_n^{\nu_n} \in C_n. \tag{12.37}$$

Thus,

(12.38) we call E, as given above by (12.34), the *algebra of convergent formal power series* (in n variables, $n \in \mathbf{N}$), being again a *unital commutative* $\mathbf{C}$-*algebra,* a subalgebra in fact of C_n, as in (12.33).

Furthermore, each one of the sets $B_\lambda \subseteq C_n$, $\lambda \in \mathbf{R}^n_+$, as in (12.34) (cf. also (12.35)), is, in effect, a (unital commutative) *Banach* $\mathbf{C}$-*algebra,* in the norm given by (12.36). The same algebras contain the polynomial algebra (12.30) as a *dense subalgebra* (see loc. cit.; p.16, Satz 1, *infra*). Furthermore,

(12.39) E is a *local* $\mathbf{C}$-*algebra, its only maximal ideal* $\mathfrak{m}$ being the set of those elements of E, which vanish at the *zero-element,* viz., equivalently, those elements in E, *whose constant term is zero.* Yet, one obtains

(12.39.1) $$E/\mathfrak{m} \cong \mathbf{C},$$

within an *isomorphism of* $\mathbf{C}$-*algebras.*

(See loc. cit.; p.27, Satz 1, *infra,* and p.210, §2). ∎

In this context, a crucial ingredient for what concerns us in the sequel, is the existence of *partial derivatives in E, with respect to the variables* (indeterminates) t_i, $1 \leqq i \leqq n$; all these are *formally defined,* while one further obtains, yet *formally,* an *indefinite integral,* for every $f \in E$ (ibid., p.9ff). Furthermore, given the $\mathbf{C}$-*algebra* E, as in (12.34), one still defines the E-*module of 1st order differentials* (or *1-forms,* or even *pfaffian forms,* yet just *pfaffians*) of E, via the aforementioned partial derivatives and the corresponding differentials of the (independent) variables, by analogy with the classical case (of $\mathcal{C}^\infty$-functions). We denote the E-module, under discussion, by

(12.40) $$\Omega(E) \equiv \Omega^1,$$

while one further defines;

(12.41) $$\Omega^p := \bigwedge_1^p \Omega^1, \quad \textit{with} \quad p \geqq 2,$$

along with the pertinent *(exterior) differentials,* by following the standard case. Thus, one obtains the next basic result;

(12.42) given the **C**-*algebra* E, as defined by (12.34) above, one has the following *exact sequence of* **C**-*vector spaces*

(12.42.1)
$$0 \longrightarrow \mathbf{C} \longrightarrow E \xrightarrow{d} \Omega^1 \longrightarrow \Omega^2 \longrightarrow \cdots$$
$$\cdots \longrightarrow \Omega^p \longrightarrow \Omega^{p+1} \longrightarrow \cdots.$$

(See loc. cit.; p.176, Satz 10). ▮ Here one also applies the fact that *the partial derivatives and* the respective *integration* are *continuous* (**C**-*linear*) *maps,* with respect to an appropriate *topology on* E, the so-called *sequential topology* (or even *sequence topology; "Folgentopologie"*. See loc. cit. p.58, and p.59, Satz 6). Yet, the latter topology makes E into a *Fréchet locally convex* (topological) **C**-*algebra* with a *unique closed maximal ideal* (given as in (12.39). See also loc. cit. p.66, Satz 7, p.74, Satz 7, and p.82, Satz 1, along with p.85, Satz 6, the Korollar and Satz 7). Thus, E is, in particular, a *Fréchet (locally convex) Q-algebra* [TA].

We call (12.42.1) the *Poincaré-de Rham (exact) sequence,* which thus is associated with the given algebra E, as in (12.34).

Thus, as a consequence of the previous discussion, we conclude that one can consider, for the case at hand, the triple

(12.43)
$$(E, d, \Omega^1),$$

as defined above, as a *differential triad* (over the *one-point space*(!) $\{\mathfrak{m}\}$, as in (12.39). Hence, one has here a *trivial sheafification over the spectrum of* E). So in conclusion;

(12.44) we can have here, once more, *with respect to the above triple* (E, d, Ω^1), as in (12.43), we call the latter, the *Grauert-Remmert differential triad,* the *differential geometric apparatus,* that has been developed so far in the preceding.

Thus, here again, one has the possibility of *doing differential geometry* (or employing, at least, its language) *not by* means of $\mathcal{C}^\infty$-*functions* (hence, when disposing a usual $\mathcal{C}^\infty$-manifold context), *but through another appropriate* (eventually more profitable or, at least, less expensive!) *algebra of functions*!

BIBLIOGRAPHY

(The items of the bibliography are given, for each one of the two Volumes of this treatise, separately; thus, for convenience, certain issues of the following list are repeated from the references of Volume I)

S. I. Andersson

1. *Vector bundle connections and liftings of partial differential operators.* In "Differential Geometric Methods in Mathematical Physics". Lecture Notes in Mathematics, No 905, pp. 119-132. Springer-Verlag, Berlin, 1982.

A. Asada

1. *Algebraic cohomology of loop spaces.* J. Fac. Sci. Shinshu Univ. 4 (1969), 1-21.
2. *Connections on differential operators.* J. Fac. Sci. Shinshu Univ. 13 (1978), 87-102.

M. F. Atiyah

1. *Complex analytic connections in fiber bundles.* Trans. Amer. Math. Soc. 85(1957), 181-207.
2. *K-Theory.* Benjamin, New York, 1967.
3. *Elliptic Operators in Compact Groups.* Lecture Notes in Mathematics, No 401. Springer-Verlag, Berlin, 1974.
4. *Geometry of Yang–Mills Fields.* Accad. Naz. Lincei (Scuola Normale Superiore), Pisa, 1979.

C. Bartocci, U. Bruzzo, D. Hernández-Ruipérez

1. *The Geometry of Supermanifolds.* Kluwer Academic Publs. Dordrecht, 1991.

J. Bernstein – V. Lunts

1. *Equivalent Sheaves and Functors.* Lecture Notes in Mathematics, No 1578. Springer-Verlag, Berlin, 1994.

E. Binz – J.Śniatycki – H. Fischer

1. *Geometry of Classical Fields.* North-Holland, Amsterdam, 1988.

R. Bkouche

1. *Couples spectraux et faisceaux associés. Applications aux anneaux de fonctions.* Bull. soc. math. France 98(1970), 253-295.

R. Bott

1. *Lectures on characteristic classes and foliations.* In "Lectures on Algebraic and Differential Topology". Lecture Notes in Mathematics, No 279, pp. 1-80. Springer-Verlag, Berlin, 1972.

N. Bourbaki

1. *Théorie des ensembles, Chap. 3 (2e éd.).* Herman, Paris, 1967.
2. *Algèbre, Chap. 1–3.* Hermann, Paris, 1970.
3. *Algèbre, Chap. 4–5 (2e éd.)* Hermann, Paris, 1959.
4. *Commutative Algebra.* Addison-Wesley, Reading, Massachusetts, 1972.

J.P. Brasselet – G. Hector – M. Saralegi

1. *Théorème de de Rham pour les variétés stratifiées.* Ann. Global Annal. Geom. 9 (1991), 211-243.

G. E. Bredon

1. *Sheaf Theory.* McGraw-Hill, New York, 1967.

R. M. Brooks

1. *Partitions of unity in F-algebras.* Math. Ann. 177(1968), 265-272.
2. *On Arens' "Weierstrass products" paper.* Math. Ann. 190(1971), 329-337.

J.-L. Brylinski

1. *Loop Spaces, Characteristic Classes and Geometric Quantization.* Birkhäuser, Boston, 1993.

L. Chambadal – J. L. Ovaert

1. *Algèbre linéaire et algèbre tensorielle.* Dunod, Paris, 1968.

K.-T. Chen

1. *Algebras of iterated path integrals and fundamental groups.* Trans. Amer. Math. Soc. 156(1971), 359-379.
2. *Iterated integrals of differential forms and loop space homology.* Ann. Math. 97(1973), 217-246.
3. *Iterated integrals, fundamental groups and covering spaces.* Trans. Amer. Math. Soc. 206(1975), 83-98.
4. *Iterated path integrals.* Bull. Amer. Math. Soc. 8(1977), 115-137.

Y. Choquet-Bruhat C. De Witt-Morette M. Dillard-Bleick

1. *Analysis, Manifolds and Physics (Revised Ed.)*. North-Holland, Amsterdam, 1982.

E. Cioranu-Verona

1. *Funcţii şi forme pe varietăţi generalizate*. St. Cerc. Mat. 26(1974), 133-143. [Romanian]

A. Colojoăra

1. *Algèbre symmetrique du dual d' un espace nucléaire – DF*. Rev. roum. math. pures appl. 23(1978), 1317-1339.
2. *Sur le théorème de Poincaré – Birkhoff – Witt*. Rev. roum. math. pures appl. 25(1980), 1327-1347.

M. Cornalba – P. Griffiths

1. *Analytic cycles and vector bundles on noncompact algebraic varieties*. Invent. Math. 26(1975), 402-406.

M. Crampin – F. A. E. Pirani

1. *Applicable Differential Geometry*. Cambridge Univ. Press, Cambridge, 1986.

C. W. Curtis

1. *Linear Algebra. An Introductory Approach*. Springer-Verlag, New York, 1984.

P. de Bartolomeis

1. *Algebre di Stein nel case reale*. Rend. Accad. Naz. dei Lincei 40 [Quaranta] 5 (1) (2) (1975) (1976); (1977), 105-144.

P. Deligne

1. *Équations différentielles à points singuliers réguliers*. Lecture Notes in Mathematics, No 163. Springer-Verlag, Berlin, 1970.

R. Dijkgraaf – E. Witten

1. *Topological gauge theories and group cohomology*. Comm. Math. Phys. (in press).

B. A. Dubrovin – A. T. Fomenko – S. P. Novikov

1. *Modern Geometry – Methods and Applications. Part I. The Geometry of Surfaces, Transformation Groups, and Fields*. Springer-Verlag, New York, 1984.

J. Dugundji

1. *Topology*. Allyn and Bacon, Boston, 1966.

T. Eguchi – P. B. Gilkey – A. J. Hanson

1. *Gravitation, Gauge Theories and Differential Geometry.* Phys. Rep. 66 (1980), 213-393.

D. Eisenbud

1 *Commutative Algebra with a View Toward Algebraic Geometry.* Springer-Verlag, New York, 1995.

C. Farsi

1. *Index Theory and Positive Scalar Curvature for Orbifolds.* Ph. D. Thesis, Univ. of Maryland, College Park, MD, 1989.
2. *K-theoretical index theorems for orbifolds.* J. Math. Oxford 43 (1992), 183-200.

H. R. Fischer

1. *Differentialkalkül für nicht-metrische Strukturen II: Differentialformen.* Arch. Math. 8 (1957), 428-443.

F. J. Flaherty

1. *The de Rham theorem for function spaces.* J. Math. Mech. 19 (1969), 317-320.
2. *A differentiable structure for* $\mathrm{Diff}(X)$ (manuscript)

M. Fragoulopoulou

1. *Uniqueness of topology for semisimple LFQ-algebras.* Proc. Amer. Math. Soc. 117 (1993), 963-969.

O. Forster

1. *Funktionentheoretische Hilfsmittel in der Theorie der kommutativen Banach-Algebren.* Jber. Deutsch. Math. -Verein. 76 (1974), 1-17.

G. Galanis

1. *Differential Geometry of Projective Limits of Fiber Bundles.* Doctoral Dissertation, Univ. of Athens, 1995. [Greek]

I. M. Gelfand – D. Raikov – G. Shilov

1. *Commutative Normed Rings.* Chelsea Publ. Co. Bronx, New York, 1964.

M. Göckeler – T. Schücker

1. *Differential Geometry, Gauge Theories, and Gravity.* Cambridge Univ. Press, Cambridge, 1987.

H. Grauert

1. *Charakterisierung der holomorph-vollständigen Räumen.* Math. Ann. 129 (1955), 233-259.
2. *Holomorphe Funktionen mit Werten in komplexen Lieschen Gruppen.* Math. Ann. 133 (1957), 450-472.

H. Grauert – R. Remmert

1. *Analytische Stellenalgebren.* Springer-Verlag, Berlin, 1971.
2. *Theory of Stein Spaces.* Springer-Verlag, Berlin, 1979.

W. Greub

1. *Multilinear Algebra (2nd Ed.).* Springer-Verlag, New York, 1978.

P. Griffiths – J. Adams

1 *Topics in Algebraic and Analytic Geometry.* Princeton Univ. Press, Princeton, New Jersey, 1974.

A. Grothendieck

1 *Produits tensoriels topologiques et espaces nucléaires.* Memoirs Amer. Math. Soc., no 16 (1955).

R. C. Gunning

1. *Lectures on Riemann Surfaces.* Princeton Univ. Press, Princeton, New Jersey, 1966.
2. *Lectures on Vector Bundles over Riemann Surfaces,* Princeton Univ. Press, Princeton, New Jersey, 1967.

R. I. Hadjigeorgiou

1. *Spectral Geometry of Topological Algebras.* Doctoral Dissertation, Univ. of Athens, 1995. [Greek]
2. *Choquet boundaries in topological algebras.* Comment. Math. 36(1996), 131-148.

R. Hartshorne

1. *Algebraic Geometry.* Springer-Verlag, New York, 1977.

G. Hector – M. Saralegi

1. *Intersection cohomology of S^1-actions.* Trans. Amer. Math. Soc. 338 (1993), 263-288.

M. Heller

1. *Algebraic foundations of the theory of differential spaces.* Demonstr. Math. 24 (1991), 349-364.
2. *Einstein algebras and general relativity.* Intern. J. Theor. Phys. 31 (1992), 277-288.
3. *Generalized Friedman's equation and its singularities.* Acta Cosmol. 19 (1993), 23-33.
4. *Structured spaces and their application to relativistic physics.* J. Math. Phys. 36 (1995), 3644-3662.
5. *Sheaves of Einstein algebras.* Intern. J. Theor. Phys. 34 (1995), 387-398.

E. Hewitt – K. A. Ross

1. *Abstract Harmonic Analysis II.* Springer-Verlag, Berlin, 1970.

F. Hirzebruch

1. *Topological Methods in Algebraic Geometry.* Springer - Verlag, Berlin, 1978.

L. Hörmander

1. *An Introduction to Complex Analysis in Several Variables.* North - Holland, Amsterdam, 1973.

T. W. Hungerford

1. *Algebra.* Springer-Verlag, Berlin, 1974.

D. Husemoller

1. *Fiber Bundles (2nd Ed.).* Springer-Verlag, New York, 1975.

M. Karoubi

1. *K-Theory.* Springer-Verlag, Berlin, 1978.

M. Karoubi – G. Leruste

1. *Algebraic Topology via Differential Geometry.* Cambridge Univ. Press, Cambridge, 1987.

Y. Katznelson

1. *An Introduction to Harmonic Analysis.* Wiley, New York, 1968.

L. Kaup – B. Kaup

1. *Holomorphic Functions of Several Variables.* W. de Gruyter, Berlin, 1983.

T. Kawasaki

1. *The signature theorem for V-manifolds.* Topology 17 (1978), 75-83.
2. *The Riemann - Roch theorem for complex V-manifolds.* Osaka J. Math 16 (1979), 151-159.
3. *The index of elliptic operators over V-manifolds.* Nagoya Math. J. 84 (1981), 135-157.

J. L. Kelley

1. *Topology.* Springer-Verlag, New York, 1975.

A. T. Kitchen

1. *Algebraic Alexander-Spanier Cohomology.* Ph. D. Thesis, Univ. of Rochester, Rochester, New York, 1971.

S. Kobayashi

1. *Differential Geometry of Complex Vector Bundles.* Princeton Univ. Press, Princeton, New Jersey, 1987.
2. *Manifolds over function algebras and mapping spaces.* Tôhoku Math. J. 41 (1989), 263-282.

S. Kobayashi – K. Nomizu

1. *Foundations of Differential Geometry, Vol. I.* Interscience Publ., New York, 1963.
2. *Foundations of Differential Geometry, Vol. II.* Interscience Publ., New York, 1969.

R. Kochendörffer

1. *Introduction to Algebra.* Wolters-Noordhoff, Groningen, 1972.

A. A. Korobov

1. *Cohomology of algebras with unit.* Soviet Math., Doklady 10 (1969), 17-20.

B. Kostant

1. *Quantization and unitary representations.* In "Lectures in Modern Analysis and Applications, II". Lecture Notes in Mathematics, No 170, pp. 87-208. Springer-Verlag, Berlin, 1970.

A. I. Kostrikin

1. *Introduction to Algebra.* Springer-Verlag, New York, 1982.

J. L. Koszul

1. *Sur certaines groupes de transformations de Lie.* Coll. Geom. Diff., Strasbourg, 1953.
2. *Lectures on Fiber Bundles and Differential Geometry.* Tata Institute, Bombay, 1960.

G. Köthe

1. *Topological Vector Spaces I.* Springer-Verlag, Berlin, 1969.

S. G. Krantz

1. *Function Theory of Several Complex Variables.* J. Wiley, New York, 1982.

E. Kunz

1. *Kähler Differentials.* Vieweg, Braunschweig, 1986.

S. Lang

1. *Algebra (2nd Ed.).* Addison - Wesley, Reading, Massachusetts, 1984.

K. Lønsted

1. *An algebraization of vector bundles on compact manifolds.* J. Pure Appl. Algebra 2 (1972), 193-207.
2. *Vector bundles over finite CW-complexes are algebraic.* Proc. Amer. Math. Soc. 38 (1973), 27-31.

A. Mallios

1. *Vector bundles and K-theory over topological algebras.* J. Math. Anal. Appl. 92 (1983), 452-506.
2. *Homotopy invariants of the spectrum of a topological algebra.* J. Math. Anal. Appl. 101 (1984), 297-307.
3. *Hermitian K-theory over topological $*$-algebras.* J. Math. Anal. Appl. 106 (1985), 454-539.
4. *Topological Algebras. Selected Topics.* North - Holland, Amsterdam, 1986.
5. *Continuous vector bundles over topological algebras, II.* J. Math. Anal. Appl. 132 (1988), 401-423.
6. *On an abstract form of Weil's integrality theorem.* Note Mat. 12 (1992), 167-202.
7. *On geometric topological algebras.* J. Math. Anal. Appl. 173 (1993), 301-322.
8. *The de Rham-Kähler complex of the Gel'fand sheaf of a topological algebra.* J. Math. Anal. Appl. 175 (1993), 143-168.
9. *Topological Algebras. Localizations and Extensions.* (Book, in preparation).

Yu. I. Manin

1. *Gauge Field Theory and Complex Geometry.* Springer - Verlag, Berlin, 1988.

C. D. Marshall

1. *The de Rham cohomology of subcartesian spaces.* J. Diff. Geom. 10 (1975), 575-588.

J. W. Milnor – J. D. Stasheff

1. *Characteristic classes.* Princeton Univ. Press, Princeton, New Jersey, 1974.

M. A. Mostow

1. *The differentiable space structures of Milnor classifying spaces, simplicial complexes, and geometric realizations.* J. Diff. Geom. 14 (1979), 255-293.

L. Nachbin

1. *Elements of Approximation Theory.* D. van Nostrand, Princeton, New Jersey, 1967.
2. *Introduction to Functional Analysis: Banach Spaces and Differential Calculus.* Marcel Dekker, New York, 1981.

M. Nakahara

1. *Geometry, Topology and Physics.* Adam Hilger, Bristol, 1990.

M. A. Naĭmark

1. *Normed Algebras.* Wolters-Noordhoff, Groningen, 1972.

M. E. Novodvorskiĭ

1. *On certain homotopic invariants of the space of maximal ideals.* Mat. Zametki 1 (1967), 487-494. [Russian]

M. Papatriantafillou

1. *On the de Rham's complex in abstract spaces* (manuscript).
2. *The de Rham's complex in function spaces* (to appear).

T. Petrie – J. Randall

1. *Connections, Definite forms, and Fourmanifols.* Oxford Univ. Press, Oxford, 1990.

W. A. Poor

1. *Differential Geometric Structures.* McGraw-Hill, New York, 1981.

E. Prugovečki

1. *Quantum Mechanics in Hilbert Space (2nd Ed.).* Academic Press, New York, 1981.
2. *Quantum Geometry. A Framework for Quantum General Relativity.* Kluwer, Dordrecht, 1992.

M. Puta

1. *Hamiltonian Mechanical Systems and Geometric Quantization.* Kluwer, Dordrecht, 1993.

H. Reiter

1. *Classical Harmonic Analysis and Locally Compact Groups.* Oxford Univ. Press, Oxford, 1968.

M. Rosenfeld

1. *Commutative F-algebras.* Pacific J. Math. 16 (1966), 159-166.

J.J. Rotman

1. *An Introduction to Algebraic Topology.* Springer-Verlag, New York, 1988.

M. Saralegi – G. Hector

1. *Formes différentielles d' intersection: un théorème de de Rham pour l' homologie d' intersection des préstratifications abstraites.* C. R. Acad. Sci. Paris 308 (1989), 25-28.

E. Sasin

1. *On some exterior algebra of the differential forms over a differential space.* Demonstr. Math. 19 (1986), 1063-1075.
2. *The de Rham cohomology of differential spaces.* Demonstr. Math. 22 (1989), 249-270.
3. *Differential spaces and singularities in differential space-times.* Demonstr. Math. 24 (1991), 601-634.

W. Sasin – K. Spallek

1. *Gluing of differentiable spaces and applications.* Math. Ann. 292 (1992), 85-102.

I. Satake

1. *On a generalization of the notion of manifold.* Proc. Nat. Acad. Sci. 42 (1956), 359-363.

H.H. Schaefer

1. *Topological Vector Spaces.* Springer-Verlag, New York, 1980.

J. Schwartz

1. *De Rham's theorem for arbitrary spaces.* Amer. J. Math. 77 (1995), 29-44.

S. A. Selesnick

1. *Watts cohomology for a class of Banach algebras and the duality of compact abelian groups.* Math. Z. 130 (1973), 313-323.
2. *Line bundles and harmonic analysis on compact groups.* Math. Z. 146 (1976), 53-67.
3. *Second quantization, projective modules, and local gauge invariance.* Intern. J. Theor. Phys. 22 (1983), 29-53.
4. *Geometrical symmetry breaking.* Nuovo Cimento 83A (1984), 175-190.
5. *Correspondence principle for the quantum net.* Intern. J. Theor. Phys. 30 (1991), 1273-1292.

R. Sikorski

1. *Abstract covariant derivative.* Colloq. Math. 18 (1967), 251-271.
2. *Differential modules.* Colloq. Math. 24 (1971), 45-70.

D. J. Simms – N. M. J. Woodhouse

1. *Lectures on Geometric Quantization.* Lecture Notes in Physics, No 53. Springer-Verlag, Berlin, 1976.

P. Shanahan

1. *The Atiyah-Singer Index Theorem. An Introduction.* Lecture Notes in Mathematics, No 638. Springer-Verlag, Berlin, 1978.

L. Smith

1. *Linear Algebra (2nd Ed.).* Springer-Verlag, New York, 1984.

J. W. Smith

1. *The de Rham theorem for general spaces.* Tôhoku Math. J. 18 (1966), 115-137.

J. Śniatycki

1. *Prequantization of charge.* J. Math. Phys. 15 (1974), 619-620.
2. *Geometric Quantization and Quantum Mechanics.* Springer-Verlag, New York, 1980.

J.-M. Souriau

1. *Structure de systèmes dynamiques.* Dunod, Paris, 1970.

K. Spallek

1. *Differezierbare Raüme.* Math. Ann. 180 (1969), 269-296.

E. H. Spanier

1 *Algebraic Topology.* Springer - Verlag, New York, 1966.

E. L. Stout

1. *The theory of Uniform Algebras.* Bogden & Quigley, Inc., Tarrytown-on-Hudson, New York, 1971.

J. N. Tavares

1. *Chen integrals, generalized loops and loop calculus* (preprint) (1993).

E. Vassiliou

1. *Connections on principal sheaves* (manuscript).
2. *On Mallios $\mathcal{A}$-connections as connections on principal sheaves.* Note Mat. (in press).
3. *Transformations of sheaf connections* (to appear).
4. *From principal connections to connections on principal sheaves.* Anal. Stiint. Univ. "Al. I. Cusa" 42(1996), 149-160.

A. Verona

1. *Le théorème de de Rham pour les préstratifications abstraites.* C. R. Acad. Sci. Paris 273 (1971), 886-889.

M. E. Verona

1. *A de Rham theorem for generalized manifolds.* Proc. Edinburg Math. Soc. 22 (1979), 127-135.

J. Wagner

1. *Faisceau structural associé à une algèbre de Banach.* Sém. P. Lelong (1968/69), Lecture Notes in Mathematics, No 116. pp. 164-167. Springer-Verlag, Berlin, 1970.

S. Warner

1. *Polynomial completeness in locally multiplicatively-convex algebras.* Duke Math. J. 23 (1956), 1-11.

C. E. Watts

1. *Alexander-Spanier cohomology and rings of continuous functions.* Proc. Nat. Acad. Sci. USA 54 (1965), 1027-1028.

A. Weil

1. *Introduction à l' étude des variétés kählériennes (nouv. éd.).* Hermann, Paris, 1971.

R. O. Wells

1. *Differential Analysis on Complex Manifolds.* Springer - Verlag, New York, 1980.

K. Whyburn

1. *Differentiable cohomology on locally compact groups.* Proc. Nat. Acad. Sci. USA 67 (1970), 45-51.

N. Woodhouse

1. *Geometric Quantization.* Oxford Univ. Press, Oxford, 1980.

Notational Index

Subject Index

(Notions already appeared in Volume I, but still applied in the present volume, are not included in the following list)

Other *Mathematics and Its Applications* titles of interest:

V.I. Istratescu: *Fixed Point Theory. An Introduction.* 1981, 488 pp.
out of print, ISBN 90-277-1224-7

A. Wawrynczyk: *Group Representations and Special Functions.* 1984, 704 pp.
ISBN 90-277-2294-3 (pb), ISBN 90-277-1269-7 (hb)

R.A. Askey, T.H. Koornwinder and W. Schempp (eds.): *Special Functions: Group Theoretical Aspects and Applications.* 1984, 352 pp. ISBN 90-277-1822-9

A.V. Arkhangelskii and V.I. Ponomarev: *Fundamentals of General Topology. Problems and Exercises.* 1984, 432 pp. ISBN 90-277-1355-3

J.D. Louck and N. Metropolis: *Symbolic Dynamics of Trapezoidal Maps.* 1986, 320 pp. ISBN 90-277-2197-1

A. Bejancu: *Geometry of CR-Submanifolds.* 1986, 184 pp.
ISBN 90-277-2194-7

R.P. Holzapfel: *Geometry and Arithmetic Around Euler Partial Differential Equations.* 1986, 184 pp. ISBN 90-277-1827-X

P. Libermann and Ch.M. Marle: *Sympletic Geometry and Analytical Mechanics.* 1987, 544 pp. ISBN 90-277-2438-5 (hb), ISBN 90-277-2439-3 (pb)

D. Krupka and A. Svec (eds.): *Differential Geometry and its Applications.* 1987, 400 pp. ISBN 90-277-2487-3

Shang-Ching Chou: *Mechanical Geometry Theorem Proving.* 1987, 376 pp.
ISBN 90-277-2650-7

G. Preuss: *Theory of Topological Structures. An Approach to Categorical Topology.* 1987, 318 pp. ISBN 90-277-2627-2

V.V. Goldberg: *Theory of Multicodimensional (n+1)-Webs.* 1988, 488 pp.
ISBN 90-277-2756-2

C.T.J. Dodson: *Categories, Bundles and Spacetime Topology.* 1988, 264 pp.
ISBN 90-277-2771-6

A.T. Fomenko: *Integrability and Nonintegrability in Geometry and Mechanics.* 1988, 360 pp. ISBN 90-277-2818-6

L.A. Cordero, C.T.J. Dodson and M. de Leon: *Differential Geometry of Frame Bundles.* 1988, 244 pp. *out of print,* ISBN 0-7923-0012-2

E. Kratzel: *Lattice Points.* 1989, 322 pp. ISBN 90-277-2733-3

E.M. Chirka: *Complex Analytic Sets.* 1989, 396 pp. ISBN 0-7923-0234-6

Kichoon Yang: *Complete and Compact Minimal Surfaces.* 1989, 192 pp.
ISBN 0-7923-0399-7

A.D. Alexandrov and Yu.G. Reshetnyak: *General Theory of Irregular Curves.* 1989, 300 pp. ISBN 90-277-2811-9

Other *Mathematics and Its Applications* titles of interest:

J.-F. Pommaret: *Partial Differential Equations and Group Theory. New Perspectives for Applications.* 1994, 473 pp. ISBN 0-7923-2966-X

Kichoon Yang: *Complete Minimal Surfaces of Finite Total Curvature.* 1994, 157 pp. ISBN 0-7923-3012-9

N.N. Tarkhanov: *Complexes of Differential Operators.* 1995, 414 pp. ISBN 0-7923-3706-9

L. Tamássy and J. Szenthe (eds.): *New Developments in Differential Geometry.* 1996, 444 pp. ISBN 0-7923-3822-7

W.C. Holland (ed.): *Ordered Groups and Infinite Permutation Groups.* 1996, 255 pp. ISBN 0-7923-3853-7

K.L. Duggal and A. Bejancu: *Lightlike Submanifolds of Semi-Riemannian Manifolds and Applications.* 1996, 308 pp. ISBN 0-7923-3957-6

D.N. Kupeli: *Singular Semi-Riemannian Geometry.* 1996, 187 pp. ISBN 0-7923-3996-7

L.N. Shevrin and A.J. Ovsyannikov: *Semigroups and Their Subsemigroup Lattices.* 1996, 390 pp. ISBN 0-7923-4221-6

C.T.J. Dodson and P.E. Parker: *A User's Guide to Algebraic Topology.* 1997, 418 pp. ISBN 0-7923-4292-5

B. Rosenfeld: *Geometry of Lie Groups.* 1997, 412 pp. ISBN 0-7923-4390-5

A. Banyaga: *The Structure of Classical Diffeomorphism Groups.* 1997, 208 pp. ISBN 0-7923-4475-8

A.N. Sharkovsky, S.F. Kolyada, A.G. Sivak and V.V. Fedorenko: *Dynamics of One-Dimensional Maps.* 1997, 272 pp. ISBN 0-7923-4532-0

A.T. Fomenko and S.V. Matveev: *Algorithmic and Computer Methods for Three-Manifolds.* 1997, 346 pp. ISBN 0-7923-4770-6

A. Mallios: *Geometry of Vector Sheaves.* An Axiomatic Approach to Differential Geometry.
Volume I: Vector Sheaves. General Theory. 1998, 462 pp. ISBN 0-7923-5004-9
Volume II: Geometry. Examples and Applications. 1998, 462 pp. ISBN 0-7923-5005-7
(Set of 2 volumes: 0-7923-5006-5)

MIX
Papier aus verantwortungsvollen Quellen
Paper from responsible sources
FSC® C105338

If you have any concerns about our products,
you can contact us on
ProductSafety@springernature.com

In case Publisher is established outside the EU,
the EU authorized representative is:
Springer Nature Customer Service Center GmbH
Europaplatz 3, 69115 Heidelberg, Germany

Printed by Libri Plureos GmbH
in Hamburg, Germany